# The
# Influenza Viruses

# THE VIRUSES

Series Editors
HEINZ FRAENKEL-CONRAT, *University of California*
*Berkeley, California*

ROBERT R. WAGNER, *University of Virginia School of Medicine*
*Charlottesville, Virginia*

THE VIRUSES: Catalogue, Characterization, and Classification
Heinz Fraenkel-Conrat

THE ADENOVIRUSES
Edited by Harold S. Ginsberg

THE BACTERIOPHAGES
Volumes 1 and 2 • Edited by Richard Calendar

THE HERPESVIRUSES
Volumes 1–3 • Edited by Bernard Roizman
Volume 4 • Edited by Bernard Roizman and Carlos Lopez

THE INFLUENZA VIRUSES
Edited by Robert M. Krug

THE PAPOVAVIRIDAE
Volume 1 • Edited by Norman P. Salzman
Volume 2 • Edited by Norman P. Salzman and Peter M. Howley

THE PARVOVIRUSES
Edited by Kenneth I. Berns

THE PLANT VIRUSES
Volume 1 • Edited by R. I. B. Francki
Volume 2 • Edited by M. H. V. Van Regenmortel and Heinz Fraenkel-Conrat
Volume 3 • Edited by Renate Koenig
Volume 4 • Edited by R. G. Milne

THE REOVIRIDAE
Edited by Wolfgang K. Joklik

THE RHABDOVIRUSES
Edited by Robert R. Wagner

THE TOGAVIRIDAE AND FLAVIVIRIDAE
Edited by Sondra Schlesinger and Milton J. Schlesinger

THE VIROIDS
Edited by T. O. Diener

# The
# Influenza Viruses

Edited by
## ROBERT M. KRUG
Memorial Sloan-Kettering Cancer Center
New York, New York

PLENUM PRESS • NEW YORK AND LONDON

Library of Congress Cataloging in Publication Data

The Influenza viruses  /  edited by Robert M. Krug.
   p.     cm. — (The Viruses)
  Includes bibliographies and index.
  ISBN 0-306-43191-2
  1. Influenza viruses. 2. Molecular biology. 3. Viral genetics. 4. Orthomyxoviridae — genetics. 5. Orthomyxoviridae — physiology. I. Krug, Robert M. II. Series.
  [DNLM: QW 168.5.07 1435]
QR201.I6I535  1989
616.2′030194 — dc20
DNLM/DLC                                           89-8702
for Library of Congress                              CIP

© 1989 Plenum Press, New York
A Division of Plenum Publishing Corporation
233 Spring Street, New York, N.Y. 10013

Printed in the United States of America

# Contributors

**Ramesh K. Akkina,** Department of Microbiology, College of Veterinary Medicine, Colorado State University, Fort Collins, Colorado 80523

**Firelli V. Alonso-Caplen,** Graduate Program in Molecular Biology, Memorial Sloan-Kettering Cancer Center, New York, New York 10021

**Thomas M. Chambers,** Department of Virology and Molecular Biology, St. Jude Children's Research Hospital, Memphis, Tennessee 38101

**P. M. Colman,** CSIRO Division of Biotechnology, Parkville 3052, Australia

**Mary-Jane Gething,** Department of Biochemistry and Howard Hughes Medical Institute, University of Texas Southwestern Medical Center, Dallas, Texas 75235-9038

**Charles J. Hackett,** Wistar Institute of Anatomy and Biology, Philadelphia, Pennsylvania 19104

**Ilkka Julkunen,** Graduate Program in Molecular Biology, Memorial Sloan-Kettering Cancer Center, New York, New York 10021

**Michael G. Katze,** Department of Microbiology, School of Medicine, University of Washington, Seattle, Washington 98195

**Robert M. Krug,** Graduate Program in Molecular Biology, Memorial Sloan-Kettering Cancer Center, New York, New York 10021

**Robert A. Lamb,** Department of Biochemistry, Molecular Biology, and Cell Biology, Northwestern University, Evanston, Illinois 60208

**Debi P. Nayak,** Department of Microbiology and Immunology, Jonsson Comprehensive Cancer Center, UCLA School of Medicine, Los Angeles, California 90024

**Peter Palese,** Department of Microbiology, Mount Sinai School of Medicine, New York, New York 10029

**Michael G. Roth,** Department of Biochemistry, University of Texas Southwestern Medical Center, Dallas, Texas 75235-9038

**Joe Sambrook,** Department of Biochemistry, University of Texas Southwestern Medical Center, Dallas, Texas 75235-9038

**J. J. Skehel,** Division of Virology, National Institute for Medical Research, London NW7 1AA, England

**Frances I. Smith,** Department of Microbiology, Mount Sinai School of Medicine, New York, New York 10029

**W. Weis,** Department of Biochemistry and Molecular Biology, Harvard University, Cambridge, Massachusetts 02138

**S. A. Wharton,** Division of Virology, National Institute for Medical Research, London NW7 1AA, England

**D. C. Wiley,** Department of Biochemistry and Molecular Biology, Harvard University, Cambridge, Massachusetts 02138

**Jonathan W. Yewdell,** Laboratory of Viral Diseases, National Institute of Allergy and Infectious Diseases, National Institutes of Health, Rockville, Maryland 20852

# Preface

Influenza virus is an important human pathogen, frequently causing widespread disease and a significant loss of life. Much has been learned about the structure of the virus, its genetic variation, its mode of gene expression and replication, and its interaction with the host immunologic system. This knowledge has the potential of leading to approaches for the control of influenza virus. In addition, research on influenza virus has led to important advances in eukaryotic molecular and cellular biology and in immunology.

A major focus of this book is the molecular biology of influenza virus. The first chapter, which serves as an introduction, describes the structure of each of the genomic RNA segments and their encoded proteins. The second chapter discusses the molecular mechanisms involved in the expression and replication of the viral genome. In addition to other subjects, this chapter deals with one of the most distinctive features of influenza virus, namely the unique mechanism whereby viral messenger RNA synthesis is initiated by primers cleaved from newly synthesized host-cell RNAs in the nucleus. Among the most significant accomplishments in influenza virus research has been the delineation of the three-dimensional structure of the two surface glycoproteins of the virus, the hemagglutinin and neuraminidase. This has provided a structural basis for mapping both the antigenic sites and the regions involved in the major biological functions of these two molecules. The current state of research on the hemagglutinin and neuraminidase is presented in the third and fourth chapters of this book. Chapter 5 describes the research on the biosynthesis, processing, and transport of these two viral glycoproteins. This research has been important not only for understanding the morphogenesis of influenza virus, but also for providing new information about the biosynthesis and transport of all cell-surface glycoproteins.

Defective-interfering virus particles were first discovered in influenza virus preparations, and the sixth chapter describes what is currently known about the generation and mechanism of action of influenza virus defective-interfering particles. The different evolutionary patterns of genetic variation of influenza A, B, and C viruses in humans, and the roles of different influenza virus genes in pathogenicity, are dealt with in Chapter 7. Influenza virus induces thymus-derived lymphocytes (T lymphocytes), which play an important role in antiviral immunity in humans. As discussed in Chapter 8, studies using influenza A viruses have provided some of the most significant findings on the nature of the antigens recognized by T lymphocytes, on the specificity of T lymphocytes, and on the process by which antigens are presented on the cell surface for recognition by T lymphocytes.

This book was put together with two overlapping objectives in mind: providing a current review of the research on influenza virus for virologists, while highlighting for a wide audience of scientists the impact that influenza virus research has had on eukaryotic molecular and cellular biology and on immunology.

Robert M. Krug

# Contents

## Chapter 2

### Expression and Replication of the Influenza Virus Genome

*Robert M. Krug, Firelli V. Alonso-Caplen, Ilkka Julkunen, and
Michael G. Katze*

## Chapter 3

### Structure, Function, and Antigenicity of the Hemagglutinin of Influenza Virus

*S. A. Wharton, W. Weis, J. J. Skehel, and D. C. Wiley*

## Chapter 6

**Structure of Defective-Interfering RNAs of Influenza Viruses and Their Role in Interference**

*Debi P. Nayak, Thomas M. Chambers, and Ramesh K. Akkina*

*Chapter 7*

**Variation in Influenza Virus Genes: Epidemiological, Pathogenic, and Evolutionary Consequences**

*Frances I. Smith and Peter Palese*

*Chapter 8*

**Specificity and Function of T Lymphocytes Induced by Influenza A Viruses**

*Jonathan W. Yewdell and Charles J. Hackett*

CHAPTER 1

# Genes and Proteins of the Influenza Viruses

ROBERT A. LAMB

## I. INTRODUCTION

This chapter describes the structure of the genes of influenza A, B, and C viruses. Influenza viruses contain a segmented single-stranded RNA genome that has been called negative stranded because the viral messenger RNA (mRNAs) are transcribed from the viral RNA segments. A great deal of new knowledge has been obtained about influenza A, B, and C viruses, since the last major multiauthored reviews of the genetics, molecular biology, and structural biology of influenza viruses (Palese and Kingsbury, 1983). The complete nucleotide sequence of the 8 RNA segments of the influenza A and B viruses has been obtained, and significant progress has been made with the sequencing of the influenza C virus genome. Other major developments include the following:

1. The three-dimensional structure of both major surface antigens, hemagglutinin and neuraminidase, has been determined from X-ray studies of crystallized proteins, and the structure of a neuraminidase–antibody complex has been obtained. In addition, the structure of the influenza virus hemagglutinin complexed with its receptor sialic acid has been elucidated, which may provide a basis for the rational design of antiviral drugs that would block viral attachment to cells.
2. In both influenza A and B viruses, previously unrecognized small integral membrane proteins, $M_2$ and NB, respectively, have been identified and extensively characterized.

ROBERT A. LAMB • Department of Biochemistry, Molecular Biology, and Cell Biology, Northwestern University, Evanston, Illinois 60208.

1

3. The influenza A virus N9 neuraminidase has also been found to exhibit hemagglutinating activity.
4. Influenza C virus glycoprotein exhibits both hemagglutinating and neuraminate-O-acetyl esterase activity.

The following sections review the structure of each of the RNA segments of influenza viruses and their encoded proteins. Emphasis is placed on topics unique to this chapter, as later chapters in this volume concern influenza virus replication, the hemagglutinin, the neuraminidase and antigenic variation. Inevitably, there is some overlap, but attempts have been made to minimize it and at the same time will make this chapter an overall survey. Little attempt is made to discuss the history of the disease, the response to infection, the ecology and epidemiology of the disease or the control of influenza through vaccination strategies or antiviral compounds. For these topics, the reader is referred to a recent review by Kilbourne (1987).

## A. Structure of the Genome

Influenza virus, when cultured in tissue culture cells or embryonated eggs, has a fairly regular appearance, when negatively stained and visualized in the electron microscope, of particles of 80–120 nm in diameter. The virion contains a lipid envelope containing surface projections or spikes radiating outward. These spikes are of two readily distinguishable types: the rod-shaped hemagglutinin and the mushroom shaped neuraminidase (Kilbourne, 1987). Inside the virus, and observable by thin sectioning of virus or by disrupting particles, are the ribonucleoprotein (RNP) structures, which contain the different RNA segments. The genetic information of influenza A and B viruses is contained in eight segments of single-stranded RNA and for influenza C virus is 7 segments of single-stranded RNA (see Lamb, 1983, and Air and Compans, 1983, for references to the early paper describing these findings). The RNA of the virus is not infectious, and the mRNAs are transcribed from the virion RNA (vRNA) by the virion-associated RNA-dependent RNA transcriptase. Thus, as by convention, mRNA is plus-stranded (Baltimore, 1971) and thus the influenza viruses are known as negative-strand RNA viruses.

The early evidence for a segmented genome of influenza viruses has been extensively reviewed (Lamb, 1983; Lamb and Choppin, 1983); a major step forward in understanding the structure of the influenza virus genome came from the electrophoretic separation of the virion RNAs on polyacrylamide gels containing 6 M urea (Bean and Simpson, 1976; Pons, 1976; Palese and Schulman, 1976; Ritchey et al., 1976; McGeoch et al., 1976) (Fig. 1). The critical study showing that the eight RNA segments of influenza A viruses were distinct was done by two-dimensional

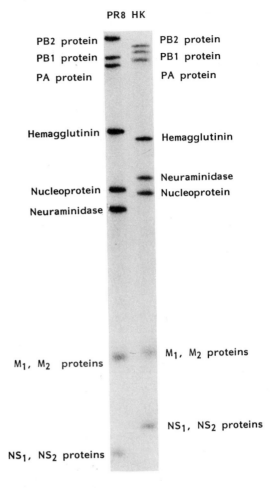

FIGURE 1. The influenza A virus RNA segments indicating their encoded gene products. The RNAs of influenza A/PR/8/34 and A/HK/8/68 viruses (PR8 and HK) were separated on a 2.6% polyacrylamide gel. RNA segments 1–8 are shown from top to bottom of the gel. The gene product assignments are discussed in the text. Under the gel electrophoresis conditions used the migration of the RNAs is dependent on the size as well as the secondary structure of the molecules. From Ritchey *et al.*, (1976) and kindly made available by Dr. Peter Palese, Mount Sinai School of Medicine, New York.

oligonucleotide fingerprinting (McGeoch *et al.*, 1976); this set the stage for a flurry of activity from several laboratories to characterize further the genome RNA segments and to determine their encoded protein products. The assignment of specific RNA segments to virus polypeptides was made in three ways:

1. A method pioneered by Palese, Schulman, and collaborators that depended on the apparent differences between strains in the electrophoretic mobilities of the RNA segments on polyacrylamide gels and the ability to distinguish proteins between strains, either by immunological methods for the hemagglutinin and neuraminidase or by differences in the electrophoretic mobilities of the polypeptides on protein gels [Recombinants were prepared between two parental strains and comparisons made of the mobilities of

TABLE I. Influenza A Virus Genome RNA Segments and Coding Assigments

| Segment | Length[a] (nucleotides) | mRNA length (nucleotides)[b] | Encoded polypeptide[b] | Nascent polypeptide length[d](aa) | Mol. wt. predicted[a] | Approx. no. molecules per virion[e] | Remarks[f] |
|---|---|---|---|---|---|---|---|
| 1 | 2341 | 2320 | PB2 | 759 | 85,700 | 30–60 | 7Me-Gppp Nm (cap) recognition of host-cell RNA; component of RNA transcriptase complex; endo-nucleolytic cleavage of host RNA? |
| 2 | 2341 | 2320 | PB1 | 757 | 86,500 | 30–60 | Catalyzes nucleotide addition; component of RNA transcriptase complex; endonucleolytic cleavage of host RNA? |
| 3 | 2233 | 2211 | PA | 716 | 82,400 | 30–60 | Component of RNA transcriptase complex; function unknown |
| 4 | 1778 | 1757 | HA | 566 | 61,468 | 500 | Major surface glycoprotein; sialic acid binding; proteolytic cleavage activation; fusion activity at acid pH; major antigenic determinant; trimer |
| 5 | 1565 | 1540 | NP | 498 | 56,101 | 1000 | Monomer binds to RNA to form coiled ribonucleoprotein; involved in switch from mRNA to template RNA synthesis; and in virion RNA synthesis |
| 6 | 1413 | 1392 | NA | 454 | 50,087 | 100 | Surface glycoprotein; neuraminidase activity; tetramer; antigenic determinant; less usual membrane orientation |

| 7 | 1027 | M$_1$ | 252 | 27,801 | 3000 | Major component of virion, probably underlies lipid bilayer; no known enzymatic activity |
|---|------|------|------|--------|------|-----------------------------------------|
|   | 316  | M$_2$ | 97  | 11,010 | 20–60 | Integral membrane protein, no cleavable signal sequence; high abundance in infected cells—M$_2$ sequence determines amantadine hydrochloride sensitivity |
|   | 276  | ?    | ? (9) | —    | —    | Spliced mRNA sequence predicts that 9 amino acid peptide could be made; no evidence for its synthesis |
| 8 | 890  | NS$_1$ | 230 | 26,815 | —  | High abundance, nonstructural protein found in nucleus, nucleolus, and cytoplasm; function unknown |
|   | 395  | NS$_2$ | 121 | 14,216 | —  | Nonstructural protein, cytoplasmic, and nuclear location; function unknown |

[a] For A/PR/8/34 strain: see Table I for references.
[b] Deduced form RNA sequence, excluding poly(A) tract.
[c] Determined by biochemical and genetic approaches (see text).
[d] Determined by nucleotide sequence analysis and protein sequencing (see text).
[e] Adapted from Compans and Choppin (1975) and Zebedee and Lamb (1988).
[f] See text for references.

the RNA segments and polypeptides between the parental types and each recombinant; from the analyses, gene assignments were made (reviewed in Palese, 1977; Lamb, 1983).]

2. A hybridization strategy dependent on base-sequence homologies between corresponding RNA segments of different strains and temperature-sensitive mutants (reviewed in Lamb, 1983)

3. A direct method using hybrid arrest of translation of individual influenza virus mRNAs using purified vRNA segments (reviewed in Lamb, 1983)

From all these studies, the gene assignment for influenza A virus is as follows: RNA segment 1 codes for PB2, 2 for PB1, 3 for PA, 4 for HA, 5 for NP, 6 for NA, 7 for $M_1$ and $M_2$ and 8 for $NS_1$ and $NS_2$. These findings are illustrated schematically in Fig. 1 and a summary of the gene assignments is shown in Table I. For the gene assignment of influenza B and C viruses the reader is referred to the individual sections below.

The enormous increase in knowledge of the genome structure of influenza viruses came with the development in 1979 of direct RNA sequencing methods and the molecular cloning of cDNA copies of the RNA segments of influenza viruses, which coupled with rapid DNA sequencing methods facilitated the determination of the complete sequence of an influenza A virus by 1982 (reviewed by Lamb, 1983). It was soon discovered that the 13 nucleotides at the 5'-terminus of an RNA segment are common to all RNA segments and that there is a common sequence of 12 nucleotides at the 3'-terminus of each segment. The postulated role of these sequences in transcriptase recognition and replication and the observation that they show partial and inverted complementarity is discussed in detail in Chapter 2 (this volume). In the following sections, the structure of each RNA segment and properties of their encoded polypeptide(s) of influenza A, B, and C viruses are described.

## II. RNA SEGMENTS 1, 2, AND 3: THE THREE POLYMERASE-ASSOCIATED PROTEINS FORM A COMPLEX TO TRANSCRIBE RNA

The three largest virion RNA segments (1–3) encode three proteins—PB1, PB2, and PA—which together form a complex and have a variety of enzymatic activities connected with the unusual mechanism of RNA transcription of influenza virus. Only the basic structural features of the genes and proteins are described in this chapter, as a detailed description of the biological roles of the proteins is to be found in Chapter 2 (this volume).

## A. Three P Proteins

The introduction of higher-resolution polyacrylamide protein gel systems during the early 1970s made it possible to show that influenza virions contained three discrete P proteins of apparent $M_r \approx$ 95,000–82,000 (Lamb, 1974; Lamb and Choppin, 1976; Inglis et al., 1976). Earlier studies had reported that one or two P proteins could be found in virions (Compans et al., 1970; Skehel and Schild, 1971; Skehel, 1972; Bishop et al., 1972). The three P proteins were found to be associated with transcriptionally active ribonucleoprotein structures and at an abundance of about 30–40 molecules of each P protein per virion particle (Inglis et al., 1976). Analysis of the RNA segment and protein electrophoretic migration differences between recombinants of strains A/PR/8/34 and A/HK/68 demonstrated that RNA segment 1 encoded the P3 protein, RNA segment 2 the P1 protein, and RNA segment 3 the P2 protein (Palese et al., 1977; Ritchey et al., 1977). However, with A/FPV/Rostock/34, a different coding assignment of polypeptides to RNA segments was obtained (Scholtissek et al., 1976). It soon became clear that the assignment of the names P1, P2, or P3, which was based solely on gel electrophoretic mobility and not biological function, was inappropriate, as it was found that the P2 protein of one strain was the functional equivalent of P3 of another strain (Almond and Barry, 1979).

A more useful nomenclature for the P proteins, based on an intrinsic property of each amino acid sequence, was suggested by Horisberger (1980). When viral polypeptides were separated on a two-dimensional gel system using isoelectric focusing (IEF) in the first dimension and sodium dodecyl sulfate–polyacrylamide gel electrophoresis (SDS–PAGE) in the second dimension, it was found that two of the P polypeptides were relatively basic and one was relatively acidic (Horisberger, 1980). Thus, the more rapidly migrating basic P protein encoded by RNA segment 1 was called PB2, the slowly migrating basic P protein coded by RNA segment 2 was called PB1 and the acidic P protein coded by RNA segment 3 was called PA (Horisberger, 1980; Ulmanen et al., 1981; Winter and Fields, 1982).

## B. PB1, PB2, and PA Gene Sequences

The complete sequence of RNA segments 1, 2, and 3 has been obtained for several human and avian viruses (Fields and Winter, 1982; Winter and Fields, 1982; Bishop et al., 1982a,b; Kaptein and Nayak, 1982; Sivasubramanian and Nayak, 1982; Roditi and Robertson, 1984; Robertson et al., 1984). RNA segments 1 and 2 are both 2341 nucleotides in length and code for proteins of 759 (PB2; $M_r \approx$ 85,700) and 757 (PB1; $M_r \approx$

86,500) amino acids, respectively. These are basic proteins at pH 6.5 having a net charge of +28. RNA segment 3 is 2233 nucleotides in length and codes for a protein (PA; $M_r \approx 82,400$) of 716 amino acids with a net charge of $-13.5$ at pH 6.5.

## C. Functions of the PB1, PB2, and PA Proteins

The known roles and other postulated activities of the individual P proteins is described in detail in Chapter 2 (this volume). Briefly, it has been found from UV crosslinking studies, photoaffinity labeling using cap analogues, and studies with specific *ts* mutants, that PB2 recognizes and binds to the $^7$mGpppGpN$_m$ type 1 cap structure at the 5' end of the host-cell RNA used to prime influenza virus transcription (Ulmanen *et al.*, 1981; Nichol *et al.*, 1981; Braam *et al.*, 1983). From studies of transcription reactions performed in the absence of specific nucleotide triphosphates coupled with experiments involving UV crosslinking, it became clear that PB1 is responsible for the chain initiation/elongation events of adding the first base to the host-cell-derived RNA primer (Braam *et al.*, 1983). It is not known which of the P proteins constitutes the endonuclease to cleave the host-cell RNA primer 10–13 nucleotides from the cap structure, nor is it known which protein is responsible for poly(A) addition to mRNAs.

## D. P Proteins Form a Complex That Migrates to the Cell Nucleus

Several groups have raised antisera to the individual P proteins after expressing the cloned cDNAs of RNA segments 1–3 in bacteria and purifying the fusion-protein antigens (Jones *et al.*, 1986; Detjen *et al.*, 1987; Akkina *et al.*, 1987). These antisera have been used to show that the P proteins in the cytoplasm and nucleus form a complex of PB1, PB2, and PA that is largely resistant to disruption by normal immunoprecipitation buffers and on sucrose gradients sediments from 11S to 22S (Detjen *et al.*, 1987; Akkina *et al.*, 1987). The stoichiometry of the P proteins in the complex seemingly varies depending on the specific antisera used to isolate the complex (Detjen *et al.*, 1987). These data could be interpreted to indicate that there is more than one type of P-protein complex, but the simpler explanation is that some components dissociate from the complex during its isolation. Almost all P proteins not associated with nucleocapsid structures are found in a complex, and there is very little free discrete soluble P protein (Detjen *et al.*, 1987). Immunofluorescence studies indicate that the P proteins are localized to the nucleus, which suggests that they are transported from the cytoplasm to the nucleus (Jones et al., 1986; Akkina et al., 1987). A biochemical examination of infected cell nuclei indicated that the P-protein complex can be isolated from the nucleus free of RNP particles (Detjen et al., 1987). It remains to be determined whether these P-protein complexes by themselves possess cap-binding and cap-dependent endonuclease activities or whether the activities are only activated when complexed with the RNA containing nucleocapsid.

To determine whether each P-protein contains a nuclear transport (karyophilic) signal, several groups have expressed the cloned cDNA to RNA segments 1–3 in eukaryotic vectors. PB2 (Jones et al., 1986) and PA (Smith et al., 1987) were expressed by vaccinia-recombinant viruses and were found to be localized to the nucleus using immunofluorescent techniques. PB1 was also found to be localized to the nucleus when expressed by an SV40 recombinant virus (Akkina et al., 1987). Thus, each of the P proteins appears to contain its own intrinsic nuclear transport signal, even though it seems likely that the P proteins are always transported to the nucleus as a complex in the influenza virus-infected cell. The nature of the amino acids constituting the karyophilic signal has not been defined further (see discussion below concerning NP and $NS_1$ proteins).

Cell lines expressing functional P proteins have been used to complement P-protein temperature-sensitive mutants and rescue virus particles at the nonpermissive temperature. When PB2 was expressed in a bovine papillomavirus vector under the control of a mouse metallothionein I promoter, it was found to relieve the block in viral mRNA synthesis exhibited by ts6, a mutant containing a cap-binding defect in the PB2 protein (Braam-Markson et al., 1985). Krystal and co-workers (1986) constructed cell lines that express only one or all three of the P proteins and have complemented temperature-sensitive mutations in the PB2 and PA proteins. Interestingly, it was found that cell lines expressing all three P proteins were several orders of magnitude more efficient at complementing ts mutants at the nonpermissive temperature in terms of titer of released virus than were cells expressing only the appropriate single P protein (Krystal et al., 1986). To explain the synergistic effect on virus growth complementation of ts mutants by having all three P proteins expressed in one cell, with some prescience it was proposed that the polymerase proteins combine to form a complex (Krystal et al., 1986). If a preformed trimer was used, it would have a significant competitive advantage over a cell expressing only a single P protein (Krystal et al., 1986).

To examine complex formation of the PB1, PB2, and PA proteins, the genes were expressed in insect cells using a baculovirus vector (St. Angelo et al., 1987). On co-infection of cells with all three vectors, although each of the P proteins was synthesized, only a complex containing the PB1 and PB2 proteins that lacked PA could be found (St. Angelo et al., 1987). Although this complex is different from that isolated from influenza virus-infected cells, this may be attributable to many trivial factors, including the low temperature needed to maintain the insect cells.

### E. Influenza B Virus P Proteins and RNA Segments

Influenza B virus encodes three P proteins having an apparent $M_r \approx$ 100,000–80,000, as determined by polyacrylamide gel electrophoresis (PAGE) (Racaniello and Palese, 1979a; Briedis et al., 1981). It can only be presumed that these P proteins are involved with recognition of capped host-cell RNAs, cap-dependent endonuclease activity, transcription, and replication because very little work on influenza B virus RNA synthesis has been reported since the description of RNA-dependent RNA polymerase activity in influenza B virions (Oxford, 1973). The best available evidence that the transcription process of influenza B virus is very similar to that of influenza A virus is the finding of additional heterogeneous sequences at the 5' end of cloned cDNAs of influenza B virus mRNAs (Briedis et al., 1982; Briedis and Lamb, 1982; Shaw et al., 1982; Shaw and Lamb, 1984; Briedis and Tobin, 1984; Kemdirim et al., 1986).

The complete nucleotide sequences of RNA segment 1 coding the PB1 protein (Kemdirim et al., 1986; DeBorde et al., 1988), RNA segment 2 encoding the PB2 protein (DeBorde et al., 1988), and RNA segment 3 encoding the PA protein (Akota-Amanfu et al., 1987; DeBorde et al., 1988) have been determined. RNA segment 1 is 2386 nucleotides in length with an open reading frame for PB1 of 752 amino acids ($M_r \approx$ 84,407). The protein has a net charge of +20 at pH 7.0. RNA segment 2 is 2396 nucleotides in length with an open reading frame for PB2 of 770 amino acids ($M_r \approx$ 88,035). The protein has a net charge of +24.5 at pH 7.0. RNA segment 3 is 2304 nucleotides in length with an open reading frame of 725 amino acids ($M_r \approx$ 83,000). The protein has a net charge of −9 at pH 6.5. A comparison of the predicted amino acid sequences of PB1 of influenza A and B viruses indicates that they share 61% amino acid identity. The position of 90% of the proline residues is conserved. This is the highest level of homology seen for any protein between the influenza A and B viruses and suggests a constraint on the evolution of this PB1 protein (Kemdirim et al., 1986). The PA proteins exhibit 38% identity between influenza A and B viruses (Akoto-Amanfu et al., 1987).

### F. Influenza C Viruses

Influenza C viruses contain three minor polypeptides designated P1, P2, and P3 (Compans et al., 1977; Petri et al., 1980). In addition, three RNA segments of 2350–2150 nucleotides have been identified (Racaniello and Palese, 1979b). It is presumed that these RNA segments encode the P proteins and that the proteins are involved in transcription and replication. To date no information is available concerning the nucleotide sequence of the RNA segments.

## III. RNA SEGMENT 4: SYNTHESIS, STRUCTURE, AND FUNCTION OF THE HEMAGGLUTININ

The influenza virus hemagglutinin (HA) is one of the most studied integral membrane proteins in virology and cell biology. It has three major functions during the influenza virus replicative cycle:

1. HA binds to a sialic acid-containing receptor on the cell surface, establishing the attachment of the infectious virus particle to the plasma membrane of the susceptible host cell.
2. HA initiates the infection process by mediating the fusion of the endocytosed virus particle with the endosomal membrane, thereby liberating the viral cores into the cytoplasm. For this event to occur, the HA of the infecting virus has to be activated by proteolytic cleavage.
3. HA is the major antigen of the virus against which neutralizing antibodies are produced and recurrent influenza epidemics are associated with changes in its antigenic structure.

The determination of the three-dimensional structure of HA (Wilson *et al.*, 1981) has provided a structural basis on which to relate a vast body of accumulated data concerning receptor binding, proteolytic activation, conformational changes leading to the fusion activity, and mapping of antigenic sites on the molecule. A detailed description of the structure and function of HA is found in Chapter 3 (this volume) and also in Wiley and Skehel (1987). Here, knowledge of the gene structure and synthesis of HA will be reviewed. Further details on the synthesis of HA and its intracellular transport can be found in Chapter 5 (this volume).

## A. General Introduction to the Hemagglutinin

The hemagglutinin was originally named because of the ability of the virus to agglutinate erythrocytes (Hirst, 1941; McClelland and Hare, 1941) by attachment to specific sialic acid glycoprotein receptors (Hirst, 1942). The HA constitutes a large percentage of the spikes detected on the surface of the virion by electron microscopy (Laver and Valentine, 1969). Further examination of HA by electron microscopy, after its removal from the virion by detergent or proteolysis, showed that the molecule consisted of more than one subunit (Laver and Valentine, 1969; Griffith, 1975; Wrigley *et al.*, 1977). Chemical crosslinking studies on HA (Wiley *et al.*, 1977), and ultimately determination of the X-ray crystallographic structure (Wiley *et al.*, 1981), showed that HA is a trimer of non-covalently linked monomers.

HA is synthesized as a single polypeptide chain ($M_r \approx 77,000$), which, depending on the virus strain, host-cell type, and growth conditions, can be proteolytically cleaved into two disulfide linked chains, $HA_1$ and $HA_2$ (Lazarowitz *et al.*, 1971, 1973a,b; Klenk *et al.*, 1972; Skehel, 1972). This cleavage of HA does not affect its antigenic or receptor-binding properties (Lazarowitz *et al.*, 1971, 1973a,b; Lohmeyer and Klenk, 1979; McCauley *et al.*, 1980), but it is essential for the virus to be infectious (Lazarowitz and Choppin, 1975; Klenk *et al.*, 1975), and hence is a critical determinant in pathogenicity (Bosch *et al.*, 1979, 1981; Rott, 1979; Deshpande *et al.*, 1987; Kawaoka *et al.*, 1984, 1985) and in the spread of infection in the organism (Rott *et al.*, 1980).

## B. RNA Segment 4 Gene Structure

The nucleotide sequence of RNA segment 4 encoding the HA has been obtained for a large number of antigenic subtypes and for many variants within a subtype in order to understand antigenic changes (Table II). The overall basic structure of the gene is very similar for all subtypes with an RNA segment ranging from 1742 to 1778 nucleotides encoding a polypeptide of 562–566 amino acids. The $HA_1$ chain is from 319 to 326 residues and $HA_2$ from 221 to 222 residues. Depending on the particular subtype and in some cases variant within a subtype, the number of amino acids lost on proteolytic cleavage between the $HA_1$ and $HA_2$ chains ranges from 1 to 6 residues (for references, see Table II: many of the details can be found in earlier reviews, e.g., Ward, 1981; Lamb, 1983). As the X-ray crystallographic structure of HA was obtained for the A/Aichi/68 (H3) strain, the gene structure of this strain will be used as an example (Verhoeyen *et al.*, 1980).

The A/Aichi/68 RNA segment 4 is 1765 nucleotides in length and contains a 5′ untranslated region of 20 nucleotides before the first AUG codon, which is followed by an open reading frame encoding a total of 566 amino acids. The hydrophobic signal peptide cleaved from HA after its insertion into the rough endoplasmic reticulum (RER) membrane is 16 amino acids in length. The mature HA contains 328 residues in the $HA_1$ chain, and 221 residues in the $HA_2$ chain and $HA_1$ and $HA_2$ are linked by a cysteine bridge. A single arginine residue is lost on proteolytic cleavage of the HA precursor, suggesting that two enzymes must be involved in the activation of HA. A trypsinlike enzyme would make the initial cleavage followed by an exopeptidase of the carboxypeptidase B type to remove the arginine from the cleavage site (Dopheide and Ward, 1978; Garten *et al.*, 1981).

The newly liberated N-terminus of $HA_2$ is hydrophobic, is highly conserved in HAs of different influenza virus strains (reviewed in Lamb,

TABLE II. Compilation of Nucleotide Sequences of Influeza Virus RNA Segments for the A, B, and C Viruses[a]

| RNA segment | Protein | Strain | References |
|---|---|---|---|
| 1 | PB2 | A/PR/8/34 | Fields and Winter (1982) |
| | | A/WSN/33 | Kaptein and Nayak (1982) |
| | | A/FPV/Rostock/34 | Roditi and Robertson (1984) |
| | | B/AA/1/66 | Deborde et al. (1988) |
| 2 | PB1 | A/PR/8/34 | Winter and Fields (1982) |
| | | A/WSN/33 | Sivasubramanian and Nayak (1982) |
| | | A/NT/60/68 | Bishop et al. (1982a) |
| | | B/Lee/40 | Kendirin et al. (1986) |
| | | B/AA/1/66 | DeBorde et al. (1988) |
| 3 | PA | A/PR/8/34 | Fields and Winter (1982) |
| | | A/NT/60/68 | Bishop et al. (1982b) |
| | | A/FPV/Rostock/34 | Robertson et al. (1984) |
| | | B/Singapore/222/79 | Akoto-Amanfu et al. (1984) |
| | | B/AA/1/66 | DeBorde et al. (1988) |
| 4 | HA | A/FPV/Rostock/34 | Porter et al. (1979) |
| | | A/Japan/305/57 | Gething et al. (1980) |
| | | A/Aichi/2/68 | Verhoeyen et al. (1980) |
| | | A/Victoria/3/75 | Min Jou et al. (1980) |
| | | A/NT/60/68 | Both and Sleigh (1980) |
| | | A/Memphis/102/72 | Sleigh et al. (1980) |
| | | A/Duck/Ukraine/63 | Fang et al. (1981) |
| | | A/WSN/33 | Hiti et al. (1981) |
| | | A/PR/8/34 | Winter et al. (1981a) |
| | | A/Bangkok/1/79 | Both and Sleigh (1981) |
| | | A/NT/60/68 | Both et al. (1983b) |
| | | A/Queensland/7/70 | Both et al. (1983b) |
| | | A/Hong Kong/107/71 | Both et al. (1983b) |
| | | A/England/42/72 | Both et al. (1983b) |
| | | A/Port Chalmers/1/73 | Both et al. (1983b) |
| | | A/Singapore/4/75 | Both et al. (1983b) |
| | | A/Mayo Clinic/1/75 | Both et al. (1983b) |
| | | A/Tokyo/1/75 | Both et al. (1983b) |
| | | A/England/864/75 | Both et al. (1983b) |
| | | A/Victoria/112/76 | Both et al. (1983b) |
| | | A/Allegheny County/29/76 | Both et al. (1983b) |
| | | A/Texas/1/77 | Both et al. (1983b) |
| | | A/Bangkok/1/79 | Both et al. (1983b) |
| | | A/Bangkok/2/79 | Both et al. (1983b) |
| | | A/Shanghai/31/80 | Both et al. (1983b) |
| | | A/England/321/77 | Hauptmann et al. (1983) |
| | | A/Seal/Mass/1/80 | Naeve and Webster (1983) |
| | | A/NJ/11/76 (Swine) | Both et al. (1983a) |
| | | A/Memphis/1/71 | Newton et al. (1983) |
| | | A/Memphis/1/71 | Newton et al. (1983) |
| | | A/Duck/5/77 | Kida et al. (1987) |
| | | A/Duck/8/80 | Kida et al. (1987) |
| | | A/Duck/33/80 | Kida et al. (1987) |
| | | A/Duck/21/82 | Kida et al. (1987) |

(continued)

TABLE II. (*Continued*)

| RNA segment | Protein | Strain | References |
|---|---|---|---|
| | | A/Duck/7/82 | Kida *et al.* (1987) |
| | | A/Duck/9/85 | Kida *et al.* (1987) |
| | | A/Duck/10/85 | Kida *et al.* (1987) |
| | | A/CK/Penn/1/83 | Kawaoka *et al.* (1984) |
| | | A/CK/Penn/1370/83 | Kawaoka *et al.* (1984) |
| | | A/CK/Md/2287/83 | Kawaoka and Webster (1985) |
| | | A/Turkey/VA/6962/83 | Kawaoka and Webster (1985) |
| | | A/CK/Penn/11709/83 | Kawaoka and Webster (1985) |
| | | A/CK/Wash/13413 | Kawaoka and Webster (1985) |
| | | A/CK/Penn/12094 | Kawaoka and Webster (1985) |
| | | A/CK/Penn/29708 | Kawaoka and Webster (1985) |
| | | A/CK/Vic/1/85 | Nestorowicz *et al.* (1987) |
| | | A/Starling/Vic/5156/85 | Nestorowicz *et al.* (1987) |
| | | A/USSR/90/77 | Concannon and Cummings (1984) |
| | | A/Eq/Miami/63 | Daniels *et al.* (1985) |
| | HA$_1$ domain only | A/Fort Warren/1/50 | Raymond *et al.* (1986) |
| | | A/England/1/51 | Raymond *et al.* (1986) |
| | | A/Fort Lenard Wood/1/52 | Raymond *et al.* (1986) |
| | | A/Queensland/34/54 | Raymond *et al.* (1986) |
| | | A/Denver/1/57 | Raymond *et al.* (1986) |
| | | A/USSR/90/77 | Raymond *et al.* (1986) |
| | | A/Lackland/3/78 | Raymond *et al.* (1986) |
| | | A/Brazil/11/78 | Raymond *et al.* (1986) |
| | | A/England/333/80 | Raymond *et al.* (1986) |
| | | A/India/6263/80 | Raymond *et al.* (1986) |
| | | A/Texas/12/82 | Raymond *et al.* (1986) |
| | | A/Texas/29/82 | Raymond *et al.* (1986) |
| | | A/Georgia/79/83 | Raymond *et al.* (1986) |
| | | A/Georgia/114/83 | Raymond *et al.* (1986) |
| | | A/Hong Kong/32/83 | Raymond *et al.* (1986) |
| | | A/Chile/1/83 | Raymond *et al.* (1986) |
| | | A/Duneidin/6/83 | Raymond *et al.* (1986) |
| | | A/Duneidin/27/83 | Raymond *et al.* (1986) |
| | | A/Victoria/7/83 | Raymond *et al.* (1986) |
| | HA | B/Lee/40 | Krystal *et al.* (1982) |
| | | B/Maryland/59 | Krystal *et al.* (1983a) |
| | | B/Hong Kong/73 | Krystal *et al.* (1983a) |
| | | B/Singapore/222/79 | Verhoeyen *et al.* (1983) |
| | HA | C/JHG/66 | Pfeifer and Compans (1984) |
| | | C/Cal/78 | Nakada *et al.* (1984b) |
| | | C/Taylor/1223/47 | Buonogurio *et al.* (1985) |
| | | C/Ann Arbor/1/50 | Buonogurio *et al.* (1985) |
| | | C/Great Lakes/1167/54 | Buonogurio *et al.* (1985) |
| | | C/Mississippi/80 | Buonogurio *et al.* (1985) |
| | | C/Yamagata/10/81 | Buonogurio *et al.* (1985) |
| | | C/England/892/83 | Buonogurio *et al.* (1985) |
| | | C/Pig/Beijing/10/81 | Buonogurio *et al.* (1985) |
| | | C/Pig/Beijing/115/81 | Buonogurio *et al.* (1985) |
| | | C/Pig/Beijing/439/82 | Buonogurio *et al.* (1985) |
| 5 | NP | A/PR/8/34 | Winter and Fields (1981) |

## TABLE II. (Continued)

| RNA segment | Protein | Strain | References |
|---|---|---|---|
| | | A/PR/8/34 | Van Rompuy et al. (1981) (with corrections see Winter and Fields, 1981) |
| | | A/NT/60/68 | Huddleston and Brownlee (1982) |
| | | A/FPV/Rostock/34 | Tomley and Roditi (1984) |
| | | A/Parrot/Ulster/33 | Steuler et al. (1985) |
| | | A/Udorn/307/72 | Buckler-White and Murphy (1986) |
| | | A/Mallard/NY/6750/78 | Buckler-White and Murphy (1986) |
| | NP | B/Singapore/222/79 | Londo et al. (1983) |
| | | B/Lee/40 | Briedis and Tobin (1984) |
| | | B/AA/1/66 | DeBorde et al. (1988) |
| | NP | C/Cal/78 | Nakada et al. (1984) |
| 6 | NA | A/PR/8/34 | Fields et al. (1981) |
| | | A/WSN/33 | Hiti and Nayak (1982) |
| | | A/NT/60/68 | Bentley and Brownlee (1982) |
| | | A/Udorn/72 | Markoff and Lai (1982) |
| | | A/Victoria/3/75 | Van Rompuy et al. (1982) |
| | | A/RI/5/57 | Elleman et al. (1982) |
| | | A/Bangkok/1/79 | Martinez et al. (1983) |
| | | A/USSR/90/77 | Concannon et al. (1984) |
| | | A/Parrot/Ulster/73 | Steuler et al. (1984) |
| | | A/Tokyo/3/67 | Lentz et al. (1984) |
| | | A/G70C/75 | Air et al. (1985) |
| | | A/Cor/16/74 | Dale et al. (1986) |
| | | A/Ken/1/81 | Dale et al. (1986) |
| | | A/Whale/Maine/1/84 | Air et al. (1987) |
| | | A/tern Australia/G70C/75 | Air et al. (1987) |
| | NA | B/Lee/40 | Shaw et al. (1982) |
| 7 | $M_1$ and $M_2$ | A/PR/8/34 | Winter and Fields (1980) |
| | | A/PR/8/34 | Allen et al. (1980) |
| | | A/Udorn/72 | Lamb and Lai (1981) |
| | | A/FPV/Rostock/34 | McCauley et al. (1982) |
| | | A/Swine/Iowa/30 | Nakajima et al. (1984) |
| | | A/Bangkok/1/79 | Ortin et al. (1983) |
| | | A/FPV/Weybridge/27 | Markushin et al. (1988) |
| | | A/Mallard/NY/6750/78 | Buckler-White et al. (1986) |
| | | A/WSN/33 | Markushin et al. (1988) |
| | | A/WSN/33 | Zebedee and Lamb (1989) |
| | | A/WS/33 | Zebedee and Lamb (1989) |
| | | A/Singapore/1/57 | Zebedee and Lamb (1989) |
| | | A/Port Chalmers/1/73 | Zebedee and Lamb (1989) |
| | | A/USSR/90/77 | Zebedee and Lamb (1989) |
| | | A/FW/1/50 | Zebedee and Lamb (1989) |
| | M | B/Lee/40 | Briedis et al. (1982) |
| | | B/Singapore/222/79 | Hiebert et al. (1986) |
| | | B/AA/1/66 | DeBorde et al. (1988) |
| 6 | M | C/JJ/50 | Yamashita et al. (1988) |
| 8 | $NS_1$ and $NS_2$ | A/Udorn/72 | Lamb and Lai (1980) |

(continued)

TABLE II. (Continued)

| RNA segment | Protein | Strain | References |
|---|---|---|---|
| | | A/FPV/Rostock/34 | Porter et al. (1980) |
| | | A/PR/8/34 | Baez et al. (1980) |
| | | A/PR/8/34 | Winter et al. (1981b) |
| | | A/Duck/Alberta/60/76 | Baez et al. (1981) |
| | | A/Alaska/6/77 | Buonogurio et al. (1984) |
| | | A/FM/1/47 | Krystal et al. (1983b) |
| | | A/FW/1/50 | Krystal et al. (1983b) |
| | | A/USSR/90/77 | Krystal et al. (1983b) |
| | | A/WSN/33 | Buonogurio et al. (1986) |
| | | A/Bellamy/42 | Buonogurio et al. (1986) |
| | | A/Maryland/2/80 | Buonogurio et al. (1986) |
| | | A/Houston/18515/84 | Buonogurio et al. (1986) |
| | | A/Houston/23284/85 | Buonogurio et al. (1986) |
| | | A/Denver/1/57 | Buonogurio et al. (1986) |
| | | A/Ann Arbor/6/60 | Buonogurio et al. (1986) |
| | | A/Berkley/1/68 | Buonogurio et al. (1986) |
| | | A/Houston/24269/85 | Buonogurio et al. (1986) |
| | | A/Turkey/Oregon/71 | Norton et al. (1987) |
| | | A/Chicken/Japan/24 | Nakajima et al. (1987) |
| | | A/Duck/England/56 | Nakajima et al. (1987) |
| | | A/Duck/Ukraine/1/63 | Nakajima et al. (1987) |
| | | A/Tern/South Africa/61 | Nakajima et al. (1987) |
| | | A/Mynah/Haneda-Thai/76 | Nakajima et al. (1987) |
| | NS$_1$ and NS$_2$ | B/Lee/40 | Briedis and Lamb (1982) |
| | | B/Yamagata/1/73 | Norton et al. (1987) |
| | | B/AA/1/66 | DeBorde et al. (1988) |
| 7 | NS$_1$ and NS$_2$ | C/Cal/78 | Nakada et al. (1985) |

[a]This listing of nucleotide sequences of influenza virus RNA segments makes no claims as to its completeness. For all the sequences available, the reader is referred to the GENBANK/EMBL data base. Papers cited are those that intrigued this reviewer and are biased toward the first descriptions of specific RNA sequences or where a nucleotide sequence illustrated a point of biological significance discussed in the review. Apologies are given to those whose papers have not been cited.

1983), and has been implicated in participating in the fusion activity (Gething et al., 1978, 1986; Richardson et al., 1980; Skehel et al., 1982; Wilson et al., 1981; Daniels et al., 1985; Doms et al., 1985). Near the C-terminus of HA$_2$ is a stretch of 27 hydrophobic residues, which acts as a stop-transfer signal during the translocation of HA across the membrane of the RER and anchors HA in the lipid bilayer, leaving a cytoplasmic tail of 10 amino acids (Sveda et al., 1982; Gething et al., 1982; Doyle et al., 1985, 1986).

## C. Three-Dimensional Structure of the Hemagglutinin

Bromelain treatment of virus released an antigenically and structurally intact trimeric ectodomain of HA (BHA) which is water soluble and contained all of HA$_1$ and the first 175 of 221 residues of HA$_2$ (Brand

and Skehel, 1972; Skehel and Waterfield, 1975; Wiley *et al.*, 1977; Waterfield *et al.*, 1979). This molecule can be crystallized and the three-dimensional structure has been determined from X-ray studies to 3 Å resolution (Wilson *et al.*, 1981).

The structure of HA is described in detail in Chapter 3 (this volume). Briefly, the molecule is an elongated cylinder 135 Å long consisting of two major components: (1) a long fibrous stem extending 76 Å from the lipid bilayer containing two antiparallel α-helices that terminate near the membrane in a compact five-stranded antiparallel β-sheet globular fold; and (2) on top of the elongated stem, a globular domain of antiparallel β-sheet structure. This region is composed entirely of residues from $HA_1$ and is connected to the $HA_2$ stem by only two antiparallel chains from $HA_1$. The N-terminal region of $HA_1$ is very close to the membrane, and it seems possible that the nascent chain may have folded into a mature form in the endoplasmic reticulum while tethered to the membrane not only by the hydrophobic C-terminal anchor domain but also by the N-terminal signal sequence (Wilson *et al.*, 1981). Cleavage by signal peptidase of the signal peptide may be a fairly slow and late event during the synthesis of the nascent chain. The hydrophobic N-terminus of $HA_2$ is separated from the C-terminus of $HA_1$ by 22 Å, indicating that a conformational change accompanied the cleavage of the two molecules. This hydrophobic fusion peptide in each monomer is tucked into the trimer interface ~35 Å from the lipid bilayer. The HA receptor-binding site is a pocket located on the distal end of the molecule and is composed of amino acids that are conserved between isolates (Wilson *et al.*, 1981; Wiley and Skehel, 1987). Confirmation that the pocket is the receptor-binding site was provided by the selection of single amino acid substitution mutants of HA with altered specificity such that they would bind sialic acid containing $\alpha_{2\rightarrow6}$ but not $\alpha_{2\rightarrow3}$ linkages to galactose and vice versa (Rogers *et al.*, 1983; Weis *et al.*, 1988). Six N-linked carbohydrate chains are attached to $HA_1$ and one chain to $HA_2$. All the carbohydrate chains are of the complex type, except for two on $HA_1$, which are of the high mannose type (Ward and Dopheide, 1980). All the sites for carbohydrate addition are on the lateral surfaces of HA, except one, which appears to stabilize the oligomeric contacts between globular units at the top of the structure (Wilson *et al.*, 1981). Although no required function of the carbohydrate has been shown, it is likely that carbohydrate addition is needed for the folding of the HA molecule into its native form. HA of some strains synthesized in the presence of the inhibitor of N-linked glycosylation, tunicamycin, fails to be recognized by some classes of monoclonal antibody and is retained in the RER (Gething *et al.*, 1986).

Determination of the three-dimensional structure of HA also permitted the mapping of the antigenic sites on the molecule. This was done by determining the location of amino acid changes in the HAs of natural influenza virus isolates and in HAs of antigenic variants selected by growth in the presence of monoclonal antibodies (Laver *et al.*, 1979, 1980,

1981; Yewdell *et al.*, 1979; Both and Sleigh, 1981; Caton *et al.*, 1982) and correlating these changes to the HA structure (Wilson *et al.*, 1981; Wiley *et al.*, 1981). The regions of antigenic variation cover much of the surface of the distal domain of the molecule, including residues surrounding the receptor-binding pocket, and they have been divided into five operational antigenic sites (Wiley *et al.*, 1981; Skehel *et al.*, 1984). The critical assumption in this work is that the site of a single amino acid change, in response to virus growth in the presence of the antibody, is the antibody-binding site. This has been tested by determining the X-ray structure of the HA of a variant virus (Knossow *et al.*, 1984). The structural changes between the variant and the parental HA were found in the local neighborhood of the amino acid substitution, indicating that small local variations in structure permit escape from neutralization by antibodies (Knossow *et al.*, 1984).

## D. Synthesis of HA, Co-translational and Post-translational Modifications, Oligomerization, and the Exocytotic Pathway

The available data indicate that the biosynthesis of HA follows the same pathway as cellular integral membrane proteins. When the mRNA for HA is translated, the signal sequence, after emerging from a protease inaccessible space in the ribosome is recognized by the signal-recognition particle (SRP) which mediates the co-translational insertion of the nascent polypeptide chain across the membrane of the RER. The elongating HA chain is glycosylated in the RER (Compans, 1973b; Hay, 1974; Klenk *et al.*, 1974; Elder *et al.*, 1979; McCauley *et al.*, 1980; Hull *et al.*, 1988). Recently, considerable attention has been focused on the time course of folding of the HA monomer and its assembly into trimers. It has been shown that trimerization of HA occurs post-translationally from a free pool of monomers within 7–10 min of synthesis (Gething *et al.*, 1986; Copeland *et al.*, 1986, 1988; Boulay *et al.*, 1988; Yewdell *et al.*, 1988). Most of the evidence suggests that oligomerization occurs in the RER (Gething *et al.*, 1986; Copeland *et al.*, 1988) and that correct folding of HA and its oligomerization are sequential and obligatory events, before HA can be transported from the RER to the *cis*-Golgi complex (Gething *et al.*, 1986; Copeland *et al.*, 1986, 1988). Many mutant HAs constructed by protein/genetic engineering fail to be transported from the RER and are blocked at different stages of the folding and/or oligomerization pathway (Gething *et al.*, 1986; Doyle *et al.*, 1985, 1986).

Trimming of the high mannose carbohydrate chains occurs in the Golgi complex. $HA_2$ is also modified in the Golgi complex by the addition of palmitic acid in an ether linkage susceptible to cleavage by hydroxylamine, but it is not clear whether this is to a serine or threonine residue or whether it is a thioether linkage to cysteine (Schmidt, 1982). The site of palmitic acid addition is not known, and it could be in the cytoplasmic tail, as found for the modification on vesicular stomatitis virus G protein (Rose *et al.*, 1984) or in the ectodomain. Sialic acid residues are added to the carbohydrate chains of HA in the *trans*-Golgi compartment (Basak *et al.*, 1985), but in an influenza virus-infected cell the sialic acid is not found on the mature HA at the cell surface or in virions (Klenk *et al.*, 1970) because of the action of the viral neuraminidase. The transport of HA to the cell surface is discussed in greater detail in Chapter 5 (this volume), particularly with respect to the targeting of HA to the apical surface of polarized epithelial cells (Rodriguez-Boulan and Sabatini, 1978; Roth *et al.*, 1979, 1983, 1987; McQueen *et al.*, 1986).

## E. Cleavage Activation

Proteolytic cleavage of the HA precursor is mediated by cellular proteases to yield the disulfide-linked $HA_1$ and $HA_2$ subunits and has been reported to occur just before or at the time of insertion of the protein into the host-cell membrane (Klenk *et al.*, 1975). Cleavage of the HA molecule into $HA_1$ and $HA_2$ is a prerequisite for virus infectivity (Klenk *et al.*, 1975; Lazarowitz *et al.*, 1975) and permits fusion of the viral envelope with the secondary endosome. A comparison of the HA amino acid sequences of virulent and aviralent H5 and H7 subtypes of influenza virus indicate that the virulent viruses contain a series of basic residues in the connecting peptide, whereas most of the avirulent strains contain only a single arginine residue (Kawaoka *et al.*, 1987; Nestorowicz *et al.*, 1987). The suggestion was originally made that the number of basic residues in the connecting peptide would correlate with cleavability, hence virulence of the virus (Bosch *et al.*, 1981). A direct test of this hypothesis was done by constructing HA molecules using genetic engineering that contained varying residues in the cleavage site (Kawaoka and Webster, 1988); the data indicate that the relationship is not straightforward. In addition to the number of basic residues critical for cleavage activation in tissue culture, unknown structural features of HA seem to be involved, which include the length of the connecting peptide affecting a structural feature of HA (Kawaoka and Webster, 1988). Cleavage activation of HA is analogous to the process by which several hormone precursors and enzymes (e.g., proinsulin, proenkephalin, prosomatostatin, progastrin, and proalbumin) are activated (reviewed by Steiner *et al.*, 1984). For many of these precursors, cleavage occurs either in the *trans*-Golgi or in secretion granules formed from the *trans*-Golgi complex (Steiner *et al.*, 1984); it is likely that this slightly acidified subcellular compartment is also the site of cleavage of those HAs that contain more than a single basic residue. Cleavage of HA causes the first of several conformational changes that the mature structure can undergo as the N-terminus of $HA_2$ is 22 Å from the C-terminus of $HA_1$ (Wilson *et al.*, 1981).

## F. Viral Entry and Membrane Fusion

Influenza viruses once attached to the cell-surface receptor by HA are thought to enter cells by a process involving endocytosis and to fuse their membranes with the membranes of endosomes, a process activated by conformational changes in HA at the acidified pH of the endosome. The pathway of viral entry has been reviewed recently (White et al., 1983); the low-pH-induced profound conformational changes that activate fusion activity of HA are described in detail in Chapter 3 (this volume) and also by Wiley and Skehel (1987) and are not described here.

## G. Influenza B Virus Hemagglutinin

The influenza B virus HA is thought to resemble that of the influenza A viruses in both structure and biochemical properties. The nucleotide sequence of RNA segment 4 of several isolates of influenza B virus has been obtained (Krystal et al., 1982, 1983b; Verhoeyen et al., 1983). The gene structure of isolate B/Lee/40 (Krystal et al., 1982) will be used here as an example. RNA segment 4 contains 1882 nucleotides and has a single open reading frame that can encode a polypeptide of 584 amino acids. By analogy to influenza A viruses, it can be deduced that $HA_1$ would be composed of 346 residues and $HA_2$ would contain 223 amino acids. Only single basic residues are found in the connecting peptides of influenza B virus HAs and not basic pairs, but in the absence of direct protein sequence data the exact C-terminus of $HA_1$ cannot be determined. A computer-assisted analysis indicates that 13 of 15 cysteine residues found in influenza A/PR/8/34 virus HA are conserved in influenza B virus HA and that the overall amino acid identity is 24% and 39% in $HA_1$ and $HA_2$, respectively (Krystal et al., 1982). When the HA sequence of A/PR/8/34 is compared with A/Aichi/2/68 (Verhoeyen et al., 1980), the $HA_1$ and $HA_2$ identity is 35% and 53%, respectively. Thus, the B HA sequence is somewhat removed from, but still homologous to, the influenza A HA sequences. Further analysis of the sequence of influenza B virus isolates B/Maryland/59, B/Hong Kong/73, and B/Lee/40 and correlation with the three-dimensional structure of influenza A virus HA suggest that the major structural features are likely to be very similar (Krystal et al., 1983a).

## H. Influenza C Virus Glycoprotein

In contrast to the two distinct glycoproteins of influenza A and B viruses, which exhibit HA and NA activities, only a single type of glycoprotein has been detected in influenza C virions. This glycoprotein displays receptor-binding activity for 9-O-acetyl-N-acetylneuraminic acid, a penetration (fusion) function, and a receptor-destroying activity, which is a neuraminate-O-acetylesterase.

The influenza C glycoprotein is synthesized as a single polypeptide chain ($M_r \approx 88,000$) cleaved by a trypsinlike protease into two subunits that are disulfide linked (Herrler *et al.*, 1979). Cleavage causes a concomitant increase in specific infectivity and is a prerequisite for viral infectivity (Herrler *et al.*, 1979). Activation of infectivity is accompanied by activation of hemolysis and fusion activities of influenza C virions (Ohuchi *et al.*, 1982). The requirement for acid pH of the fusion activity suggests that, like influenza A viruses, influenza C virus uncoating involves endocytosis and uptake into secondary endosomes followed by fusion of the viral envelope with the endosomal membrane.

The complete nucleotide sequence of RNA segment 4 of several isolates of influenza C virus has been obtained (Pfeifer and Compans, 1984; Nakada *et al.*, 1984b; Buonogurio *et al.*, 1985). The gene structure of isolate influenza C/JHB/1/66 will be used as an example (Pfeifer and Compans, 1984). The viral RNA segment is 2073 nucleotides in length and can encode a polypeptide of 655 amino acids with a predicted molecular weight of 72,063. There are eight sites for the potential addition of N-linked carbohydrate. The influenza C glycoprotein shares structural features with HA of influenza A and B viruses, including three hydrophobic domains thought to act as a signal sequence, a fusion peptide of 17 hydrophobic residues and a membrane anchorage domain. Although the fusion peptide has considerable homology to the similar domain of the influenza A and B viruses, the remainder of the protein lacks direct sequence homology to these glycoproteins.

It has long been known that there is a major difference between influenza C virus and the influenza A and B viruses, because influenza A and B viruses destroy their own receptors on erythrocytes without affecting the receptor for influenza C virus (Hirst, 1950). Conversely, influenza C viruses bind to receptors on erythrocytes and possess a receptor-destroying enzyme (RDE) (Hirst, 1950) that does not affect receptors for influenza A and B viruses (Kendal, 1975; Meier-Ewert *et al.*, 1978). Further evidence that the RDE of influenza C virus is not a neuraminidase was obtained by finding that it did not liberate neuraminic acid from any NA substrate and was not inhibited by soluble neuraminic acid-containing glycoproteins (Kendal, 1975). Attempts at defining the specificity of the RDE led to the isolation of a rat serum inhibitor of hemagglutination (O'Callaghan *et al.*, 1977) that contains sialic acid. Inhibition can be abolished by prior treatment with certain neuraminidases (Herrler *et al.*, 1985a; Kitame *et al.*, 1985). Recently, it was shown that the receptor for influenza C virus is 9-O-acetyl-N-acetylneuraminic acid (Rogers *et al.*, 1986) and that the RDE is a neuraminate-O-acetylesterase (Herrler *et al.*, 1985b). Evidence that the influenza C glycoprotein contains the esterase activity was demonstrated by expressing the RNA segment 4 product using an SV40-based vector and showing that the expressed protein exhibited both receptor-binding (hemagglutinin) and receptor-destroying (esterase) activity (Vlasak *et al.*, 1987).

## IV. RNA SEGMENT 5: THE NUCLEOCAPSID PROTEIN FORMS THE STRUCTURAL MONOMER UNIT OF THE RIBONUCLEOPROTEIN PARTICLES

The nucleocapsid protein (NP) is the major structural protein that interacts with the viral RNA segment, to form the ribonucleoprotein particles (RNP). Associated with each RNP are the three transcriptase-associated polypeptides: PB1, PB2, and PA. A calculation based on stoichiometry of NP in RNPs suggests that each NP molecule can interact with 20 nucleotides (Compans and Choppin, 1975).

### A. Structure of RNPs and Involvement in RNA Synthesis

The RNPs of influenza virus were observed in the electron microscope, after ether disruption of virions, as 9- to 10-nm-diameter particles of variable length, averaging 60 nm (Hoyle *et al.*, 1961). Later measurements on detergent-disrupted virions indicated that the RNPs were of 15-nm diameter (Pons *et al.*, 1969; Compans *et al.*, 1972) and that they have a periodic distribution of alternate deep and shallow grooves along the strand, suggesting that the RNP is a double helix of a strand folded back on itself with a loop at one end (Schulze *et al.*, 1970; Compans *et al.*, 1972; Murti *et al.*, 1988). In other studies, RNP forms with loops at both ends have been observed, and it is clear that the nucleocapsid helices are right-handed (Heggeness *et al.*, 1982). The RNP structures have been shown to contain the individual genome RNA segments (Duesberg, 1969; Pons, 1971; Compans *et al.*, 1972; Caliguiri and Gerstein, 1978). Unlike the structure of the best studied ribonucleoprotein, that of tobacco mosaic virus (TMV) (Klug, 1980), which has its RNA internal and well protected, the influenza virus RNP is thought to have its RNA bound to the outside of the structure, since it can be displaced by polyvinylsulfate (Pons *et al.*, 1969; Goldstein and Pons, 1970) and is susceptible to digestion with RNase without disrupting the RNP structure (Duesberg, 1969; Kingsbury and Webster, 1969; Pons *et al.*, 1969; Murti *et al.*, 1980). RNA crosslinking experiments on intact RNPs indicate that the 3' and 5' end of each virion RNA segment is base paired, providing further evidence for the RNPs being folded back on themselves (Hsu *et al.*, 1987). Each RNP contains a promoter for transcriptase activity (Abrahams, 1979; Pons and Rochavansky, 1979); and each of the RNPs has transcriptase activity (Plotch *et al.*, 1981; Ulmanen *et al.*, 1981). High-resolution immunoelectron microscopy of ribonucleoproteins using antisera to PB1, PB2, and PA suggests that there is single polymerase binding site located at, or very close to, the end of each RNP (Murti *et al.*, 1988). The role of the NP subunit in transcription has not been identified because of the difficulty in dissociating the polymerase protein complex (PB1 + PB2 + PA), NP, and RNA from one another and reconstituting any activity. This type of approach was very successful in elucidating the roles of the polypeptides associated with the RNA polymerase of vesicular stomatitis virus (VSV) (Emerson and Yu, 1975). Monoclonal antibodies to NP inhibit transcription (Van Wyke *et al.*, 1980), but this may be because of steric hindrance rather than blocking an enzymatic role of NP. The analysis of *ts* mutants of NP indicate that NP is involved in viral RNA replication (Barry and Mahy, 1979; Murphy *et al.*, 1982; Thiery and Danos, 1982; Mahy, 1983). Studies of viral RNA replication *in vitro* have shown that NP is required for antitermination during RNA synthesis and for elongation in virion RNA synthesis (Beaton and Krug, 1986; Shapiro and Krug, 1988). It has also been shown that a *ts* mutant in NP was *ts* in template RNA synthesis, but not in mRNA synthesis *in vitro* (Shapiro and Krug, 1988). These findings suggest that there are two populations of NP in the infected cell: nucleocapsid-associated and non-nucleocapsid associated. The role of NP in viral RNA replication is discussed in greater detail in Chapter 2 (this volume).

As might be expected, the RNPs, which display the enzymatic activities for transcription, are infectious when introduced into cells by DEAE-mediated transfection (Hirst and Pons, 1973), although at a low level. This is presumably because of the difficulty of getting eight active and ribonuclease-sensitive RNPs into a single cell and also because the normal viral entry route, via the endocytic pathway, has been bypassed. Higher levels of biological activity of transfected RNPs have been observed in the marker rescue of *ts* mutants (Hirst and Pons, 1973), probably by supplying an enzymatic activity in *trans*.

## B. Properties of the Nucleocapsid Protein

The nucleotide sequence of RNA segment 5 has been obtained for several human and avian strains (e.g., Winter and Fields, 1981; Huddleston and Brownlee, 1982; Tomley and Roditi, 1984; Steuler *et al.*, 1985; Buckler-White and Murphy, 1986). The RNA segment is 1565 nucleotides in length with, in the mRNA sense, a 5′ noncoding region of 45 nucleotides and a 3′ noncoding region of 26 nucleotides. The open reading frame of 1494 nucleotides codes for 498 amino acids, giving a predicted protein of $M_r \approx 56,101$. The protein is rich in arginine residues and has a net positive charge of +14 at pH 6.5. There are no clusters of basic residues, which might have been predicted for the interaction of NP with the acidic phosphate residues of RNA; therefore, the RNA is probably associated with many regions of the NP molecule to neutralize the phosphate residue charges (Winter and Fields, 1981). The NP protein sequence contains few cysteine residues, as expected for a protein that is intra-cellular, and most likely lacks disulfide bonds. NP has been found to be phosphorylated at up to one serine residue per molecule (Kamuta and Watanabe, 1977; Privalsky and Penhoet, 1981; Petri and Dimmock, 1981), but it is not clear what percentage of NP molecules are phosphory-lated or whether phosphorylation is essential for function.

NP is the type-specific antigen of influenza viruses used to dis-tinguish the A, B, and C viruses; it was originally thought to be anti-genically invariant among influenza A viruses. However, with influenza A viruses, minor antigenic differences between NP of different strains have been observed, both with polyclonal antibodies (Schild et al., 1979) and with monoclonal antibodies (Van Wyke et al., 1980). Analysis of the nucleotide sequences and of the predicted amino acid sequences of influenza A virus NP has led to the proposal that there are two classes of NP protein: one avian type and one human type (Buckler-White and Murphy, 1986).

## C. Nuclear Transport and Karyophilic Sequences

Influenza virus transcription and replication occur in the host-cell nucleus (Herz et al., 1981; Jackson et al., 1982; Jones et al., 1986; Shapiro et al., 1987); a mechanism must exist for the transport of RNPs to the nucleus from the cytoplasm. The movement of NP can be followed from the cytoplasm to the nucleus (Briedis et al., 1981), and RNPs have been isolated from purified nuclei (Krug, 1972; Krug and Etkind, 1973). To examine whether the nuclear targeting of NP is a property inherent of NP, the gene for NP has been expressed in eukaryotic cells from an SV40 vector (Lin and Lai, 1983) and in established cell lines (Ryan et al., 1986). In both cases, NP was shown by fluorescence analysis to accumulate in the nucleus. To identify the sequence responsible for the nuclear ac-cumulation of NP, a variety of nested deletions of the NP gene were made, inserted into a vector, and the plasmids microinjected into Xenopus oocytes; the cellular location of the expressed NP protein was then examined. These studies indicated that the region of NP consisting of residues 327–345 was necessary for accumulation (retention) in the nucleus (Davey et al., 1985). This was further supported by experiments in which these residues of NP were shown to be sufficient to direct and to retain chimpanzee α-globin in the nucleus (Davey et al., 1985). The pre-cise boundaries of the karyophilic sequence of NP have not been further defined experimentally, although it is interesting to note that in aligning the influenza A and B virus NP sequences, the region of residues 336–345 is the second most highly conserved stretch of 10 amino acids between the A- and B-type NP sequences. These sequences differ only at residue 338 (Phe to Tyr); thus, this region may be conserved because of its karyo-philic function (Davey et al., 1985). The sequence of residues 336–345 is

N-Ala-Ala-Phe-Glu-Asp-Leu-Arg-Val-Leu-Ser-C, which bears little resemblance to the basic karyophilic sequence of SV40 T antigen (Kalderon et al., 1984b), yet it does have some similarity to the karyophilic sequence in yeast $\alpha_2$-protein (Hall et al., 1984). One peculiarity of these experiments using Xenopus oocyte is that, whereas NP showed nuclear accumulation mimicking the results found in mammalian or avian cells, the $NS_1$ protein expressed in a similar manner did not (Davey et al., 1985). Further discussion of nuclear targeting and retention signals is to be found in the section of the $NS_1$ and $NS_2$ polypeptides.

## D. RNA–Protein Interactions

The mechanism of assembly of the nucleocapsid is poorly understood. Both virion RNA (− strand) and template (+ strand) RNAs are found associated with NP molecules, whereas the viral mRNAs (+ strand) are not encapsidated (Pons, 1971; Hay et al., 1977); therefore, there must be a selection mechanism to prevent NP from associating with mRNAs. Although influenza mRNAs may not necessarily contain NP molecules, this should not be interpreted as implying that the mRNAs are naked in the cytoplasm. They are most likely associated with proteins such as those of the mRNP or nuclear RNP complexes found with cellular pre-mRNAs or mRNAs and some viral mRNAs (reviewed by Dreyfuss, 1986). Such an association has not been explored for influenza virus mRNAs.

The nucleocapsid protein synthesized from cloned NP DNA in bacteria is capable of binding to influenza virion RNA, to other single-stranded RNAs of nonviral origin, to single-stranded DNA, and to dextran sulfate (Kingsbury et al., 1987). This association of NP with single-stranded RNA has some properties similar to that of authentic nucleocapsids: (1) the RNA is susceptible to RNase digestion; (2) the protein is resistant to release by detergents and urea; and (3) the complexes have a density of 1.34 g/cm³ in CsCl (Kingsbury et al., 1987). However, these structures differ in their morphological characteristics from native RNPs. Although the association of NP with RNA lacks selectivity, this may be because the assay system used failed to represent adequately the environment of the intracellular compartment where nucleocapsids are assembled—most likely a specific region of the nucleus (Kingsbury et al., 1987). It is interesting to note the similarities with TMV, which, in vivo, self-assembles its RNA and coat protein. Under some pH conditions, the coat protein will associate with heterologous synthetic polyribonucleotides (Fraenkel-Conrat and Singer, 1964; Matthews, 1966), but at other pH values the coat protein forms disks; under these conditions, TMV disks will only associate with TMV RNA and not other RNAs (Butler and Klug, 1971). Later it was shown that the specificity in initiation on the viral RNA is brought about by the presence of a unique site on the viral

RNA that interacts with the first protein disk during assembly (Zimmern, 1977; Zimmern and Butler, 1977). This enucleation site is internal in the TMV RNA and is not located at the end of the RNA molecule, as might have been expected.

Influenza virus must form RNPs with all eight RNA segments; if there is a common nucleation site, it might well lie in the conserved 3'- and 5'-terminal nucleotides of the vRNA segments. As template (+ strand) RNAs are covered in NP and mRNAs are not, it is tempting to speculate that this putative nucleation site lies in the 13 common 5' vRNA nucleotides whose complement is lacking in mRNAs. The assembly of viral RNPs is also discussed in Chapter 2 (this volume).

The current scheme of influenza virus assembly, which is based on evidence derived from examination of thin sections using the electron microscope, is that RNPs recognize the viral $M_1$ protein at a patch on the infected cell plasma membrane (reviewed in Choppin and Compans, 1975) that already contains HA and NA. This order of events necessitates a protein–protein interaction between $M_1$ and NP (or perhaps $M_2$ and NP). When native RNPs were separated into different size classes on low percentage acrylamide gels, it was observed on subsequent analysis of the polypeptide composition of these RNPs that some $M_1$ protein was associated with each RNP (Rees and Dimmock, 1981). It was not clear, however, whether this represented a specific or nonspecific association. The $M_1$ protein is difficult to solubilize with high salt (Lamb et al., 1985) and in the presence of detergent $M_1$ aggregates and can adhere to other proteins. Thus, genuine interactions between $M_1$ and RNPs will be difficult to demonstrate using conventional methodologies.

## E. NP Is a Major Target for Cytotoxic T Lymphocytes

The nucleocapsid protein is the major target antigen for cross-reactive cytotoxic T lymphocytes (CTL) generated against influenza A viruses. The NP molecule is recognized in association with a class I molecule of the major histocompatibility complex (MHC) by CTL that are cross-reactive to all influenza A subtypes. This system has become the prototype to further define the nature of the antigen recognized by CTL and the mechanism(s) by which antigens are presented at the cell surface.

It has been shown that NP is recognized by specific T-cell clones (Townsend et al., 1984a) and that in vitro, polyclonal murine cultures prepared from repeated stimulation with infected cells yielded a predominantly NP-specific CTL population (Townsend and Skehel, 1984b). The NP protein expressed in transfected mouse cells from cloned DNA under the control of the SV40 early region promoter (Townsend et al., 1984b) served as a target for cross-reactive CTL. Recombinant vaccinia viruses that express NP have been shown to act as a CTL target in both murine (Yewdell et al., 1985) and human (McMichael et al., 1986) systems.

Recent work to define the mechanism by which a nonmembrane anchored (and largely nuclear localized protein) can be expressed at the cell surface, so that it can be recognized by CTL, has been done by expressing truncated portions of the NP gene (Townsend *et al.*, 1985). Both N-terminal and C-terminal fragments of NP expressed in L cells are active as target molecules for CTL at the cell surface (Townsend *et al.*, 1985). The target has been further defined by showing that synthetic peptides to NP added exogeneously to the target cell surface can also be recognized by CTL further supporting the concept of antigen processing before presentation at the cell surface (Townsend *et al.*, 1986). The mechanism involved in CTL recognition of NP and other internal influenza viral proteins is discussed in detail in Chapter 8 (this volume).

## F. Influenza B Virus NP Gene and Protein

The influenza B virus NP protein is immunologically unrelated to influenza A virus, yet performs the same structural role in influenza B virus RNPs. The gene sequence of RNA segment 5 has been obtained for two strains of influenza B virus: B/Singapore/222/79 (Londo *et al.*, 1983), and B/Lee/40 (Briedis and Tobin, 1984). Comparison of the influenza B/Singapore/222/79 NP amino acid sequence with that of the influenza A/PR/8/34 NP sequence showed 47% direct homology in the aligned regions, a similar distribution of basic amino acid residues and an added 50 amino acids at its N-terminal end (Londo *et al.*, 1983). The B/NP protein contains 560 amino acids and has a charge of +21 at pH 7. The RNA segment contains 1841 (B/Lee/40) or 1839 (B/Singapore/222/79) nucleotides with the extra nucleotides occurring in the 5' noncoding region.

## G. Influenza C Virus NP Gene and Protein

The complete sequence of RNA segment 5 of influenza C virus (C/California/78) was determined without definitive evidence for the coding assignment of the gene (Nakada *et al.*, 1984). The predicted protein sequence was thought to be that of NP based on its size, lack of specialized features such as signal or anchorage domains for membrane targeting, and small regions of strong homology to NP of the A and B viruses (Nakada *et al.*, 1984). The gene contains 1809 nucleotides and can code for a protein of 565 amino acids ($M_r \approx 63{,}525$) with a charge of +24 at pH 7. Two regions of high homology to the NPs of influenza A and B viruses have been found (Nakada *et al.*, 1984a), and it is very interesting to note that one of these regions is implicated as the karyophilic sequence of the A and B virus NP proteins (Davey *et al.*, 1985). If one considers the nature of the amino acid differences and makes allowances for type of residue, the similarity is even more striking. Although there is no evidence to show that influenza C virus NP protein does accumulate in the nucleus, it would be surprising if it did not.

## V. RNA SEGMENT 6: THE NEURAMINIDASE

RNA segment 6 codes for the NA integral membrane glycoprotein. NA is a major spike glycoprotein of the virion and is important both for its biological activity in removing sialic acid from glycoproteins and as a major antigenic determinant that undergoes antigenic variation. The gene structure, protein structure, and three-dimensional crystal structure of NA have been extensively reviewed (Colman and Ward, 1985); see Chapter 4 (this volume), which is devoted to the subject of neuraminidase. To avoid repetition, only the salient features of NA, necessary for completeness of a discussion of the gene structure of influenza viruses, are included.

### A. Neuraminidase Function

Neuraminidase (acylneuraminyl hydrolase, EC 3.2.1.18) catalyzes the cleavage of the α-ketosidic linkage between a terminal sialic acid and an adjacent D-galactose or D-galactosamine (Gottschalk, 1957). The functional role of NA in virus replication has not been established with great certainty, although it appears to be related to the ability of the virus to free itself from sialic acid-containing structures. NA activity may permit transport of the virus through mucin present in the respiratory tract, enabling the virus to find its way to the target epithelial cells. The removal of sialic acid from the HA by NA activity also seems to facilitate proteolytic cleavage of HA (Schulman and Palese, 1977; Sugiura and Ueda, 1980; Nakajima and Sugiura, 1980); such a mechanism may be important for the neurovirulence of some strains. NA activity destroys the HA receptor (Burnet and Stone, 1947) on the host cell, aiding in the elution of progeny virus particles from infected cells and preventing progeny virus particles from reabsorbing to the infected cell (Palese and Schulman, 1974). The removal of sialic acid by NA activity from the complex carbohydrate chains of HA and NA has been shown to be necessary to prevent self-aggregation of the virus (Palese et al., 1974).

### B. Structure of Neuraminidase

The NA is a tetramer ($M_r \approx 220,000$) consisting of four identical disulfide-linked subunits (Varghese et al., 1983) that can be released from the virion membrane with detergent. When viewed in the electron microscope, these detergent-solubilized molecules appear as a mushroom-shaped structure containing a stalk and head, with the molecules aggregated by the hydrophobic membrane anchorage region near the end of the stalk to form rosettes (Laver and Valentine, 1969; Wrigley et al., 1973). From the gene sequence, it has become clear that this hydrophobic region is very near the N-terminus of the molecule and anchors the NA in membranes with six N-terminal amino acids, forming a cytoplasmic tail (Fields et al., 1981). NA molecules lacking the hydrophobic domain and part of the stalk can be isolated by pronase or trypsin digestion of virions or infected cells; these head molecules retain their enzymatic and antigenic properties (Mayron et al., 1961; Noll et al., 1962; Wilson and Rafelson, 1963; Lazdins et al., 1972; Wrigley et al., 1977; Blok et al., 1982c). The head molecules have been crystallized and by a clever mathematical manipulation of information derived from crystals of NA of strains RI/5$^+$/57 and Tokyo/3/67, the structure of the Tokyo/3/67 N2 NA was determined at 2.9 Å resolution (Varghese et al., 1983) (see Chapter 4, this volume).

## C. General Deductions from the Gene Sequence of Influenza A Virus NA

The complete nucleotide sequence and deduced amino acid sequence of influenza A virus RNA segment 6 that encodes NA has been obtained for several N1 subtypes (Fields et al., 1981; Hiti and Nayak, 1982; Steuler et al., 1984), two N2 subtypes (Markoff and Lai, 1982; Lentz et al., 1984), and examples of each of the N7, N8, and N9 subtypes (Air et al., 1985; Dale et al., 1986). In addition, the complete amino acid sequence for the head molecule of A/Tokyo/67 (N2) was determined by protein chemistry (Ward et al., 1982). Taking the PR/8/34 (N1) strain sequence (Fields et al., 1981) as an example, it was found that RNA segment 6 is 1413 nucleotides in length with a 5' noncoding region of 20 nucleotides and a 3' noncoding region of 31 nucleotides. There is a single open reading frame of 1362 nucleotides coding for 453 amino acids ($M_r$ = 50,087). Five possible sites for N-linked glycosylation are found, although with strain PR/8/34 there is no evidence as to how many are used. With the Tokyo/67 (N2) strain, direct evidence indicates that one potential site on the head of the molecule is not used (Ward et al., 1982) and that the oligosaccharide chains are of both the high-mannose and complex types (Ward et al., 1982).

The NA molecule contains only one major hydrophobic region located near the N-terminus, which extends from residues 7 to 35, sufficient to anchor the protein in the lipid bilayer (Fields et al., 1981). On the basis of the nucleotide sequence information alone, it was originally suggested that NA was inserted and anchored in the membrane by its N-terminus, which remains intracellular. This has been confirmed by protein sequence studies on the intact and protease cleaved head molecules (Blok *et al.*, 1982c).

## D. Influenza A Virus Strain Similarities and Differences

Alignment of the predicted amino acid sequences of the N1, N2, N7, N8, and N9 subtypes highlights several important features of the structure of NA (Dale *et al.*, 1986). In the head region of the molecule (Tokyo N2 residues 74–469), there is extensive amino acid homology between the N1 and N8 molecules (55% overall) and the N7 and N9 molecules (57% overall). The cysteine residues are in virtually identical positions in all strains, strongly suggesting a similar tertiary structure. One carbohydrate site is conserved in all the strains (Dale *et al.*, 1986) (Asn at residue 146 in the Tokyo sequence). The oligosaccharide attached at this position of NA in the Tokyo strain is of the complex type and is serologically cross-reactive with chick cell host antigen and unique from the other NA sugar chains, in that it contains significant amounts of *N*-acetylgalactosamine (Ward *et al.*, 1983). The stalk regions show the greatest variability even within subtypes, both in amino acid sequence and in length, ranging from 62 to 82 amino acids (Blok and Air, 1982a,b; Dale *et al.*, 1986). The stalk contains a varying number of potential sites for the addition of N-linked carbohydrates, and it is known that the stalk is extensively glycosylated (Lazdins *et al.*, 1972). While the amino acid composition of the N-terminal hydrophobic anchorage domain varies between subtypes, the N-terminal cytoplasmic tail, which has the sequence

N-Met-Asn-Pro-Asn-Glu-Lys-C

is conserved, i.e., the N-terminal Met is not removed (Blok *et al.*, 1982c). It is tempting to speculate that the conservation of NA cytoplasmic tail sequences is functional, perhaps during the initial translocation of NA across the rough endoplasmic reticulum (RER) membrane during its synthesis or in interacting with another viral component during assembly of the virion.

The N9 subtype NA is different from other NAs, as it is also able to exhibit hemagglutinating activity (Laver *et al.*, 1984). The predicted amino acid sequence of the N9 NA does not suggest an immediate reason as to why the N9 NA protein should bind sialic acid tightly such that in the presence of red blood cells (RBCs) hemagglutination is exhibited (Air *et al.*, 1985). The inhibitor 2-deoxy-2,3-dehydro-*N*-acetylneuraminic acid completely abolishes NA activity, but not the hemagglutinating activity of the N9 neuraminidase (Laver *et al.*, 1984), suggesting, but not proving, that the NA and HA activities reside in separate sites on the molecule. Recent analysis of the X-ray three-dimensional structure of the N9 NA (Colman *et al.*, 1987) and analysis of the effect of monoclonal antibodies on NA and HA activities (Webster *et al.*, 1987) indicate that both activities reside in separate sites. However, the crystallographic structure does not indicate a second sialic acid-binding pocket of the type found on NA at the top of the molecule or in HA (Wilson et al., 1981). The biological advantage of a second molecule exhibiting HA activity is not understood. However, these finding have interesting evolutionary implications because the paramyxovirus glycoprotein HN has both HA and NA activity on the same polypeptide (Scheid et al., 1972) and the HN glycoprotein like NA has an N-terminal signal anchor (Hiebert et al., 1985). In 1974, Scheid and Choppin (1974), described experiments suggesting that both activities reside at the same site on the HN glycoprotein, whereas Portner (1981) provided evidence from monoclonal antibody-inhibition experiments that indicate the presence of two independent active sites. The similar membrane orientation of HN of paramyxoviruses and NA of myxoviruses and the observation that the N9 NA hemagglutinates but has a very similar primary structure to other NAs is intriguing both from a structural viewpoint and as to how the gene evolved. Determination of the three dimensional structure of a paramyxovirus HN molecule should provide significant insight into the mechanism of biological activities of these proteins.

## E. N-Terminal Signal Anchor

Neuraminidase is a member of a group of integral membrane proteins, class II (Garoff, 1985), that have an uncleaved N-terminal extended signal domain that both targets the protein to the membrane during synthesis and anchors the protein in the bilayer. Other members of the group include transferrin receptor, sucrose–isomaltase complex, intestinal brush-border aminopeptidase, HLA-DR invariant chain, asialoglycoprotein receptor, and the HA–NA of paramyxoviruses (Schneider et al., 1984; Hunziker et al., 1986; Maroux and Louvard, 1976; Chiacchia and Drickamer, 1984; Claesson et al., 1983; Strubin et al., 1984; Spiess et al., 1985; Spiess and Lodish, 1985, 1986; Hiebert et al., 1985).

The N-terminal hydrophobic domain of NA was shown to be necessary for targeting the protein to membranes by making mutants with deletions in this region (Davis et al., 1983; Markoff et al., 1984; Sivasubramanian and Nayak, 1987). That the N-terminal hydrophobic domain of NA is sufficient to act as a signal sequence was shown by transferring the region to a HA molecule that lacked its own signal sequence (Bos et al., 1984). It was found that the chimeric HA was glycosylated, indicating that it had been translocated across the membrane, and thus had its signal function restored. However, the effect on transmembrane topology of placing the NA hydrophobic domain at the N-terminus of HA while retaining the normal HA anchor at the C-terminus was not investigated. Results obtained from genetic manipulation of the transferrin N-terminal hydrophobic domain indicate that this region is both necessary and sufficient to act as an uncleaved signal anchor (Zerial *et al.*, 1986, 1987).

## F. Three-Dimensional Structure of NA

The structure of the NA heads derived after pronase treatment, as determined from an X-ray-generated electron-density map, indicates that NA is a tetramer with a box-shaped head (100 Å × 100 Å × 60 Å) that, based on the electron microscopic data, is attached to the membrane by a slender stalk (Varghese *et al.*, 1983). Electron density was obtained from just beyond the point of pronase cleavage. Each monomer of NA is composed of six topologically identical β-sheets arranged in the manner of the blades of a propeller. The tetrameric enzyme has circular four-fold symmetry stabilized in part by metal ions bound on the symmetry axis (Varghese *et al.*, 1983). The amino acids involved in the active site of NA have been identified based on locating the substrate, sialic acid, in the molecule. This was done by obtaining an electron-density difference map between NA crystals soaked in 0.5 mM sialic acid and the native crystals (Colman *et al.*, 1983). The substrate-binding site was found to be a large pocket on the top surface of each subunit rimmed by charged residues (Colman *et al.*, 1983; see also Chapter 4, this volume), and contains nine acidic and six basic residues that are conserved in all influenza A and B virus NA sequences (see Table 1). Mutagenesis of a conserved tryptophan residue found close to the sialic acid binding site resulted in a loss of enzymatic activity (Lentz and Air, 1986), although it has not been distinguished whether this is due to a loss of binding of substrate or to an inability to hydrolyze the substrate. The trivial explanation that loss of activity was the result of the mutation causing aberrant folding of the molecule was largely eliminated by demonstrating full reactivity with monoclonal antibodies, implying that the mutant retained a native structure (Lentz and Air, 1986).

## G. Influenza B Virus Neuraminidase

The complete sequence of RNA segment 6 of influenza B virus, which codes for NA, has also been determined (Shaw *et al.*, 1982). The NA gene of influenza B/Lee/40 virus is approximately 10% larger than that of A/PR/8/34 NA and contains 466 amino acids ($M_r = 51,721$) with four potential glycosylation sites. The predicted amino acid composition determined from the nucleotide sequence is in very close agreement with the amino acid analysis of purified influenza B/Lee/40 NA (Laver and Baker, 1972). There is only a single uncharged potential membrane-spanning sequence extending from residues 4–34, and this N-terminal region anchors B/NA in membranes (Shaw *et al.*, 1982). A comparison of sequences from influenza A (N1 and N2 subtypes) and B neuraminidase shows conserved residues with crucial structural and functional roles (Shaw *et al.*, 1982; Varghese *et al.*, 1983; Wei *et al.*, 1987). Many of the cysteine residues can be aligned in all sequences, and catalytic charged residues are conserved (Colman *et al.*, 1983). The cross-reactive carbohydrate attachment site (residue 146 in Tokyo N2) is conserved. A number of glycine residues found in tight bends of the N2 structure are also conserved as are several of the residues around the substrate binding site including a tryptophan at position 364 that is necessary for enzymatic activity (Wei *et al.*, 1987). It seems quite likely that all influenza virus NA will show the basic structural features of blades of a propeller, even though there is only 30% direct amino acid homology in the head region. The stalk region of B/NA is predicted to contain two carbohydrate chains, and analysis of the head structure indicates a loss of about 80 amino acids and 50% of the carbohydrate from the molecule (Lazdins *et al.*, 1972), consistent with a stalk containing two carbohydrate chains.

## H. Influenza B Virus RNA Segment 6

Unlike the situation found with any other influenza A or B virus RNA segment known to date, in which the first AUG codon from the 5′ end of the mRNA would be expected to initiate the NA polypeptide, influenza B virus NA initiates protein synthesis at the second AUG codon from the 5′ end of the mRNA. Furthermore, the B/NA sequence is unusual in that the first AUG codon, which is separated from the second by four nucleotides, is followed by an open reading frame of 100 amino acids that overlaps the NA reading frame by 292 nucleotides (Fig. 2).

Initiation of translation begins at the 5′-proximal AUG codon in 90% of eukaryotic mRNAs that have been sequenced (Kozak, 1983a) and, together with other circumstantial evidence (Kozak, 1983b), the hypothesis was proposed that eukaryotic ribosomes find the AUG initiator codon by "scanning" the 5′ end of the mRNA (Kozak, 1978, 1983a). The model was recently revised (Kozak, 1984a,b) as the following information became

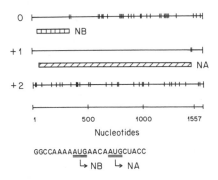

FIGURE 2. Schematic representation of the termination codons in all three reading frames on the influenza B virus RNA segment 6 to show the open reading frames for NB and NA. The nucleotide sequence surrounding the two AUG initiation codons is shown below. Data from Shaw *et al.* (1982); diagram modified from Lamb *et al.* (1983).

available. First, there is a growing list of exceptional cases in which one or more AUG codons occur upstream from the start of the protein coding region. Second, the distribution of nucleotides flanking functional initiator codons in eukaryotic mRNAs shows a striking bias—an A occurs in 80% of mRNAs examined at position −3 of the AUG and a G at position +4 in 40–60% of cases, i.e., the consensus context for initiation of translation is AXXAUGG (Kozak, 1983a). As more data have been surveyed, the optimal context is now proposed to be CCₐCCAUGG, and site-specific mutagenesis experiments add further weight to this proposal (Kozak, 1984b, 1986a). If the first AUG is in poor context, it is suggested that the ribosome may slip by to search for a downstream AUG in a "stronger" context (Kozak, 1984b). Third, there is a growing body of evidence suggesting that there may be polycistronic mRNAs in eukaryotes with termination of translation occurring after the first AUG codon and reinitiation at the second downstream initiation AUG codon (Kozak, 1984b; Liu et al., 1984; Johansen et al., 1984; Haarr et al., 1984; Khalili et al., 1987).

The first and second AUG codons on the B/NA mRNA lie within the context

$$\text{5'-CAAAA}AUG\text{AACA}AUG\text{CU-3'}$$

Thus, from the context rules, it would seem that the first AUG is in a better environment than the second AUG codon (Kozak, 1986b). These data made it very interesting to investigate whether the B/NA mRNA would be bicistronic and, as described below, stimulated the search for another protein product (NB) derived from the 100-amino acid open reading frame.

## I. NB Glycoprotein of Influenza B Virus

The nucleotide sequence of influenza B virus RNA segment 6 predicts that the 100-amino acid NB protein would contain an $NH_2$-terminal methionine residue, seven cysteine residues, 18 isoleucine residues, and four potential glycosylation sites. By selecting appropriate radioactively labeled precursors, the NB polypeptide was identified ($M_r \sim 18,000$ on polyacrylamide gels) and found to be abundantly expressed in influenza B virus-infected cells (Shaw et al., 1983). When HeLa cells were infected with B/Lee/40 or B/MB/50 strains of influenza B virus, NB was found to migrate differently on polyacrylamide gels between the two strains, reflecting its virus specificity and eliminating the possibility that it is a polypeptide whose synthesis is induced in the host cell (Shaw et al., 1983). To investigate the glycosylation of NB, infected cells were labeled with [³H]glucosamine, and a glycosylated polypeptide was detected that co-migrated with NB labeled with [³H]isoleucine. Furthermore, treatment of [³⁵S]cysteine-labeled infected cell lysates with endoglycosidase

H, to digest high-mannose carbohydrate chains, resulted in a polypeptide of $M_r \approx 11{,}000$, which co-migrates on gels with unglycosylated NB synthesized in vitro (Shaw et al., 1983). To confirm that NB was translated from the 5' end of the RNA segment 6 derived mRNA, hybrid-arrest translation experiments were performed with appropriate restriction fragments. To rule out the unlikely possibility that NB was merely a fragment of NA, the proteins were shown to be different by tryptic peptide mapping (Shaw et al., 1983). NB is synthesized in cells infected with all strains of influenza B virus examined (Shaw and Choppin, 1984).

Analysis of the mRNA that encodes NB and NA by sedimenting influenza B virus mRNAs on sucrose gradients and translating the fractions in vitro suggested that a single size class of mRNA coded for both NA and NB (Shaw et al., 1983). To exclude the possibility that two mRNA transcripts, similar in size, were derived from influenza B virus RNA segment 6, either by differential transcription or by subsequent processing, further experiments were performed. Neither the data derived from nuclease S1 mapping nor primer extension sequencing on the mRNA indicated another mRNA species (Shaw et al., 1983). Therefore, the data suggest that the mRNA is bicistronic, but a subtle modification giving rise to separate mRNA pools (such as methylation) has not been completely eliminated. However, this post-transcriptional modification would also have to occur when the B/NB/NA mRNA is transcribed under the control of the SV40 late promoter and polyadenylation signal, because both NB and NA are synthesized using this vector (Williams and Lamb, 1986). In addition, both NB and NA are translated from RNA, which has been transcribed from cloned B/NB/NA DNA using bacteriophage T7 polymerase. This result strongly suggests the lack of a eukaryotic post-transcriptional modification that alters the ability to translate NB or NA (M. A. Williams and R. A. Lamb, unpublished observations).

Examination of the predicted amino acid sequence of NB (Fig. 3) indicates that NB contains a region of 22 uncharged amino acids (residues 19–40). This region has a hydropathic index of >2, a value normally found for proteins that interact with membranes (Kyte and Doolittle, 1982). NB was shown to associate with the same subcellular fractions as HA and NA and to have properties of an integral membrane protein as solubilization of NB required 2% Triton X-100 and 0.5 M KCl (Williams and Lamb, 1986). There are four potential sites (Asn-X-Ser/Thr) for the addition of N-linked carbohydrate, which are distributed such that two are on each side of the hydrophobic domain (Fig. 3). As the addition of asparagine N-linked carbohydrate to an integral membrane protein is asymmetrical, occurring in the lumen of endoplasmic reticulum (ER) vesicles, a maximum of two sites on NB could be used, depending on the orientation of the protein in membranes.

To examine whether NB is expressed at the infected cell surface and to determine its topography in membranes, intact cells were treated with

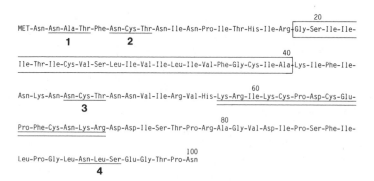

FIGURE 3. Predicted amino acid sequence of influenza virus B/Lee/40 NB protein. The amino acid sequence of NB is predicted from the nucleotide sequence of the NB/NA mRNA (data derived from Shaw *et al.*, 1982). The four potential glycosylation sites are underlined and numbered with respect to their position from the $NH_2$-terminal, the membrane-spanning hydrophobic domain is boxed, and the sequence of an oligopeptide used to make antisera to NB is indicated by double underlines. From Williams and Lamb (1986).

proteinase K. Some NB was found in a trimmed form of $M_r \approx 10,000$ when precipitated with an antisera to a COOH-terminal specific peptide. The large shift in mobility from NB ($M_r \approx 18,000$) to trimmed NB ($M_r \approx 10,000$) would be expected if proteinase K removed a small N-terminal peptide containing carbohydrate residues. The specificity of the COOH-terminal antisera made it likely that the $NH_2$-terminus of NB containing the sugar chains is exposed at the cell surface (Williams and Lamb, 1986) (Fig. 4). As an alternative approach to show that NB is expressed at the cell surface, intact cells were treated with endoglycosidase F to examine for the loss of carbohydrate from NB. Immunoprecipitation demonstrated a major NB-specific band of $M_r \approx 11,500$, identical in size to NB synthe-

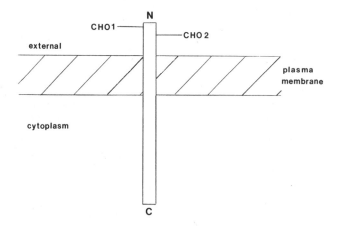

FIGURE 4. Schematic diagram of NB indicating its orientation in membranes. From Williams and Lamb (1986).

sized *in vitro* or NB immunoprecipitated from infected cell lysates and treated with endoglycosidase F. Subsequent digestion of the endo-glycosidase F-treated intact cells with proteinase K showed that the $M_r \approx$ 11,500 NB band was reduced in size to $M_r \approx$ 10,000, consistent with removing ~18 $NH_2$-terminal amino acids from the cell surface after re-moval of the carbohydrate (Williams and Lamb, 1986). Further examina-tion of the carbohydrate complexity of NB indicated that, in pulse-chase experiments, NB never acquired endoglycosidase H resistance in either MDCK or CV1 cells. These data suggest that NB contains carbohydrate side chains of the high-mannose type (Williams and Lamb, 1986). In influ-enza B-virus infected MDCK cells and to a slightly lesser extent in CV1 cells, a diffuse region of radioactive material ($M_r \approx$ 35,000 to $\approx$50,000) was observed after precipitation with NB-specific antisera. This hetero-geneously migrating material was lost after endoglycosidase F treatment of the cell surface and, concomitantly, a large increase in the amount of deglycosylated NB ($M_r \approx$ 11,500) was observed. Originally it was thought possible that the high-molecular-weight NB represented either (1) NB that interacts with itself through carbohydrate moieties, or (2) higher heteroglycans (Williams and Lamb, 1986). Recent data indicate that the heterogeneous higher-molecular-weight form of NB is due to the addition of polylactosamine to both carbohydrate chains of the high mannose form of the N-linked carbohydrate of NB (Williams and Lamb, 1988). Polylactosaminoglycan is a carbohydrate moiety characterized by a vari-able number of repeating units of galactose $\beta 1 \rightarrow$ 4-$N$-acetylglucosamine $\beta 1 \rightarrow$ 3 (Gal$\beta 1 \rightarrow$ 4-GlcNAc$\beta 1 \rightarrow$ 3) attached to a (mannose)$_3$(GlcNAc)$_2$ core oligosaccharide.

The precise sites of carbohydrate addition in NB were determined by site-specific mutagenesis and expression in eukaryotic cells. The Asn residue in each of the four Asn-X-Ser/Thr sequences was changed to a Ser residue by oligonucleotide-directed site-specific mutagenesis of the cloned cDNA of RNA segment 6. Each of the mutants was expressed in CV1 cells using an SV40 late-region vector. The phenotype of the mu-tants that had lost a glycosylation site was predicted to result in ex-pression of a NB-specific polypeptide of intermediate gel electrophoretic mobility (i.e., $M_r \approx$ 14,000 to 15,000). The data indicated that mutants at sites 1 and 2 (numbered from the N-terminus) expressed major new NB-specific bands of $M_r \approx$ 14,000 to 15,000, whereas mutants in sites 3 and 4 expressed an NB specific band of $M_r \approx$ 18,000, the wild-type size. A double mutant in both sites 1 and 2 resulted in the expression of a major NB-specific $M_r \approx$ 10,500 band. Thus, these data indicate that the $NH_2$-terminal Asn residues at positions 3 and 7 of the NB glycoprotein, which are exposed at the cell surface, are both used for the addition of carbohy-drate. This genetic approach also confirmed the orientation of NB in membranes because as discussed above the addition of carbohydrate is asymmetric and the orientation of an integral membrane protein is abso-lute and maintained, regardless of its final destination in the transport pathway.

It is not known whether the single uncleaved hydrophobic domain of NB can be defined as an extended signal anchor that interacts with the signal recognition particle (SRP) to insert NB co-translationally into membranes and then acts as a stop-transfer sequence to anchor NB in membranes. If NB does interact with SRP, it is interesting to note that ~78% of the NB polypeptide chain would be completed before the interaction with SRP occurs, since by the time the hydrophobic domain is exposed outside the ribosome a further 40 or so amino acids will have been synthesized but not yet exited from the ribosome (Walter *et al.*, 1984). It is also possible that SRP may interact with the hydrophobic domain posttranslationally, as found with a few specific proteins (Zimmermann and Meyer, 1986). Alternatively, NB may insert into membranes independently of SRP with the hydrophobic domain acting as an insertion sequence (Anderson *et al.*, 1983).

The role of NB in influenza B virus infections is not known. Although NB is readily detected in infected cells, it has not been detected in purified virions (Shaw *et al.*, 1983). However, the caveat has to be added that this may be dependent on the sensitivity of the methods used. The absence of NB in any great amount from virions as compared to HA and NA suggests that there must be an exclusion mechanism for NB before or during the budding process. It will be of great interest to determine the cellular distribution of NB on the cell surface with respect to the budding process. The cell-surface localization of NB makes it unlikely that the protein is involved in the transcription or replication of RNA. A role in organizing proteins at the cell surface to form patches of viral proteins or in the budding process would seem more likely. There is no evidence for a role of NB in the immune response to influenza B virus infection in humans, although immunoprecipitation studies using hyperimmune mouse sera demonstrate that NB is produced during productive infections of mice (Shaw and Choppin, 1984; Shaw *et al.*, 1985).

## J. Effect of Mutations and Deletions in the Bicistronic mRNA for NB and NA

The influenza B virus NB/NA mRNA is a naturally occurring mRNA; from examination of the mRNA nucleotide sequence in conjunction with the rules of the modified scanning hypothesis (Kozak, 1986b), it would be expected that only NB and not NA should be synthesized in infected cells. However, NB and NA accumulate in a 0.6 : 1 ratio, suggesting that approximately 60% of ribosome preinitiation complexes scanning the NB/NA mRNA do not initiate protein synthesis at the first AUG codon but continue scanning until reaching the second AUG codon four nucleotides downstream (Williams and Lamb, 1989).

To determine the importance of the influenza B virus mRNA nucleotide sequences in allowing initiation at the second 5' proximal AUG codon, changes in the 5'-terminal region of the mRNA were made, including deletions, insertions and site-specific mutations and the recombinant DNA molecules were then expressed in eukaryotic cells (Williams and Lamb, 1989). A C residue at position −3 of the first AUG codon caused a threefold decrease in NB accumulation and a twofold increase in NA accumulation. A C residue at position +2 of the first AUG codon had almost no effect on either NB or NA synthesis, whereas a C residue at +4 of the first AUG codon caused nearly a twofold decrease in accumulation of both NB and NA. A G residue at position +4 of the first AUG codon did not increase the accumulation of NB and caused a slight decrease in accumulation of NA. Elimination of the NA AUG codon by conversion of the second AUG codon to ACG caused a very small decrease in NB accumulation and elimination of the NB AUG codon by conversion of the first AUG codon to ACC caused a twofold decrease in NA accumulation. Thus, these effects of the single nucleotide changes on NB and NA synthesis are different from those that might be predicted on the basis of previous studies (Kozak, 1986b), as only small changes in synthesis (two- to threefold), and not 10-fold, were found. In addition, the relative importance of the G residue at the +4 position of the first AUG codon for efficient initiation at this AUG codon was not observed. These data suggest that nucleotides outside the −3 and +4 region with respect to the NB AUG codon are important in determining the ratio of initiation events for NB and NA. Deletion of 40 of the 46 influenza virus-specific 5'-untranslated region nucleotides caused a 10-fold decrease in NA synthesis and suggest that an unrecognized feature of the entire region, e.g., a specific secondary structure, also has a major role in the initiation of protein synthesis at the second AUG codon (Williams and Lamb, 1989).

## K. Influenza C Virus

Influenza C viruses do not contain an RNA segment that encodes a class two integral membrane protein that displays neuraminidase activity. The receptor-destroying activity of influenza C virus is a property of HA, which exhibits esterase activity (see above).

## VI. RNA SEGMENT 7 OF INFLUENZA A VIRUS: STRUCTURE AND SYNTHESIS OF THE MEMBRANE PROTEIN ($M_1$) AND AN INTEGRAL MEMBRANE PROTEIN ($M_2$) FROM UNSPLICED AND SPLICED mRNAs

RNA segment 7 of influenza A virus encodes two known proteins, the membrane protein $M_1$, which is the most abundant polypeptide in the virion and an integral membrane protein $M_2$ that is abundantly expressed at the cell surface. Three RNA transcripts have been identified that are derived from RNA segment 7: a co-linear transcript encoding the membrane protein $M_1$, a spliced mRNA encoding the $M_2$ protein, and an alternatively spliced mRNA (mRNA$_3$) that has the potential to encode a small peptide, $M_3$. There is no evidence for the existence of $M_3$.

## A. RNA Segment 7 and Its mRNAs

The complete sequence of RNA segment 7 has been obtained for several strains, including PR/8/34 (Winter and Fields, 1980; Allen et al., 1980), Udorn/72 (Lamb and Lai, 1981), FPV/Rostock/34 (McCauley et al., 1982), Bangkok/1/79 (Ortin et al., 1983), Mallard/NY/6750/78 (Buckler-White et al., 1986), and WS/33 (Zebedee and Lamb, 1989). The RNA 7 segment of all strains is 1027 nucleotides in length; the largest open reading frame for the Udorn strain extends from an AUG codon at nucleotides 26–28 to a termination codon at nucleotides 782–784 and encodes the membrane protein ($M_1$) (252 amino acids, $M_r = 27,801$). The predicted amino acid sequence agrees with the available amino acid composition and partial amino acid sequences (Both and Air, 1979; Robertson et al., 1979).

RNA segment 7 also contains a second reading frame that could code for a maximum of 97 amino acids and that overlaps the $M_1$ reading frame by 68 nucleotides (Winter and Fields, 1980; Allen et al., 1980; Lamb and Lai, 1981). This open reading frame encodes the influenza A virus $M_2$ protein.

Analysis of the RNA transcripts in influenza virus infected cells derived from RNA segment 7 indicated that in addition to the mRNA encoding the $M_1$ protein (~1000 nucleotides), there were other poly-adenylated RNA segment 7 specific transcripts. Nuclease $S_1$ mapping analysis and direct primer extension nucleotide sequencing of these small mRNAs indicated that a mRNA of ~350 nucleotides, designated the $M_2$ mRNA, was lacking 689 nucleotides found in the $M_1$ mRNA (Lamb et al., 1981). The 51-nucleotide leader sequence of the $M_2$ mRNA is derived from the 5' end of the co-linear $M_1$ mRNA transcript and the 271 nucleotides of the body region of the $M_2$ mRNA is 3'-co-terminal with the $M_1$ mRNA. The leader sequence of the $M_2$ mRNA contains the AUG initiation codon and codons for eight subsequent amino acids (which would be shared with $M_1$), and in the body region of the $M_2$ mRNA translation occurs in the +1 reading frame: the $M_1$ and $M_2$ protein reading frames overlap by 14 amino acids (Lamb et al., 1981) (Fig. 5). A second interrupted mRNA (designated M mRNA$_3$) was also identified in infected cells as being derived from RNA segment 7 (Lamb et al., 1981; Inglis and Brown, 1981). M mRNA$_3$ has a 5'-leader sequence of 11 virus-specific nucleotides shared with the $M_1$ and $M_2$ mRNAs, followed by a body region of 271 nucleotides identical to that of the $M_2$ mRNA. The first AUG codon of the M mRNA$_3$ transcript is in a reasonably weak context (AAACGA<u>AUG</u>G), and this open reading frame would only code

FIGURE 5. Model for the arrangement of $M_1$ mRNA, $M_2$ mRNA and mRNA$_3$ and their coding regions. Thin lines at the 5' and 3' terminals of the mRNAs represent noncoding regions. Cross hatched areas represent the coding regions of the mRNAs. In the region 740–1004, $M_2$ mRNA is translated in a reading frame different from that used for $M_1$. No evidence has yet been obtained

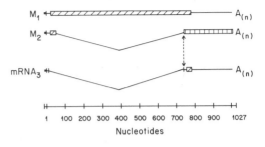

that mRNA$_3$ is translated, but its open reading frame which corresponds to the last nine residues of $M_1$ protein is indicated. The introns in the mRNAs are shown by the V-shaped lines; the filled in black boxes at the 5' end of the mRNAs represent heterogeneous nucleotides derived from cellular mRNAs that are covalently linked to the viral sequences. Redrawn from data in Lamb et al. (1981).

for 9 amino acids that would be identical to the C-terminal region of the membrane protein $M_1$. Subsequent AUG codons on M mRNA$_3$ are in even weaker contexts for the initiation of protein synthesis.

The nucleotides at the 5' and 3' splice junctions of the interrupted $M_2$ mRNA and M mRNA$_3$ have similarities to the consensus sequences of spliced eukaryotic mRNAs (Mount, 1982) but a well-defined pyrimidine tract in the intron before the 3' splice site is lacking. However, it was thought most likely that the interrupted mRNAs are probably produced by alternate splicing of the co-linear $M_1$ mRNA (Lamb et al., 1981). To establish that the influenza virus mRNA junctions can be used by the cellular splicing apparatus in the nucleus, the cloned RNA segment 7 DNA was inserted into a Simian virus 40 (SV40) vector under the control of the late region promoter and polyadenylation signal (Lamb and Lai, 1982). As shown schematically in Fig. 6, mRNAs were isolated from the SV40 recombinant virus-infected cells that contained uninter-

FIGURE 6. Schematic diagram of mRNAs and coding regions synthesized from an SV40 recombinant virus that expresses an unspliced mRNA encoding influenza virus $M_1$ protein and a spliced mRNA that uses the same junctions as influenza virus mRNA$_3$. For details, see the text and Lamb and Lai (1982). Redrawn from Lamb and Lai (1982).

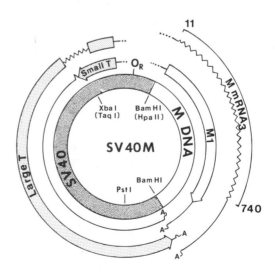

rupted influenza virus M sequences, an interrupted mRNA with the M mRNA$_3$ junctions, and a mRNA chimera containing SV40 5' leader sequences and M-specific 3' body sequences on either side of the interrupted region (Lamb and Lai, 1982). However, it is interesting that the M$_2$ mRNA splice junction was not used. Nonetheless, these data indicate that when influenza virus sequences are transcribed from a DNA template by RNA polymerase II, they are recognized and used by cellular splicing enzymes. Thus, these data derived from use of the SV40 expression vectors strongly suggest that cellular enzymes in the nucleus splice influenza viral mRNAs when they are transcribed from an RNA template (mRNA) by the viral transcriptase.

## B. Membrane Protein (M$_1$)

The membrane protein (M$_1$), also known as the matrix protein, is the most abundant protein in the virion (Compans et al., 1970; Schulze, 1970; Haslam et al., 1970; Skehel and Schild, 1971). It is now dogma that the membrane protein underlines the lipid bilayer and adds rigidity to the bilayer thus providing structural integrity to the virion envelope. However, direct experimental evidence for the location of M$_1$ and its function have proved difficult to obtain, and the seemingly plausible role assigned M$_1$ is based on indirect evidence. Electron microscopic examination of virions shows an electron-dense layer beneath the lipid bilayer, attributed to the M$_1$ protein (Apostolov and Flewett, 1969; Compans and Dimmock, 1969; Bachi et al., 1969). Proteolytic digestion of virions removes the spikes (HA and NA) but leaves M$_1$ protein undigested (Compans et al., 1970), and lipid extraction of fixed virions results in a spikeless shell that can be observed by electron microscopy (Schulze, 1972). Iodination of virions under various conditions indicates that M$_1$ protein, while not on the surface, is external to the nucleocapsid (Stanley and Haslam, 1971; Rifkin et al., 1972), as supported by fluorescent transfer experiments (Lenard et al., 1974). Finally, M$_1$ is the only protein in the virion in sufficient quantity to form a continuous shell beneath the lipid bilayer (Compans et al., 1970; Schulze, 1972).

It has been speculated that the M$_1$ protein may recognize the cytoplasmic tails of HA and NA (and perhaps M$_2$) and form a domain on the inner surface of the plasma membrane, providing a site for binding of ribonucleoprotein segments during virion assembly (Choppin et al., 1972). Surprisingly, no data have yet been obtained either to support or refute this attractive hypothesis, but the ability to express individual genes in various combinations using eukaryotic vectors should make this possibility more readily testable.

The M$_1$ protein requires detergent plus 0.5 M KCl (Lamb et al., 1985) to be solubilized, but it is not an integral membrane protein. However, M$_1$ does interact with lipid, as found by reconstituting lipid and purified

$M_1$ protein (Bucher *et al.*, 1980; Gregoriades, 1980; Gregoriades and Frangione, 1981). The predicted amino acid sequence indicates there is one region of 37 amino acids in the middle of the molecule possessing 16 hydrophobic amino acids and only two charged amino acids. This region may be involved in hydrophobic interactions with either protein or lipid and may account for the solubility of $M_1$ protein in chloroform/methanol (Gregoriades, 1973).

## C. Influenza Virus $M_2$ Protein Structure and Subcellular Localization

The amino acid sequence of the $M_2$ protein as predicted from the nucleotide sequence of its mRNA is shown in Fig. 7. The initiation codon and subsequent eight $NH_2$-terminal amino acids of $M_1$ and $M_2$ are predicted to be in common, as they are encoded before the 5′ splice junction of the $M_2$ mRNA. The remaining 88 amino acids of $M_2$, encoded after the 3′ splice junction, are predicted to be translated from the second open reading frame, and this has been confirmed using antisera to synthetic peptides (Lamb *et al.*, 1985).

Inspection of the predicted amino acid sequence of $M_2$ indicated the presence of three cysteine residues but only a single internal methionine residue. The $M_2$ protein ($M_r \approx 15{,}000$) was identified in influenza virus-infected cells when the radioisotopic precursor was switched from

(Met)Ser-Leu-Leu-Thr-Glu-Val-Glu-Thr-Pro-Ile-Arg-Asn-Glu-Trp-Gly-(16)
SP1
Cys-Arg-Cys-Asn-Asp-Ser-Ser-Asp-Pro-Leu-Val-Val-Ala-Ala-Ser-Ile-(32)

Ile-Gly-Ile-Leu-His-Leu-Ile-Leu-Trp-Ile-Leu-Asp-Arg-Leu-Phe-Phe-(48)

Lys-Cys-Ile-Tyr-Arg-Phe-Phe-Glu-His-Gly-Leu-Lys-Arg-Gly-Pro-Ser-(64)

Thr-Glu-Gly-Val-Pro-Glu-Ser-Met-Arg-Glu-Glu-Tyr-Arg-Lys-Glu-Gln-(80)
SP2
Gln-Ser-Ala-Val-Asp-Ala-Asp-Asp-Ser-His-Phe-Val-Ser-Ile-Glu-Leu-(96)

Glu. (97)

FIGURE 7. Predicted amino acid sequence of the $M_2$ protein of influenza A/Udorn/72 showing salient features. The sequence is derived from the nucleotide sequence of the $M_2$ mRNA (Lamb and Lai, 1980; Lamb *et al.* 1981). The underlined regions (SP1 and SP2) show the oligo peptides to which antisera were raised in rabbits to confirm the predicted sequence. The hydrophobic membrane spanning domain is boxed. A potential glycosylation site that is not used is indicated by a double-dotted underline. The arrows indicate potential sites of trypsin cleavage when $M_2$ is expressed at the cell surface, and both sites are thought to be used. From Lamb *et al.*, (1985).

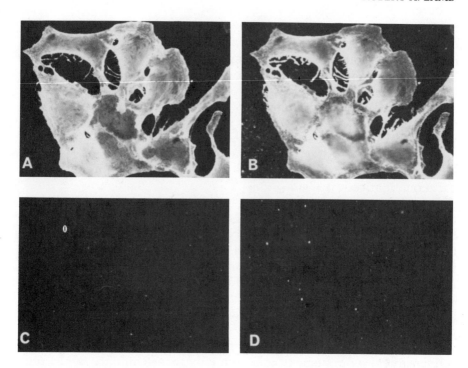

FIGURE 8. Detection of $M_2$ at the cell surface by indirect immunofluorescence using antibody to the $NH_2$-terminal peptide (SP1). Infected and uininfected CV1 cells were fixed in 2% formaldehyde for 30 min and were incubated sequentially with affinity-purified IgG to the $NH_2$-terminal peptide, SP1 and fluoresceine isothiocyanate-conjugated goat antirabbit IgG or a mixture of monoclonal antibodies (ascites fluid) to influenza virus HA, and tetramethylrhodamine isothiocyanate conjugated goat antimouse IgG. (A) infected cells showing staining with rabbit anti-$M_2$ $NH_2$-terminal peptide and fluoresceine goat antirabbit. (B) same field of cells as in A, showing staining with mouse anti-HA and rhodamine goat antimouse. (C) uninfected cells stained as in A. (D) same field of uninfected cells stained as in B. Exposure times for C and D were manually adjusted to be the same as for A and B, respectively. From Lamb et al. (1985).

[$^{35}$S]methionine to [$^{35}$S]cysteine (Lamb and Choppin, 1981). The $M_2$ protein exhibits different mobilities on polyacrylamide gels between influenza virus strains (Lamb and Choppin, 1981; Lamb et al., 1985; Zebedee et al., 1985); this observation was exploited in the genetic assignment of $M_2$ to RNA segment 7 using recombinant viruses of strains PR8 and HK (Lamb and Choppin, 1981). Biochemical evidence, involving hybrid-arrest translation experiments, also indicated that $M_2$ was encoded by RNA segment 7 (Lamb and Choppin, 1981). Tryptic peptide map analysis showed that $M_2$ is a discrete protein from $M_1$ and eliminated the possibility of $M_2$ being a proteolytic degradation product of $M_1$ (Lamb and Choppin, 1981).

The predicted amino acid sequence of $M_2$ indicates that it contains a region of 19 uncharged amino acids (residues 25–43) (see Fig. 7). This domain has a hydropathic index (Kyte and Doolittle, 1982) of >2, a value normally found for regions of proteins that interact with membranes, and it is long enough to span a lipid bilayer. The subcellular localization of $M_2$ was examined and evidence was obtained indicating that $M_2$ is an integral membrane protein (Lamb et al., 1985).

To facilitate experiments and to confirm indirectly the $NH_2$-terminal sequence of $M_2$ and the use of the second open reading frame on RNA segment 7, three antisera specific to $M_2$ were made. Antisera were generated to synthetic oligopeptides (SP1, residues 2–10 and SP2, residues 69–79 of the Udorn strain) such that site specific antisera to the $NH_2$-terminal and COOH-terminal sides of the hydrophobic domain were available. The third antisera (DM2) was to gel-purified denatured WSN $M_2$ polypeptide (Lamb et al., 1985). The SP1 $NH_2$-terminal specific antiserum is capable of immunoprecipitating $M_2$ from all strains of human and avian influenza viruses tested from H1 to H13 subtypes (Zebedee et al., 1985). It was shown both biochemically and by immunofluorescence (Fig. 8) that $M_2$ is expressed abundantly at the cell surface and, in addition to the HA and NA, is a third virus-specific infected cell membrane protein (Lamb et al., 1985). Trypsin treatment of infected cells and immunoprecipitation using site-specific antisera indicate that a minimum of 18 $NH_2$-terminal amino acids of $M_2$ are exposed at the cell surface. Treatment of intracellular microsomal vesicles with protease (trypsin or proteinase K), followed by immunoprecipitation with the SP2 antiserum indicated that the $M_2$ protein contains an extensive region of ~56 amino acids exposed on the cytoplasmic side of the infected-cell membrane (Zebedee et al., 1985). A model for the orientation of $M_2$ at the plasma membrane is shown in Fig. 9.

An investigation of the properties of $M_2$ indicates that in CV1 cells infected with the WSN strain, $M_2$ is synthesized by 2 hr p.i. together with the other polypeptides (Zebedee et al., 1985). However, in other cell types

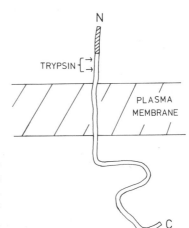

FIGURE 9. Schematic diagram indicating the orientation of the $M_2$ integral membrane protein. The cross-hatched region at the extracellular N-terminal of $M_2$ is a region of amino acids that is conserved in all influenza A viruses examined. From Lamb et al. (1985).

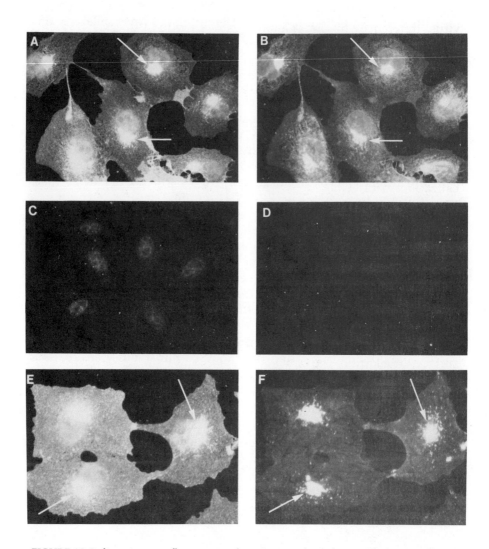

FIGURE 10. Indirect immunofluorescence detects $M_2$ in the Golgi complex. CV1 cells were mock infected or infected with the WSN strain of influenza virus; at 5 hr p.i. they were fixed in 2% formaldehyde for 30 min and permeabilized in acetone at $-20°C$ for 5 min. The upper panels show the same field of virus-infected cells stained sequentially with purified rabbit IgG to gel-denatured $M_2$ protein and fluorescein isothiocyanate (FITC)-conjugated goat anti-rabbit IgG (A), or after staining with a mixture of monoclonal antibodies (ascites fluid) to influenza virus HA and tetramethylrhodamine isothiocyanate-conjugated goat anti-mouse IgG (B). (C, D) Mock-infected cells stained in the same manner as A and B, respectively. The bottom panels show the pattern of $M_2$ staining in WSN-infected cells, using antiserum to the gel-denatured $M_2$ (as in A) and FITC-conjugated goat anti-rabbit IgG (E) tetramethyl-rhodamine isothiocyanate-conjugated WGA (F). Arrows indicate the Golgi complex. Exposure times for C and D were manually adjusted to be the same as for A and B, respectively.

(e.g., chicken embryo fibroblasts and L cells) infected with fowl plague virus, differences in the amount of spliced mRNA species 10 (which presumably encodes $M_2$) have been observed (Inglis and Brown, 1984). The A/Udorn/72 strain of influenza virus in CV1 or HeLa cells seems to synthesize or accumulate less $M_2$ than occurs in other strains. This differs from the amount of $M_2$ synthesized in the same cell type when infecting with the closely related A/Hong Kong/75 strain (Lamb and Choppin, 1981; Lamb et al., 1985). Thus, the control of splicing of the mRNAs in the nucleus or the ability of the $M_2$ mRNA to be translated requires further investigation.

The intracellular pattern of $M_2$ protein distribution includes localization to the Golgi apparatus (Fig. 10), suggesting that $M_2$ is transported to the cell surface by a pathway similar or identical to that of other integral membrane proteins. The half-time transport of $M_2$ in CV1 cells infected with the WSN strain is approximately 30–40 min, similar to that observed for the rate of HA transport to the cell surface (Gething and Sambrook, 1981; Sveda and Lai, 1981). Ironically, for the WSN strain, it has been calculated that HA is transported very slowly (~3 hr) (McQueen et al., 1986). (This calculation is based on the indirect method of measuring endoglycosidase H sensitivity of HA and may be biased in that some molecules may be aberrantly terminated in transport. The more direct method of measuring rate of HA transport is to measure arrival at the cell surface. In addition, the data of McQueen and co-workers were derived using HA expressed by an SV40 vector, and the level of expression may be low, thus affecting the rate of trimerization and ultimately the transport of HA.)

The predicted amino acid sequence of $M_2$ indicates the presence of a potential site (Asn-Asp-Ser, residues 20–22) for N-linked glycosylation located within the region of $M_2$ exposed at the cell surface but close to the lipid bilayer surface. However, the available evidence indicates that the site is not used: (1) The $M_2$ protein synthesized in the presence of tunicamycin, the inhibitor of N-linked glycosylation, does not have a changed electrophoretic mobility; (2) $M_2$ synthesized in vitro in the absence of membranes has the same electrophoretic mobility as $M_2$ synthesized in vivo, and it is possible that the site of N-linked sugar attachment in $M_2$ cannot be used because of stearic hindrance between the dolichol–lipid enzyme complex and the lipid bilayer, since the Asn at residue 20 is within 4 amino acids of the membrane (Zebedee et al., 1985); and (3) the site for N-linked glycosylation of the $M_2$ protein is not conserved in the sequence of $M_2$ for many avian strains of influenza virus (Fig. 11), suggesting that it is dispensable.

## D. Expression of $M_2$ in Eukaryotic Cells from Cloned cDNA

To determine whether the transport of $M_2$ to the cell surface is independent of other influenza virus-specific polypeptides, a cDNA clone of

```
                         1         10        20        30        40        50        60        70        80        90     97
A/Udorn/72               M S L L T E V E T P I R N E W G C R C N D S S D P L V V A A S I I G I L H L I L W I L D R L F F K C I Y R F F E H G L K R G P S T E G V P E S M R E E Y R K E Q Q S A V D A D D S H F V S I E L E
A/WSN/33                 *  * * *  * * * *  * * * I *  N *  *  *  R * K Y  *  *  *  * *  N  *  V  G  N
A/WS/33                  *  * * *  * * * *  * * * I *  N *  *  *  R * K Y  *  *  *  * *  N  *  V  G  N
A/PR/8/34                *  * * *  * * * G  * * A I *  N *  *  *  R * K Y  *  G  K  * *  *  *  *  G  *
A/FW/1/50                *  * * *  * * * *  * * * * *  * *  *  *  L * K *  *  *  *  * *  *  *  *  *  N
A/Singapore/1/57         *  * * *  * * * *  * * * * *  * *  *  *  * * K *  *  *  *  * *  *  *  *  *  *
A/Pt Chalmers/1/73       *  * * *  * * * *  * * * * *  * *  *  *  * * * *  *  *  *  * *  *  *  *  *  *
A/USSR/90/77             *  * * *  * * * *  * * * * *  * *  *  *  L * K *  *  *  *  * *  N  *  *  *  N
A/Bangkok/1/79           *  * * *  * * * *  * * * * *  * *  *  *  * * K *  *  *  *  * *  N  *  *  *  *

A/FPV/Weybridge/27       *  T * G  E  S * S *  * * * I *  * *  F  *  R L K Y  *  *  *  Q *  *  *  V  G  N
A/FPV/Rostock/34         *  T * G  E  * * * *  * * I I *  * *  *  N  R L K Y  *  *  *  Q *  *  *  V  G  N
A/mallard/NY/78          *  T * G  E  K * S *  * * * I *  * *  *  *  R L K Y  *  *  *  Q *  *  *  V  G  N
A/swine/Iowa/30          P  T * *  *  * * * *  H I T I *  K F  *  *  R L K Y  P  *  D  Q K  *  A  V  G  N

A/Bel/42                 *  * * *  *  * * * *  * * * * *  N *  *  *  R L K Y  *  |
A/HK/8/68                *  * * *  *  * * * *  * * * * *  * *  *  *|
H10-A/chick/Germany/49   *  T * G  E  * * S *  * * * I *  * *  *  *|
H4-A/duck/Czech/56       *  T * G  E  * Y S G  * * * I *  * *  *  *|
H3-A/duck/Ukraine/63     *  * * G  E  * * S *  * * * I *  * *  *  *|
H6-A/turkey/MA/65        *  T S G  E  K * S *  * * I F T  * V  *  *|
H9-A/turkey/WI/66        *  T * G  E  K * S *  * * * I *  * *  *  *|
H2-A/duck/Germany/73     *  T * G  E  K Y S *  * * * I *  * *  *  *|
H1-A/duck/Alberta/76     *  T * G  E  K * S *  * * * I *  * *  *  *|
H13-A/gull/MD/77         *  * * *  E  * * S *  * * * I *  * *  *  *|
```

FIGURE 11. The predicted amino acid sequence of the $M_2$ protein of human, swine, and avian strains of influenza A virus. Only the A/Udorn/72 sequence is shown in full. Identity is indicated either by a blank or a star. Amino acid changes are indicated. (The data are compiled from the references listed in Table II and Zebedee and Lamb, 1988, 1989b.) Note that for strains listed in the bottom sector, the sequence of $M_2$ is incomplete and the vertical line indicates the end of the available data. From Zebedee (1989).

the $M_2$ mRNA was expressed under the control of the SV40 late region promoter and polyadenylation signal (Zebedee *et al.*, 1985). The cDNA to $M_2$ mRNA was prepared by isolating poly(A) containing RNAs from influenza virus-infected cells enriching for small mRNAs on sucrose gradients, and pooling fractions yielding $M_2$ on *in vitro* translation. To increase the yield of complete $M_2$ cDNA clones, first strand cDNA synthesis was performed using oligo(dT) as a primer for reverse transcriptase and second strand synthesis was primed with the synthetic dodecamer d(5'-AGCAAAAGCAGG-3'), which is complementary to the 3' end of all influenza virus RNA segments. The $M_2$ protein expressed by the SV40 vector became associated with the Golgi complex and was found on the surface of vector-infected cells (Zebedee *et al.*, 1985), indicating that its cell surface expression is not dependent on other influenza virus proteins. The cDNA to the $M_2$ mRNA described here has also been expressed in recombinant vaccinia virus vectors under the control of the 7.5K promoter (Smith *et al.*, 1987) and this recombinant virus was used to examine the expression of $M_2$ in mouse cells of different haplotypes (see Chapter 8, this volume).

## E. Characterization of a Monoclonal Antibody to the $M_2$ Extracellular Domain

A monoclonal antibody to $M_2$ prepared using polyacrylamide gel purified $M_2$ as antigen has been obtained (Zebedee and Lamb, 1988). The antibody-binding site was located to the extracellular $NH_2$-terminal of $M_2$, as shown by the loss of recognition after proteolysis at the infected cell surface, which removes 18 $NH_2$-terminal residues, and by the finding that the antibody recognizes $M_2$ in cell-surface fluorescence. The epitope was further defined to involve residues 11 and 14 by comparison of the predicted amino acid sequences of $M_2$ from several avian and human strains and the ability of the $M_2$ protein to be recognized by the antibody (Zebedee and Lamb, 1988). Further use of this antibody reagent is described below.

## F. Initial Interaction of $M_2$ with Membranes

It has been proposed that there are two distinct mechanisms for the integration of newly synthesized polypeptides into cellular membranes (Blobel, 1980). The first mechanism of integration is specified by an insertion sequence in the polypeptide chain whereby the protein proceeds unassisted into exposed cellular membranes and results in the insertion of a hairpin-loop domain of the polypeptide chain into the lipid bilayer. The protein may extend into the hydrophilic milieu on the other side of the membrane (Engelmann and Steitz, 1981). The other mechanism is mediated by a signal sequence and is dependent on specific interactions of the signal sequence with components at the membrane that effect the translocation of the polypeptide from the cytoplasmic side of the membrane to the lumen of a vesicle, (Blobel, 1980). A great deal of evidence has been accumulated to indicate that a signal sequence, after emerging from a protease inaccessible space in the ribosome, is recognized by the signal recognition particle (SRP) (Walter et al., 1981). This recognition causes high-affinity binding of SRP to the ribosome and in many cases an arrest of chain elongation (Walter and Blobel, 1981; Anderson et al., 1983). The interaction of SRP with the SRP receptor (or docking protein) on the endoplasmic reticulum membrane (Gilmore et al., 1982; Meyer et al., 1982) is then proposed to cause a displacement of SRP with a concomitant release of the elongation arrest (Gilmore and Blobel, 1983), and the co-translational translocation of the nascent polypeptide chain across the membrane proceeds. If the signal sequence is N-terminal, it is normally cleaved by the signal peptidase. However, of membrane proteins examined so far, a small group have internal noncleavable signal sequences, e.g., sarcoplasmic reticulum calcium ATPase (Reithmeier and MacLennon, 1981), opsin (Friedlander and Blobel, 1985), influenza virus NA (Fields et al., 1981) and paramyxovirus HN (Hiebert et al., 1985). For membrane-bound polypeptides translocation is proposed to be interrupted by a stop-transfer sequence, thereby yielding precisely specified asymmetrical integration of the polypeptide chain into the membrane (reviewed by Blobel, 1980).

The structure of influenza virus $M_2$ as an integral membrane protein is unusual. The $M_2$ protein does not contain a cleavable signal sequence (Lamb et al., 1985) and the single hydrophobic domain in the polypeptide must act either as an extended signal/anchorage (stop-transfer) domain or as an insertion sequence. Most proteins that have extended signal/anchorage domains (e.g., influenza virus NA, paramyxovirus HN, sucrase isomaltase, transferrin receptor, HLA-DR invariant chain, asialoglycoprotein receptor (reviewed in Wickner and Lodish, 1985) are orientated in membranes with the N-terminus on the cytoplasmic side of the membrane, whereas $M_2$ has its C-terminus exposed to the cytoplasm (Lamb et al., 1985; Zebedee et al., 1985).

It was recently shown that in vitro integration of $M_2$ into membranes requires the co-translational presence of SRP (Hull et al., 1988). This is of interest because the minimum length of a ribosome-bound polypeptide that could potentially be recognized by SRP is apparently equal to the length of the signal sequence plus the 40–50 residues of a nascent chain buried in a protease and, by inference, SRP-inaccessible groove in the large ribosomal subunit. Because $M_2$ protein has 23 amino acid residues that precede the hydrophobic transmembrane spanning sequence, most of the 97 residue $M_2$ protein will be synthesized before emergence of the functional signal from the ribosome. As the $M_2$ protein contains a single hydrophobic domain that performs the dual role of acting as an SRP-dependent signal sequence and as a stop-transfer sequence, a series of deletion mutants were constructed within the $M_2$ hydrophobic domain in an attempt to delineate the signal and anchor functions (Hull et al., 1988). Deletion of as few as two residues from the hydrophobic segment of $M_2$ markedly decreased the efficiency of membrane integration, whereas deletion of six residues completely eliminated integration. $M_2$ proteins containing deletions that eliminate stable membrane association failed to be recognized by SRP, indicating that they lack a functional signal sequence. These data thus indicate that the signal sequence that initiates membrane integration of $M_2$ resides within the transmembrane spanning segment of the polypeptide (Hull et al., 1988).

## G. Is $M_2$ a Structural Component of Virions?

Quantitation of the molar amounts of $M_2$ synthesized relative to HA 3–6 hr p.i., based on [$^3$H]phenylalanine or [$^{35}$S]cysteine labeling and the known sequences of HA and $M_2$, indicate an $M_2$ to HA ratio of 1 : 1.5 (Lamb et al., 1985). As approximately $5 \times 10^6 - 1 \times 10^7$ molecules of HA are expressed at the surface of influenza virus-infected CV1 cells (strain A/Jap/305) (Gething and Sambrook, 1982), it can be inferred that a great number of molecules of $M_2$ are expressed at the surface of influenza virus-infected cells. In order to understand the mechanism by which $M_2$ exerts its function, particularly with respect to the effect of amantadine hydrochloride on virus assembly and its linkage with $M_2$, it became very important to determine whether $M_2$ is in virions. There is no doubt that $M_2$ is underrepresented in virions as compared with its relative abundance in infected cells, and a mechanism must exist that prevents $M_2$ incorporation into virions during maturation and budding. In retrospect, although $M_2$ is underrepresented in virions compared with the infected cell, experiments originally performed to determine whether $M_2$ is in virions (cited in Lamb et al., 1985) were not done at a sensitivity necessary to detect 100 molecules or less per virion, and it should be remembered that the critically important transcriptase-associated P proteins of influenza virus are only found in an abundance of approximately 40 molecules per virion (Table I). To quantitate the amount of $M_2$ associated with virions, use was made of the $M_2$-specific monoclonal antibody described above (Zebedee and Lamb, 1988). A quantitative immunoblot procedure involving a comparison of the amount of $M_2$ detected in purified virions with known amounts of purified $M_2$ protein as well as a direct analysis of virions labeled to high specific activity with [$^{35}$S]cysteine, indicated that in the virion preparations used there are 14–68 molecules of $M_2$ per virion (Zebedee and Lamb, 1988). Thus, although the amount of $M_2$ found in virions is small, these calculations would be an overestimation if the virion preparations were contaminated by copurification of plasma membrane or cytoplasmic vesicles containing $M_2$ from the infected cells.

## H. Function of $M_2$ in Infected Cells

The finding that $M_2$ is abundantly expressed at the plasma membrane of influenza A virus-infected cells led to the suggestion that a possible role for $M_2$ in the virus life cycle is in the assembly of the virus particle (Lamb et al., 1985). It will be of interest to examine the cellular distribution of $M_2$ compared with HA and NA, especially in regions of budding virus particles. Influenza virus buds from the apical surface of polarized epithelial cells (Rodriguez-Boulan and Sabatini, 1978; Roth et al., 1979), and $M_2$ is expressed at the apical surface of polarized epithelial cells (Hughey, P., Compans, R. W., Zebedee, S. L., and Lamb, R. A., unpublished observations). Thus, it is possible that $M_2$ aligns itself on either side of a budding virus particle and assists in pinching off the bud from the membrane.

The question of whether the $M_2$ monoclonal antibody described above that recognizes the extracellular region of $M_2$, could affect the growth of influenza A virus in a plaque assay has recently been addressed. It was found that the size, but not the number of plaques for certain strains of influenza A virus was restricted. Strains sensitive to the effect of the $M_2$ antibody on virus growth include A/FW/1/50, A/Singapore/1/57, A/HK/8/68, A/USSR/90/77, and A/Udorn/305/72, whereas A/WSN/33 and A/PR/8/34 are resistant to the antibody restriction on growth (Zebedee and Lamb, 1988). This plaque size reduction is a specific effect for the $M_2$ antibody, as shown from an analysis of recombinants of defined genome composition and by finding that competition with an $NH_2$-terminal peptide prevents the antibody restriction of virus growth (Zebedee and Lamb, 1988). Because the $M_2$ protein of both sensitive and resistant strains is recognized by the $M_2$ antibody in immunofluorescence, immunoprecipitation, and immunoblot experiments, the observed difference in growth restriction cannot be explained by differences in antibody binding at the cell surface. To examine the mechanism of $M_2$ antibody growth restriction further, viruses that grew to a normal plaque size in the presence of the antibody were isolated from a plaque titration (Zebedee and Lamb, 1989). Most of the variant viruses were not conventional antigenic variants, as their $M_2$ protein was still recognized by the antibody. Genetic analysis of reassortant influenza viruses prepared from the antibody-resistant variants and an antibody-sensitive parent virus indicated that $M_2$ antibody growth restriction was linked to RNA segment 7. Analysis of the RNA segment 7 nucleotide sequence of these variants predicted amino acid changes in the C-terminal region of $M_2$ (3 variants) or the N-terminal region of the $M_1$ protein (5 variants) (Zebedee and Lamb, 1989). As changes in the variant viruses occur in both the $M_1$ and $M_2$ proteins as second site mutations outside the $M_2$ antibody recognition site, it suggests that antibody binding to the $M_2$ protein of a sensitive influenza virus interferes with an interaction of the cytoplasmic domain of $M_2$ with the $M_1$ protein. One possibility is that one of the major roles of $M_2$ is in chaperoning $M_1$, either in transport to the cell surface or in the formation of a virus particle (Zebedee and Lamb, 1989).

## I. Effect of Amantadine Hydrochloride on $M_2$

Amantadine (1-aminoadamantane) hydrochloride displays a specific anti-influenza A virus action (Davies et al., 1964) and has been used in the prophylaxis and treatment of influenza A infections (reviewed by Oxford and Galbraith, 1980). Many studies have been done to elucidate the inhibitory mechanism of amantadine (reviewed in Schulman, 1982) and have led to the conclusion that the drug exerts its effect after the virus is bound to the cell but before uncoating. The picture is complicated by the fact that (1) different laboratories and experiments from the same group have used widely varying concentrations (5 μM–1 mM) of the drug, (2) different strains of influenza A virus are inhibited at varying concentrations, and (3) some strains are only inhibited if the drug is present before infection. One of the more detailed studies showed that with fowl plague virus, primary transcription was inhibited in vivo (but not in vitro) and, as could be expected, every further step of replication was blocked (Skehel et al., 1977). Much of this early work on examining the block in replication was done with concentrations of the drug (>200 μM), at which it is now recognized that amantadine acts as a lysosomaltropic reagent with an amine effect similar to that of $NH_4Cl$ or chloroquine, causing a rise in the pH value of endosomes and lysosomes. A few studies using low concentrations (0.1–25 μM) of amantadine showed that mutants resistant to the effect of the drug (at that concentration) could be isolated at high frequency ($10^3$–$10^4$); the amantadine resistance was mapped through the isolation and characterization of recombinant viruses to RNA segment 7 encoding $M_1$ and $M_2$ (Lubeck et al., 1978; Hay et al., 1979) or to RNA segment 4, encoding the HA (Scholtissek and Faulkner, 1979; Hay and Zambon, 1984).

It now seems clear that there are at least two very different effects of amantadine, depending on concentration. At high concentrations of the drug (500 μM), resistant mutants have been isolated that have a raised pH of activation of HA-mediated fusion with endosomal membranes, and amino acid changes in HA occur at positions compatible with a lowering of the energy barrier to the conformational transition to a fusion active state such that fusion occurs at the higher pH (Daniels et al., 1985). At low concentrations (0.1–10 μM), the picture is still complicated. It has been reported that with some strains, e.g., A/Singapore/1/57, replication is inhibited at an early, presynthetic stage when the drug is present prior to infection, and primary transcription is inhibited (Hay and Zambon, 1984). However, with the avian virus strains A/FPV/Rostock/34 and A/FPV/Weybridge/27, polypeptides are synthesized at normal levels in the presence of the drug, but virus assembly is inhibited. The small amount (10% of control) of virus released contains significantly diminished amounts of HA, whereas all the other polypeptides are found in normal amounts (Hay and Zambon, 1984). Thus, with A/FPV/Rostock/34 and A/FPV/Weybridge/27 the action of the drug appears to be to prevent HA incorporation into virions. The drug must be present before synthesis of virus proteins in order to affect their incorporation into virus particles, suggesting an effect on the production, localization or interactions of virus components rather than disruption of a previral structure (Hay et al., 1985).

Recently a study has been done to characterize low-concentration amantadine-resistant mutants at a molecular level (Hay et al., 1985). Nucleotide sequencing studies showed that in 90 individual isolates of amantadine-resistant mutants that mapped genetically to RNA segment 7, amino acid substitutions occurred at only four amino acids in the membrane-spanning domain of the $M_2$ protein (residues 25–43) (Lamb et al., 1985). No changes were found in the coding regions of $M_1$ and only a few of the mutants had changes in HA (Hay et al., 1985). Although resistance to the drug maps to the 20 amino acid hydrophobic region of $M_2$, a genetic analysis shows there is an influence of HA on amantadine sensitivity between virus strains (Hay et al., 1985), which implies that the drug may interfere with interactions between these two virus proteins. The amino acid changes in $M_2$ found in the resistant mutants all reduce the hydrophobicity of the membrane-spanning domain and include changes at $M_2$ residue 27, 30, and 31. More unexpectedly mutants were also frequently found which introduced a charged group (glutamic acid) at $M_2$ residue 34. However, it has previously been observed that substitution of a charged group does not prevent anchorage of hydrophobic domains in membranes (Adams and Rose, 1985).

These data indicate the $M_2$ has a crucial role in virus replication. The strain dependent differences (i.e., the inhibition occurring at the initiation of infection or at assembly) may not necessarily indicate that there is more than one effect of the drug at low concentrations. Strains inhibited by amantadine at the stage of initiation of infection are also affected by the drug if it is added later in infection. A role of $M_2$ in the initiation of infection is only easily explainable if $M_2$ is a structural component of virions. The strain specific differences of the effect of the drug may reflect very different amounts of $M_2$ in virions.

It would be imprudent to propose a single hypothesis that incorporates all the known phenomenon, including the effect of amantadine on uncoating in some strains and assembly in others. However, some order can be achieved if the following ideas are considered. It is known that $M_2$ and HA expressed from vectors can be transported to the cell surface independently (Zebedee et al., 1985; Gething et al., 1982) and that HA is biologically active by all criteria measured, but in vivo fusion in endosomes cannot be measured. However, it is possible that during influenza virus infection there is an association of $M_2$ and HA (either directly or indirectly via $M_1$) that is necessary to bring about a change in HA. This $M_2$-mediated change in HA may be needed for budding or needed later in the virus particle so that HA can cause fusion in endosomes. This putative association could occur in some strains before the vesicles transporting $M_2$ and HA reach the plasma membrane and in others could occur at the plasma membrane. It has already been shown that HA and $M_2$ share the usual ER to Golgi to plasma membrane transport pathway (Zebedee et al., 1985). It is also known that amantadine intercalates into lipid bilayers (Jain et al., 1976); thus, it is possible that amantadine may cause a perturbation of $M_2$ by altering a property of the $M_2$ hydrophobic domain. It is of added interest that if this hydrophobic region of $M_2$ forms an α-helix, then all the mutations occur in residues on the same face of the α-helix (Hay et al., 1985). An association of amantadine with $M_2$ may alter properties of $M_2$ such that a hydrophilic association of $M_2$ with HA is unaffected but the transport of HA is retarded. Alternatively, amantadine may prevent a hydrophobic association of $M_2$ with HA (either directly or indirectly via $M_1$) in the lipid bilayer, which is necessary to bring about a change in the structure of HA. In either case, the transport of HA or budding may be affected. If $M_2$ becomes incorporated into virions in some strains and HA has not "matured" through the change in structure, but this change can also occur in endosomes, then the drug interaction with $M_2$ may prevent the change occurring in endosomes during the next round of infection. In all cases, the amino acid changes in the resistant mutants would prevent the interaction of the drug but still allow $M_2$ to associate with HA.

## J. Evolution of the Nucleotide Sequence of RNA Segment 7 and the Amino Acid Sequences of $M_1$ and $M_2$

A comparison of the available sequences of RNA segments 7 [A/PR/8/34 (Winter and Fields, 1980; Allen et al., 1980); A/Udorn/72 (Lamb and Lai, 1981); A/swine/Iowa/30 (Nakajima et al., 1984); A/FPV/Rostock/34 (McCauley et al., 1982); A/Bangkok/1/79 (Ortin et al., 1983); A/Mallard/NY/6750/78 (Buckler-White et al., 1986); A/WSN/33, A/WS/33, A/Singapore/1/57, A/Fort Warren/1/50, A/Port Chalmers/1/73 (Zebedee and Lamb, 1989)] indicates that the sequence is highly conserved. Conservation of the amino acid sequence of $M_1$ between strains is not unexpected as the $M_1$ protein is antigenically similar between strains (Schild, 1972), and only limited differences have been found by peptide-mapping procedures (Brand et al., 1977; Dimmock et al., 1977). There is less conservation in the 97 amino acids of the $M_2$ protein between sequences of different strains. However, the changes observed preserve the basic structural feature of the uncleaved hydrophobic membrane-spanning domain. In addition the 9 $NH_2$-terminal amino acids of $M_2$ shared with $M_1$ are conserved in all strains of influenza A virus examined, except in the reported sequence of A/swine/Iowa/30 (Fig. 11). Interestingly, the potential site for N-linked glycosylation (residues 21–23), which the available data indicate is not used in strains A/WSN/33, A/Udorn/72, and A/PR/8/34, has been lost from the sequence of A/Mallard/NY/6750/78 and all other avian strains tested (Fig. 11) totally eliminating the possibility of N-linked glycosylation for these strains. Thus, $M_2$ is one of a small group of integral membrane proteins that are transported through the Golgi apparatus and expressed at the cell surface that are not glycosylated.

## K. RNA Segment 7 of Influenza B Virus

RNA segment 7 of influenza B virus also codes for the $M_1$ protein of this virus (Raccaniello and Palese, 1979; Briedis et al., 1981). The complete nucleotide sequence of a cloned full-length DNA copy of genome RNA segment 7 of influenza B/Lee/40 virus has been determined (Briedis et al., 1982). RNA segment 7 of B/Lee/40 contains 1191 nucleotides. After a 5'-noncoding region of 24 nucleotides, there is an open reading frame that extends from the first possible initiation codon for protein synthesis at nucleotides 25–27 to a termination codon at nucleotides 769–771. This region could code for a protein of 248 amino acids, compatible in size with the influenza B virus $M_1$ protein; 63 amino acids, without any deletions or insertions, are conserved between influenza A and B viruses in the $M_1$ protein, suggesting conservation of structural features. A tract of five adenine residues occurs at nucleotides 1171–1175. This probably represents the polyadenylation site for the virus mRNA(s), as is seen in influenza A virus (Robertson et al., 1981). Polyadenylation at this site would lead to an $M_1$, mRNA of ~1175 nucleotides. From nucleotide 513 there is a second gene region, in the +2 reading frame (with respect to $M_1$), extending until a termination codon at nucleotides 1095–1097 (Fig. 12).

The second coding region overlaps that of the influenza $M_1$ protein by 86 amino acids. This region could code for a maximum of 195 amino acids in the +2 reading frame, depending on whether the mRNA is

FIGURE 12. Schematic representation of the open reading frames and termination codons in the cloned B-M-DNA of influenza B virus RNA segment 7. The 0 reading frame codes for the membrane protein of 247 amino acids. The +2 reading frame contains 195 amino acids. Redrawn from data provided in Briedis et al. (1982).

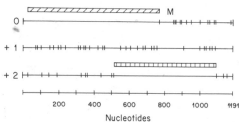

spliced and depending on the location of the initiating methionine residue. No obvious homology was detected in the amino acid sequence of
$M_2$ of influenza A virus and the second open reading frame of RNA
segment 7 of influenza B virus (Briedis *et al.*, 1982). The identification of a
second protein has not been reported and in the author's laboratory
straightforward approaches to search for a spliced mRNA or polypeptide
produced from this overlapping reading have been unsuccessful. To obtain further information, Hiebert and colleagues (1986) have investigated
whether the second open reading frame observed with B/Lee/40 (Briedis
*et al.*, 1982) is conserved in another isolate of influenza B virus, B/Sin-
gapore/222/79. It was found that the second open reading frame was
maintained in the 39 years separating the two isolates and displays 86%
amino acid homology with the B/Lee/40 sequence (27 changes). The
deduced amino acid sequence of the $M_1$ protein indicates that B/Sin-
gapore/222/79 contains 248 amino acids, 243 of which are identical in
the B/Lee/40 $M_1$ protein. The conservation of the second open reading
frame between two viruses isolated 39 years apart would suggest that the
second open reading frame is conserved because its gene product is essential to the virus (Hiebert *et al.*, 1986).

## L. Membrane Protein of Influenza C Virus

The membrane protein of influenza C virus has a $M_r \approx 28,000$ on
polyacrylamide gels (Compans *et al.*, 1977; Sugawara *et al.*, 1983; Yokota
*et al.*, 1983). Nucleotide sequence determination of a molecular clone of
RNA segment 6 of influenza C/JJ/50 indicates that the RNA segment is
1180 nucleotides in length and contains a single open reading frame that
could encode a polypeptide of 374 amino acids ($M_r \approx 41,700$) (Yamashita
*et al.*, 1988). However, hybrid-selection translation experiments of
mRNAs isolated from influenza C virus-infected cells resulted in synthesis of the membrane protein ($M_r \approx 28,000$) (Yamashita *et al.*, 1988).
Analysis of the mRNA structure indicated that the predominant mRNA
contains an interrupted region from nucleotides 753 to 981, such that on
translation of the mRNA a protein ($M_r \approx 26,970$) would be synthesized,
which is the expected size of the membrane protein (Yamashita *et al.*,

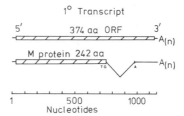

FIGURE 13. Schematic representation of influenza C
virus RNA segment 6 derived mRNAs. The influenza
C virus membrane protein is thought to be translated
from a spliced mRNA. Drawn from data provided in
Yamashita *et al.* (1988).

1988). The 3' junction of the interrupted mRNA is not a close approximation to the usual consensus sequence junction of spliced mRNAs (Mount,
1982), but it seems reasonable to assume that, like influenza A and B
viruses, influenza C virus has a spliced mRNA derived from the RNA
segment that encodes the membrane protein. What is different with the
influenza C viruses is that instead of the primary RNA transcript encoding the membrane protein, it is encoded by a spliced mRNA (Yamashita
et al., 1988). Small quantities of a co-linear RNA transcript derived from
influenza C virus RNA segment 6 have also been identified in infected
cells. A protein translated from an unspliced mRNA would contain an
additional 132 amino acids on the C-terminus of the membrane protein
(Fig. 13). This region of additional amino acids contains two regions that
are of sufficient length and hydrophobicity to act as membrane-spanning
domains. Thus, it will be interesting to determine whether additional
proteins are synthesized from RNA segment 6 of influenza C viruses and,
if so, to determine their subcellular distribution.

## VII. INFLUENZA A VIRUS RNA SEGMENT 8: UNSPLICED AND SPLICED mRNAs ENCODE THE NONSTRUCTURAL PROTEINS $NS_1$ AND $NS_2$

RNA segment 8 of influenza A virus encodes two proteins $NS_1$ and
$NS_2$, that are only found in infected cells. $NS_1$ is encoded by a colinear
mRNA transcript whereas $NS_2$ is encoded by an interrupted mRNA. The
finding of a spliced $NS_2$ mRNA was the first evidence for splicing with an
RNA virus that did not use a DNA intermediate in its replication.

### A. RNA Segment 8 and Its mRNAs

The complete nucleotide sequence of RNA segment 8 has been determined for a large number of human and avian strains (e.g., Lamb and Lai,

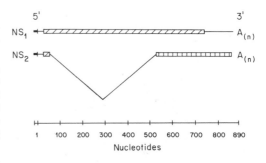

FIGURE 14. Schematic representation for the arrangement of the influenza A virus $NS_1$ and $NS_2$ mRNAs.
The thin lines at the 5' and 3' termini
represent noncoding regions. The
cross-hatched rectangles represent
the coding regions of the two
mRNAs. In the region 529–861, the
$NS_2$ mRNA is translated in a reading
frame different from that for $NS_1$.
The V-shaped line in the $NS_2$ mRNA
represents the intron. The small box at the 5' end of the mRNAs represents the heterogeneous
nucleotides derived from cellular mRNAs. Redrawn from data in Lamb and Lai (1980).

1980; Porter et al., 1980; Baez et al., 1980, 1981; Buonogiorio et al., 1984; 1986; Krystal et al., 1983b; Norton et al., 1987). RNA segment 8 is 890 nucleotides in length, and the open reading frame coding for the $NS_1$ protein ranges from 202–237 amino acids depending on the virus strain: the $NS_1$ protein of one strain, A/Turkey Oregon/71, contains only 124 amino acids. There is a second open reading frame found at the 5' end of the virion RNA segment 8 that is used in part to encode $NS_2$. The prediction that RNA segment 8 encodes two proteins ($NS_1$ and $NS_2$) using overlapping reading frames was made on the basis of biochemical and genetic evidence (Lamb et al., 1978; Lamb and Choppin, 1979; Inglis et al., 1979) and nucleotide sequencing confirmed this data. It had been shown that $NS_2$ is a unique influenza virus-encoded polypeptide on the basis of its peptide composition and strain-specific differences in migration on polyacrylamide gels (Lamb et al., 1978; Lamb and Choppin, 1979). It was suggested that one RNA segment must code for two polypeptides (Lamb et al., 1978) because of the evidence at that time for nine virus-coded polypeptides ($M_2$ had not been recognized) and the existence of only eight influenza virus RNA segments. This led to the finding that influenza virus RNA segment 8 codes for both $NS_1$ and $NS_2$, as shown in studies with recombinant viruses in which $NS_1$ and $NS_2$ reassort together and from experiments in which hybridization of virion RNA segment 8 to total virus mRNA specifically prevented the synthesis of both $NS_1$ and $NS_2$ in vitro (Lamb and Choppin, 1979; Inglis et al., 1979).

The estimated sizes of $NS_1$ ($M_r \approx 25,000$), $NS_2$ ($M_r \approx 11,000$) and RNA segment 8 (890 nucleotides) indicated that $NS_1$ and $NS_2$ would have to be translated from overlapping reading frames (Lamb and Choppin, 1979; Inglis et al., 1979). The finding that $NS_2$ is translated from a separate mRNA (~370 nucleotides without poly(A) residues) (Lamb and Choppin, 1979; Inglis et al., 1979) raised several interesting possibilities. Using cloned DNA derived from RNA segment 8, the mRNAs for $NS_1$ and $NS_2$ were mapped on RNA segment 8 using the S1 nuclease technique (Lamb et al., 1980) and by blot hybridization experiments (Inglis et al., 1980). These data, together with the results from hybrid-arrested translation experiments indicated (1) the body of the $NS_1$ mRNA (~850 nucleotides) maps from 0.05–0.95 units of the cloned cDNA of RNA segment 8; (2) $NS_1$ terminates translation at ~0.75 map units, and (3) the body of the $NS_2$ mRNA (~340 nucleotides) maps from 0.59 to 0.95 units. This suggested that the two mRNAs are 3' co-terminal, share the same polyadenylation site, and that the $NS_1$ and $NS_2$ coding regions must be in different reading frames, which overlap by ~150 nucleotides (Lamb et al., 1980). To analyze the 5' nucleotides of the $NS_2$ mRNA, the nucleotide sequence of the A/Udorn/72 RNA segment 8 cloned DNA and the sequence of the $NS_2$ mRNA was obtained (Lamb and Lai, 1980). This sequencing showed that the $NS_2$ mRNA contained an interrupted region of 473 nucleotides; the first 56 virus-specific nucleotides at the 5' end of the $NS_2$ mRNA are the same as those found at the 5' end of the $NS_1$ mRNA, and then $NS_2$ mRNA continues at nucleotide 529. The leader sequence of the NS$_2$ mRNA contains the AUG initiation codon for protein synthesis and coding information for nine amino acids which are common to NS$_1$ and NS$_2$. The ~340 nucleotide body region of the NS$_2$ mRNA can be translated in the +1 reading frame, and the sequence of strain A/Udorn/72 indicates that NS$_1$ and NS$_2$ overlap by 70 amino acids that are translated from different reading frames (Fig. 14). The 5' and 3' junctions of the NS$_2$ mRNA are a good match with those used in spliced mRNAs (Mount, 1982) and, although at the time of the discovery several formal possibilities apart from splicing had to be considered, all the available evidence indicates that the NS$_2$ mRNA is generated by splicing the NS$_1$ co-linear transcript (Lamb and Lai, 1980; see Krug, 1983).

To establish that the influenza virus mRNA junctions can be used by the cellular splicing apparatus, the cloned NS DNA was inserted into a SV40 late region vector that lacked its own splice sites. As shown schematically in Fig. 15 both unspliced and spliced NS$_1$ and NS$_2$ mRNAs were obtained that were translated to yield the NS$_1$ and NS$_2$ polypeptides (Lamb and Lai, 1984). The steady-state amounts of the NS$_1$ and NS$_2$ mRNAs expressed by the SV40 late region vector (approximately 1 : 1 ratio) differs from that observed in influenza virus infected cells (10 : 1 ratio), but this may well be due to an effect of such factors as the additional 5' and 3' sequences derived from SV40, thereby altering stability or splicing efficiency or different retention times in the nucleus. As described in great detail in Chapter 2, it has been difficult to show a precursor-product relationship between NS$_1$ and NS$_2$ by using *in vitro* splicing extracts (Plotch and Krug, 1986). It is not known how the extent of splic-

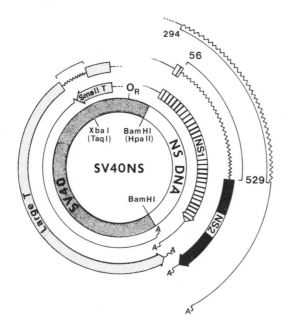

FIGURE 15. Schematic diagram of an SV40 recombinant virus containing the cloned NS DNA and indicating the mRNAs produced in infected cells and their open reading frames. Zig-zag lines in the mRNAs represent introns and the dashes at the 5' end of the mRNAs represent varying positions of the 5' ends. From Lamb and Lai (1984).

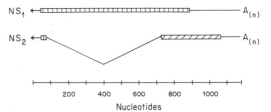

FIGURE 16. Schematic representation of the arrangement of the influenza B virus $NS_1$ and $NS_2$ mRNAs. Redrawn from data in Briedis and Lamb (1982).

ing of $NS_1$ (and $M_1$) is controlled. Both the unspliced and the spliced mRNAs encode proteins that are required for viral replication. Hence some mechanism must have evolved to preserve the unspliced precursor, while at the same time producing sufficient amounts of the spliced product. Several mechanisms can be envisaged for this control: (1) *trans*, i.e., a viral protein that regulates the extent of splicing; or (2) *cis*, i.e., by sequence elements in the promoter. In influenza virus-infected cells, the extent of splicing is regulated to give rise to a 10:1 ratio of unspliced to spliced mRNAs. In the SV40 recombinant NS or M virus-infected cells some of this control is lost, perhaps because these mRNA precursors contain SV40-specific 5' and 3' regions in addition to influenza virus specific sequences (Lamb and Lai, 1982, 1984). However, some possible evidence for a *trans* factor has been obtained because the ratio of unspliced to spliced RNA segment 8 derived transcripts is altered with certain *ts* mutants with lesions in RNA segment 8 of influenza virus and it has been suggested that the production of the $NS_2$ mRNA is regulated by virus-specific products (Smith and Inglis, 1985). More recently, evidence has been obtained for a *cis* acting mechanism in which some sequence element starting 23 nucleotides downstream of the 5' splice site have been found to be inhibitory for splicing (Plotch and Krug, 1986). Using *in vitro* splicing extracts it has been found that the $NS_1$ mRNA forms 55S complexes containing the U1, U2, U4, U5, and U6 snRNAs but that no catalysis occurs (Agris, Nemeroff, and Krug, personal communication). This indicates that the 3' splice site and branchpoints can interact with these snRNPs, but that some other sequence element blocks splicing. A further discussion of splicing of influenza virus mRNA can be found in Chapter 2.

## B. $NS_1$ Protein

The $NS_1$ polypeptide was first identified in an examination of the polypeptides synthesized in influenza A virus infected cells by Lazarowitz and co-workers (1971) who observed the abundant synthesis of a $M_r \approx 26,000$ polypeptide, designated $NS_1$, that was not found in purified virions. Several other reports confirmed the presence of $NS_1$ in infected cells (Skehel, 1972; Compans, 1973a; Krug and Etkind, 1973). Although a great deal has been learned about the structure of the gene and polypeptide in the 15 years that have elapsed since the original description of $NS_1$, remarkably little has been learned about its function during the replicative cycle of influenza virus. It has been suggested that $NS_1$ is involved in the shutoff of host cell protein synthesis or the synthesis of vRNA (Lazarowitz et al., 1971; Compans, 1973a; Wolstenholme et al., 1980; Koennecke et al., 1981). Two ts mutants of A/Udorn/72 with lesions in $NS_1$ (ts ICR 1629, residue 132 Ala changed to Thr, and ts SPC 45, residue 62 Lys changed to Asn) show the phenotype, as measured by polypeptide synthesis, of a failure of infection to progress beyond primary transcription/translation (Shimizu et al., 1982), which is consistent with a lack of vRNA synthesis. Data with other ts mutants of RNA segment 8 indicate similar phenotypes (Wolstenholme et al., 1980; Koennecke et al., 1981), but it has not been determined whether these mutations are in the $NS_1$ or $NS_2$ coding regions. $NS_1$ is found in infected cells associated with polysomes and also in the nucleus and nucleolus (Lazarowitz et al., 1971; Compans, 1973a; Krug and Etkind, 1973; Krug and Soeiro, 1975). $NS_1$ has been shown to migrate to the nucleus shortly after its synthesis in the cytoplasm (Briedis et al., 1981). With some strains of influenza A virus (e.g., PR8), $NS_1$ appears to accumulate in the nucleus early in infection and only later can accumulation in the nucleolus be observed (Young et al., 1983a). With certain strains late in infection, $NS_1$ forms electron-dense paracrystalline inclusion bodies (Morrongiello and Dales, 1977; Shaw and Compans, 1978), which contain a mixture of cellular RNA species (Yoshida et al., 1981). It is not clear if the $NS_1$ inclusions have a functional significance in influenza virus replication, but they do provide a source of semipure $NS_1$ protein for use as an antigen (Shaw and Compans, 1978). The only known post-translational modification to $NS_1$ is phosphorylation, with phosphate covalently linked to one or two threonine residues per molecule (Privalsky and Penhoet, 1977, 1978, 1981; Almond and Felsenreich, 1982).

The first nucleotide sequences to be obtained for RNA segment 8 (Lamb and Lai, 1980; Porter et al., 1980) indicated that $NS_1$ was 230–237 amino acids in length depending on the virus strain. The predicted amino acid sequence agrees well with the amino acid composition of $NS_1$ (Shaw and Compans, 1978) and the tryptic peptides of $NS_1$ can readily be correlated with the predicted sequence (Lamb et al., 1978; Lamb and Choppin, 1979). As more nucleotide sequences became available, it became clear that $NS_1$ ranged from 202 amino acids for the A/FM/1/47 strain to 237 amino acids for the A/Udorn/72 strain. These "early" termination codons made it possible to use the $NS_1$ mRNAs as "test genes" (strains A/Cambridge/46 and A/Texas/1/68) to assay for ochre and amber suppressor transfer RNA activities introduced into mammalian cells (Young et al., 1983b). Recently, it has been found that the A/Turkey/Oregon/71

NS$_1$ protein contains only 124 amino acids (Norton *et al.*, 1987), yet the virus forms plaques in tissue culture and is highly infectious in turkeys (Beard and Helfer, 1972; Norton *et al.*, 1987). This suggests the C-terminal half of NS$_1$ is dispensable for its function. An influenza virus host range mutant CR43-3, derived by recombination from the A/Alaska/6/77 and the cold-adapted and temperature-sensitive A/Ann Arbor/6/60 viruses have also been characterized and found to contain an in-frame deletion from amino acids 66-77 inclusive in the NS$_1$ protein (Buonagurio *et al.*, 1984). The restricted host range phenotype of the mutant virus was found to be that it forms plaques in primary chick kidney cells, but not in Madin–Darby canine kidney cells (Maassab and De Borde, 1983). Although in CR43-3 it is possible that one NS$_1$ chain may complement another, it is also possible that NS$_1$ may be semifunctional as a 124-amino acid polypeptide that contains a 12-amino acid deletion. If a biochemical assay for NS$_1$ activity can be identified in the future, the structural domains of this unusual protein can be readily determined by molecular genetics. It has been proposed (Winter *et al.*, 1981b) that the NS$_1$ and NS$_2$ cistrons may not have overlapped on the influenza virus progenitor RNA segment 8 but were contiguous cistrons and that mutations in the stop codon for the NS$_1$ polypeptide resulted in continued translation overlapping the NS$_2$ reading frame. The present data can be interpreted as supporting this hypothesis.

The A/Turkey/Oregon/71 NS$_1$ protein of 124 residues and the CR43-3 NS$_1$ protein both accumulate in the nucleus, suggesting that the karyophilic signal necessary to direct NS$_1$ from the cytoplasm to the nucleus must be between amino acids 2–65 and 78–124. With SV40 T antigen it has been shown that the basic sequence from residues 127–132 [Pro-Lys-Lys-Lys-Arg-Lys] is both sufficient and necessary to direct proteins to the nucleus (Kalderon *et al.*, 1984a,b). It has been pointed out that in NS$_1$, four out of five of these basic residues occur at positions 17–21 and 35–46 for the A/PR/8/34 strain (Greenspan *et al.*, 1985) and that one or more of these may be a karyophilic sequence. These determinations are not always simple, i.e., Hall and colleagues (Hall *et al.*, 1984) have found that yeast $\alpha_2$-protein has more than one nuclear targeting signal. To understand the nature of these signals, it is necessary to determine whether the signals are independent at the same step in transport, cooperative at the same step or independent at different steps. Recently, it has been shown that NS$_1$ contains two independent signals for nuclear localization, using deletion mutants of NS$_1$ (Greenspan *et al.*, 1988). One signal contains the stretch of basic amino acids Asp-Arg-Leu-Arg-Arg (codons 34–38) and the other has only been defined to residues 203–237, a region which contains basic residues (Greenspan *et al.*, 1988). For these types of experiments, the NS$_1$ protein has been expressed from cloned cDNA using eukaryotic and prokaryotic vectors (Young *et al.*, 1983; Lamb and Lai, 1984; Greenspan *et al.*, 1985, 1988; Portela *et al.*, 1985; 1986). Expression in bacteria provided a convenient source of large amounts of pure $NS_1$ against which polyclonal antisera has been raised, and the antisera used to confirm the presence of $NS_1$ in the nucleus when expressed from the SV40-based eukaryotic vector in the absence of the other influenza virus-specific polypeptides (Greenspan et al., 1985).

## C. $NS_2$ Protein

A small polypeptide of observed $M_r \approx 11,000$ was observed in influenza A virus infected cells but not in purified virions (Skehel, 1972; Follett et al., 1974; Minor and Dimmock, 1975; Krug and Etkind, 1973; Stephenson et al., 1977; Lamb and Choppin, 1978). After it was shown that $NS_2$ exhibited strain-specific differences in its mobility on polyacrylamide gels and that $NS_2$ had a tryptic peptide composition that was distinct from that of any of the other virally encoded polypeptides, it became clear that $NS_2$ is a discrete viral polypeptide (Lamb et al., 1978). This led to the suggestion that one virion RNA segment coded for two proteins (Lamb et al., 1978). Hybrid arrest of in vitro translation experiments, analysis of recombinants with defined genome composition and sequence analysis of the $NS_1$ and $NS_2$ mRNAs all indicate that $NS_2$ $M_r \approx$ 14,216 is encoded by RNA segment 8 (Lamb and Choppin, 1979; Inglis et al., 1979; Lamb and Lai, 1980).

The function of $NS_2$ in infected cells is not known, and the results concerning its cellular location are conflicting. Using antisera raised to $NS_2$ expressed in bacteria to perform immunofluorescence on cells expressing $NS_2$ either during an influenza virus infection or from a eukaryotic vector, suggests $NS_2$ is located in the nucleoplasm (Greenspan et al., 1985). However, earlier cell fractionation experiments suggest $NS_2$ is a soluble cytoplasmic protein (Mahy et al., 1980; Briedis et al., 1981) and data obtained using a vaccinia virus recombinant that expresses $NS_2$ indicated that $NS_2$ was located to the cytoplasm (Smith et al., 1987). It is possible that the subcellular localizations are correct and these, of course, are not mutually exclusive. However, the experimental approaches tried to date have their complications. The cell fractionation experiments may be affected by the fact that $NS_2$ is a small protein and may rapidly leak out of the nucleus. Some of the immunofluorescence data clearly shows that $NS_2$ localizes in the nucleus but there is diffuse cytoplasmic staining (Greenspan et al., 1985). Other immunofluorescence data clearly shows $NS_2$ localized to the cytoplasm (Smith et al., 1987). If $NS_2$ in the nucleus is bound in a complex, it could be expected to fluoresce much more intensely than soluble $NS_2$ in the cytoplasm, and therefore it is very difficult to quantitate the relative nuclear and cytoplasmic fluorescence intensities. In the author's laboratory, influenza virus infected CV1 cells have been treated with 2% glutaraldehyde for 15' and then the membranes permeabilized with 0.5% Triton-X100. It has been observed that ~75% of the $NS_2$ polypeptide is released from the cell in a soluble fraction after centrifugation (100,000g for 30 min) (Zebedee and Lamb, un-published observations). As a considerable amount of $NS_2$ escaped from fixation by the glutaraldehyde, it would suggest that some $NS_2$ is soluble in the cytoplasm.

## D. RNA Segment 8 of Influenza B Virus: Unspliced and Spliced mRNAs Code for $NS_1$ and $NS_2$

RNA segment 8 of influenza B virus also encodes two nonstructural proteins $NS_1$ (apparent $M_r \approx 40,000$) and $NS_2$ (apparent $M_r \approx 11,500$) (Briedis et al., 1981). $NS_1$ was identified as a polypeptide synthesized only in infected cells and encoded by RNA segment 8 (Oxford, 1973; Choppin et al., 1975; Almond et al., 1979; Racaniello and Palese, 1979). $NS_2$ was recognized in influenza B-virus infected cells (Lamb and Choppin, 1979; Almond et al., 1979) and shown to be a discrete virus-specific polypeptide by strain-specific differences in mobility and tryptic peptide mapping (Lamb and Choppin, 1979; Briedis et al., 1981). $NS_2$ is encoded by a separate small mRNA as shown by sucrose-gradient separation of in-fected cell mRNAs followed by in vitro translation of the fractions, and $NS_1$ and $NS_2$ were shown to be encoded by RNA segment 8 by hybrid arrest of translation experiments (Briedis et al., 1981). From the nu-cleotide sequence of cloned cDNA made to RNA segment 8 of influenza B/Lee/40, it was determined that RNA segment 8 codes for $NS_1$ and $NS_2$ with an overall structure similar to that of the influenza A viruses (Briedis and Lamb, 1982). RNA segment 8 is 1096 nucleotides in length and contains a large open reading frame of 281 amino acids which en-codes the $NS_1$ polypeptide $M_r = 32,026$). Sequencing of the $NS_2$ mRNA showed that it contained an interrupted sequence of 655 nucleotides (Fig. 16). The first ~75 virus-specific nucleotides at the 5' end of the $NS_2$ mRNA are the same as found at the 5' end of the $NS_1$ mRNA. This region contains the initiation codon for protein synthesis and coding informa-tion for 10 amino acids common to the two proteins. The ~350-nucleotide body region of the $NS_2$ mRNA can be translated in the +1 reading frame, and the sequence indicates that the $NS_1$ and $NS_2$ protein-coding regions overlap by 52 amino acids translated from different read-ing frames. The sequences at the 5' and 3' splice junctions have sim-ilarities to the consensus sequences for splicing signals (Mount, 1982). $NS_2$ contains 122 amino acids, one more than that of influenza A virus. Between the influenza A and B viruses, the organization of the $NS_1$ and $NS_2$ mRNAs and the sizes of the $NS_2$ mRNA and protein are conserved despite the larger size of the influenza B virus RNA segment, $NS_1$ mRNA, and $NS_1$ protein (Briedis and Lamb, 1982). The structure of the B/Yamagata/1/73 strain RNA segment 8 shows little variation compared with that of the B/Lee/40 strain (Norton et al., 1987) (60 nucleotide changes, resulting in 22 $NS_1$ amino acid differences and 3 $NS_2$ amino acid

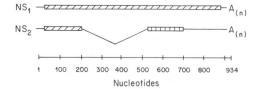

FIGURE 17. Schematic representation of the influenza C virus RNA segment 7 derived NS$_1$ and NS$_2$ mRNAs. Redrawn from data derived by Nakada *et al.* (1986).

changes). However, a laboratory variant of an influenza B virus (Yamagata clone 302), encodes a truncated NS$_1$ protein of 110 amino acids derived from the NS$_1$ reading frame and 17 carboxy-terminal residues from the +1 reading frame. This occurred because of a deletion of 13 bases between positions 374 and 386 of RNA segment 8. The NS$_2$ protein coding sequences and the splice junctions are unaffected by the changes. The truncated NS$_1$ protein accumulates in the nucleus and the virus replicates at a wild-type rate (Norton *et al.*, 1987). Thus, it appears that the NS$_1$ polypeptides of both influenza A and B viruses can tolerate a large deletion in the C-terminal region without affecting infectivity of the virus in tissue culture. Like influenza A virus NS$_1$ and NS$_2$ proteins, the function of the influenza B virus counterparts remains to be determined.

## E. Influenza C Virus RNA Segment 7: Unspliced and Spliced mRNAs Code for NS$_1$ and NS$_2$

The nonstructural proteins of influenza C virus have proven to be more difficult to examine experimentally, partly because of the poorer growth properties of the virus as compared with influenza A and B viruses. There have been several reports of the identification of nonstructural proteins. Yokota and co-workers (1983) identified five nonstructural polypeptides in influenza C virus infected cells, but on tryptic peptide analysis only a polypeptide of $M_r \approx 29,000$ had a unique peptide map. The other polypeptides ranging in size from $M_r \approx 27,500$ to 14,000 were determined to be proteolytic fragments of the M protein. However, Petri and co-workers (1980) and Nakada *et al.*, (1985) identified two nonstructural polypeptides of $M_r \approx 24,000$ and 14,000 in C/Johannesburg/1/66 infected chick kidney cells. Once the RNA segment 7 was cloned and the nucleotide sequence obtained, the pattern of expression became somewhat clearer.

The complete nucleotide sequence of RNA segment 7 of influenza C/California/78 indicates that it contains 934 nucleotides with a long open reading frame, starting at nucleotide 27 and terminating at nucleotide 884, and can encode 286 amino acids (Nakada *et al.*, 1984). Hybrid arrest translation experiments with the cloned cDNA and mRNAs isolated from infected cells indicate that RNA segment 7 encodes a $M_r \approx 28,500$ polypeptide which was also identified in infected cells (Nakada et al., 1984). Examination of the RNA segment 7 specific mRNAs using the primer-extension method, indicated that there is a second RNA species, smaller than a co-linear transcript (Nakada et al., 1986). Direct sequencing of the primer-extension product and the virion RNA segment 7 indicate that the second RNA transcript contains an interrupted region from nucleotide 213 to 527 (Fig. 17). The sequences around the 5' and 3' junctions conform to the established splicing signals. The resulting spliced mRNA is such that translation could initiate at the same AUG codon as $NS_1$ and continue for 62 amino acids in the $NS_1$ reading frame. After the splice junction, translation would continue in the $+1$ reading frame for 59 amino acids and, therefore, the mRNA encodes 122 amino acids (Nakada et al., 1986). The second open reading frame of the influenza C virus RNA segment 7 is completely overlapped by that of the $NS_1$ protein as shown in Fig. 17.

To determine whether the spliced mRNA is expressed in influenza C virus-infected cells, a peptide was synthesized to the C-terminal region of the $+1$ reading frame region of the mRNA and antisera raised to the synthetic peptide (Nakada et al., 1986). The antisera was used to immunoprecipitate a $M_r \approx 14,000$ polypeptide from infected cells that corresponded in size to a polypeptide that is synthesized in vitro from influenza C virus-infected cell mRNAs and in hybrid-arrest translation experiments was shown to be RNA segment 7 specific (Nakada et al., 1986). Thus, an unspliced mRNA encoding $NS_1$ (286 amino acids) and a spliced mRNA encoding $NS_2$ (122 amino acids) are derived from influenza C virus RNA segment 7, but their function in the replicative cycle of the virus is not known.

ACKNOWLEDGMENTS. I am very grateful to Dr. Reay Paterson, Dr. Mark Williams, and Dr. Suzanne Zebedee for their thoughtful comments on both the original and then the later updated version of this review. However, the blame for all errors and subjective interpretations of scientific papers rests with me. It is with pleasure that I acknowledge the assistance of Sandy Getowicz and Carolyn Jenkins for typing the text and the endless references. To keep this review to a reasonable size, it is inevitable that many citations of older papers and some current papers were omitted. Many of these references appear in other chapters of this book that specialize on a particular topic, but to those authors who feel slighted I apologize. Research in this author's laboratory was supported by research grants AI-20201 and AI-23173 from the National Institutes of Health. Finally, I should like to thank Dr. Reay Paterson for forebearance during the writing of this review.

# REFERENCES

Abrahams, G., 1979, The effect of ultraviolet radiation on the primary transcription of influenza virus messenger RNAs, Virology 97:177–182.

Adams, G. A., and Rose, J. K., 1985, Incorporation of a charged amino acid into the membrane-spanning domain blocks cell surface transport but not membrane anchoring of a viral glycoprotein, *Mol. Cell Biol.* **5:**1442–1448.

Air, G. M., and Compans, R. W., 1983, Influenza B and influenza C viruses, in: *Genetics of Influenza Viruses* (P. Palese and D. W. Kingsbury, eds.), pp. 280–304, Springer-Verlag, Vienna.

Air, G. M., Ritchie, L. R., Laver, W. G., and Colman, P. M., 1985, Gene and protein sequence of an influenza neuraminidase with hemagglutinating activity, *Virology* **145:**117–122.

Air, G. M., Webster, R. G., Colman, P. M., and Laver, W. G., 1987, Distribution of sequence differences in influenza N9 neuraminidase of tern and whale viruses and crystallization of the whale neuraminidase complexed with antibodies, *Virology* **160:**346–354.

Akkina, R. K., Chambers, T. M., Londo, D. R., and Nayak, D. P. 1987, Intracellular localization of the viral polymerase proteins in cells infected with influenza virus and cells expressing PB1 protein from cloned cDNA, *J. Virol.* **61:**2217–2224.

Akoto-Amanfu, E., Sivasubramanian, N., and Nayak, D. P., 1987, Primary structure of the polymerase acidic (PA) gene of influenza B virus (B/Sing/222/79), *Virology* **159:**147–153.

Allen, H., McCauley, J., Waterfield, M., and Gething, M.-J., 1980, Influenza virus RNA segment 7 has the coding capacity for two polypeptides, *Virology* **107:**548–551.

Almond, J. W., and Barry, R. D., 1979, Genetic recombination between two strains of fowl plague virus: Construction of genetic maps, *Virology* **92:**407–415.

Almond, J. W., and Felsenreich, V., 1982, Phosphorylation of the nucleoprotein of an avian influenza virus, *J. Gen. Virol.* **60:**295–305.

Almond, J. W., Haymerle, H. A., Felsenreich, V. D., and Reeve, P., 1979, The structural and infected cell polypeptides of influenza B virus, *J. Gen. Virol.* **45:**611–621.

Anderson, D. J., Mostov, K. E., and Blobel, G., 1983, Mechanisms of integration of de novo-synthesized polypeptides into membranes: Signal recognition particle is required for integration into microsomal membranes of calcium ATPase and of lens MP26 but not cytochrome $b_5$, *Proc. Natl. Acad. Sci. USA* **80:**7249–7253.

Apostolov, K., and Flewett, T. H., 1965, Internal structure of influenza virus, *Virology* **26:**506–508.

Bachi, T., Gerhard, W., Lindermann, J., and Muhlethaler, K., 1969, Morphogenesis of influenza A virus in Ehrlich ascites tumor cells as revealed by thin sectioning and freeze-etching, *J. Virol.* **4:**769–776.

Baez, M., Taussig, R., Zazra, J. J., Young, J. F., Palese, P., Reisfeld, A., and Skalka, A. M., 1980, Complete nucleotide sequence of the influenza A/PR/8/34 virus NS gene and comparison with the NS genes of the A/Udorn/72 and A/FPV/Rostock/34 strains, *Nucl. Acids* **8:**5845–5858.

Baez, M., Zazra, J. J., Elliott, R. M., Young, J. F., and Palese, P., 1981, Nucleotide sequence of the influenza A/duck/Alberta/60/76 virus NS RNA: Conservation of the $NS_1/NS_2$ overlapping gene structure in a divergent influenza virus RNA segment, *Virology* **113:**397–402.

Baltimore, D., 1971, Expression of animal virus genomes, *Bacteriol. Rev.* **35:**235–241.

Barry, R. D., and Mahy, B. W. J., 1979, The influenza virus genome and its replication, *Br. Med. Bull.* **35:**39–46.

Basak, S., Tomana, M., and Compans, R. W., 1985, Sialic acid is incorporated into influenza hemagglutinin glycoproteins in the absence of viral neuraminidase, *Virus Res.* **2:**61–68.

Bean, W. J., Jr., and Simpson, R. W., 1976, Transcriptase activity and genome composition of defective influenza virus, *J. Virol.* **18:**365–369.

Beard, C. W., and Helfer, D. H., 1972, Isolation of two turkey influenza viruses in Oregon, *Avian Dis.* **16:**1133–1136.

Beaton, A. R., and Krug, R. M., 1986, Transcription antitermination during influenza virus template RNA synthesis requires the nucleocapsid protein and the absence of a 5' capped end, *Proc. Natl. Acad. Sci. USA* **83:**6282–6286.

Bentley, D. R., and Brownlee, G. G., 1982, Sequence of the N2 neuraminidase from influenza virus A/NT/60/68, *Nucl. Acids Res.* **10**:5033–5042.

Bishop, D. H. L., Huddleston, J. A., and Brownlee, G. G., 1982a, The complete sequence of RNA segment 2 of influenza A/NT/60/68 and its encoded P1 protein, *Nucl. Acids Res.* **10**:1335–1343.

Bishop, D. H. L., Jones, K. L., Huddleston, J. A., and Brownlee, G. G., 1982b, Influenza A virus evolution: Complete sequences of influenza A/NT/60/68 RNA segment 3 and its predicted acidic P polypeptide compared with those of influenza A/PR/8/34, *Virology* **120**:481–489.

Bishop, D. H. L., Roy, P., Bean, W. J., and Simpson, R. W., 1972, Transcription of the influenza ribonucleic acid genome by a virion polymerase. III. Completeness of the transcription process, *J. Virol.* **10**:689–697.

Blobel, G., 1980, Intracellular protein topogenesis, *Proc. Natl. Acad. Sci. USA* **77**:1496–1500.

Blok, J., and Air, G. M., 1980, Comparative nucleotide sequences at the 3′-end of the neuraminidase gene from eleven influenza A viruses, *Virology* **107**:50–60.

Blok, J., and Air, G. M., 1982a, Block deletions in the neuraminidase genes from some influenza A viruses of the N1 subtype, *Virology* **118**:229–234.

Blok, J., and Air, G. M., 1982b, Sequence variation at the 3′ end of the neuraminidase gene from 39 influenza type A viruses, *Virology* **121**:211–229.

Blok, J., Air, G. M., Laver, W. G., Ward, C. W., Lilley, G. G., Woods, E. F., Roxburgh, C. M., and Inglis, A. S., 1982, Studies on the size, chemical composition, and partial sequence of the neuraminidase (NA) from type A influenza viruses show that the N-terminal region of the NA is not processed and serves to anchor the NA in the membrane, *Virology* **119**:109–121.

Bos, J. J., Davis, A. R., and Nayak, D. P., 1984, NH$_2$-terminal hydrophobic region of influenza virus neuraminidase provides the signal function in translocation, *Proc. Natl. Acad. Sci. USA* **81**:2327–2331.

Bosch, F., Orlich, M., Klenk, H.-D., and Rott, R., 1979, The structure of the hemagglutinin, a determinant for the pathogenicity of influenza viruses, *Virology* **95**:197–207.

Both, G. W., and Air, G. M., 1979, Nucleotide sequence coding for the N-terminal region of the matrix protein of influenza virus, *Eur. J. Biochem.* **96**:363–372.

Both, G. W., and Sleigh, M. J., 1980, Complete nucleotide sequence of the haemagglutinin gene from a human influenza virus of the Hong Kong subtype, *Nucl. Acids Res.* **8**:2561–2575.

Both, G. W., and Sleigh, M. J., 1981, Conservation and variation in the hemagglutinins of Hong Kong subtype influenza viruses during antigenic drift, *J. Virol.* **39**:663–672.

Both, G. W., Cheng, H. S., and Kilbourne, E. D., 1983a, Hemagglutinin of swine influenza virus: A single amino acid change pleiotropically affects viral antigenicity and replication, *Proc. Natl. Acad. Sci. USA* **80**:6996–6700.

Both, G. W., Sleigh, M. J., Cox, N. J., and Kendal, A. P., 1983b, Antigenic drift in influenza virus H3 hemagglutinin from 1968 to 1980: Multiple evolutionary pathways and sequential amino acid changes at key antigenic sites, *J. Virol.* **48**:52–60.

Boulay, F., Doms, R. W., Webster, R., and Helenius, A., 1988, Post-translational oligomerization and cooperative acid-activation of mixed influenza hemagglutinin trimers, *J. Cell Biol.* **106**:629–639.

Braam-Markson, J., Jaudon, C., and Krug, R. M., 1985, Expression of a functional influenza viral cap-recognizing protein by using a bovine papilloma virus vector, *Proc. Natl. Acad. Sci. USA* **82**:4326–4330.

Braam, J., Ulmanen, I., and Krug, R. M., 1983, Molecular model of a eucaryotic transcription complex: Functions and movements of influenza P proteins during capped RNA-primed transcription, *Cell* **34**:609–618.

Brand, C. M., and Skehel, J. J., 1972, Crystalline antigen from the influenza virus envelope, *Nature New Biol.* **238**:145–147.

Brand, C. M., Stealy, V. M., and Rowe, J., 1977, Peptide mapping of [$^{125}$I]-labeled influenza virus proteins and matrix proteins as markers in recombination, *J. Gen. Virol.* **36:**385–394.

Briedis, D. J., Conti, G., Munn, E. A., and Mahy, B W. J., 1981, Migration of influenza virus-specific polypeptides from cytoplasm to nucleus of infected cells, *Virology* **111:**154–164.

Briedis, D. J., and Lamb, R. A., 1982, Influenza B virus genome: Sequences and structural organization of RNA segment 8 and the mRNAs coding for the NS$_1$, and NS$_2$ proteins, *J. Virol.* **42:**186–193.

Briedis, D. J., and Tobin, M., 1984, Influenza B virus genome: Complete nucleotide sequence of the influenza B/Lee/40 virus genome RNA 5 encoding the nucleoprotein and comparison with the B/Singapore/22/79 nucleoprotein, *Virology* **133:**448–455.

Briedis, D. J., Lamb, R. A., and Choppin, P. W., 1981, Influenza B virus RNA segment 8 codes for two nonstructural proteins, *Virology* **112:**417–425.

Briedis, D. J., Lamb, R. A., and Choppin, P. W., 1982, Sequence of RNA segment 7 of the influenza B virus genome: Partial amino acid homology between the membrane proteins (M$_1$) of influenza A and B viruses and conservation of a second open reading frame, *Virology* **116:**581–588.

Bucher, D. J., Kharitonenkow, L. G., Zakomiridin, J. A., Griboriev, V. B., Klimenko, S. M., and Davis, J. F., 1980, Incorporation of influenza M-protein into liposomes, *J. Virol.* **36:**586–590.

Buckler-White, A. J., and Murphy, B. R., 1986, Nucleotide sequence analysis of the nucleoprotein gene of an avian and a human influenza virus strain identifies two classes of nucleoproteins, *Virology* **155:**345–355.

Buckler-White, A. J., Naeve, C. W., and Murphy, B. R., 1986, Characterization of a gene coding for M proteins which is involved in host range restriction of an avian influenza A virus in monkeys, *J. Virol.* **57:**697–700.

Buonagurio, D. A., Krystal, M., Palese, P., DeBorde, D. C., and Maassab, H. F., 1984, Analysis of an influenza A virus mutant with a deletion in the NS segment, *J. Virol.* **49:**418–425.

Buonagurio, B. A., Nakada, S., Desselberger, U., Krystal, M., and Palese, P., 1985, Noncumulative sequence changes in the hemagglutinin genes of influenza C virus isolates, *Virology* **146:**221–232.

Buonagurio, D. A., Nakada, S., Parvin, J. D., Krystal, M., Palese, P., and Fitch, W. M., 1986, Evolution of human influenza A viruses over 50 years: Rapid and uniform rate of change in the NS gene, *Science* **232:**980–982.

Burnet, F. M., and Stone, J. D., 1947, The receptor-destroying enzyme of *V. cholerae*, *Aust. J. Exp. Biol. Med. Sci.* **25:**227–233.

Butler, P. J. G., and Klug, A., 1971, Assembly of the particle of tobacco mosaic virus from RNA and disks of protein, *Nature New Biol.* **229:**47–50.

Caliguiri, L. A., and Gerstein, H., 1978, Subclasses of ribonucleoproteins in influenza virus-infected cells, *Virology* **90:**119–132.

Caton, A. J., Brownlee, G. G., Yewdell, J. W., and Gerhard, W., 1982, The antigenic structure of the influenza virus A/PR/8/34 hemagglutinin (H1 subtype), *Cell* **31:**417–427.

Chiacchia, K. B., and Drickamer, K., 1984, Direct evidence for the transmembrane orientation of the hepatic glycoprotein receptors, *J. Biol. Chem.* **259:**15440–15446.

Choppin, P. W., Compans, R. W., Scheid, A., McSharry, J. J., and Lazarowitz, S. G., 1972, Structure and assembly of viral membranes, in: *Membrane Research* (C. F. Fox, ed.), pp. 163–179, Academic, Orlando, FL.

Choppin, P. W., Lazarowitz, S. G., and Goldberg, A. R., 1975, Studies on proteolytic cleavage and glycosylation of the haemagglutinin of influenza A and B viruses, in: *Negative Strand Viruses*, Vol. 1 (B. W. J. Mahy and R. D. Barry, eds.), pp. 105–119, Academic, London.

Claesson, L., Larhammar, D., Rask, L., and Peterson, P. A., 1983, cDNA clone for the human invariant class of class II histocompatibility antigens and its implications for protein structure, *Proc. Natl. Acad. Sci. USA* **80:**7395–7399.

Colman, P. M., and Ward, C. W., 1985, Structure and diversity of the influenza virus neuraminidase, *Curr. Top. Microbiol. Immunol.* **114:**177–255.

Colman, P. M., Varghese, J. N., and Laver, W. G., 1983, Structure of the catalytic and antigenic sites in influenza virus neuraminidase, *Nature (Lond.)* **303:**41–44.

Colman, P. M., Laver, W. G., Varghese, J. N., Backer, A. T., Tulloch, P. A., Air, G. M., and Webster, R. G., 1987, The three-dimensional structure of a complex of influenza virus neuraminidase and an antibody, *Nature (Lond.)* **326:**358–363.

Compans, R. W., 1973a, Influenza virus proteins. II. Association with components of the cytoplasm, *Virology* **51:**56–70.

Compans, R. W., 1973b, Distinct carbohydrate components of influenza virus glycoproteins in smooth and rough cytoplasmic membranes, *Virology* **55:**541–545.

Compans, R. W., and Dimmock, N. J., 1969, An electron microscopic study of single cycle infection of chick embryo fibroblasts by influenza virus, *Virology* **39:**499–515.

Compans, R. W., and Choppin, P. W., 1975, Reproduction of myxoviruses, in: *Comprehensive Virology*, Vol. IV (H. Fraenkel-Conrat and R. R. Wagner, eds.), pp. 179–252, Plenum, New York.

Compans, R. W., Klenk, H. D., Caliguiri, L. A., and Choppin, P. W., 1970, Influenza virus proteins. I. Analysis of polypeptides of the virion and identification of spike glycoproteins, *Virology* **42:**880–889.

Compans, R. W., Content, J., and Duesberg, P. H., 1972, Structure of the ribonucleoprotein of influenza virus, *J. Virol.* **10:**795–800.

Compans, R. W., Bishop, D. H. L., and Meier-Ewert, H., 1977, Structural components of influenza C virions, *J. Virol.* **21:**658–665.

Concannon, P., and Cummings, I. W., 1984, Nucleotide sequence of the influenza virus A/USSR/90/77 hemagglutinin gene, *J. Virol.* **49:**276–278.

Concannon, P., Kwolek, C. J., and Salser, W. A., 1984, Nucleotide sequence of the influenza virus A/USSR/90/77 neuraminidase gene, *J. Virol.* **50:**654–656.

Copeland, C. S., Doms, R. W., Bolzau, E. M., Webster, R. G., and Helenius, A., 1986, Assembly of influenza hemagglutinin trimers and its role in intracellular transport, *J. Cell Biol.* **103:**1179–1191.

Copeland, C. S., Zimmer, K.-P., Wagner, K. R., Healey, G. A., Mellman, I., and Helenius, A., 1988, Folding, trimerization, and transport are sequential events in the biogenesis of influenza virus hemagglutinin, *Cell* **53:**197–209.

Dale, B., Brown, R., Miller, J., Tyler White, R., Air, G. M., and Cordell, B., 1986, Nucleotide and deduced amino acid sequence of the influenza neuraminidase genes of two equine serotypes, *Virology* **155:**460–468.

Daniels, R. S., Downie, J. C., Hay, A. J., Knossow, M., Skehel, J. J., Wang, M. L., and Wiley, D. C., 1985, Fusion mutants of the influenza virus hemagglutinin glycoprotein, *Cell* **40:**431–439.

Davey, J., Dimmock, J., and Colman, A., 1985, Identification of the sequence responsible for the nuclear accumulation of the influenza virus nucleoprotein in *Xenopus* oocytes, *Cell* **40:**667–675.

Davies, W. L., Grunert, R. R., Haff, R. F., McGahen, J. W., Neumayer, E. M., Paulshock, M., Watts, J. C., Wood, T. R., Herrman, E. C., and Hoffman, C. E., 1964, Antiviral activity of 1-Adamantanamine (Amantadine), *Science* **144:**862–863.

Davis, A. R., Bos, T. J., and Nayak, D. P., 1983, Active influenza virus neuraminidase is expressed in monkey cells from cDNA cloned in simian virus 40 vectors, *Proc. Natl. Acad. Sci. USA* **80:**3976–3980.

DeBorde, D. C., Donabedian, A. M., Herlocher, M. L., Naeve, C. W., and Maassab, H. F., 1988, Sequence comparison of wild-type and cold-adapted B/Ann Arbor/1/66 Influenza virus genes, *Virology* **163:**429–443.

Deshpande, K. L., Fried, V. A., Ando, M., and Webster, R. G., 1987, Glycosylation affects cleavage of an H5N2 influenza virus hemagglutinin and regulates virulence, *Proc. Natl. Acad. Sci. USA* **84**:36–40.

Detjen, B. M., St. Angelo, C., Katze, M. G., and Krug, R. M., 1987, The three influenza virus polymerase (P) proteins not associated with viral nucleocapsids in the infected cell are in the form of a complex, *J. Virol.* **61**:16–22.

Dimmock, N. J., Carver, A. S., and Webster, R. G., 1980, Categorization of nucleoproteins and matrix proteins from type A influenza viruses by peptide mapping, *Virology* **103**:350–356.

Doms, R. W., Helenius, A., and White, J., 1985, Membrane fusion activity of the influenza virus hemagglutinin: The low pH-induced conformational change, *J. Biol. Chem.* **260**:2973–2931.

Dopheide, T. A., and Ward, C. W., 1978, The carboxyl-terminal sequence of the heavy chain of a Hong Kong influenza hemagglutinin, *Eur. J. Biochem.* **85**:393–398.

Doyle, C., Roth, M. G., Sambrook, J., and Gething, M.-J., 1985, Mutations in the cytoplasmic domain of the influenza hemagglutinin affect different stages of intracellular transport, *J. Cell Biol.* **100**:704–714.

Doyle, C., Sambrook, J., and Gething, M.-J., 1986, Analysis of progressive deletions of the transmembrane and cytoplasmic domains of influenza hemagglutinin, *J. Cell. Biol.* **103**:1193–1204.

Dreyfuss, G., 1986, Structure and function of nuclear and cytoplasmic ribonucleoprotein particles, *Annu. Rev. Cell. Biol.* **2**:459–498.

Duesberg, P., 1969, Distinct subunits of the ribonucleoprotein of influenza virus, *J. Mol. Biol.* **42**:485–499.

Elder, K. T., Bye, J. M., Skehel, J. J., Waterfield, M. D., and Smith, A. E., 1979, In vitro synthesis, glycosylation and membrane insertion of influenza virus hemagglutinin, *Virology*, 95:343–350.

Elleman, T. C., Azad, A. A., and Ward, C. W., 1982, Neuraminidase gene from the early Asian strain of human influenza virus, A/RI/5⁻/57 (H$_2$N$_2$), *Nucl. Acids. Res.* **10**:7005–7016.

Emerson, S. U., and Yu, Y.-H., 1975, Both NS and L proteins are required for *in vitro* RNA synthesis by vesicular stomatitis virus, *J. Virol.* **15**:1348–1356.

Engelman, D. M., and Steitz, T. A., 1981, The spontaneous insertion of proteins into and across membranes: The helical hairpin hypothesis, *Cell* **23**:411–422.

Fang, R., Min Jou, W., Huylebroeck, D., Devos, R., and Fiers, W., 1981, Complete structure of A/duck/Ukraine/63 influenza hemagglutinin gene: Animal virus as progenitor of human H$_3$ Hong Kong 1968 influenza hemagglutinin, *Cell* **25**:315–323.

Fields, S., and Winter, G., 1982, Nucleotide sequences of influenza virus segments 1 and 3 reveal mosaic structure of a small viral RNA segment, *Cell* **28**:303–313.

Fields, S., Winter, G., and Brownlee, G. G., 1981, Structure of the neuraminidase in human influenza virus A/PR/8/34, *Nature (Lond.)* **290**:213–217.

Follett, E. A. C., Pringle, C. R., Wunner, W. H., and Skehel, J. J., 1974, Virus replication in enucleate cells: Vesicular stomatitis virus and influenza virus, *J. Virol.* **13**:394–399.

Fraenkel-Conrat, H., and Singer, B., 1964, Reconstitution of tobacco mosaic virus. IV. Inhibition by enzymes and other proteins and use of polynucleotides, *Virology* **23**:354–362.

Friedlander, M., and Blobel, G., 1985, Bovine opsin has more than one signal sequence, *Nature (Lond.)* **318**:338–343.

Garoff, H., 1985, Using recombinant DNA techniques to study protein targeting in the eucaryotic cell, *Annu. Rev. Cell Biol.* **1**:403–445.

Garten, W., Bosch, F. X., Linder, D., Rott, R., and Klenk, H.-D., 1981, Proteolytic activation of the influenza virus hemagglutinin: The structure of the cleavage site and the enzymes involved in cleavage, *Virology* **115**:361–374.

Gething, M.-J., Bye, J., Skehel, J. J., and Waterfield, M. D., 1981, Cloning and DNA sequence of double stranded copies of haemagglutinin genes from H2 and H3 strains elucidates antigenic shift and drift in human influenza virus, *Nature* 287:301–306.

Gething, M.-J., Doms, R. W., York, D., and White, J. M., 1986, Studies on the mechanism of membrane fusion: Site-specific mutagenesis of the hemagglutinin of influenza virus, *J. Cell. Biol.* **102**:11–23.

Gething, M.-J., McCammon, K., and Sambrook, J., 1986, Expression of wild-type and mutant forms of influenza hemagglutinin: The role of folding in intracellular transport, *Cell* **46**: 939–950.

Gething, M.-J., and Sambrook, J., 1981, Cell-surface expression of influenza haemagglutinin from a cloned DNA copy of the RNA gene, *Nature (Lond.)* **293**:620–625.

Gething, M.-J., and Sambrook, J., 1982, Construction of influenza virus haemagglutinin genes that code for intracellular and secreted forms of the protein, *Nature (Lond.)* **300**: 598–603.

Gething, M.-J., White, J. M., and Waterfield, M. D., 1978, Purification of the fusion protein of Sendai virus: Analysis of the $NH_2$-terminal sequence generated during precursor activation, *Proc. Natl. Acad. Sci. USA* **75**:2737–2740.

Gilmore, R., and Blobel, G., 1983, Transient involvement of signal recognition particle and its receptor in the microsomal membrane prior to protein translocation, *Cell* **35**:677–685.

Gilmore, R., Walter, P., and Blobel, G., 1982, Protein translocation across the endoplasmic reticulum. 2. isolation and characterization of the signal recognition particle, *J. Cell. Biol.* **95**:470–477.

Goldstein, E. A., and Pons, M. W., 1970, The effect of polyvinyl-sulfate on the ribonucleo-protein of influenza virus, *Virology* **41**:382–384.

Gottschalk, A., 1957, The specific enzyme of influenza virus and *Vibrio cholerae, Biochim. Biophys. Acta* **23**:645–646.

Greenspan, D., Krystal, M., Nakada, S., Arnheiter, H., Lyles, D. S., and Palese, P., 1985, Expression of influenza virus $NS_2$ nonstructural protein in bacteria and localization of $NS_2$ in infected eucaryotic cells, *J. Virol.* **54**:833–843.

Greenspan, D., Palese, P., and Krystal, M., 1988, Two nuclear location signals in the influenza virus $NS_1$ nonstructural protein, *J. Virol.* **62**:3020–3026.

Gregoriades, A., 1973, The membrane protein of influenza virus: Extraction from virus and infected cells with acidic chloroform-methanol, *Virology* **54**:369–383.

Gregoriades, A., 1980, Interaction of influenza M protein with viral lipids and phosphatidylcholine vesicles, *J. Virol.* **36**:470–479.

Gregoriades, A., and Frangione, B., 1981, Insertion of influenza M protein into the viral lipid bilayer and localization of site of insertion, *J. Virol.* **40**:323–328.

Griffith, I. P., 1975, The fine structure of influenza virus, in: *Negative Strand Viruses*, Vol. 1, (B. W. J. Mahy and R. D. Barry, eds.), pp. 121–132, Academic, London.

Guo, Y. J., Jin, F. G., Wang, P., Wang, M., and Zhu, J. M., 1983, Isolation of influenza C virus from pigs and experimental infection of pigs with influenza C virus, *J. Gen. Virol.* **64**: 177–182.

Haarr, L., Marsden, H. S., Preston, C. M., Smiley, J. R., Summers, W. C., and Summers, W. P., 1985, Utilization of internal AUG codons for initiation of protein synthesis directed by mRNAs from normal and mutant genes encoding herpes simplex virus-specified thymidine kinase, *J. Virol.* **56**:512–519.

Hall, M. N., Hereford, L., and Herskowitz, I., 1984, Targeting of *E. coli* β-galactosidase to the nucleus in yeast, *Cell* **36**:1057–1065.

Haslam, E. A., Hampson, A. W., Radiskevics, I., and White, D. O., 1970, The polypeptides of influenza virus. II. Interpretations of polyacrylamide gel electrophoresis patterns, *Virology* **42**:555–565.

Hauptmann, R., Clark, L. D., Mountford, R. C., Bachmayer, H., and Almond, J. W., 1983, Nucleotide sequence of the haemagglutinin gene of influenza virus. A/England/321/77, *J. Gen. Virol.* **64**:215–220.

Hay, A., 1974, Studies on the formation of the influenza virus envelope, *Virology* **60**:398–418.

Hay, A. J., and Zambon, M. C., 1984, Multiple actions of amantadine against influenza viruses, in: *Antiviral Drugs and Interferon: The Molecular Basis of Their Activity* (Y. Becker, ed.), pp. 301–315, Martinus Nijhoff, Boston.

Hay, A. J., Lomniczi, B., Bellamy, A. R., and Skehel, J. J., 1977, Transcription of the influenza virus genome, *Virology* **83**:337–355.

Hay, A. J., Kennedy, N. C. T., Skehel, J. J., and Appleyard, G., 1979, The matrix protein gene determines amantadine-sensitivity of influenza viruses, *J. Gen. Virol.* **42**:189–191.

Hay, A. J., Wolstenholme, A. J., Skehel, J. J., and Smith, M. H., 1985, The molecular basis of the specific anti-influenza action of amantadine, *EMBO J.* **4**:3021–3024.

Heggeness, M. H., Smith, P. R., Ulmanen, I., Krug, R. M., and Choppin, P. W., 1982, Studies on the helical nucleocapsid of influenza virus, *Virology* **118**:466–470.

Herrler, G., Compans, R. W., and Meier-Ewert, H., 1979, A precursor glycoprotein in influenza C virus, *Virology* **99**:49–56.

Herrler, G., Geyer, R., Mueller, H.-P., Stirm, S., and Klenk, H.-D., 1985a, Rat alpha-1 macroglobulin inhibits hemagglutination by influenza C virus, *Virus. Res.* **2**:183–192.

Herrler, G., Rott, R., Klenk, H.-D., Mueller, H.-P., Skukla, A. K., and Schauer, R., 1985b, The receptor-destroying enzyme of influenza C virus is neuraminate-O-acetyl esterase, *EMBO J.* **4**:1503–1506.

Herz, C., Stavnezer, E., Krug, R. M., and Gurney, T., Jr., 1981, Influenza virus, an RNA virus, synthesizes its messenger RNA in the nucleus of infected cells, *Cell* **26**:391–400.

Hiebert, S. W., Paterson, R. G., and Lamb, R. A., 1985, Hemagglutinin–neuraminidase protein of the paramyxovirus simian virus 5: Nucleotide sequence of the mRNA predicts an N-terminal membrane anchor, *J. Virol.* **54**:1–6.

Hiebert, S. W., Williams, M. A., and Lamb, R. A., 1986, Nucleotide sequence of RNA segment 7 of influenza B/Singapore/222/79: Maintenance of a second large open reading frame, *Virology* **155**:747–751.

Hirst, G. K., 1941, Agglutination of red cells by allantoic fluid of chick embryos infected with influenza virus, *Science* **94**:22–23.

Hirst, G. K., 1942, The quantitative determination of influenza virus and antibodies by means of red cell agglutination, *J. Exp. Med.* **75**:47–64.

Hirst, G. K., 1950, The relationship of the receptors of a new strain of virus to those of the mumps–NDV–influenza group, *J. Exp. Med.* **91**:177–184.

Hirst, G. K., and Pons, M. W., 1973, Mechanism of influenza virus recombination. II. Virus aggregation and its effect on plaque formation by so-called noninfectious virus, *Virology* **56**:620–631.

Hiti, A. L., Davis, A. R., and Nayak, D. P., 1981, Complete sequence analysis shows that the hemagglutinins of the H0 and H2 subtypes of human influenza virus are closely related, *Virology* **111**:113–124.

Hiti, A. L., and Nayak, D. P., 1982, Complete nucleotide sequence of the neuraminidase gene of human influenza virus A/WSN/33, *J. Virol.* **41**:730–734.

Horisberger, M. A., 1980, The large P proteins of influenza A viruses are composed of one acidic and two basic polypeptides, *Virology* **107**:302–305.

Hoyle, L., Horne, R. W., and Waterson, A. P., 1961, The structure and composition of the myxoviruses. II. Components released from the influenza virus particle by ether, *Virology* **13**:448–459.

Hsu, M.-T., Parvin, J. D., Gupta, S., Krystal, M., and Palese, P., 1987, Genomic RNAs of influenza viruses are held in a circular conformation in virions and in infected cells by a terminal panhandle, *Proc. Natl. Acad. Sci. USA* **84**:8140–8144.

Huddleston, J. A., and Brownlee, G. G., 1982, The sequence of the nucleoprotein gene of human influenza A virus, strain A/NT/60/68, *Nucl. Acids Res.* **10**:1029–1038.

Hull, D. J., Gilmore, R., and Lamb, R. A., 1988, Integration of a small integral membrane protein, $M_2$, of influenza virus into the endoplasmic reticulum: analysis of the internal signal-anchor domain of a protein with an ectoplasmic $NH_2$ terminus, *J. Cell Biol.* **106**: 1489–1498.

Hunziker, W., Spiess, M., Semenza, G., and Lodish, H. F., 1986, The sucrase–isomaltase complex: Primary structure, membrane orientation and evolution of a stalked, intrinsic brush border protein, *Cell* **46:**227–234.

Inglis, S. C., and Brown, C. M., 1981, Spliced and unspliced RNAs encoded by virion RNA segment 7 of influenza virus, *Nucl. Acids. Res.* **9:**2727–2740.

Inglis, S. C., and Brown, C. M., 1984, Differences in the control of virus mRNA splicing during permissive or abortive infection with influenza A (fowl plague) virus, *J. Gen. Virol.* **65:**153–164.

Inglis, S. C., Carroll, A. R., Lamb, R. A., and Mahy, B. W. J., 1976, Polypeptides specified by the influenza virus genome. 1. Evidence for eight distinct gene products specified by fowl plague virus, *Virology* **74:**489–503.

Inglis, S. C., Barrett, T., Brown, C. M., and Almond, J. W., 1979, The smallest genome RNA segment of influenza virus contain two genes that may overlap, *Proc. Natl. Acad. Sci. USA* **76:**3790–3794.

Inglis, S. C., Gething, M. J., and Brown, C. M., 1980, Relationship between the messenger RNAs transcribed from two overlapping genes of influenza virus, *Nucl. Acids. Res.* **8:**3575–3589.

Jackson, D. A., Caton, A. J., McCready, S. J., and Cook, P. R., 1982, Influenza virus RNA is synthesized at fixed sites in the nucleus, *Nature (Lond.)* **296:**366–368.

Jain, M. K., Wu, Y.-H., Morgan, T. K., Biggs, M. S., and Murray, R. K., 1976, Phase transition in a lipid bilayer. II. Influence of adamantane derivatives, *Chem. Phys. Lipids* **17:**71–78.

Johansen, H., Schumperli, D., and Rosenberg, M., 1984, Affecting gene expression by altering the length and sequence of the 5' leader, *Proc. Natl. Acad. Sci. USA* **81:**7698–7702.

Jones, I. M., Reay, P. A., and Philpott, K. L., 1986, Nuclear location of all three influenza polymerase proteins and a nuclear signal in polymerase PB2, *EMBO J.* **5:**2371–2376.

Kalderon, D., Richardson, W. D., Markham, A. F., and Smith, A. E., 1984a, Sequence requirements for nuclear localization of SV40 large-T antigen, *Nature (Lond.)* **311:**33–38.

Kalderon, D., Roberts, B. L., Richardson, W. D., and Smith, A. E., 1984b, A short amino acid sequence able to specify nuclear location, *Cell* **39:**499–509.

Kamata, T., and Watanabe, Y., 1977, Role for nucleocapsid protein phosphorylation in the transcription of influenza virus genome, *Nature (Lond.)* **267:**460–462.

Kaptein, J. S., and Nayak, D. P., 1982, Complete nucleotide sequence of the polymerase 3 gene of human influenza virus A/WSN/33, *J. Virol.* **42:**55–63.

Kawaoka, Y., Naeve, C. W., and Webster, R. G., 1984, Is virulence of H5N2 influenza viruses in chickens associated with loss of carbohydrate from the hemagglutinin? *Virology* **139:**303–316.

Kawaoka, Y., and Webster, R. G., 1985, Evolution of the A/Chicken/Pa/83 (H5N2) influenza virus, *Virology* **146:**130–137.

Kawaoka, Y., and Webster, R. G., 1988, Sequence requirements for cleavage activation of influenza virus hemagglutinin expressed in mammaliar cells, *Proc. Natl. Acad. Sci. USA* **85:**324–328.

Kemdirim, S., Palefsky, J., and Briedis, D. J., 1986, Influenza B virus PB1 protein: Nucleotide sequence of the genome RNA segment predicts a high degree of structural homology with the corresponding influenza A virus polymerase protein, *Virology* **152:**126–135.

Kendal, A. P., 1975, A comparison of influenza C virus with prototype myxoviruses: Receptor destroying activity (neuraminidase) and structural polypeptides, *Virology* **65:**87–89.

Khalili, K., Brady, J., and Khoury, G., 1987, Translational regulation of SV40 early mRNA defines a new viral protein, *Cell* **48:**639–645.

Kilbourne, E. D., 1987, *Influenza,* Plenum, New York.

Kingsbury, D. W., Jones, I. M., and Murti, K. G., 1987, Assembly of influenza ribonucleoprotein in vitro using recombinant nucleoprotein, *Virology* **156:**396–403.

Kingsbury, D. W., and Webster, R. G., 1969, Some properties of influenza virus nucleocapsids, *J. Virol.* **4:**219–225.

Kitame, F., Nakamura, K., Saito, A., Sindhara, H., and Homma, M., 1985, Isolation and characterization of influenza C virus inhibitor in rat serum, *Virus Res.* **3:**231–244.

Klenk, H.-D., Compans, R. W., and Choppin, P. W., 1970, An electron microscopic study of the presence or absence of neuraminic acid in enveloped viruses, *Virology* **42:**1158–1162.

Klenk, H.-D., Rott, R., and Becht, H., 1972, On the structure of the influenza virus envelope, *Virology* **47:**579–591.

Klenk, H.-D., Wollert, W., Rott, R., and Scholtissek, C., 1974, Association of influenza virus proteins with cytoplasmic fractions, *Virology* **57:**28–41.

Klenk, H.-D., Rott, R., Orlich, M., and Blodorn, J., 1975, Activation of influenza A viruses by trypsin treatment, *Virology* **68:**426–439.

Klug, A., 1980, The assembly of tobacco mosaic virus structure and specificity, *Harvey Lect.* **74:**141–172.

Knossow, M., Daniels, R. S., Douglas, A. R., Skehel, J. J., and Wiley, D. C., 1984, Three-dimensional structure of an antigenic mutant of the influenza virus hemagglutinin, *Nature (Lond.)* **311:**678–680.

Koennecke, I., Boschek, C. B., and Scholtissek, C., 1981, Isolation and properties of a temperature-sensitive mutant (ts412) of the influenza A virus recombinant with a ts lesion in the gene coding for the nonstructural protein, *Virology* **110:**16–25.

Kozak, M., 1978, How do eucaryotic ribosomes select initiation regions in messenger RNA? *Cell* **15:**1109–1123.

Kozak, M., 1981, Mechanism of mRNA recognition by eukaryotic ribosomes during initiation of protein synthesis, *Curr. Top. Microb. Immunol.* **93:**81–123.

Kozak, M., 1983a, Comparison of initiation of protein synthesis in procaryotes, eucaryotes, and organelles. *Microbiol. Rev.* **47:**1–45.

Kozak, M., 1983b, Translation of insulin-related polypeptides from mRNAs with tandemly reiterated copies of the ribosome binding site, *Cell* **34:**971–978.

Kozak, M., 1984a, Compilation and analysis of sequences upstream from the translational start site in eukaryotic mRNAs, *Nucl. Acids. Res.* **12:**857–872.

Kozak, M., 1984b, Point mutations close to the AUG initiator codon affect the efficiency of translation of rat preproinsulin in vivo, *Nature (Lond.)* **308:**241–246.

Kozak, M., 1986a, Point mutations define a sequence flanking the AUG initiator codon that modulates translation by eukaryotic ribosomes, *Cell* **44:**283–292.

Kozak, M., 1986b, Bifunctional messenger RNAs in eukaryotes, *Cell* **47:**481–483.

Krug, R. M., 1972, Cytoplasmic and nucleoplasmic viral RNP's in influenza virus-infected MDCK cells, *Virology* **50:**103–113.

Krug, R. M., 1983, Transcription and replication of influenza viruses, in: *Genetics of Influenza Viruses* (P. Palese and D. W. Kingsbury, eds.), pp. 70–98, Springer-Verlag, Vienna.

Krug, R. M., and Etkind, P. R., 1973, Cytoplasmic and nuclear specific proteins in influenza virus-infected MDCK cells, *Virology* **56:**334–348.

Krug, R. M., and Soeiro, R., 1975, Studies on the intranuclear localization of influenza virus-specific proteins, *Virology* **64:**378–387.

Krystal, M., Elliott, R. M., Benz, E. W., Jr., Young, J. F., and Palese, P., 1982, Evolution of influenza A and B virus: Conservation of structural features in the hemagglutinin genes, *Proc. Natl. Acad. Sci. USA* **79:**4800–4804.

Krystal, M., Young, J. F., Palese, P., Wilson, I. A., Skehel, J. J., and Wiley, D. C., 1983a, Sequential mutations in hemagglutinins of influenza B virus isolates: Definition of antigenic domains, *Proc. Natl. Acad. Sci. USA* **80:**4527–4531 (including correction: *Proc. Natl. Acad. Sci. USA* **81:**1261; 1984).

Krystal, M., Buonagurio, D., Young, J. F., and Palese, P., 1983b, Sequential mutations in the NS genes of influenza virus field strains, *J. Virol.* **45:**547–554.

Krystal, M., Li, R., Lyles, D., Pavlakis, G., and Palese, P., 1986, Expression of the three influenza virus polymerase proteins in a single cell allows growth complementation of virus mutants, *Proc. Natl. Acad. Sci. USA* **83:**2709–2713.

Kyte, J., and Doolittle, R. F., 1982, A simple method for displaying the hydropathic character of a protein, *J. Mol. Biol.* **157:**105–132.

Lamb, R. A., 1974, *Aspects of the Structure and Replication of Sendai virus*, Doctoral dissertation, The University of Cambridge, England.

Lamb, R. A., 1983, The influenza virus RNA segments and their encoded proteins, in: *Genetics of Influenza Viruses* (P. Palese and D. W. Kingsbury, eds.), pp. 26–69, Springer-Verlag, Vienna.

Lamb, R. A., and Choppin, P. W., 1976, Synthesis of influenza virus proteins in infected cells: Translation of viral polypeptides, including three P polypeptides, from RNA produced by primary transcription, *Virology* 74:504–519.

Lamb, R. A., and Choppin, P. W., 1979, Segment 8 of the influenza virus genome is unique in coding for two polypeptides, *Proc. Natl. Acad. Sci. USA* 76:4908–4912.

Lamb, R. A., and Choppin, P. W., 1981, Identification of a second protein (M$_2$) encoded by RNA segment 7 of influenza virus, *Virology* 112:729–737.

Lamb, R. A., and Choppin, P. W., 1983, The gene structure and replication of influenza virus, *Annu. Rev. Biochem.* 52:467–506.

Lamb, R. A., and Lai, C.-J., 1980, Sequence of interrupted and uninterrupted mRNAs and cloned DNA coding for the two overlapping nonstructural proteins of influenza virus, *Cell* 21:475–485.

Lamb, R. A., and Lai, C.-J., 1981, Conservation of the influenza virus membrane protein (M$_1$) amino acid sequence and an open reading frame of RNA segment 7 encoding a second protein (M$_2$) in H1N1 and H3N2 strains, *Virology* 112:746–751.

Lamb, R. A., and Lai, C.-J., 1982, Spliced and unspliced messenger RNAs synthesized from cloned influenza virus M DNA in an SV40 vector: Expression of the influenza virus membrane protein (M$_1$), *Virology* 123:237–256.

Lamb, R. A., and Lai, C.-J., 1984, Expression of unspliced NS$_1$ mRNA, spliced NS$_2$ mRNA, and a spliced chimera mRNA from cloned influenza virus NS DNA in an SV40 vector, *Virology* 135:139–147.

Lamb, R. A., Etkind, P. R., and Choppin, P. W., 1978, Evidence for a ninth influenza viral polypeptide, *Virology* 91:60–78.

Lamb, R. A., Choppin, P. W., Chanock, R. M., and Lai, C.-J., 1980, Mapping of the two overlapping genes for polypeptides NS$_1$ and NS$_2$ on RNA segment 8 of influenza virus genome, *Proc. Natl. Acad. Sci. USA* 77:1857–1861.

Lamb, R. A., Lai, C.-J., and Choppin, P. W., 1981, Sequences of mRNAs derived from genome RNA segment 7 of influenza virus: Colinear and interrupted mRNAs code for overlapping proteins, *Proc. Natl. Acad. Sci. USA* 78:4170–4174.

Lamb, R. A., Shaw, M. W., Briedis, D. J., and Choppin, P. W., 1983, The nucleotide sequence of the neuraminidase of influenza B virus reveals two overlapping reading frames, in: *The Origin of Pandemic Influenza Viruses* (W. G. Laver, ed.), pp. 77–86, Elsevier, New York.

Lamb, R. A., Zebedee, S. L., and Richardson, C. D., 1985, Influenza virus M$_2$ protein is an integral membrane protein expressed on the infected-cell surface, *Cell* 40:627–633.

Laver, W. G., and Baker, N., 1972, Amino acid composition of polypeptides from influenza virus particles, *J. Gen. Virol.* 17:61–67.

Laver, W. G., and Valentine, R. C., 1969, Morphology of the isolated hemagglutinin and neuraminidase subunits of influenza virus, *Virology* 38:105–119.

Laver, W. G., Gerhard, W., Webster, R. G., Frankel, M. E., and Air, G., 1979, Antigenic drift in type A influenza virus: Peptide mapping and antigenic analysis of A/PR/8/34 (HON1) variants selected with monoclonal antibodies, *Proc. Natl. Acad. Sci. USA* 76: 1425–1429.

Laver, W. G., Air, G. M., Dopheide, T. A., and Ward, C. W., 1980, Amino acid sequence changes in the haemagglutin of A/Hong Kong (H3N2) influenza virus during the period 1968–77, *Nature (Lond.)* 283:454–457.

Laver, W. G., Air, G. M., and Webster, R. G., 1981, The mechanisms of antigenic drift in influenza virus. Amino acid sequence changes in an antigenically active region of Hong Kong (H3N2) influenza virus hemagglutinin, *J. Mol. Biol.* 145:339–361.

Laver, W. G., Colman, P. M., Webster, R. G., Hinshaw, V. S., and Air, G. M., 1984, Influenza virus neuraminidase with hemagglutinin activity, *Virology* 137:314–323.

Lazarowitz, S. G., and Choppin, P. W., 1975, Enhancement of infectivity of influenza A and

B viruses by proteolytic cleavage of the hemagglutinin polypeptide, *Virology* **68:**440–454.

Lazarowitz, S. G., Compans, R. W., and Choppin, P. W., 1971, Influenza virus structural and nonstructural proteins in infected cells and their plasma membranes, *Virology* **46:**830–843.

Lazarowitz, S. G., Compans, R. W., and Choppin, P. W., 1973a, Proteolytic cleavage of the hemagglutinin polypeptide of influenza virus: Function of the uncleaved polypeptide HA, *Virology* **52:**199–212.

Lazarowitz, S. G., Goldberg, A. R., and Choppin, P. W., 1973b, Proteolytic cleavage by plasmin of the HA polypeptide of influenza virus. Host cell activation of serum plasminogen, *Virology* **56:**172–180.

Lazdins, I., Haslam, E. A., and White, D. O., 1972, The polypeptides of influenza virus. VI. Composition of the neuraminidase, *Virology* **49:**758–765.

Lenard, J. C., Wong, C. Y., and Compans, R. W., 1974, Association of the internal membrane protein with the lipid bilayer in influenza virus. A study with the fluorescent probe 12-(9-anthroyl)-stearic acid, *Biochim. Biophys. Acta* **332:**341–349.

Lentz, M. R., and Air, G. M., 1986, Loss of enzyme activity in a site-directed mutant of influenza neuraminidase compared to expressed wild-type protein, *Virology* **148:**74–83.

Lentz, M. R., Air, G. M., Laver, W. G., and Webster, R. G., 1984, Sequence of the neuraminidase gene of influenza virus A/Tokyo/3/67 and previously uncharacterized monoclonal variants, *Virology* **135:**257–265.

Lin, B. C., and Lai, C. J., 1983, The influenza virus nucleoprotein synthesized from cloned DNA in a simian virus 40 vector is detected in the nucleus, *J. Virol.* **45:**434–438.

Liu, C.-C., Simonsen, C. C., and Levinson, A. D., 1984, Initiation of translation at internal AUG codons in mammalian cells, *Nature (Lond.)* **309:**82–85.

Lohmeyer, J., Talenz, L. T., and Klenk, H.-D., 1979, Biosynthesis of the influenza virus envelope in abortive infection, *J. Gen. Virol.* **42:**73–88.

Londo, D. R., Davis, A. R., and Nayak, D. P., 1983, Complete nucleotide sequence of the nucleoprotein gene of influenza B virus, *J. Virol.* **47:**642–648.

Lubeck, M. D., Schulman, J. L., and Palese, P., 1978, Susceptibility of influenza A viruses to amantadine is influenced by the gene coding for M protein, *J. Virol.* **28:**710–716.

Maassab, H. F., and DeBorde, D. C., 1983, Characterization of an influenza A host range mutant, *Virology* **130:**342–350.

Mahy, B. W. J., 1983, Mutants of influenza virus, in: *Genetics of Influenza Viruses* (P. Palese and D. W. Kingsbury, eds.), pp. 192–254, Springer-Verlag, Vienna.

Mahy, B. W. J., Barrett, T., Briedis, D. J., Brownson, J. M., and Wolstenholme, A. J., 1980, Influence of the host cell on influenza virus replication, *Phil. Trans. R. Soc. Lond.* **B288:**349–357.

Markoff, L., and Lai, C. J., 1982, Sequence of the influenza A/Udorn/72 (H3N2) virus neuraminidase gene as determined from cloned full-length DNA, *Virology* **119:**288–297.

Markoff, L., Lin, B.-C., Sveda, M. M., and Lai, C.-J., 1984, Glycosylation and surface expression of the influenza virus neuraminidase requires the N-terminal hydrophobic region, *Mol. Cell Biol.* **4:**8–16.

Markushin, S., Ghiasi, H., Sokolov, N., Shilov, A., Sinitsin, B., Brown, D., Klimov, A., and Nayak, D., 1988, Nucleotide sequence of RNA segment 7 and the predicted amino acid sequence of $M_1$ and $M_2$ proteins of FPV/Weybridge (H7N7) and WSN (H1N1) influenza viruses, *Virus Res.* **10:**263–272.

Maroux, S., and Louvard, D., 1976, On the hydrophobic part of aminopeptidase and maltases which bind the enzyme to the intestinal brush membrane, *Biochim. Biophys. Acta* **419:**189–195.

Martinez, C., Del Rio, L., Portela, A., Domingo, E., and Ortin, J., 1983, Evolution of the influenza virus neuraminidase gene during drift of the N2 subtype, *Virology* **130:**539–545.

Matthews, R. E. F., 1966, Reconstitution of turnip yellow mosaic virus RNA with TMV protein subunits, *Virology* **30**:82–96.

Mayron, L. W., Robert, B., Winzler, R. J., and Rafelson, M. E., 1961, Studies on the neuraminidase of influenza virus. 1. Separation and some properties of the enzyme from asian and PR8 strains, *Arch. Biochem. Biophys.* **92**:475–483.

McCauley, J., Skehel, J. J., Elder, K., Gething, M.-J., Smith, A., and Waterfield, M., 1980, Haemagglutinin biosynthesis, in: *Structure and Variation in Influenza Viruses* (G. Laver and G. Air, eds.), pp. 97–104, Elsevier/North-Holland, New York.

McCauley, J. W., Mahy, B. W. J., and Inglis, S. C., 1982, Nucleotide sequence of fowl plague virus RNA segment 7, *J. Gen. Virol.* **58**:211–215.

McClelland, L., and Hare, R., 1941, The adsorption of influenza virus by red cells and a new in vitro method of measuring antibodies for influenza virus, *Can. J. Public Health* **32**:530–538.

McGeoch, D. J., Fellner, P., and Newton, C., 1976, The influenza virus genome consists of eight distinct RNA species, *Proc. Natl. Acad. Sci. USA* **73**:3045–3049.

McMichael, A. J., Michie, C. A., Gotch, G. M., Smith, G. L., and Moss, G., 1986, Recognition of influenza A virus nucleoprotein by human cytotoxic T lymphocytes, *J. Gen. Virol.* **67**:719–726.

McQueen, N., Nayak, D. P., Stephens, E. B., and Compans, R. W., 1986, Polarized expression of a chimeric protein in which the transmembrane and cytoplasmic domains of the influenza virus hemagglutinin have been replaced by those of the vesicular stomatitis virus G protein, *Proc. Natl. Acad. Sci. USA* **83**:9318–9322.

Meier-Ewert, H., Compans, R. W., Bishop, D. H. L., and Herrler, G., 1978, Molecular analysis of influenza C virus, in: *Negative Strand Viruses and the Host Cell* (B. W. J. Mahy and R. D. Barry, eds.), pp. 127–133, Academic, Orlando, Florida.

Meyer, D. I., Krause, E., and Dobberstein, B., 1982, Secretory protein translocation across membranes—The role of the "docking protein," *Nature (Lond.)* **297**:647–650.

Min Jou, W., Verhoeyen, M., Devos, R., Saman, E., Fang, R., Huylebroeck, D., Fiers, W., Threlfall, G., Barber, C., Carey, N., and Emtage, S., 1980, Complete structure of the hemagglutinin gene from the human influenza A/Victoria 3/75 (H3H2) strain as determined from cloned DNA, *Cell* **19**:683–696.

Minor, P. D., and Dimmock, N. J., 1975, Inhibition of synthesis of influenza virus proteins: Evidence for two host-cell dependent events during multiplication, *Virology* **67**:114–123.

Morrongiello, M. P., and Dales, S., 1977, Characterization of cytoplasmic inclusions formed during influenza/WSN virus infection of chick embryo fibroblasts, *Intervirology* **8**:281–293.

Mount, S. M., 1982, A catalogue of splice junction sequences, *Nucl. Acids Res.* **10**:459–472.

Murphy, B. R., Markoff, L. J., Hosier, N. T., Massicot, J. G., and Chanock, R. M., 1982, Production and level of genetic stability of an influenza A virus temperature-sensitive mutant containing two genes with ts mutations, *Infect. Immun.* **37**:235–242.

Murti, K. G., Bean, W. J., Jr., Webster, R. G., 1980, Helical ribonucleoproteins of influenza virus: An electron microscope analysis, *Virology* **104**:224–229.

Murti, K. G., Webster, R. G., and Jones, I. M., 1988, Localization of RNA polymerases of influenza viral ribonucleoproteins by immunogold labeling, *Virology* **164**:562–566.

Naeve, C. W., and Webster, R. G., 1983, Sequence of the hemagglutinin gene from influenza virus A/Seal/Mass/1/80, *Virology* **129**:298–308.

Nakada, S., Creager, R. S., Krystal, M., and Palese, P., 1984a, Complete nucleotide sequence of the influenza C/Calif/78 virus nucleoprotein gene, *Virus Res.* **1**:433–441.

Nakada, S., Creager, R. S., Krystal, M., Aaronson, R. P., and Palese, P., 1984b, Influenza C virus hemagglutinin: Comparison with influenza A and B virus hemagglutinins, *J. Virol.* **50**:118–124.

Nakada, S., Graves, P. N., Desselberger, U., Creager, R. S., Krystal, M., and Palese, P., 1985, Influenza C virus RNA 7 codes for a nonstructural protein, *J. Virol.* **56**:221–226.

Nakada, S., Graves, P. N., and Palese, P., 1986, The influenza C virus NS gene: Evidence for a spliced mRNA and a second NS gene product (NS$_2$ protein), *Virus Res.* **4**:263–273.

Nakajima, K., Nobusawa, E., and Nakajima, S., 1984, Genetic relatedness between A/Swine/Iowa/15/30 (H1N1) and human influenza viruses, *Virology* **139**:194–198.

Nakajima, S., and Sugiura, A., 1980, Neurovirulence of influenza virus in mice. II. Mechanism of virulence as studied in a neuroblastoma cell line, *Virology* **101**:450–457.

Nestorowicz, A., Kawaoka, Y., Bean, W. J., and Webster, R. G., 1987, Molecular analysis of the hemagglutinin genes of Australian H7N7 influenza viruses: Role of passerine birds in maintenance or transmission?, *Virology* **160**:411–418.

Newton, S. E., Air, G. M., Webster, R. G., and Laver, W. G., 1983, Sequence of the hemagglutinin gene of influenza virus A/Memphis/1/71 and previously uncharacterized monoclonal antibody-derived variants, *Virology* **128**:495–501.

Nichol. T., Penn, C. R., and Mahy, B. W. J., 1981, Evidence for the involvement of influenza A (fowl plague Rostock) virus protein P2 in ApG and mRNA primed *in vitro* RNA synthesis, *J. Gen. Virol.* **57**:407–413.

Noll, H., Aoyagi, T., and Orlando, J., 1962, The structural relationship of sialidase to the influenza virus surface, *Virology* **18**:154–157.

Norton, G. P., Tanaka, T., Tobita, K., Nakada, S., Buonaburio, D. A., Greenspan, D., Krystal, M., and Palese, P., 1987, Infectious influenza A and B virus variants with long carboxyl terminal deletions in the NS$_1$ polypeptides, *Virology* **156**:204–213.

O'Callaghan, R. J., Loughlin, S. M., Labat, D. D., and Howe, C., 1977, Properties of influenza C virus grown in cell culture, *J. Virol.* **24**:875–882.

Ohuchi, M., Ohuchi, R., and Mifune, K., 1982, Demonstration of hemolytic and fusion activities of influenza C virus, *J. Virol.* **42**:1076–1079.

Ortin, J., Martinez, C., Del Rio, L., Davila, M., Lopez-Galindez, C., Villanueva, N., and Domingo, E., 1983, Evolution of the nucleotide sequence of influenza virus RNA segment during drift of the H3N2 subtype, *Gene* **23**:233–239.

Oxford, J. S., 1973, Polypeptide composition of influenza B viruses and enzymes associated with the purified virus particle, *J. Virol.* **12**:827–835.

Oxford, J. S., and Galbraith, A., 1980, Antiviral activity of amantadine: A review of laboratory and clinical data, *Pharmacol. Ther.* **11**:181–262.

Palese, P., 1977, The genes of influenza virus, *Cell* **10**:1–10.

Palese, P., and Kingsbury, D. W., 1983, *Genetics of Influenza Viruses*, Springer-Verlag, Vienna.

Palese, P., and Schulman, J. L., 1974, Isolation and characterization of influenza virus recombinants with high and low neuraminidase activity: Use of 2-3′ methoxyphenyl-N-acetylneuraminic acid to identify cloned populations, *Virology* **57**:227–237.

Palese, P., and Schulman, J. L., 1976, Differences in RNA patterns of influenza A viruses, *J. Virol.* **17**:876–884.

Palese, P., Tohita, K., Ueda, M., and Compans, R. W., 1974, Characterization of temperature sensitive influenza virus mutants defective in neuraminidase, *Virology* **61**:397–410.

Palese, P., Ritchey, M. B., and Schulman, J. L., 1977, Mapping of the influenza virus genome. II. Identification of the P1, P2 and P3 genes, *Virology* **76**:114–121.

Petri, T., and Dimmock, N. J. 1981, Phosphorylation of influenza virus nucleoprotein in vivo, *J. Gen. Virol.* **57**:185–190.

Petri, T., Herrler, G., Compans, R. W., and Meier-Ewert, H., 1980, Gene products of influenza C virus, *FEMS Microbiol. Lett.* **9**:43–47.

Pfeifer, J. B., and Compans, R. W., 1984, Structure of the influenza C glycoprotein gene as determined from cloned DNA, *Virus Res.* **1**:281–296.

Plotch, S. J., and Krug, R. M., 1986, *In vitro* splicing of influenza viral NS$_1$ mRNA and NS$_2$-β-globin chimeras: Possible mechanisms for the control of viral mRNA splicing, *Proc. Natl. Acad. Sci. USA* **83**:5444–5448.

Plotch, S. J., Bouloy, M., Ulmanen, I., and Krug, R. M., 1981, Initiation of influenza viral RNA transcription by capped RNA primers: A unique cap (m$^7$GpppXm)-dependent virion endonuclease generates 5' terminal RNA fragments that prime transcription, *Cell* **23**:847–858.

Pons, M. W., 1971, Isolation of influenza virus ribonucleoprotein from infected cells. Demonstration of the presence of negative stranded RNA in viral RNP, *Virology* **46**:149–160.

Pons, M. W., 1976, A re-examination of influenza single and double-stranded RNAs by gel electrophoresis, *Virology* **69**:789–792.

Pons, M. W., and Rochovansky, O. M., 1979, Ultraviolet inactivation of influenza virus RNA in vitro and in vivo, *Virology* **97**:183–189.

Pons, M. W., Schulze, I. T., and Hirst, G. K., 1969, Isolation and characterization of the ribonucleoprotein of influenza virus, *Virology* **39**:250–259.

Portela, A., Melero, J. A., Luna de la, S., and Ortin, J., 1986, Construction of cell lines that regulate by temperature the amplification and expression of influenza virus non-structural protein genes, *EMBO J.* **5**:2387–2392.

Portela, A., Melero, J. A., Martinez, C., Domingo, E., and Ortin, J., 1985, A primer vector system that allows temperature dependent gene amplification and expression in mammalian cells: Regulation of the influenza virus $NS_1$ gene expression, *Nucl. Acids Res.* **13**:7959–7977.

Porter, A. G., Barber, C., Carey, N. H., Hallewell, R. A., Threlfall, G., and Emtage, J. S., 1979, Complete nucleotide sequence of an influenza virus hemagglutinin gene from cloned DNA, *Nature (Lond.)* **282**:471–477.

Porter, A. G., Smith, J. C., and Emtage, J. S., 1980, Nucleotide sequence of influenza virus RNA segment 8 indicates that coding regions for $NS_1$ and $NS_2$ proteins overlap, *Proc. Natl. Acad. Sci. USA* **77**:5074–5078.

Portner, A., 1981, The HN glycoprotein of Sendai virus: Analysis of site(s) involved in hemagglutinating and neuraminidase activities, *Virology* **115**:375–384.

Privalsky, M. L., and Penhoet, E. E., 1977, Phosphorylated protein component present in influenza virions, *J. Virol.* **24**:401–405.

Privalsky, M. L., and Penhoet, E. E., 1978, Influenza virus proteins: identity, synthesis, and modification analyzed by two-dimensional gel electrophoresis, *Proc. Natl. Acad. Sci. USA* **75**:3625–3629.

Privalsky, M. L., and Penhoet, E. E., 1981, The structure and synthesis of influenza virus phosphoproteins, *J. Biol. Chem.* **256**:5368–5376.

Racaniello, V. R., and Palese, P., 1979a, Influenza B virus genome: Assignment of viral polypeptides to RNA segments, *J. Virol.* **29**:361–373.

Racaniello, V. R., and Palese, P., 1979b, Isolation of influenza C virus recombinants, *J. Virol.* **32**:1006–1014.

Raymond, F. L., Caton, A J., Cox, N. J., Kendal, A. P., and Brownlee, G. G., 1986, The antigenicity and evolution of influenza H1 haemagglutinin, from 1950–1957 and 1977–1983: Two pathways from one gene, *Virology* **148**:275–287.

Rees, P. J., and Dimmock, N. J., 1981, Electrophoretic separation of influenza virus ribonucleoproteins, *J. Gen. Virol.* **53**:125–132.

Reithmeier, R. A. F., and MacLennan, D. H., 1981, the $NH_2$ terminus of the $(Ca^{2+} + Mg^{2+})$-Adenosine triphosphatase is located on the cytoplasmic surface of the sarcoplasmic reticulum membrane, *J. Biol. Chem.* **256**:5957–5960.

Richardson, C. D., Scheid, A., and Choppin, P. W., 1980, Specific inhibition of paramyxovirus and myxovirus replication by oligopeptides with amino acid sequences similar to those at the N-termini of the $F_1$ or $HA_2$ viral polypeptides, *Virology* **105**:205–222.

Rifkin, D. B., Compans, R. W., and Reich, E., 1972, A specific labeling procedure for proteins on the outer surface of membranes, *J. Biol. Chem.* **247**:6432–6437.

Ritchey, M. B., Palese, P., and Schulman, J. L., 1976, Mapping of the influenza virus genome. III. Identification of genes coding for nucleoprotein, membrane protein, and nonstructural protein, *J. Virol.* **20**:307–313.

Ritchey, M. B., Palese, P., and Schulman, J. L., 1977, Differences in protein patterns of influenza A viruses, *Virology* **76:**122–128.

Robertson, B. H., Bhown, A. S., Compans, R. W., and Bennett, J. C., 1979, Structure of the membrane protein of influenza virus. 1. Isolation and characterization of cyanogen bromide cleavage products, *J. Virol.* **30:**759–766.

Robertson, J. S., Schubert, M., and Lazzarini, R. A., 1981, Polyadenylation sites for influenza mRNA, *J. Virol.* **38:**157–163.

Robertson, J. S., Robertson, M.E.St.C., and Roditi, I. J., 1984, Nucleotide sequence of RNA segment 3 of the avian influenza A/FPV/Rostock/34 and its comparison with the corresponding segment of human strains A/PR/8/34 and A/NT/60/68, *Virus Res.* **1:**73–80.

Roditi, I. J., and Robertson, J. S., 1984, Nucleotide sequence of the avian influenza virus A/fowl plague/Rostock/34 segment 1 encoding the PB2 polypeptide, *Virus Res.* **1:**65–71.

Rodriguez-Boulan, E., and Sabatini, D. D., 1978, Asymmetric budding of viruses in epithelial monolayers: A model system for study of epithelial polarity, *Proc. Natl. Acad. Sci. USA* **75:**5071–5075.

Rogers, G. N., Paulson, J. C., Daniels, R. S., Skehel, J. J., Wilson, I. A., and Wiley, D. C., 1983, Single amino acid substitutions in influenza hemagglutinin change receptor-binding specificity, *Nature (Lond.)* **304:**76–78.

Rogers, G. N., Herrler, G., Paulson, J. C., and Klenk, H.-D., 1986, Influenza C virus uses 9-*O*-acetyl-*N*-acetylneuraminie acid as a high affinity receptor determinant for attachment to cells, *J. Biol. Chem.* **261:**5947–5951.

Rose, J. K., Adams, G. A., and Gallione, C. J., 1984, The presence of cysteine in the cytoplasmic domain of the vesicular stomatitis virus glycoprotein is required for palmitate addition, *Proc. Natl. Acad. Sci. USA* **81:**2050–2054.

Roth, M. G., Fitzpatrick, J. P., and Compans, R. W., 1979, Polarity of influenza and vesicular stomatitis virus maturation in MDCK cells. Lack of requirement for glycosylation of viral glycoproteins, *Proc. Natl. Acad. Sci. USA* **76:**6430–6434.

Roth, M. G., Compans, R. W., Giusti, L., Davis, A. R, Nayak, D. P., Gething, M.-J., and Sambrook, J., 1983, Influenza virus hemagglutinin expression is polarized in cells infected with recombinant SV40 viruses carrying cloned hemagglutinin DNA, *Cell* **33:**435–443.

Roth, M. G., Gundersen, D., Patil, N., and Rodriguez-Boulan, E., 1987, The large external domain is sufficient for the correct sorting of secreted or chimeric influenza virus hemagglutinins in polarized monkey kidney cells, *J. Cell. Biol.* **104:**769–782.

Rott, R., 1979, Molecular basis of infectivity and pathogenicity of myxovirus, *Arch. Virol.* **59:**285–298.

Rott, R., Reinacher, M., Orlich, M., and Klenk, H.-D., 1980, Cleavability of hemagglutinin determines spread of avian influenza viruses in the chorioallantoic membrane of chicken embryo, *Arch. Virol.* **65:**123–133.

Ryan, K. W., MacKow, E. R., Chanock, R. M., and Lai, C.-J., 1986, Functional expression of influenza A viral nucleoprotein in cells transformed with cloned DNA, *Virology* **154:**144–154.

Scheid, A., and Choppin, P. W., 1974, The hemagglutinin and neuraminidase protein of a paramyxovirus: Interaction with neuraminic acid in affinity chromatography, *Virology* **62:**125–133.

Scheid, A., Caliguiri, L. A., Compans, R. W., and Choppin, P. W., 1972, Isolation of paramyxovirus glycoproteins. Association of both hemagglutinating and neuraminidase activities with the larger SV5 glycoprotein, *Virology* **50:**640–652.

Schild, G. C., 1972, Evidence for a new type-specific structural antigen of the influenza virus particle, *J. Gen. Virol.* **15:**99–103.

Schild, G. C., Oxford, J. S., and Newman, R. W., 1979, Evidence for antigenic variation in influenza A nucleoprotein, *Virology* **93:**569–573.

Schmidt, M. F. G., 1982, Acylation of viral spike glycoproteins: A feature of enveloped RNA viruses, *Virology* **116:**327–338.

Schneider, C., Owen, M. J., Banville, D., and Williams, J. G., 1984, Primary structure of human transferrin receptor deduced from the mRNA sequence, *Nature (Lond.)* **311:** 675–678.

Scholtissek, C., and Faulkner, G. P., 1979, Amantadine-resistant and -sensitive influenza A strains and recombinants, *J. Gen. Virol.* **44:**807–815.

Scholtissek, C., Harms, E., Rohde, W., Orlich, M., and Rott, R., 1976, Correlation between RNA fragments of fowl plague virus and their corresponding gene functions, *Virology* **74:**332–344.

Schulman, J. L., 1982, Influenza viruses and amantadine hydrochloride, in: *Chemotherapy of Viral Infections* (P. E. Came and L. A. Caliguiri, eds.), pp. 137–146, Springer-Verlag, New York.

Schulman, J. L., and Palese, P. J., 1977, Virulence factors of influenza A viruses: WSN virus neuraminidase required for productive infection in MDBK cells, *J. Virol.* **24:**170–176.

Schulze, I. T., 1970, The structure of influenza virus. I. The polypeptides of the virion, *Virology* **42:**890–904.

Schulze, I. T., 1972, The structure of influenza virus. II. A model based on the morphology and composition of subviral particles, *Virology* **47:**181–196.

Shaw, M. W., and Choppin, P. W., 1984, Studies on the synthesis of the influenza B virus NB glycoprotein, *Virology* **139:**178–184.

Shaw, M. W., and Compans, R. W., 1978, Isolation and characterization of cytoplasmic inclusions from influenza A virus-infected cells, *J. Virol.* **25:**605–615.

Shaw, M. W., and Lamb, R. A., 1984, A specific sub-set of host-cell mRNAs prime influenza virus mRNA synthesis, *Virus Res.* **1:**455–467.

Shaw, M. W., Lamb, R. A., Erickson, B. W., Briedis, D. J., and Choppin, P. W., 1982, Complete nucleotide sequence of the neuraminidase gene of influenza B virus, *Proc. Natl. Acad. Sci. USA* **79:**6817–6821.

Shaw, M. W., Choppin, P. W., and Lamb, R. A., 1983, A previously unrecognized influenza B virus glycoprotein from a bicistronic mRNA that also encodes the viral neuraminidase, *Proc. Natl. Acad. Sci. USA* **80:**4879–4883.

Shaw, M. W., Stoeckle, M. Y., Krah, D. L., and Choppin, P. W., 1985, The NB glycoprotein of influenza B virus, in: *Vaccines 85: Molecular and Chemical Basis of Resistance to Parasitic, Bacterial, and Viral Diseases* (R. Lerner, R. M. Chanock, and F. Brown, eds.), pp. 309–314, Cold Spring Harbor Press, Cold Spring Harbor, New York.

Shapiro, G. I., and Krug, R. M., 1988, Influenza virus RNA replication *in vitro:* synthesis of viral template RNAs and virion RNAs in the absence of an added primer, *J. Virol.* **62:** 2285–2290.

Shapiro, G. I., Gurney, T., Jr., and Krug, R. M., 1987, Influenza virus gene expression: Control mechanisms at early and late times of infection and nuclear-cytoplasmic transport of virus-specific RNAs, *J. Virol.* **61:**764–773.

Shimizu, K., Mullinix, M. G., Chanock, R. M., and Murphy, B. R., 1982, Temperature-sensitive mutants of influenza A/Udorn/72 (H3N2) virus. I. Isolation of temperature-sensitive mutants some of which exhibit host-dependent temperature sensitivity, *Virology*, **117:**38–44.

Sivasubramanian, N., and Nayak, D. P., 1982, Sequence analysis of the polymerase 1 gene and the secondary structure prediction of polymerase 1 protein of human influenza virus A/WSN/33, *J. Virol.* **44:**321–329.

Sivasubramanian, N., and Nayak, D. P., 1987, Mutational analysis of the signal-anchor domain of influenza virus neuraminidase, *Proc. Natl. Acad. Sci. USA* **84:**1–5.

Skehel, J. J., 1972, Polypeptide synthesis in influenza virus-infected cells, *Virology* **49:**23–36.

Skehel, J. J., and Schild, G. C., 1971, The polypeptide composition of influenza A viruses, *Virology* **44:**396–408.

Skehel, J. J., and Waterfield, M. D., 1975, Studies on the primary structure of the influenza virus hemagglutinin, *Proc. Natl. Acad. Sci. USA* **72**:93–97.

Skehel, J. J., Hay, A. J., and Armstrong, J. A., 1977, On the mechanism of inhibition of influenza virus replication by amantadine hydrochloride, *J. Gen. Virol.* **38**:97–110.

Skehel, J. J., Bayley, P., Brown, E., Martin, S., Waterfield, M. D., White, J., Wilson, I., and Wiley, D. C., 1982, Changes in the conformation of influenza virus hemagglutinin at the pH optimum of virus-mediated membrane fusion, *Proc. Natl. Acad. Sci. USA* **79**: 968–972.

Skehel, J. J., Stevens, D. J., Daniels, R. S., Douglas, A. R., Knossow, M., Wilson, I. A., and Wiley, D. C., 1984, A carbohydrate side chain on hemagglutinins of Hong Kong influenza viruses inhibits recognition by a monoclonal antibody, *Proc. Natl. Acad. Sci. USA* **81**:1779–1783.

Sleigh, M. J., Both, G. W., Brownlee, G. G., Bender, V. J., and Moss, B. A., 1980, The haemagglutinin gene of influenza A virus: Nucleotide sequence analysis of cloned DNA copies, in: *Structure and Variation in Influenza Viruses* (G. Laver and G. M. Air, eds.), pp. 69–79, Elsevier/North-Holland, New York.

Smith, D. B., and Inglis, S. C., 1985, Regulated production of an influenza virus spliced mRNA mediated by virus-specific products, *EMBO J.* **4**:2313–2319.

Smith, G. L., Levin, J. Z., Palese, P., and Moss, B., 1987, Synthesis and cellular location of the ten influenza polypeptides individually expressed by recombinant vaccinia viruses, *Virology* **160**:336–345.

Spiess, M., and Lodish, H. F., 1985, The sequence of a second human asialoglycoprotein receptor: Conservation of two receptors during evolution, *Proc. Natl. Acad. Sci. USA* **82**:6465–6469.

Spiess, M., and Lodish, H. F., 1986, An internal signal sequence: the asialoglycoprotein receptor membrane anchor, *Cell* **44**:177–185.

Spiess, M., Schwartz, A. L., and Lodish, H. F., 1985, Sequence of human asialoglycoprotein receptor cDNA. An internal signal sequence for membrane insertion, *J. Biol. Chem.* **260**:1979–1982.

St. Angelo, C., Smith, G. E., Summers, M. D., and Krug, R. M., 1987, Two of the three influenza viral polymerase proteins expressed by using baculovirus vectors form a complex in insect cells, *J. Virol.* **61**:361–365.

Stanley, P. M., and Haslam, E. A., 1971, The polypeptides of influenza virus. V. Localization of polypeptides in the virion by iodination techniques, *Virology* **46**:764–773.

Steiner, D. F., Docherty, K., and Carroll, R., 1984, Golgi/Granule processing of peptide hormone and neuropeptide precursors: A mini review, *J. Cell. Biochem.* **24**:121–130.

Stephenson, J. R., Hay, A. J., and Skehel, J. J., 1977, Characterization of virus-specific messenger RNAs from avian fibroblasts infected with fowl plague virus, *J. Gen. Virol.* **36**:237–248.

Steuler, H., Rohde, W., and Scholtissek, C., 1984, Sequence of the neuraminidase gene of an avian influenza virus (A/parrot/Ulster/73, H7N1), *Virology* **135**:118–124.

Steuler, H., Schroder, B., Burger, H., and Scholtissek, C., 1985, Sequence of nucleoprotein gene of influenza A/Parrot/Ulster/73, *Virus Res.* **3**:35–40.

Strubin, M., Mach, B., and Long, E. O., 1984, The complete sequence of the mRNA for the HLA-DR associated invariant chain reveals a polypeptide with an unusual transmembrane polarity, *EMBO J* **3**:869–872.

Sugawara, K., Nakamura, K., and Homma, M., 1983, Analyses of structural polypeptides of seven different isolates of influenza C virus, *J. Gen. Virol.* **64**:579–587.

Sugiura, A., and Ueda, M., 1980, Neurovirulence of influenza viruses in mice. 1. Neurovirulence of recombinants between virulent and avirulent virus strains, *Virology* **101**: 440–449.

Sveda, M. M., and Lai, C.-J., 1981, Functional expression in primate cells of cloned DNA coding for the hemagglutinin surface glycoprotein of influenza virus, *Proc. Natl. Acad. Sci. USA* **78**:5488–5492.

Sveda, M. M., Markoff, L. J., and Lai, C.-J., 1982, Cell surface expression of the influenza virus hemagglutinin requires the hydrophobic carboxyterminal sequences, *Cell* **30:** 649–656.

Thierry, F., and Danos, O., 1982, Use of specific single-stranded DNA probes cloned in M13 to study the RNA synthesis of four temperature-sensitive mutants of HK/68 influenza virus, *Nucl. Acids. Res.* **10:**2925–2938.

Tomley, F. M., and Roditi, J. J., 1984, Nucleotide sequence of RNA segment 5, encoding the nucleoprotein of influenza A/FPV/Rostock/34, *Virus Res.* **1:**625–630.

Townsend, A. R. M., and Skehel, J. J., 1984, The influenza A virus nucleoprotein gene controls the induction of both subtype specific and cross-reactive cytotoxic T cells, *J. Exp. Med.* **160:**552–563.

Townsend, A. R. M., Skehel, J. J., Taylor, P. M., and Palese, P., 1984a, Recognition of influenza A virus nucleoprotein by an H2-restricted cytotoxic T-cell clone, *Virology* **133:**456–459.

Townsend, A. R. M., McMichael, A. J., Carter, N. P., Huddleston, J. A., and Brownlee, G. G., 1984b, Cytotoxic T cell recognition of the influenza nucleoprotein and hemagglutinin expressed in transfected mouse L cells, *Cell* **39:**13–25.

Townsend, A. R. M., Gotch, F. M., and Davey, J., 1985, Cytotoxic T cells recognize fragments of the influenza nucleoprotein, *Cell* **42:**457–467.

Townsend, A. R. M., Rothbard, J., Gotch, F. M., Bahadur, G., Wraith, D., and McMichael, A. J., 1986, The epitopes of influenza nucleoprotein recognized by cytotoxic T lymphocytes can be defined by short synthetic peptides, *Cell* **44:**959–968.

Ulmanen, I., Broni, B. A., and Krug, R. M., 1981, Role of two of the influenza virus core P proteins in recognizing cap 1 structures (m$^7$GpppNm) on mRNAs and in initiating viral RNA transcription, *Proc. Natl. Acad. Sci. USA* **78:**7355–7359.

Ulmanen, I., Broni, B. A., and Krug, R. M., 1983, Influenza virus temperature-sensitive cap (m$^7$GpppNm)-dependent endonuclease, *J. Virol.* **45:**27–35.

Van Rompuy, L., Min Jou, W., Huylebroeck, D., Devos, R., and Fiers, W., 1981, Complete nucleotide sequence of the nucleoprotein gene from the human influenza strain A/PR/8/34 (HONI), *Eur. J. Biochem.* **116:**347–353.

Van Rompuy, L., Min Jou, W., Huylebroeck, D., and Fiers, W., 1982, Complete nucleotide sequence of a human influenza virus neuraminidase gene of subtype N2 (A/Victoria/3/75), *J. Mol. Biol.* **161:**1–11.

Van Wyke, K. L., Hinshaw, V. S., Bean, W. J., and Webster, R. G., 1980, Antigenic variation of influenza A virus nucleoprotein detected with monoclonal antibodies, *J. Virol.* **35:** 24–30.

Van Wyke, K. L., Yewdell, J. W., Reck, L. J., and Murphy, B. R., 1984, Antigenic characterization of influenza A virus matrix protein with monoclonal antibodies, *J. Virol.* **49:**248–252.

Vargese, J. N., Laver, W. G., and Colman, P. M., 1983, Structure of the influenza virus glycoprotein antigen neuraminidase at 2.9A resolution, *Nature (Lond.)* **303:**35–40.

Verhoeyen, M., Fang, R., Min Jou, W., Devos, R., Huylebroeck, D., Saman, E., and Fiers, W., 1980, Antigenic drift between the haemagglutinin of the Hong Kong influenza strains A/Aichi/2/68 and A/Victoria/3/75, *Nature (Lond.)* **286:**771–776.

Verhoeyen, M., Van Rompuy, L., Min Jou, W., Huylebroeck, D., and Fiers, W., 1983, Complete nucleotide sequence of the influenza B/Singapore/222/79 virus hemagglutinin gene and comparison with the B/Lee/40 hemagglutinin, *Nucl. Acids. Res.* **11:**4703–4712.

Vlasak, R., Krystal, M., Nacht, M., and Palese, P., 1987, The influenza C virus glycoprotein (HE) exhibits receptor-binding (hemagglutinin) and receptor-destroying (esterase) activities, *Virology* **160:**419–425.

Walter, P., Ibrahimi, I., and Blobel, G., 1981, Translocation of proteins across the endoplasmic reticulum. I. Signal recognition protein (SRP) binds to in-vitro assembled polysomes synthesizing secretory protein, *J. Cell Biol.* **91:**545–550.

Walter, P., Gilmore, R., and Blobel, G., 1984, Protein translocation across the endoplasmic reticulum, *Cell* **38:**5–8.

Ward, C. W., 1981, Structure of the influenza virus hemagglutinin, *Curr. Top. Microb. Immunol.* **94/95:**1–74.

Ward, C. W., and Dopheide, T. A. A., 1980, The Hong Kong (H3) hemagglutinin. Complete amino acid sequence and oligosaccharide distribution for the heavy chain of A/Memphis/102/72, in: *Structure and Variation in Influenza Virus* (G. Laver and G. Air, eds.), pp. 27–38, Elsevier/North-Holland, New York.

Ward, C. W., Elleman, T. C., and Azad, A. A., 1982, Amino acid sequence of the pronase-released heads of neuraminidase subtype N2 from the Asian strain A/Tokyo/3/67 of influenza virus, *Biochem. J.* **207:**91–95.

Ward, C. W., Murray, J. M., Roxburgh, C. M., and Jackson, D. C., 1983, Chemical and antigenic characterization of the carbohydrate side chains of an Asian (N2) influenza virus neuraminidase, *Virology* **126:**370–375.

Waterfield, M. D., Espelie, K., Elder, K., and Skehel, J. J., 1979, Structure of the haemagglutinin of influenza virus, *Br. Med. Bull.* **35:**57–63.

Webster, R. G., Air, G. M., Metzger, D. W., Colman, P. M., Varghese, J. N., Baker, A. T., and Laver, W. G., 1987, Antigenic structure and variation in influenza virus N9 neuraminidase, *J. Virol.* **61:**2910–2916.

Wei, X., Els, M. C., Webster, R. G., and Air, G. M., 1987, Effects of site-specific mutation on structure and activity of influenza virus B/Lee/40 neuraminidase, *Virology* **156:**253–258.

Weis, W., Brown, J. H., Cusaek, S., Paulson, J. C., Skehel, J. J., and Wiley, D. C., 1988, Structure of the influenza virus haemagglutinin complexed with its receptor, sialic acid, *Nature (Lond.)* **333:**426–431.

White, J., Kielian, M., and Helenius, A., 1983, Membrane fusion proteins of enveloped animal viruses, *Q. Rev. Biophys.* **16:**151–195.

Wickner, W. T., and Lodish, H. F., 1985, Multiple mechanisms of protein insertion into and across membranes, *Science* **230:**400–407.

Wiley, D. C., and Skehel, J. J., 1987, The structure and function of the hemagglutinin membrane glycoprotein of influenza virus, *Annu. Rev. Biochem.* **56:**365–394.

Wiley, D. C., Skehel, J. J., and Waterfield, M. D., 1977, Evidence from studies with a cross-linking reagent that the hemagglutinin of influenza virus is a trimer, *Virology* **79:**446–448.

Wiley, D. C., Wilson, I. A., and Skehel, J. J., 1981, Structural identification of the antibody binding sites of the Hong Kong influenza hemagglutinin and their involvement in antigenic variation, *Nature (Lond.)* **298:**373–378.

Williams, M. A., and Lamb, R. A., 1986, Determination of the orientation of an integral membrane protein and sites of glycosylation by oligonucleotide-directed mutagenesis: Influenza B virus NB glycoprotein lacks a cleavable signal sequence and has an extracellular $NH_2$-terminal region, *Mol. Cell Biol.* **6:**4317–4328.

Williams, M. A., and Lamb, R. A., 1988, Polylactosaminoglycan modification of a small integral membrane glycoprotein, influenza B virus NB, *Mol. Cell Biol.* **8:**1186–1196.

Williams, M. A., and Lamb, R. A., 1989, Effect of mutations and deletions in a bicistronic mRNA on the synthesis of influenza B virus NB and NA glycoproteins, *J. Virol.* **63:**28–35.

Wilson, I. A., Skehel, J. J., and Wiley, D. C., 1981, The hemagglutinin membrane glycoprotein of influenza virus; structure at 3 Å resolution, *Nature (Lond.)* **289:**366–373.

Wilson, V. W., and Rafelson, M. E., 1963, Isolation of neuraminidase from influenza virus, *Biochem. Prep.* **10:**113–117.

Winter, G., and Fields, S., 1980, Cloning of influenza cDNA into M13: the sequence of the RNA segment encoding the A/PR/8/34 matrix protein, *Nucl. Acids. Res.* **8:**1965–1974.

Winter, G., and Fields, S., 1981, The structure of the gene encoding the nucleoprotein of human influenza virus A/PR/8/34, *Virology* **114:**423–428.

Winter, G., and Fields, S., 1982, Nucleotide sequence of human influenza A/PR/8/34 segment 2, *Nucl. Acids. Res.* **10:**2135–2143.

Winter, G., Fields, S., and Brownlee, G. G., 1981a, Nucleotide sequence of the haemagglutinin of a human influenza virus H1 subtype, *Nature (Lond.)* **292:**72–75.

Winter, G., Fields, S., Gait, M. J., and Brownlee, G. G., 1981b, The use of synthetic oligodeoxynucleotide primers in cloning and sequencing segment 8 of influenza virus. (A/PR/8/34), *Nucl. Acids. Res.* **9:**237–245.

Wolstenholme, A. J., Barrett, T., Nichol, S. T., and Mahy, B. W. J., 1980, Influenza virus-specific RNA and protein synthesis in cells infected with temperature-sensitive mutants defective in the genome segment encoding nonstructural proteins, *J. Virol.* **35:**1–7.

Wrigley, N. G., Skehel, J. J., Charlwood, P. A., and Brand, C. M., 1973, The size and shape of influenza virus neuraminidase, *Virology* **51:**525–529.

Wrigley, N. G., Laver, W. G., and Downie, J. C., 1977, Binding of antibodies to isolated hemagglutinin and neuraminidase molecules of influenza virus observed in the electron microscope, *J. Mol. Biol.* **109:**405–421.

Yamashita, M., Krystal, M., and Palese, P., 1988, Evidence that the matrix protein of influenza C virus is coded for by a spliced mRNA, *J. Virol.* **62:**3348–3355.

Yewdell, J. W., Webster, R. G., and Gerhard, W., 1979, Antigenic variation in three distinct determinants of an influenza type A hemagglutinin molecule, *Nature (Lond.)* **279:**246–248.

Yewdell, J. W., Bennick, J. R., Smith, G. L., and Moss, B., 1985, Influenza A virus nucleoprotein is a major target antigen for cross-reactive anti-influenza A virus cytotoxic T lymphocytes, *Proc. Natl. Acad. Sci. USA* **82:**1785–1789.

Yewdell, J. W., Yellen, A., and Bachi, T., 1988, Monoclonal antibodies localize events in the folding, assembly and intracellular transport of the influenza virus hemagglutinin glycoprotein, *Cell* **52:**843–852.

Yokota, M., Nakamura, K., Suganara, K., and Homma, M., 1983, The synthesis of polypeptides in influenza C virus-infected cells, *Virology* **130:**105–117.

Yoshida, T., Shaw, M., Young, J. F., and Compans, R. W., 1981, Characterization of the RNA associated with influenza A cytoplasmic inclusions and the interaction of $NS_1$ protein with RNA, *Virology* **110:**87–97.

Young, J. F., Desselberger, U., Palese, P., Ferguson, B., Shatzman, A. R., and Rosenberg, M., 1983a, Efficient expression of influenza virus $NS_1$ nonstructural protein in *Escherichia coli, Proc. Natl. Acad. Sci. USA* **80:**6105–6109.

Young, J. F., Capecchi, M., Laski, F. A., Rajbhandary, U. L., Sharp, P. A., and Palese, P., 1983b, Measurement of suppressor transfer RNA activity, *Science* **221:**873–875.

Zebedee, S. L., 1989, Structural and functional analysis of the influenza A virus $M_2$ protein, Doctoral dissertation, Northwestern University, Evanston, Illinois.

Zebedee, S. L., and Lamb, R. A., 1988, Influenza A virus $M_2$ protein: Monoclonal antibody restriction of virus growth and detection of $M_2$ in virions, *J. Virol.* **62:**2762–2772.

Zebedee, S. L., and Lamb, R. A., 1989a, Growth restriction of influenza A virus by $M_2$ protein antibody is genetically linked to the $M_1$ protein, *Proc. Natl. Acad. Sci. (USA)* **86:**1061–1065.

Zebedee, S. L., and Lamb, R. A., 1989b, Nucleotide sequences of influenza A virus RNA segment 7: A comparison of five isolates, *Nucl. Acids. Res.* **17:**2870.

Zebedee, S. L., Richardson, C. D., and Lamb, R. A., 1985, Characterization of the influenza virus $M_2$ integral membrane protein and expression at the infected-cell surface from cloned cDNA, *J. Virol.* **56:**502–511.

Zerial, M., Melancon, P., Schneider, C., and Garoff, H., 1986, The transmembrane segment of the human transferrin receptor functions as a signal peptide, *EMBO J* **5:**1543–1550.

Zerial, M., Huylebroeck, D., and Garoff, H., 1987, Foreign transmembrane peptides replacing the internal signal sequence of transferrin receptor allows its translocation and membrane binding, *Cell* **48:**147–155.

Zimmermann, R., and Meyer, D. I., 1986, 1986: A year of new insights into how proteins cross membranes, *Trends Biochem. Sci.* **11:**512–515.

Zimmern, D., 1977, The nucleotide sequence at the origin for assembly on tobacco mosaic virus RNA, *Cell* **11:**463–482.

Zimmern, D., and Butler, P. J. G., 1977, The isolation of tobacco mosaic virus RNA fragments containing the origin for viral assembly, *Cell* **11:**455–462.

CHAPTER 2

# Expression and Replication of the Influenza Virus Genome

ROBERT M. KRUG, FIRELLI V. ALONSO-CAPLEN,
ILKKA JULKUNEN, AND MICHAEL G. KATZE

## I. INTRODUCTION

The genome of influenza A viruses is composed of eight segments of negative polarity (see Chapter 1). In infected cells, these virion RNAs (vRNAs) are both transcribed into messenger RNAs (mRNAs) and replicated. The distinctive feature of influenza viral mRNA synthesis is that it is primed by 5' capped ($m^7$GpppNm-containing) fragments derived from newly synthesized host-cell RNA polymerase II transcripts (Bouloy et al., 1978, 1979; Krug et al., 1979; Plotch et al., 1979, 1981; Krug, 1981, 1983; Herz et al., 1981). The mRNA chains are elongated until a stretch of uridine residues is reached 17–22 nucleotides before the 5' ends of the vRNAs, where transcription terminates and polyadenylate [poly(A)] is added to the mRNAs (Hay et al., 1977a; Robertson et al., 1981). For replication to occur, an alternative type of transcription is required that results in the production of full-length copies of the vRNAs. The full-length transcripts, or template RNAs, are initiated without a primer and are not terminated at the poly(A) site used during mRNA synthesis (Hay et al., 1977a, 1982). The second step in replication is the copying of the template RNAs into vRNAs. This synthesis also occurs without a primer, since the vRNAs contain 5' triphosphorylated ends (Young and Content, 1971). The three types of virus-specific RNAs—mRNAs, template

ROBERT M. KRUG, FIRELLI V. ALONSO-CAPLEN, and ILKKA JULKUNEN • Graduate Program in Molecular Biology, Memorial Sloan-Kettering Cancer Center, New York, New York 10021.    MICHAEL G. KATZE • Department of Microbiology, School of Medicine, University of Washington, Seattle, Washington 98195.

89

RNAs, and vRNAs— are all synthesized in the nucleus of infected cells (Herz *et al.*, 1981; Jackson *et al.*, 1982; Shapiro *et al.*, 1987). During the early phase of infection, the synthesis of these three types of virus-specific RNAs is coupled (Hay *et al.*, 1977a; G. L. Smith and Hay, 1982; Shapiro *et al.*, 1987), whereas during the later phase of infection essentially only vRNAs are synthesized (Shapiro *et al.*, 1987).

In the nucleus, the viral mRNAs undergo at least some of the same processing steps as cellular and DNA virus-encoded mRNA precursors. Internal adenosine residues of influenza viral mRNAs are methylated (Krug *et al.*, 1980; Narayan *et al.*, 1987), and two of the viral mRNAs, those coding for the M1 protein (membrane protein) and for the NS1 protein (nonstructural protein 1), are themselves spliced to form smaller mRNAs (Lamb and Lai, 1980, 1982, 1984; Lamb *et al.*, 1981). Both the internal methylation and the splicing of influenza viral mRNAs are almost certainly catalyzed by host nuclear enzymes. Of the RNAs shown to undergo splicing, the influenza virus M1 and NS1 mRNAs are the only known examples that are not DNA-directed and that are not synthesized by RNA polymerase II. Because both the unspliced (NS1 and M1) and spliced (NS2 and M2) mRNAs code for proteins, the extent of splicing is controlled such that the ratio of unspliced to spliced mRNA is about 10:1 (Lamb *et al.*, 1980, 1981). How this splicing is controlled may be relevant to the regulated splicing of retrovirus RNAs and to one or more types of alternative splicing that operate to produce cellular mRNAs.

During influenza virus infection, protein synthesis is maintained at high levels; a dramatic switch from cellular to viral protein synthesis occurs despite the presence of high levels of functional cellular mRNAs in the cytoplasm of infected cells (Katze *et al.*, 1986a, 1988). The overall high level of protein synthesis is maintained via the action of an unidentified viral gene product(s) that suppresses the protein kinase (P68) that phosphorylates the α-subunit of the eIF-2 translation-initiation factor (Katze *et al.*, 1986a, 1988). This influenza viral gene product apparently acts in a manner similar to that of adenovirus-encoded VA1 RNA (Siekierka *et al.*, 1985; Schneider *et al.*, 1985; O'Malley *et al.*, 1986; Kitajewski *et al.*, 1986; Katze *et al.*, 1987). The switch from cellular to influenza virus protein synthesis results in large part from a block in cellular mRNA translation at both initiation and elongation (Katze *et al.*, 1986b).

Various steps in viral gene expression and replication might be candidates as targets for antiviral agents. In fact, a natural inhibitor of influenza virus replication has already been described, the interferon (IFN)-induced Mx protein. In mouse cells, the antiviral state induced by IFNα/β is controlled by the host gene Mx (Haller *et al.*, 1980; Haller, 1981; Staehli and Haller, 1987). Only mouse cells that possess this gene develop an efficient antiviral state against influenza virus after exposure to IFNα/β, whereas the antiviral state against other viruses is independent of the Mx gene. The mouse Mx gene product is a 72,000-molecular-weight protein that accumulates in the nucleus (Horisberger *et al.*, 1983;

Dreiding *et al.*, 1985). In mouse cells expressing the IFN-induced Mx protein, primary transcription—the viral mRNA synthesis catalyzed by the inoculum viral RNA transcriptase—is inhibited (Krug *et al.*, 1985). The antiviral activity of the Mx protein does not require other IFN-induced proteins: when the cloned DNA encoding the mouse Mx protein was transfected into Mx-negative cells, an antiviral state specifically directed against influenza virus was established (Staehli *et al.*, 1986a). These areas of virus gene expression and replication are discussed in detail in this chapter.

## II. VIRAL mRNA SYNTHESIS

Influenza virus employs a novel mechanism for the synthesis of its viral mRNAs. Viral mRNA synthesis requires initiation by host-cell primers, specifically capped (m$^7$GpppNm-containing) RNA fragments derived from host-cell RNA polymerase II transcripts (Bouloy *et al.*, 1978; Plotch *et al.*, 1979, 1981; Krug, 1981). Because this occurs in the nucleus of infected cells (Herz *et al.*, 1981), viral mRNA synthesis requires continuous functioning of the cellular RNA polymerase II and is therefore inhibited by α-amanitin (Mark *et al.*, 1979). Host-cell primers are generated by a viral cap-dependent endonuclease that cleaves the capped cellular RNAs 10–13 nucleotides from their 5′ ends preferentially at a purine residue (Plotch *et al.*, 1981). Priming does not require hydrogen bonding between the capped primer fragments and the 3′ ends of the vRNA templates (Krug *et al.*, 1980; Krug, 1983). Rather, priming requires the presence of a 5′-methylated cap structure (Plotch *et al.*, 1979, 1981; Bouloy *et al.*, 1979). In fact, each of the methyl groups in the cap, the 2′-*O*-methyl on the penultimate base as well as the 7-methyl on the terminal G, strongly increases priming activity (Bouloy *et al.*, 1980). Transcription is initiated by the incorporation of a G residue onto the 3′ end of the resulting fragments, directed by the penultimate C residue of the vRNAs (Plotch *et al.*, 1981). Viral mRNA chains are then elongated until a stretch of 5–7 uridine (U) residues is reached 17–22 nucleotides before the 5′ ends of the vRNAs, where transcription terminates and poly(A) is added to the mRNAs (Hay *et al.*, 1977a,b; Robertson *et al.*, 1981).

Viral mRNA synthesis is catalyzed by viral nucleocapsids (Inglis *et al.*, 1976; Plotch *et al.*, 1981), which consist of the individual vRNAs associated with four viral proteins: the nucleocapsid (NP) protein and the three P (PB1, PB2, and PA) proteins (Inglis *et al.*, 1976; Ulmanen *et al.*, 1981). The P proteins are responsible for viral mRNA synthesis, and some of their roles have been determined by analyses of the *in vitro* reaction catalyzed by virion nucleocapsids. Ultraviolet (UV)-light-induced crosslinking experiments showed that the three P proteins are in the form of a complex that starts at the 3′ ends of the vRNA templates and moves down the templates in association with the elongating mRNAs during transcription (Braam *et al.*, 1983). The PB2 protein in this complex recognizes and binds to the cap of the primer RNA (Ulmanen *et al.*, 1981; Blass *et al.*, 1982; Braam *et al.*, 1983). This was verified by the *in vitro* behavior of the nucleocapsids from PB2 temperature-sensitive (*ts*) mutants (Ulmanen *et al.*, 1983). The cap-dependent endonuclease catalyzed by these nucleocapsids and the binding of capped primer fragments to these nucleocapsids were *ts in vitro*. The PB2 protein shows homology to the 25,000 molecular weight cap-binding protein that is involved in an early step in the initiation of protein synthesis in eukaryotes (Rychlik *et al.*, 1987). The UV-crosslinking experiments also showed that the PB1 protein, which is initially found at the first residue (a G residue) added onto the primer, moves as part of the P-protein complex to the 3' ends of the growing viral mRNA chains, indicating that it most likely catalyzes each nucleotide addition (Braam *et al.*, 1983). On the basis of the relative positions of PB1 and PB2 on the nascent chains, it was concluded that the P-protein complex most likely has the PB1 protein at its leading edge and the PB2 protein at its trailing edge (Braam *et al.*, 1983). Figure 1 presents a model of the functions and movements of the P proteins during viral mRNA synthesis.

Immunoelectron microscopy has provided strong support for this model. Using monospecific antisera against the individual proteins and

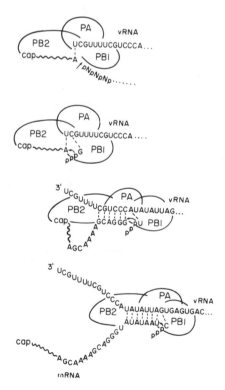

FIGURE 1. Model of the functions and movements of the three P proteins during capped RNA-primed viral mRNA synthesis. The sequence shown is that of the vRNA and mRNA coding for the NP protein. From Braam *et al.* (1983).

indirect immunogold labeling, it was shown that each of the P proteins is found essentially at only one position on each virion nucleocapsid, i.e., at one end (Murti et al., 1988) (Fig. 2). Furthermore, binding of an antibody to one P protein (e.g., PB2) inhibited the binding of an antibody to a second P protein (e.g., PB1), strongly suggesting that all three P proteins are associated with each other at the same end of each nucleocapsid.

Although PA is in the P-protein complex, no specific role for PA in viral mRNA synthesis has been found. In fact, experiments with ts virus mutants with defects in the PA gene indicate that the principal role of the PA protein in infected cells is not in viral mRNA synthesis, but rather in viral RNA replication. After a temperature shiftup, template and vRNA synthesis, but not viral mRNA synthesis, was immediately inhibited (Krug et al., 1975; Mahy et al., 1981; Mowshowitz, 1981). By contrast, with ts mutants containing defects in either the PB1 or PB2 protein, viral mRNA synthesis was inhibited immediately after a shiftup.

It is likely that the complex of the three P proteins is assembled first and that it is this complex, rather than a particular one of the P proteins, that recognizes and binds to the 3' ends of the vRNAs to initiate mRNA synthesis. Infected cells contain a pool of P proteins not associated with viral nucleocapsids, and the P proteins in this pool are largely, if not totally, in the form of a complex with each other (Detjen et al., 1987; Akkina et al., 1987). Thus, for example, when the cytoplasmic and nuclear extracts from infected cells were depleted of nucleocapsids by centrifugation and then subjected to immunoprecipitation with either an anti-PB1 or an anti-PB2 antiserum, all three P proteins (and little or no NP protein) were precipitated (Detjen et al., 1987) (Fig. 3). The immunoprecipitated complexes were enriched in the P protein against which the antiserum was directed. Because little or no free P protein was detected, this enrichment indicates that either breakdown of the P-protein complexes occurs to various degrees during immunoprecipitation or several types of P complexes containing different ratios of the P proteins exist, or that both take place. It should also be kept in mind that the P proteins are involved in viral RNA replication as well as viral mRNA synthesis, so that the nonnucleocapsid P-protein complexes identified in infected cell extracts may participate in either mRNA synthesis or replication, or both.

The formation of P-protein complexes also occurs in the absence of other influenza virus gene products. Each of the P genes was inserted into a baculovirus vector under the control of the polyhedrin promoter; insect cells were infected simultaneously with all three baculovirus P recombinants (St. Angelo et al., 1987). Under relatively gentle conditions of immunoprecipitation [0.5% Triton X-100 and no sodium dodecyl sulfate (SDS)], an anti-PB1 antiserum precipitated not only the PB1 protein, but the PB2 and PA proteins as well (Krug et al., 1987) (Fig. 4). In the absence of the PB1 protein, neither PB2 or PA was precipitated by PB1 antiserum (St. Angelo et al., 1987). PB2 antiserum also precipitated all three P pro-

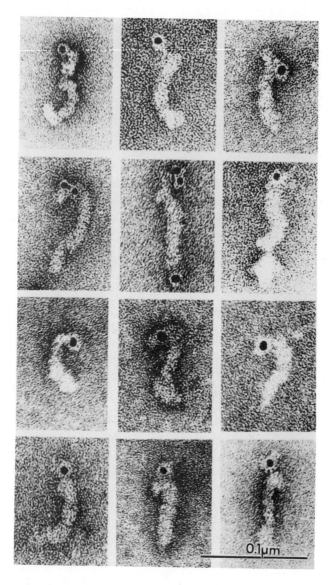

FIGURE 2. Immunogold labeling of purified influenza virion nucleocapsids with monoclonal anti-PB1 antibodies. The antibodies were purified by chromatography on protein A-
Sepharose before use in immunogold labeling. Virion nucleocapsids adsorbed to parlodion-
coated grids were floated first on a solution of anti-PB1 antibody, followed by goat anti-
mouse antibodies conjugated with 5-nm gold particles. The samples were negatively stained
with 1% aqueous uranyl acetate and viewed in a Philips EM301 electron microscope.
Arrows denote an exceptional case in which the nucleocapsid appears labeled at both ends.
Electron micrographs were by G. Murti. From Murti *et al.* (1988).

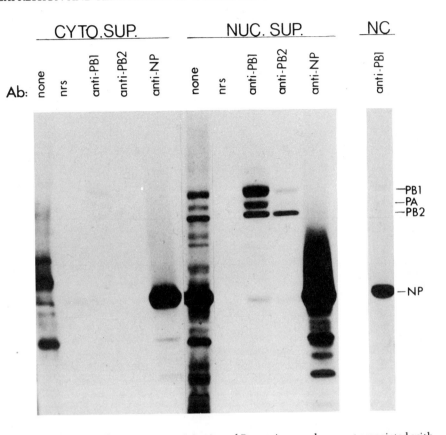

FIGURE 3. Detection by immunoprecipitation of P-protein complexes not associated with viral nucleocapsids. BHK-21 cells were labeled with [$^{35}$S]methionine 2.5–3.5 hr p.i. The high-speed supernatant from the cytoplasm and nucleus was immunoprecipitated with normal rabbit serum (nrs), anti-PB1 antiserum, anti-PB2 antiserum, or pooled anti-NP monoclonal antibodies. Each immunoprecipitation reaction contained equal amounts of radioactivity (10$^6$ cpm), corresponding to cell-equivalent amounts of the cytoplasmic and nuclear samples. The immunoprecipitated proteins were analyzed on 8% polyacrylamide gels containing 4 M urea. The lanes denoted as "none" show the pattern of labeled proteins in the high-speed supernatant from the cytoplasm and nucleus (3 × 10$^4$ cpm) without immunoprecipitation. The last lane shows the results of immunoprecipitating glycerol gradient-purified intracellular nucleocapsids (7.5 × 10$^4$ cpm) with anti-PB1 antiserum. The positions of the nucleocapsid-associated proteins are indicated. From Detjen et al. (1987).

teins. These results indicate that the three P proteins expressed in insect cells did form a complex with each other and that the ability to form a complex is an intrinsic property of these three P proteins that does not require the participation of other influenza virus gene products. When immunoprecipitation was carried out under harsher conditions (i.e., in the presence of 0.1% SDS), the PA protein was not found in the complex (St. Angelo et al., 1987), indicating that the PA protein was less stably

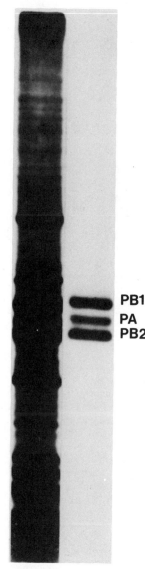

**Anti**

**E     PB1**

PB1
PA
PB2

FIGURE 4. Formation of a complex containing the three P proteins in baculovirus-infected insect cells. Cells infected with all three baculovirus P recombinants were labeled with [$^{35}$S]methionine 38–39 hr p.i. and were fractionated into cytoplasmic and nuclear extracts. The proteins in the nuclear extract were resolved on a 8% polyacrylamide gel containing 4 M urea (lane E). An equivalent aliquot of the nuclear extract was adjusted to 50 mM Tris–HCl, 5 mM EDTA, 0.5% Triton X-100, and 50 mM NaCl and immunoprecipitated with an anti-PB1 antiserum. The immune complexes isolated on protein A Sepharose were then solubilized by heating at 100°C in gel-loading buffer; labeled proteins were resolved on a 8% polyacrylamide gel containing 4 M urea (anti-PB1 lane). From Krug *et al.* (1987).

associated with the other two P proteins. By contrast, under these harsher immunoprecipitation conditions, the PA protein remained associated with the P-protein complex found in influenza virus-infected cell extracts (Detjen *et al.*, 1987), suggesting that there may be subtle differences between the P-protein complexes formed in the two systems.

Further information about the mechanism of viral mRNA synthesis and about the role of each of the P proteins awaits the development of a better *in vitro* system, one in which the individual P proteins can be added separately and in various combinations to viral templates, i.e., viral nucleocapsids lacking the P proteins, in order to reconstitute activity in one or more steps of viral mRNA synthesis. Probably the greatest difficulty in developing such a system will be the preparation of viral nucleocapsids lacking the P proteins. With another negative-strand virus, vesicular stomatitis virus (VSV), investigators have succeeded in selectively dissociating the polymerase proteins from virion nucleocapsids (Emerson and Yu, 1975). The resulting nucleocapsids were shown to retain their ability to serve as templates when the polymerase proteins were added back. Despite the efforts of many investigators, it has not yet been possible to carry out similar experiments with influenza virus. The selective dissociation of the P proteins without disruption of the nucleocapsid template has not been achieved. This may necessitate the assembly of nucleocapsids *de novo* using vRNA and NP protein, an endeavor with its own inherent problems (see Section III).

A recent paper (Szewczyk *et al.*, 1988) reported reconstitution of the influenza virus transcriptase simply by renaturing the NP and P proteins isolated by denaturing SDS gel electrophoresis. The proteins in virion nucleocapsids were separated by SDS gel electrophoresis and were then blotted onto a polyvinylidine difluoride membrane. The NP and three P proteins were eluted from these membranes with Triton X-100 at pH 9. These four proteins were then mixed with influenza vRNA in the presence of the bacterial enzyme thioredoxin to renature the proteins. This mixture was assayed for RNA synthesis in the presence of a primer, either a capped RNA, alfalfa mosaic virus (AlMV) RNA 4 (containing a cap 0 $m^7GpppG$ rather than a cap 1 $m^7GpppGm$ structure that is normally required), or the dinucleotide ApG, which at high concentrations has been shown to initiate viral mRNA synthesis (McGeoch and Kitron, 1975; Plotch and Krug, 1977). Some RNA complementary to vRNA was synthesized. The RNA products were apparently of relatively large size (~900–1800 nucleotides in length), but the gels used for analysis showed only a broad smear of product RNAs (Szewczyk *et al.*, 1988), rather than the discrete mRNA species observed by others (Plotch and Krug, 1977, 1978). These results (Szewczyk *et al.*, 1988) suggest that the complicated structure of active influenza virion transcriptase complexes re-formed at least to a certain extent in this mixture of four proteins and vRNA: the NP protein was able to bind to vRNA to form a functional helical template (see Section III), the three P proteins formed complexes with each other, and these complexes bound to the 3' ends of the reformed helical NP–vRNA structures and were able to initiate and elongate mRNA chains. This is a surprising and potentially exciting result. A key question concerns the efficiency of reformation of active transcriptase complexes. Szewczyk and co-workers report that they reconstituted 10–30% of the activity present in their original virion nucleocapsids. The problem is that their original nucleocapsids exhibit extremely low activity—only about 0.1% of that obtained by other workers (Ulmanen *et al.*, 1983). Consequently, relative to virion nucleocapsids with good transcriptase activity, these workers have achieved only about 0.03% reconstitution. Further work is needed to determine whether this method for reconstitution can be made efficient enough to yield important new information about influenza viral mRNA synthesis.

## III. TEMPLATE RNA SYNTHESIS

The first step in the replication of influenza vRNA is the switch from viral mRNA synthesis to the synthesis of template RNAs, the full-length copies of vRNA that then serve as templates for vRNA synthesis. This switch requires (1) a change from the capped RNA-primed initiation of transcription used during mRNA synthesis to unprimed initiation: and (2) antitermination at the poly(A) site, 17–22 nucleotides form the 5' ends of vRNAs, used during mRNA synthesis (Hay *et al.*, 1977a, 1982). Because the switch from mRNA to template RNA synthesis requires protein synthesis *in vivo* (Barrett *et al.*, 1979; Hay *et al.*, 1982), it is likely that one or more newly synthesized virus-specific proteins are needed for either unprimed synthesis or antitermination, or both.

To identify these proteins and determine their roles, an *in vitro* system was established that catalyzes the synthesis of template RNA as well as viral mRNA (Beaton and Krug, 1984, 1986). Nuclear extracts obtained from virus-infected HeLa cells proved suitable *in vitro* systems. In these experiments, M13 single-stranded DNA specific for transcripts copied off the NS vRNA (the smallest vRNA) was used to measure the NS1 mRNA and NS template RNA synthesized by the nuclear extracts. This assay included a digestion with RNase T2, which removed the poly(A) and the 5'-capped primer-donated region from the NS1 mRNA. As a result, the NS template RNA was about 20 nucleotides larger than the NS1 mRNA and therefore had a slower mobility than the NS1 mRNA during gel electrophoresis.

Nuclear extracts prepared from HeLa cells at several times postinfection were used. Initially, using extracts collected at 4 hr postinfection (p.i.), it was found that only low levels of NS1 mRNA and NS template RNA were synthesized in the absence of an added primer (Beaton and Krug, 1986) (Fig. 5, lane 1). However, the addition of a high concentration (0.4 mM) of the dinucleotide ApG, which had been shown to act as a primer for viral mRNA synthesis catalyzed by virion nucleocapsids (McGeoch and Kitron, 1975; Plotch and Krug, 1977), greatly stimulated the synthesis of both NS1 mRNA and NS template RNA catalyzed by these nuclear extracts (Beaton and Krug, 1976) (Fig. 5, lane 2). Consequently, these nuclear extracts contained the factor(s) that causes antitermination at the poly(A) site used during viral mRNA synthesis, but they were deficient in unprimed initiation of template RNA synthesis and in

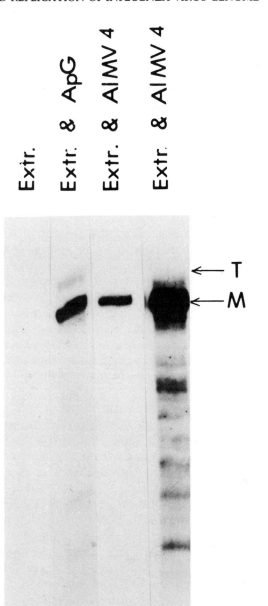

FIGURE 5. Synthesis of template RNAs *in vitro* by infected cell nuclear extracts. A nuclear extract from influenza virus-infected HeLa cells, 4 hr p.i., was incubated under RNA synthesis conditions (Beaton and Krug, 1986) in the absence of a primer (lane 1), in the presence of 0.4 mM apG (lane 2), or in the presence of A1MV RNA 4 containing a m⁷GpppGm cap (lane 3). Lane 4 shows a longer exposure of lane 3. The RNA products were analyzed for NS1 mRNA (M) and NS template RNA (T) using RNase T2 digestion as described in the text (Beaton and Krug, 1986). From Beaton and Krug (1986).

the capped primers needed for viral mRNA synthesis. The addition of ApG circumvented the inefficient unprimed initiation and therefore allowed the analysis of the mechanism of antitermination. In contrast to ApG, the addition of a capped RNA primer, AIMV RNA 4, stimulated the synthesis of only NS1 mRNA; little or no NS template RNA was synthesized (Fig. 5, lane 3). Consequently, viral RNA transcripts that initiated with a capped primer were not antiterminated by the nuclear factor(s) that antiterminated the ApG-initiated viral transcripts.

In contrast to the 4-hr extracts, extracts collected at 6 hr p.i. synthesized template RNA in the absence of ApG (Shapiro and Krug, 1988). Maximal template RNA-synthesizing activity in the absence of ApG occurred with extracts collected at 6 hr p.i. Much lower activity was present both before and after 6 hr. In many 6-hr extracts, template RNA was the predominant transcription product in the absence of ApG (Fig. 6). The addition of ApG to such an extract caused only small; (two- to threefold) stimulation of template RNA synthesis, whereas mRNA synthesis was strongly stimulated, so that mRNA synthesis predominated over template RNA synthesis in the presence of ApG. The synthesis of template RNA in the absence of ApG suggested that at least some of these template RNAs were initiated without a primer *in vitro*. However, attempts to prove that unprimed initiation occurred *in vitro* were frustrated by the presence in the nuclear extracts of enzymes that transferred the labeled $\gamma$-phosphate of ATP to the $\alpha$-position of the four ribonucleoside triphosphates. As a consequence, with $[\gamma\text{-}^{32}P]$-ATP or $[\gamma\text{-}^{35}S]$-ATP as the precursor, most of the label was incorporated into internal positions of the NS template RNAs, and unequivocal evidence for the incorporation of pppA at the 5′ ends of the NS template RNAs could not be obtained. Clearly, it will be necessary to use a better assay for initiation and/or to purify the components needed for template RNA synthesis so that they are free of the phosphate-exchange enzymes.

These nuclear extracts (both the 4-hr and 6-hr extracts) contained the factor(s) that antiterminated at the poly(A) site used during viral mRNA synthesis. In the presence of ApG, the assay for template RNA synthesis measured only the antitermination step and not the unprimed initiation step in template RNA synthesis (Beaton and Krug, 1986; Shapiro and Krug, 1988). The antitermination factor(s) could be separated from the viral nucleocapsids by ultracentrifugation. The nucleocapsids pelleted by ultracentrifugation synthesized NS1 viral mRNA but little or no NS template RNA (Fig. 7, lane 1), indicating the virtual absence of the antitermination factor in the pellet. The supernatant fraction by itself displayed little activity in either viral mRNA or template RNA synthesis (Beaton and Krug, 1986). When the supernatant was added to the pellet in the presence of ApG, NS template RNA synthesis was restored (Fig. 7, lane 2), indicating that the supernatant fraction contained the factor(s) required for antitermination at the poly(A) site.

The first approach for determining the identify of this factor was

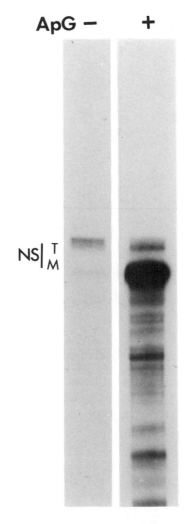

FIGURE 6. Effect of ApG on template RNA and mRNA synthesis catalyzed by a nuclear extract prepared from HeLa cells at 6 hr p.i. RNA synthesis was carried out in the absence (−) and presence (+) of ApG. RNA products were analyzed for NS1 mRNA (M) and NS template RNA (T) using RNase $T_2$ digestion, as described in the text (Beaton and Krug, 1986). From Shapiro and Krug (1988).

antibody-depletion experiments (Beaton and Krug, 1986). The supernatant was depleted of individual virus-specific proteins by incubation with protein A–Sepharose containing an antiserum directed against an individual virus-specific protein. The protein A–Sepharose was then removed by centrifugation; the resulting supernatant was then added to the pellet fraction in the presence of ApG. After incubation with protein A-Sepharose containing pooled monoclonal antibodies directed against the viral NP protein, the supernatant lost its ability to antiterminate, as NS1 mRNA, and not NS template RNA, was synthesized (Fig. 7, lane 4). A control experiment indicated that the NP protein had been quantitatively removed from the supernatant (Beaton and Krug, 1986). By contrast, antitermination activity was retained in supernatants that had been incubated with protein A-Sepharose alone (Fig. 7, lane 3) or with protein A–

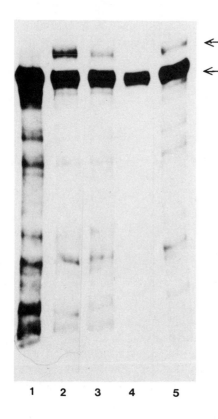

FIGURE 7. Antibody depletion of nonnucleocapsid NP protein molecules eliminated template RNA synthesis *in vitro*. The supernatant fraction from a nuclear extract from infected cells, 4 hr p.i., was either untreated (lane 2), incubated with protein A–Sepharose alone (lane 3), incubated with protein A–Sepharose containing pooled NP monoclonal antibodies (lane 4), or incubated with protein A–Sepharose containing pooled NS1 monoclonal antibodies (lane 5). Each of these supernatants was incubated with the nuclear pellet in the presence of ApG and the amount of synthesis of NS1 mRNA (M) and NS template RNA (T) determined (Beaton and Krug, 1986). Lane 1 shows the products made by the nuclear pellet alone in the presence of ApG. From Beaton and Krug (1986).

Sepharose containing pooled monoclonal antibodies directed against the NS1 protein (Fig. 7, lane 5) (Beaton and Krug, 1986), the predominant viral nonstructural protein that accumulates in the nucleus (Lazarowitz *et al.*, 1971; Krug and Etkind, 1973; Hay and Skehel, 1975; Krug and Soeiro, 1975; Greenspan *et al.*, 1988). Depletion with an antiserum directed against nonstructural protein 2 (NS2), which also accumulates in the nucleus (Greenspan et al., 1985), also did not eliminate template RNA-synthesizing activity (Shapiro and Krug, 1988). These results strongly suggest that NP protein molecules free of nucleocapsids are required for antitermination during template RNA synthesis.

Verification of the role of NP protein in template RNA synthesis was obtained using the WSN strain ts mutant ts56 containing a defect only in the NP protein (Shapiro and Krug, 1988). Several ts mutants carrying a defect in the NP gene have been shown to have a defect in virus-specific RNA synthesis (Krug et al., 1975; Sugiura et al., 1975; Scholtissek and Bowles, 1975; Scholtissek, 1978; Mahy et al., 1981). The synthesis of both template RNA (Shapiro and Krug, 1988) and vRNA (Scholtissek, 1978; Mahy et al., 1981; Shapiro and Krug, 1988), but not of viral mRNA (Krug et al., 1975; Scholtissek, 1978; Mahy et al., 1981; Shapiro and Krug, 1988), was drastically inhibited in vivo immediately after a temperature shiftup, indicating that the NP protein is required for template RNA and vRNA synthesis in vivo. To verify that the NP protein acted directly on template RNA synthesis, a nuclear extract was prepared from HeLa cells infected by the ts56 viral mutant at the permissive temperature 33°C (Shapiro and Krug, 1988). The activity of this extract in viral mRNA and template RNA synthesis at both 33°C and 39.5°C (nonpermissive temperature) was determined (Fig. 8). In both the absence and presence of

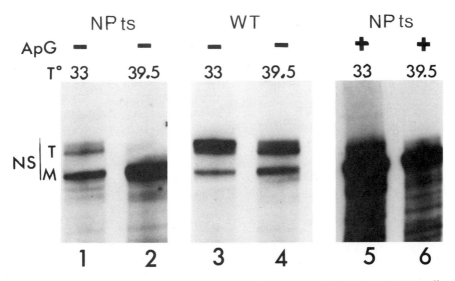

FIGURE 8. Requirement of the NP protein for template RNA synthesis in vitro. HeLa cells were infected with the ts56 virus mutant at 33°C. At 8 hr p.i., a nuclear extract (NPts) was prepared and used to catalyze RNA synthesis in the absence of ApG at 33°C (lane 1) and at 39.5°C (lane 2) and in the presence of ApG at 33°C (lane 5) and at 39.5°C (lane 6). A nuclear extract from cells infected with wild-type virus (WT) (6 hr p.i. at 37°C) was used to catalyze RNA synthesis in the absence of ApG at 33°C (lane 3) and at 39.5°C (lane 4). RNA products were analyzed for NS1 mRNA (M) and NS template RNA (T), as described in the text (Beaton and Krug, 1986). From Shapiro and Krug (1988).

ApG, this extract synthesized template RNA at 33°C (lanes 1 and 5). By contrast, at 39.5°C this extract did not synthesize any detectable template RNA in both the absence and presence of ApG (lanes 2 and 6). No template RNA was observed in lane 6, even with this dark exposure of the gel. The synthesis of mRNA-size RNA was not decreased at 39.5°C (lanes 2 and 6). In fact, in the absence of ApG, an increased amount of mRNA-size RNA synthesis at 39.5°C was seen relative to that at 33°C (lanes 1 and 2). These results indicated that the NP made by the *ts*56 mutant was inactive in antitermination at the nonpermissive temperature. However, this defect in NP did not decrease the activity at 39.5°C in mRNA synthesis catalyzed by nucleocapsids containing the same NP protein. In contrast to the *ts*56 extracts, the nuclear extracts from cells infected with wild-type WSN virus synthesized similar amounts of template RNA at 33°C and 39.5°C in the absence (lanes 3 and 4) and in the presence (data not shown) of ApG. Thus, the NP protein is required for antitermination and hence template RNA synthesis.

These results indicate that there are two populations of NP molecules, one associated with nucleocapsids and one free of nucleocapsids, and that the latter population is required for antitermination during template RNA synthesis. Most likely, the nonnucleocapsid NP proteins act by binding to the nascent template RNAs. It has been found that template RNAs in the infected cell are in the form of nucleocapsids containing NP (Pons, 1971; Krug, 1972; Hay *et al.*, 1977a), and it can be presumed that the template RNAs synthesized *in vitro* also become coated with NP to form nucleocapsids. Although antitermination occurs at a position about 20 nucleotides before the 5' ends of the vRNA templates, it is likely that NP actually first binds to the nascent template RNAs at, or close to, their 5' ends. The only sequence common to the eight influenza virus complementary RNAs is the 12-nucleotide-long sequence at their 5' ends (Skehel and Hay, 1978; Robertson, 1979). In addition, transcripts initiated with a capped primer fragment cannot be antiterminated in the presence of NP (Beaton and Krug, 1986), strongly suggesting that the capped primer sequence preceding the common 5' sequence of the viral transcripts blocks the binding of NP. Binding of the NP protein to the 5' end would thus serve as the site for initiation of nucleocapsid assembly. The ensuing addition of NP molecules would permit readthrough when the termination site is reached.

The mutation in the *ts*56 NP protein has been mapped: a single base change (G to A at position 988 in the nucleotide sequence) resulted in the substitution of an asparagine for a serine at position 314 in the amino acid sequence (M. Krystal and P. Palese, personal communication). As a result of this mutation, the NP protein was inactive in antitermination at the nonpermissive temperature (Shapiro and Krug, 1988), a function that most likely requires binding to specific sequence(s) in nascent plus-strand transcripts. Similarly, because vRNA synthesis in infected cells was inhibited at the nonpermissive temperature (Shapiro and Krug, 1988), the mutant NP protein was presumably unable to bind to sequence(s) in nascent minus-strand transcripts (see Section IV). By contrast, the mutation did not affect mRNA synthesis catalyzed by nucleocapsids containing the mutant NP protein (Shapiro and Krug, 1988). Whether the NP protein that is already associated with nucleocapsids has a specific role in viral mRNA synthesis has not been established. One group has presented evidence that dissociation of the NP protein from virion nucleocapsids resulted in polymerase–RNA complexes that catalyzed very limited elongation of mRNA chains (Kawakami and Ishihama, 1983; Kato et al., 1985). This finding is consistent with the possibility that the NP protein associated with virion nucleocapsids removes secondary structure constraints in the vRNA that inhibit mRNA chain elongation. The mutation in the ts56 NP protein did not affect this or any other role that the nucleocapsid-associated NP protein molecules might have in viral mRNA synthesis. A contributing factor to the lack of an effect of this NP mutation on viral mRNA synthesis might be that the prior binding of NP to vRNA stabilizes the NP protein at the nonpermissive temperature. Another ts mutant in the NP protein, the fowl plague virus ts81 mutant, has the same in vivo phenotype as the WSN ts56 mutant (i.e., a ts defect in vRNA synthesis) and has an amino acid substitution at position 332, close to the substituted position (amino acid 314) in ts56 (Mandler and Scholtissek, 1989). This provides further evidence that this region of the NP protein is involved in binding to nascent template RNA and vRNA chains.

It is not known how the addition of NP molecules to the nascent transcripts would permit readthrough at the termination site. Termination has been presumed to result from "stuttering" or reiterative copying of the tract of the 5–7 uridines that occurs 17–22 nucleotides before the 5' ends of the vRNAs. Unlike the vRNAs of several other negative-strand viruses (Schubert et al., 1980; Gupta and Kingsbury, 1982), the influenza vRNAs do not have a common sequence preceding the U tract that might serve as the signal for poly(A) addition. Rather, because the 5' and 3' termini of the vRNAs have at least partially inverted complementary sequences (Skehel and Hay, 1978; Robertson, 1979; Desselberger et al., 1980), a panhandle was predicted to form at the ends of the vRNAs. Using psoralen as a crosslinking reagent, it was shown that the vRNAs in virion nucleocapsids were in the form of circular structures (Hsu et al., 1987). Nuclease protection experiments were consistent with the presence in the psoralen-crosslinked vRNAs of a terminal panhandle of about 15 nucleotides. This panhandle apparently extended up to the uridine tract and could provide a physical barrier against transcription. Perhaps the addition of NP molecules to the nascent transcripts forces the panhandle open, enabling the protein complex to transcribe the 5' ends of the vRNAs. It has also been proposed that formation of the panhandle by itself could operate as a cis signal for the switch between viral mRNA and template RNA synthesis (Hsu et al., 1987). The evidence cited in support of this proposal was that vRNA was found to be predominantly in the circular (or panhandle) form at early times of infection, whereas at later times a lower percentage of the vRNA was in this form (Hsu *et al.*, 1987). However, the synthesis of both viral mRNA and template RNA occurs at early and not at late times (see Section V), so that there is no correlation between the switch from mRNA to template RNA synthesis and the fraction of the vRNA that could be isolated in circular form. Also, this hypothesis ignores the clearly documented requirement of nonnucleocapsid NP protein molecules for the switch from viral mRNA to template RNA synthesis (Beaton and Krug, 1986; Shapiro and Krug, 1988).

## IV. VIRION RNA SYNTHESIS

The nucleus is also the site of vRNA synthesis (Shapiro *et al.*, 1987). It has recently been shown that nuclear extracts from infected cells also catalyzed vRNA synthesis *in vitro* (Shapiro and Krug, 1988). To assay for vRNA synthesis, the poly (A)⁻ products of *in vitro* synthesis were hybridized to filters containing mRNA-sense M13 DNA specific for either M or NS vRNA. The hybridized RNA was eluted and analyzed by gel electrophoresis. Both M and NS vRNAs were synthesized *in vitro* by nuclear extracts collected at 6, 7, and 8 hr p.i. NS vRNA synthesis by a 7-hr extract is shown in Fig. 9, lane 1 (Shapiro and Krug, 1988). Extracts collected at these three time points displayed similar activities in M and NS vRNA synthesis. By contrast, template RNA-synthesizing activity decreased markedly with extracts collected later than 6 hr p.i. (see Section III). These results reflect the situation *in vivo* in which template RNA synthesis shut down soon after its peak rate of synthesis was achieved, whereas vRNA synthesis continued at essentially maximal rate (see Section V).

In contrast to the situation with the synthesis of transcripts complementary to vRNA, the nucleocapsids present in the nuclear pellet fraction did not synthesize a vRNA-sense RNA of discrete size in the absence of the supernatant fraction (Shapiro and Krug, 1988) (Fig. 9, lane 2). A heterogeneous array of RNAs of small size was made, as detected by analysis of the products on high percentage gels; this array varied between experiments. The supernatant also lacked activity (lane 3). When the nucleocapsids in the pellet were combined with the supernatant, vRNA synthesis was restored (lane 4). Antibody-depletion experiments indicated that NP molecules in the supernatant were required for vRNA synthesis (Shapiro and Krug, 1988). Thus, supernatant that had been incubated with protein A Sepharose containing pooled monoclonal antibodies directed against the viral NP lost its ability to support vRNA synthesis in the presence of the nucleocapsids in the pellet fraction (lane 6). By contrast, activity was retained in supernatants that had been incubated with protein A Sepharose containing normal rabbit antiserum (lane 5), anti-

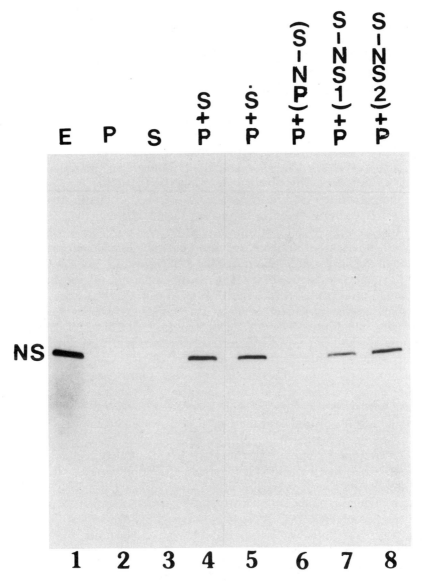

FIGURE 9. Antibody depletion of nonnucleocapsid NP molecules eliminated vRNA synthesis *in vitro*. A 7-hr infected HeLa cell nuclear extract was separated into nucleocapsids and a supernatant fraction. RNA synthesis was catalyzed by the unfractionated extract (E), the nucleocapsids alone (P), the supernatant alone (S), or the supernatant plus the nucleocapsids (S + P). Samples of the supernatant were incubated with protein A Sepharose containing normal rabbit antiserum (Ṡ), pooled NP monoclonal antibodies (S − NP), NS1 antiserum (S − NS1), or NS2 antiserum (S − NS2). Each of these supernatant samples was added to the nuclear nucleocapsids to carry out RNA synthesis *in vitro* (lanes 5, 6, 7, and 8, respectively). The poly(A)⁻ RNA products were analyzed for NS vRNA, as described in the text (Shapiro and Krug, 1988). From Shapiro and Krug (1988).

NS1 antibody (lane 7), or anti-NS2 antibody (lane 8). In the presence of supernatant lacking the NP (lane 6), a vRNA-sense RNA of discrete size was not made, as was the case with the pellet alone (lane 2). Again, shorter RNAs of consistent sizes were not detected by analysis on high-percentage gels. Thus, in the absence of NP, vRNA-sense RNA(s) of discrete size(s) was not made.

These results indicate that elongation of vRNA chains most likely ceased at any point at which NP was not available, resulting in termination at multiple sites. Presumably, these terminations also occurred at sites close to the 5' ends of the vRNA chains; it is conceivable that initiation of vRNA would be effectively blocked in the absence of NP molecules. Indeed, the 13-nucleotide-long sequence found at the 5' ends of the eight influenza vRNAs is a likely candidate for the site at which NP molecules initially bind to nascent vRNA chains.

In apparent contrast, during the synthesis of transcripts complementary to vRNA, discrete RNA species, i.e., mRNA-size transcripts, were made in the absence of nonnucleocapsid NP molecules (Beaton and Krug, 1986; Shapiro and Krug, 1988). These mRNA-sized transcripts terminate about 20 nucleotides before the 5' ends of the vRNA templates. In the presence of NP molecules, some of the transcripts did not terminate at this site, yielding full-length copies of template RNAs. Clearly, NP is required for this antitermination step. However, as already discussed, it is likely that NP actually first binds to the nascent template RNAs at or close to their 5' ends. If it is indeed necessary for NP to bind initially to the 5' ends of transcripts that will become full-length template RNAs, it is likely that, as is the case with nascent vRNA chains, the elongation of template RNAs would cease at any point at which NP was not available, resulting in termination at multiple sites when NP was depleted. A different enzyme system, i.e., a different set of nucleocapsids with their associated P proteins, which is independent of nonnucleocapsid NP molecules, would then be responsible for the synthesis of mRNA-size transcripts, which would consequently not be direct precursors to the encapsidated template RNAs. Because discrete vRNA-sense RNAs were not synthesized in the absence of nonnucleocapsid NP protein (Shapiro and Krug, 1988), an enzyme system independent of nonnucleocapsid NP molecules apparently does not exist for the copying of template RNA into vRNA-sense RNA.

Consequently, at least two types of P-protein complexes can be postulated to exist in infected cells. One type, which is also present in virion nucleocapsids, uses capped primer fragments to initiate mRNA synthesis and therefore requires the participation of the PB2 protein, but apparently not the PA protein, as discussed in a previous section of this review (see Fig. 1). The second type of complex would initiate the synthesis of either template RNA or vRNA chains without a primer and would presumably involve the action of the PA, but not of the PB2, protein. It is not known why elongation by the second type of P-protein complex requires the binding of NP protein molecules to the nascent chains.

The assembly of nucleocapsids containing either template RNAs or vRNAs is most likely coupled with the synthesis of these RNAs. A similar coupling has been observed with another negative-strand RNA virus, VSV (Patton *et al.*, 1983; Peluso and Moyer, 1983, 1984). An important question is whether influenza viral nucleocapsids can be correctly assembled independently of viral RNA synthesis, i.e., by mixing NP protein with full-length vRNA. It has been reported that an NP fusion protein (containing 32 heterologous amino acids) synthesized in bacteria was capable of binding to both viral and nonviral single-stranded RNAs (Kingsbury *et al.*, 1987). These RNAs were generated by SP6 polymerase transcription of plasmids containing viral or nonviral sequences. No specificity of the NP protein for viral sequences was observed. The RNA–protein complexes had some similarities to viral nucleocapsids. In particular, after glutaraldehyde fixation, they had a density of 1.34 g/cm$^3$ in CsCl, the same as that of virion nucleocapsids (Krug, 1972). Although the RNA–protein complexes exhibited some helical morphology in the electron microscope (Kingsbury *et al.*, 1987), the compact tightly coiled structure of virion nucleocapsids (Compans *et al.*, 1972; Murti *et al.*, 1988) was not evident. These experiments were repeated in the laboratory of the present authors, using a nonfusion NP protein expressed with a baculovirus vector (St. Angelo, 1988). This NP protein did form complexes with SP6 RNA polymerase-generated viral RNAs and with authentic vRNAs. After glutaraldehyde fixation, these complexes banded in CsCl at a density of 1.34 g/cm$^3$, indicating that they had the same RNA to protein ratio as virion nucleocapsids. However, the helical morphology of these complexes in the electron microscopy differed from that of authentic virion nucleocapsids. Also, these complexes did not sediment in sucrose gradients with S values of 35–70S, like the S values of authentic viral nucleocapsids. These complexes thus differed in important respects from authentic virion nucleocapsids. Perhaps, as suggested from the *in vitro* studies of template RNA and vRNA, the formation of authentic viral nucleocapsids requires the initial binding of NP protein to the 5' ends of nascent template RNA or vRNA chains.

It should be emphasized that the *in vitro* studies of influenza viral RNA replication have not ruled out the possibility that other nonnucleocapsid proteins in addition to NP participate in template RNA or vRNA synthesis, or both. The results make it unlikely that the NS1 and NS2 protein participate directly in the elongation of template RNA and vRNA chains. Removal of all detectable NS1 or NS2 protein from the nuclear supernatant by immunoaffinity chromatography did not eliminate the synthesis of template RNA or vRNA (Beaton and Krug, 1986; Shapiro and Krug, 1988). It is conceivable, however, that the NS1 and/or NS2 protein might act indirectly in elongation, e.g., by modifying the NP protein so that it becomes active in elongation. In addition, because it is not clear how much unprimed initiation occurred in the *in vitro* extracts that have been employed (Shapiro and Krug, 1988), these experiments did not clearly address the possibility that the NS1 and/or NS2 protein participate in unprimed initiation. Other experiments, however, make it unlikely that either the NS1 or NS2 proteins participate in template RNA synthesis but do not rule out their participation in vRNA synthesis. In cells infected with a *ts* virus mutant with a defect in the NS protein(s), vRNA synthesis, but apparently not template RNA synthesis, was inhibited at the nonpermissive temperature (Wolstenholme *et al.*, 1980). Clearly, further experiments are needed to determine whether virus-specific nonnucleocapsid proteins in addition to NP participate in either template RNA or vRNA synthesis, or both.

## V. REGULATION OF VIRAL GENE EXPRESSION IN INFECTED CELLS

Influenza virus infection is divided into two distinct phases (Shapiro *et al.*, 1987). During the early phase (before 2.5 hr p.i. of BHK-21 cells infected with the WSN strain), the synthesis of specific vRNAs, viral mRNAs, and viral proteins was coupled (Hay *et al.*, 1977a; G. L. Smith and Hay, 1982; Shapiro *et al.*, 1987). The first event detected after primary transcription was the synthesis of template RNAs, presumably copied off the parental vRNAs. Approximately equimolar amounts of each of the template RNAs were made. The peak rate of template RNA synthesis occurred early (1.5 hr p.i. in BHK-21 cells) and then sharply declined. Specific template RNAs were selectively transcribed into vRNAs. Specifically, the NS and NP vRNAs were preferentially synthesized early, whereas the synthesis of M vRNA (encoding the M1 membrane protein and the M2 protein) was delayed. The rate of synthesis of a particular vRNA correlated with, and therefore most likely determined, the rate of synthesis of the corresponding mRNA and of its encoded protein. Thus, the NS1 and NP mRNAs and proteins were preferentially synthesized at early times, whereas the synthesis of the M1 mRNA and protein were delayed. Thus, the control of viral protein synthesis during the early phase is predominantly a direct consequence of the regulation of vRNA synthesis, i.e., the selective copying of specific template RNAs into vRNAs. It will be important to determine the mechanism by which the synthesis of specific vRNAs is turned on during the early phase.

Early reports suggested that the relationships between the syntheses of vRNAs, viral mRNAs and viral proteins that occurred during the first phase of infection continued at later times (Hay *et al.*, 1977a; G. L. Smith and Hay, 1982). However, a recent study showed that these relationships change dramatically during the second phase of infection (Shapiro *et al.*, 1987). This study employed single-stranded M13 DNAs specific for various influenza viral genomic segments to analyze the synthesis of virus-specific RNAs in infected cells. It was shown that the rate of synthesis of all the vRNAs remained at, or near, maximum during the second phase (Fig. 10), whereas the rate of synthesis of all the viral mRNAs decreased

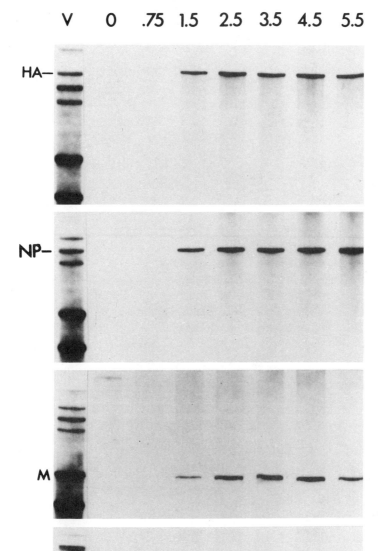

FIGURE 10. Time course of vRNA synthesis in infected cells. Infected BHK-21 cells were labeled with [³H]uridine for 15 min at the times indicated (hours postinfection). The total poly(A) − RNA from the cells in one 100-mm dish was annealed to filters containing the indicated mRNA-sense M13 DNA. Labeled RNAs eluted from the filters were electrophoresed on 5% acrylamide gels. Lane V, pattern of vRNAs (5′ end labeled). From Shapiro *et al.* (1987).

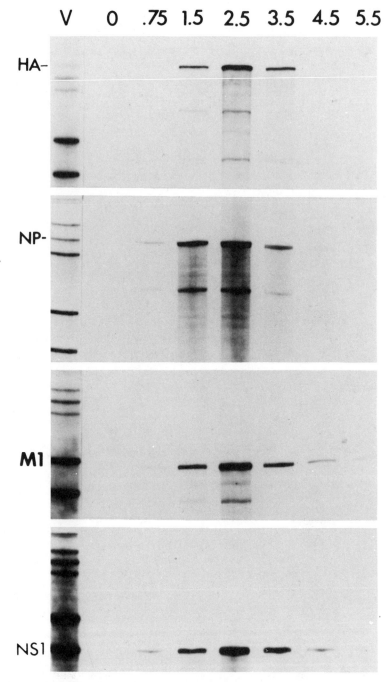

FIGURE 11. Time course of viral mRNA synthesis in infected cells. Infected BHK-21 cells were labeled with [³H]uridine for 15 min at times indicated (hours postinfection). The total poly(A)⁺ RNA from the cells in one 100-mm dish was enzymatically deadenylated with RNase H and annealed to filters containing the indicated vRNA-sense M13 DNAs. The hybridized RNAs were eluted and electrophoresed on 5% denaturing acrylamide gels, which were fluorographed. Lane V, pattern of vRNAs (5′ end labeled). From Shapiro *et al.* (1987).

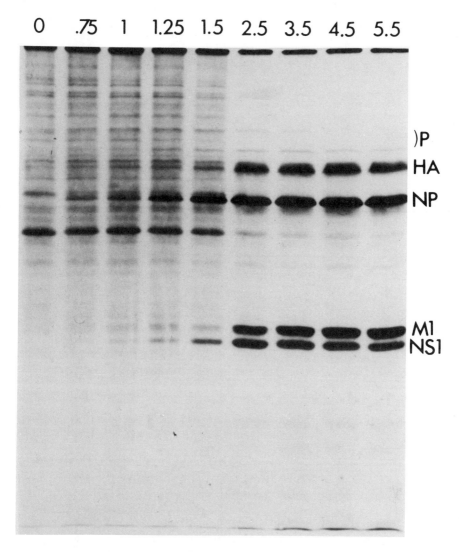

FIGURE 12. Time course of viral protein synthesis in infected cells. At the times indicated (hours postinfection), infected BHK-21 cells were labeled for 15 min with [$^{35}$S]methionone; the labeled proteins were electrophoresed on an SDS–14% polyacrylamide (SDS) gel. From Shapiro *et al.* (1987).

dramatically (Fig. 11). All the viral mRNAs behaved similarly. They had a peak rate of synthesis at the same time, 2.5 hr p.i. in BHK-21 cells, and the subsequent reduction in their rates of synthesis was identical. By 4.5 hr in BHK-21 cells, the rate of synthesis of all the viral mRNAs was 5% the maximum rate. Thus, vRNA and viral mRNA synthesis was not coupled during this second phase. In addition, viral mRNA and protein synthesis was not coupled, as the synthesis of all the viral proteins continued at maximum levels during the second phase (Fig. 12). Previously

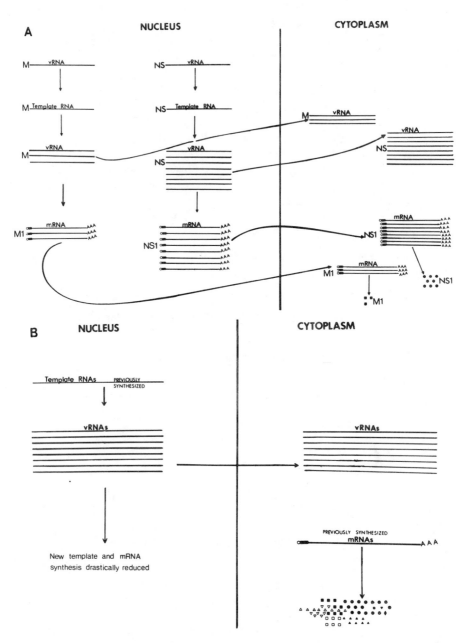

FIGURE 13. Relationships between the synthesis of template RNAs, vRNAs, viral mRNAs, and viral proteins during the early (A) and late (B) phase of infection. The structure (■) at the 5′ ends of the viral mRNAs represents the sequences derived from host-cell-capped primers. From Shapiro *et al.* (1987).

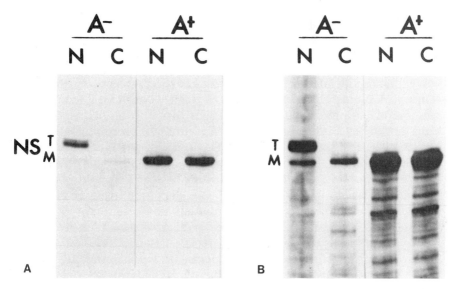

FIGURE 14. Template RNAs are sequestered in the nucleus of infected cells. Infected BHK-21 cells were labeled with [³H]uridine for 30 min (A) or 60 min (B), starting at 1.5 hr p.i. The poly(A)⁻ and poly (A)⁺ RNAs from the nuclei (lanes N) and cytoplasm (lanes C) were analyzed for NS template RNA (T) and NS mRNA (M), respectively. From Shapiro *et al.* (1987).

synthesized viral mRNAs were undoubtedly used to direct viral protein synthesis. Figure 13 diagrams the relationships between the syntheses of template RNAs, vRNAs, viral mRNAs, and viral proteins during the two phases of virus infection (Shapiro *et al.*, 1987).

The synthesis of the three types of virus-specific RNAs has been shown to occur in the nucleus (Herz *et al.*, 1981; Jackson *et al.*, 1982; Shapiro *et al.*, 1987). By using [³H]uridine-labeling periods of 30 and 60 min, it was demonstrated that viral mRNAs and vRNAs were efficiently transported to the cytoplasm (Shapiro *et al.*, 1987). This was true for the vRNAs synthesized during both the early and late phases of infection. By contrast, template RNAs were not transported (Shapiro *et al.*, 1987) (Fig. 14). Thus, the template RNAs, which were synthesized only at early times, remained in the nucleus to direct vRNA synthesis throughout infection. The nuclear–cytoplasmic relationships of vRNAs, template RNAs, and viral mRNAs during both phases of infection are diagrammed in Fig. 13 (Shapiro *et al.*, 1987).

A significant part of the control mechanism of influenza virus-infected cells is directed at the preferential synthesis of the NP and NS1 proteins early and at delaying the synthesis of the M1 protein. The NP and NS1 proteins are synthesized early presumably because they are needed for template RNA and vRNA synthesis, or both. As noted in the two previous sections of this chapter, NP molecules not associated with nucleocapsids have been shown to be required for antitermination during template RNA synthesis and for the elongation of vRNA chains. However, it has not yet been established that the NS1 protein is involved in template RNA or vRNA synthesis, or both. It is conceivable that the synthesis of the M1 protein is delayed because this protein may be involved in the transition between the early and late phases of viral infection, i.e., in stopping the transcription of vRNA into viral mRNA. The membrane (M) protein of another negative-strand RNA virus, VSV, has been implicated in the shutdown of viral RNA transcription (Clinton *et al.*, 1978; Carroll and Wagner, 1979; De *et al.*, 1982; Pal *et al.*, 1985); the influenza viral M1 protein has been shown to inhibit viral RNA transcription *in vitro* (Zvonarjev and Glendon, 1980). Perhaps the influenza virus M1 protein in the infected cell interacts selectively with the nucleocapsids containing vRNAs to inhibit the transcription of vRNA into mRNA but does not interact with the nucleocapsids containing template RNAs, as the transcription of template RNA into vRNA continues. In addition, such a selective association of the M1 protein could be involved in the selective transport of vRNAs, but not of template RNAs, from the nucleus. This hypothesis would predict that some M1 protein would be in the nucleus, which has been observed by several investigators (Hay and Skehel, 1975; Oxford and Schild, 1975; Gregoriades, 1977). Since all four proteins associated with nucleocapsids (NP and the three P proteins) possess nuclear localization signals (Davey *et al.*, 1985a,b; Portela *et al.*, 1985; Jones *et al.*, 1986; G. L. Smith *et al.*, 1987; Akkina *et al.*, 1987), it is reasonable that some process, e.g., an interaction with the M1 protein, would be needed to transport viral nucleocapsids actively out of the nucleus into the cytoplasm. It has been observed that the nuclear transport of some M1 protein that occurs in influenza virus-infected cells does not occur in cells in which the M1 protein is expressed by itself with a vaccinia virus vector (G. L. Smith *et al.*, 1987). This suggested that the transport of M1 to the nucleus might require other influenza virus proteins, consistent with the possibility that the M1 protein might interact with viral nucleocapsids in the nucleus that are in the process of being transported to the cytoplasm.

## VI. INTERFERON-INDUCED Mx PROTEIN, A SPECIFIC INHIBITOR OF INFLUENZA VIRUS REPLICATION

The IFN-induced Mx protein in mouse cells mediates selective resistance to influenza virus (see Section I). The discovery of this gene stemmed from an observation by Lindenmann (1962) that mice of the inbred strain A2G were resistant to influenza virus, whereas other inbred strains died after inoculation of this virus. This resistance was traced to a dominant gene (designated Mx$^+$) present in the A2G strain (Lindenmann, 1964; Haller, 1981). The other mouse strains were designated Mx$^-$. The Mx$^+$ mice were resistant to infection only by influenza virus and not by any other viruses tested (Lindenmann *et al.*, 1963; Lindenmann and Klein, 1966; Haller, 1981). The development of resistance to influenza virus in Mx$^+$ mice required the presence of IFN (Haller *et al.*, 1979, 1980; Haller, 1981), indicating that the Mx$^+$ gene required induction by IFN.

Macrophages, hepatocytes, and embryo fibroblasts obtained from Mx$^+$ and Mx$^-$ mice exhibited the phenotype of the whole animal (Arnheiter *et al.*, 1980; Haller *et al.*, 1980; Haller, 1981; Arnheiter and Staehli, 1983). Only the Mx$^+$ cells developed an efficient antiviral state against influenza virus after exposure to IFN α/β, whereas the antiviral state against other viruses was independent of the Mx gene. IFN γ did not induce the Mx gene (Staehli *et al.*, 1984). The Mx$^+$ gene product was shown to be a 72,000-molecular-weight protein that accumulates in the nucleus (Horisberger *et al.*, 1983; Dreiding *et al.*, 1985). Definitive evidence that this 72,000-molecular-weight protein is the Mx protein was obtained using specific antisera. These antisera were prepared by injecting Mx$^-$ BALB/C mice with extracts from the spleens obtained from congenic Mx$^+$ mice (BALB A2G-*Mx*) treated with IFNα/β (Staehli *et al.*, 1985).

The availability of both the congenic Mx$^+$ mice and the Mx-specific antisera enabled Staehli *et al.* (1986a) to clone the cDNA encoding the Mx protein. Sequence analysis of this cDNA indicated that the Mx protein contains 631 amino acids (with a calculated molecular weight of 72,037). The protein is extremely hydrophilic and contains a large number of charged amino acids: 8.6% lysine, 5.7% arginine, 9.4% glutamic acid, and 6.6% aspartic acid. Many of these charged amino acids are clustered. For example, the segment from positions 76–107 has 17 charged amino acids, and the carboxy-terminus is extremely high in arginines and lysines. There are 11 cysteine residues, indicating that the Mx protein has the potential for forming a large number of intra- and intermolecular disulfide bonds. The cDNA was shown to encode a functional Mx protein. A vector containing the cDNA under the control of the Simian virus 40 (SV40) early promoter was transfected into Mx$^-$ NIH 3T3 cells, and cells expressing the Mx protein were isolated. Cells containing large amounts of the Mx protein in the nucleus were protected against infection by influenza virus, but not by VSV (Staehli *et al.*, 1986a) (Fig. 15). Cells containing either lower amounts of the Mx protein or no Mx protein were not protected against influenza virus. This indicates that the Mx protein does not require other IFN-induced gene products in order to establish resistance to influenza virus. Interestingly, these investigators were unable to establish cell lines in which 100% of the cells expressed the Mx protein. Even after several cycles of cloning, only a certain proportion (about 30%) of the resulting cells expressed the Mx protein, and the level of Mx protein varied among expressing cells. The reasons for this phenomenon are not known.

The murine Mx gene has been mapped to chromosome 16 (Staehli *et al.*, 1986b; Reeves *et al.*, 1988) and has been shown to consist of 14 exons spanning at least 55 kilobases (kb) of DNA (Hug *et al.*, 1988). The promot-

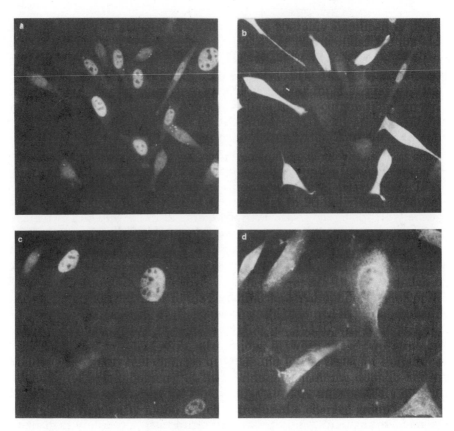

FIGURE 15. Synthesis of viral proteins in Mx-transformed 3T3 cells after infection with influenza virus and VSV. NIH 3T3 cells were transformed with a plasmid containing the Mx cDNA under the control of the SV40 early promoter (plus a plasmid containing neomycin resistance) and G418-resistant cells expressing the Mx protein were selected. These cells were infected with influenza virus (a and b) or VSV (c and d). After 4 hr, the cells were analyzed by double indirect immunofluorescence for synthesis of recombinant Mx protein (a and c) and, simultaneously, for synthesis of influenza virus proteins (b) or VSV G protein (d). The results shown are from the same experiment shown in Fig. 8 of Staehli *et al.* (1986) but represent a different microscopic field of that experiment.

er region is composed of the 5' flanking region 140 base pairs (bp) upstream of the cap site; this promoter is inducible not only by IFN but also by Newcastle disease virus (Hug *et al.*, 1988). Two types of defects have been identified in Mx$^-$ mouse strains: (1) a point mutation in exon 10 that converts a lysine codon to a TAA termination codon; and (2) a large deletion, extending from within intron 8 to within intron 11, that also generates a TAA termination codon 8 triplets downstream from the deletion (Staehli *et al.*, 1988). These defects would be expected to result in a truncated Mx protein lacking about 40% of the carboxy-terminus of the wild-type Mx protein. Such truncated proteins have not been detected, presumably because they are present at very low levels. The Mx-specific mRNAs synthesized in Mx$^-$ cells are apparently unstable and accumulate at a level about 15-fold lower than that of wild-type Mx$^+$ mRNA (Staehli *et al.*, 1988).

Most laboratory-inbred mouse strains are Mx$^-$. Only the A2G and SL/NiA strains have been found to be Mx$^+$ (Staehli and Haller, 1987). The absence of the Mx$^+$ gene from most inbred strains could be attributable to the fortuitous presence of an Mx$^-$ allele in founder animals. One study has examined the frequency of the Mx$^-$ allele in wild mice (Haller *et al.*, 1987). Because the Mx$^+$ allele is dominant, both homozygous (Mx$^+$/Mx$^+$) and heterozygous (Mx$^+$/Mx$^-$) wild mice would be expected to produce the Mx protein and to be resistant to influenza virus. About 75% of the wild mice examined had this phenotype, indicating that 25% of the wild mice were homozygous for the Mx$^-$ allele. The Mx$^+$ allele would be expected to provide selective advantage, i.e., resistance to influenza virus epidemics. Although no influenza virus strain has been isolated from wild mice, the mouse is susceptible to infection by influenza viruses isolated from other species (Staehli and Haller, 1987). It is therefore not clear how homozygous Mx$^-$ mice are able to exist in the wild. Even if the Mx protein has a function(s) in addition to its antiviral activity against influenza virus, homozygous Mx$^-$ mice would still be expected to be at a selective disadvantage.

Homologues to the murine Mx protein have been identified in other species. Proteins that exhibit immunological cross-reactivity with the murine Mx protein have been observed in rat, human, bovine, goat, and hamster cells induced with type I ($\alpha/\beta$) IFN (Staehli and Haller, 1985; Mortier and Haller, 1987; Horisberger and Hochkeppel, 1987; Horisberger, 1988; Meier *et al.*, 1988). In contrast to the murine Mx protein, most of these homologues were found to be localized in the cytoplasm. Only in rat cells has an Mx homologue been identified that is localized in the nucleus (Meier *et al.*, 1988). Three Mx homologues were identified in rat cells. One homologue was localized in the nucleus, whereas the other two were apparently cytoplasmic. As was the case for the murine Mx protein, IFN $\gamma$ did not induce the synthesis of rat, human, and bovine homologues (Staehli and Haller, 1985; Horisberger and Hochkeppel, 1987; Horisberger, 1988; Meier *et al.*, 1988), the effect of IFN $\gamma$ on the induction of the goat and hamster homologues has not been reported. In rat and bovine cells, IFN $\alpha/\beta$, but not IFN $\gamma$, protected the cells against influenza virus (Horisberger, 1988; Meier *et al.*, 1988), so that the induction of the Mx homologue correlated with the establishment of the antiviral state against influenza virus. This was not the case with human cells, in which IFN $\gamma$ protected the cells against influenza virus but did not induce the Mx homologue(s) (Staehli and Haller, 1985). Consequently, in human cells, other IFN-induced products can apparently establish an antiviral state against influenza virus in the absence of the Mx homologue(s), indicating that the human Mx homologue(s) is not necessary for the antiviral state. In fact, only in the rat system has direct evidence been obtained for the requirement of Mx homologue(s) for the antiviral state against influenza virus. When rat cells were injected with a murine Mx monoclonal antibody that recognizes all three rat Mx homologues, the ability of IFN $\alpha/\beta$ to establish an antiviral state against influenza virus was severely inhibited (Arnheiter and Haller, 1988). Because a large fraction of the microinjected antibody was transported to the nucleus, it could be concluded that the antibody reacted with the nuclear rat Mx homologue as well as presumably with the two cytoplasmic Mx homologues. Therefore, at least one of the rat Mx homologues is most likely needed for the antiviral state against influenza virus.

One of the most important issues concerning the Mx protein is the determination of the mechanism by which it selectively inhibits the replication of influenza viruses but not other viruses. Initially, it was reported that in macrophages obtained from Mx$^+$ mice, IFN $\alpha/\beta$ led to the inhibition of the translation of apparently functional viral mRNAs in the cytoplasm but did not affect the synthesis of viral mRNAs in the nucleus (Meyer and Horisberger, 1984). However, this observation has not been confirmed. Rather, it was shown that in Mx$^+$ mouse embryo fibroblasts treated with IFN $\alpha/\beta$, viral mRNA synthesis in the nucleus was severely inhibited (Krug *et al.*, 1985). Steady-state levels of viral mRNAs were measured by Northern analysis using single-stranded M13 probes specific for five viral mRNAs, those coding for the NS1, HA, and three P proteins (Fig. 16). In Mx$^+$ cells, IFN treatment resulted in the absence of detectable amounts of the HA and P mRNAs and in an extremely large reduction in the amount of the NS1 mRNA, both in the absence (compare lanes 2 and 4) and in the presence of the protein synthesis inhibitor anisomycin (compare lanes 1 and 3). In the presence of anisomycin, viral mRNA synthesis is restricted to that catalyzed by the inoculum transcriptase (primary transcription). Similar results were obtained by measuring viral mRNA levels by *in vitro* translation of poly A(+) RNA from infected cells using wheat germ extracts; when viral RNA transcription was restricted to primary transcription, only a greatly reduced amount of NS1 mRNA and an even lower amount of the M1 mRNA was detected (Krug *et al.*, 1985). The absence of most viral mRNAs and the very large reduction in the levels of the NS1 and M1 mRNAs most likely reflected an inhibition of viral mRNA synthesis, because the rate of viral mRNA synthesis catalyzed by the inoculum transcriptase, measured by *in vitro* RNA synthesis catalyzed by permeabilized cells, was severely inhibited (Krug *et al.*, 1985). By contrast, IFN treatment of Mx$^-$ cells had little or no effect on either the steady-state level (Fig. 16, lanes 5–8) or the rate of synthesis of viral mRNAs made by the inoculum transcriptase. No Mx-specific effect acting directly on viral protein synthesis in the cytoplasm was detected. Thus, the NS1 and M1 viral mRNAs that continued to be synthesized in the IFN-treated Mx$^+$ cells were translated *in vivo*; the amount of this translation was similar to the amount of translation observed when the poly A(+)

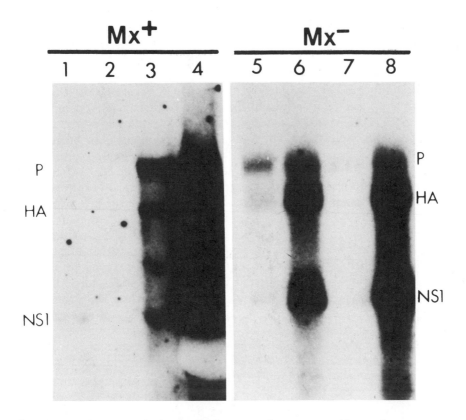

FIGURE 16. Reduction in the levels of influenza viral mRNAs in IFN-treated Mx+ mouse embryo fibroblasts. Mx+ and Mx− cells were treated with IFNα/β for 18 hr at 37°C and infected with influenza virus in the presence (lanes 1 and 5) or absence (lanes 2 and 6) of 100 μM anisomycin. Other Mx+ and Mx− cells were pretreated with medium lacking IFN and were infected with influenza virus in the presence (lanes 3 and 7) or absence (lanes 4 and 8) of 100 μM anisomycin. At 5 hr p.i., total poly(A)+ RNA was isolated, enzymatically dead-enylated, and resolved by electrophoresis on a 1% agarose–formaldehyde gel. The poly(A)+ RNA from 2 × 10⁷ cells was applied to each lane. The RNA was transferred to nitrocellulose filters hybridized to ³²P-labeled single-stranded M13 probes specific for the three P (PB1, PB2, and PA), HA, and NS1 mRNAs. The blot for the Mx+ cell RNAs was exposed to X-ray film about five times longer than the Mx− blot to detect the NS1 mRNA in IFN-treated Mx+ cell. Shorter exposure of the blot from Mx+ cells showed that the pattern of viral mRNA produced in the absence of anisomycin and IFN (lane 4) is the same as that produced in Mx− cells in the absence of anisomycin and IFN (lane 8). Equal amounts of poly(A)+ RNA were present in the four lanes of the Mx+ blot (lanes 1–4), as shown by equivalent levels of hybridization to a nick-translated cDNA clone of β-actin to the RNA in each lane (data not shown). From Krug et al. (1985).

RNA from these cells was assayed in wheat germ extracts (Krug et al., 1985). These results provided strong evidence that the Mx protein inhibited viral mRNA synthesis catalyzed by the inoculum transcriptase. Confirmation of these results came from studies on rat cells, the only other cell system for which evidence has been obtained that the antiviral state induced by IFN $\alpha/\beta$ against influenza virus is mediated by an Mx-related protein. In rat cells, as in Mx$^+$ mouse embryo fibroblasts, IFN $\alpha/\beta$ inhibited the accumulation of viral mRNAs synthesized by the inoculum transcriptase (Meier et al., 1988).

If, as indicated by the above evidence, the Mx protein inhibits influenza virus primary transcription that occurs in the nucleus, it might be expected that the Mx protein would have to be in the nucleus to inhibit influenza virus replication. The nuclear localization signal of the murine Mx protein has been shown to be composed of its 19 carboxy-terminal amino acids (Noteborn et al., 1987). When an Mx cDNA lacking the sequences encoding these carboxy-terminal amino acids was introduced into cells by transient transfection or by microinjection, the expressed truncated Mx protein was predominantly in the cytoplasm as analyzed by immunofluorescence. Relative to that observed with the complete Mx protein, the truncated protein afforded less protection against subsequent infection by influenza virus. Virus replication was assayed by immunofluorescence using a polyclonal antiserum directed against several virus proteins. The truncated Mx protein itself seemed less protective, because reduced protection against influenza virus was also observed in the few cells in which the truncated protein was predominantly in the nucleus. The residual protection observed in most cells in which the truncated Mx protein was predominantly cytoplasmic could be attributable to the cytoplasmic protein. However, because some nuclear fluorescence was present in these cells as well, it could not be ruled out that the residual nuclear protein was responsible for the protection. A better approach may be to determine whether the nonmurine Mx homologues naturally found in the cytoplasm afford protection against influenza virus. For example, in rat cells, it is the nuclear Mx-related protein and/or one or both cytoplasmic Mx-related proteins that afford protection? Current evidence already argues against the cytoplasmic Mx-related protein in human cells being responsible for protection against influenza virus: (1) IFN $\gamma$ did not induce this protein but resulted in protection against influenza virus (Staehli and Haller, 1985); and (2) microinjection of human cells with murine Mx monoclonal antibody that recognizes the human homologue did not block the establishment of the antiviral state by IFN $\alpha/\beta$ (Arnheiter and Haller, 1988).

The most direct way to establish the mechanism by which the murine Mx protein inhibits influenza virus replication will be to determine the activity of the purified Mx protein in in vitro systems that synthesize virus-specific RNAs and proteins. To this end, we (in collaboration with Dr. Jon Condra and Dr. Richard Colonno at Merck, Sharp & Dohme) have independently prepared a cDNA clone encoding the murine Mx protein and have inserted this DNA clone into a baculovirus vector under the control of the polyhedrin promoter. Large amounts of the Mx protein were synthesized and accumulated in the nucleus. We are currently endeavoring to purify the baculovirus-expressed Mx protein. Undoubtedly, other laboratories are taking a similar approach.

## VII. REGULATED SPLICING OF THE VIRAL NS1 AND M1 mRNAs

Like cellular polymerase II transcripts, influenza viral mRNAs are synthesized in the nucleus and contain 5′-terminal methylated cap structures (m⁷GpppNm). In addition, influenza viral mRNAs undergo two nuclear processing steps that polymerase II transcripts undergo: methylation of internal A residues and splicing (Krug et al., 1980; Narayan et al., 1987; Lamb and Lai, 1980, 1982, 1984; Lamb et al., 1981). Influenza virus mRNAs have been used to ascertain important aspects of these two nuclear processing reactions.

The role of the methylation of internal A residues in the metabolism of cellular polymerase II transcripts has not been established. Methylation occurs in the consensus sequences Gm⁶AC and Am⁶AC (Dimock and Stoltzfus, 1977; Schibler et al., 1977; Wei and Moss, 1977; Canaani et al., 1979), and only a specific subset of the available GAC and AAC sequences in several polymerase II transcripts has been shown to be methylated (Beemon and Keith, 1977; Aloni et al., 1979; Canaani et al., 1979; Horowitz et al., 1984; Kane and Beemon, 1985). It has been postulated that internal m⁶A residues might play a role in splicing (Chen-Kiang et al., 1973; Stoltzfus and Dane, 1982; Kane and Beemon, 1985; Zeitlin and Efstratiadis, 1984). However, the results obtained with influenza viral mRNAs have not provided support for this postulate. The entire population of influenza viral mRNA molecules contains an average of 3 m⁶A residues per molecule (Krug et al., 1980), and these m⁶A residues are found in the consensus AAC and GAC sequences (Narayan et al., 1987). The distribution of these m⁶A residues among the individual viral mRNAs is puzzling (Narayan et al., 1987). The HA and NA (neuraminidase) mRNAs contain the most m⁶A residues (8 and 7 residues per molecule, respectively), whereas the M1 and NS1 mRNAs, the two mRNAs that are spliced, as well as two of the P mRNAs contain fewer m⁶A residues (1–3 residues per molecule). Consequently, there is no obvious correlation between m⁶A content and splicing.

Two of the influenza A viral mRNAs are spliced. The NS1 and M1 mRNAs are spliced to form mRNAs coding for two other proteins, NS2 and M2, respectively (Lamb and Lai, 1980, 1982, 1984; Lamb et al., 1981). Except for a short region at their 5′ ends (corresponding to the 5′ exon), the spliced NS2 and M2 mRNAs are translated in the +1 reading frame relative to their unspliced precursor. The M1 mRNA is also spliced to form another mRNA, mRNA$_3$, which has a coding potential for only nine amino acids. The splice junctions are similar to those found in polymerase II transcripts, and most of these splice junctions were used when the NS1 and M1 genes were expressed using DNA vectors (Lamb and Lai, 1982, 1984) (see later in this section). Consequently, it can be concluded that splicing of the NS1 and M1 mRNAs is catalyzed by host-cell nuclear enzymes. Because both the unspliced (NS1 and M1) and spliced (NS2 and M2) mRNAs code for proteins, the extent of splicing is regulated such that some of the unspliced precursor is preserved at the same time that a sufficient amount of the spliced product is produced. In influenza virus-infected cells, this regulation results in a steady-state amount of the spliced mRNAs that is only about 10% of that of the unspliced mRNAs (Lamb *et al.*, 1980, 1981). This type of splicing regulation is not restricted to influenza virus. It also occurs with retroviruses (Varmus and Swanstrom, 1983). Full-length retrovirus RNA synthesized by RNA polymerase II serves as genomic RNA for progeny virus and as mRNA for several proteins. In addition, some of the full-length RNA is spliced to form subgenomic mRNA(s) that encodes other protein(s). Consequently, as is true for influenza virus, the extent of splicing of the full-length retroviral RNA is regulated. It is not known whether this type of splicing regulation occurs in other systems as well. It is conceivable that regulation of the extent of splicing could play a role both in the control of the steady-state levels of some cellular mRNAs and in some types of alternative splicing. Particularly because of their small size (~900–1000 nucleotides long), the influenza virus NS1 and M1 mRNAs are good models for elucidating the mechanism(s) by which the extent of splicing of a pre-mRNA is regulated.

Two types of mechanisms can be proposed for the control of the extent of splicing of the NS1 and M1 mRNAs. A *trans* mechanism would require the action of a virus-specific protein to either decrease or increase the rate of NS1 and M1 mRNA splicing, depending on whether the NS1 and M1 mRNAs were good or poor substrates, respectively, for splicing. Alternatively, in a *cis* mechanism, sequence element(s) in NS1 and M1 mRNA would decrease the rate of splicing. If the decreased rate of splicing were slower than the rate of transport of the unspliced mRNA from the nucleus to the cytoplasm, this would result in a decrease in the extent of splicing.

Most of the recent studies to discriminate between these two mechanisms has focused on the splicing of NS1 mRNA. One approach has been to examine the splicing of NS1 mRNA *in vitro* using HeLa cell nuclear extracts (Plotch and Krug, 1986; Agris *et al.*, 1989). Studies of the *in vitro* splicing of other pre-mRNAs have demonstrated a great deal about the mechanism of splicing. Splicing has been shown to occur in two steps (Padgett *et al.*, 1984; Ruskin *et al.*, 1984). The first step is cleavage at the 5' splice site, to generate the 5' exon and a lariat form of the intron attached to the 3' exon. Subsequently, the 5' and 3' exons are ligated to form the mature mRNA and to release the intron lariat. During the initial phase of the reaction, the pre-mRNA substrate is assembled into ribonucleoprotein complexes, or spliceosomes, containing small nuclear ribonucleoproteins (snRNPs) (Brody and Abelson, 1985; Frendeway and Keller, 1985; Grabowski et al., 1985; Bindereif and Green, 1986; Perkins et al., 1986). The largest complexes found in mammalian systems sediment at about 50–60S is sucrose gradients (Frendeway and Keller, 1985; Grabowski et al., 1985; Bindereif and Green, 1986; Perkins et al., 1986). Analysis of the snRNP composition of 50–60S spliceosomes by affinity selection of biotinylated pre-mRNA on streptavidin–agarose beads indicated that the U1, U2, U4, U5, and U6 snRNPs are present in these spliceosomes (Grabowski and Sharp, 1986; Bindereif and Green, 1987). The same snRNPs have been identified in the 50–60S spliceosomes by other methods (Konarska and Sharp, 1986, 1987; Frendeway et al., 1987; Zillman et al., 1988). The association of the U1 snRNPs with the 50–60S spliceosomes appears to be more labile than that of the other snRNPs (Grabowski and Sharp, 1986; Bindereif and Green, 1987; Zillman et al., 1988). Of the five snRNPs found in spliceosomes, at least four, the U1, U2, U4, and U6 snRNPs, have been shown to be required for splicing (for a recent review, see Maniatis and Reed, 1987). The U1 and U2 snRNPs interact with the 5' splice and intron branch point, respectively. The 50–60S spliceosomes contain the products of the first step in splicing, 5' exon and lariat-3' exon, indicating that these spliceosomes are functional structures (Brody and Abelson, 1985; Frendeway and Keller, 1985; Grabowski et al., 1985; Bindereif and Green, 1986; Perkins et al., 1986).

The 5' and 3' splice junctions of NS1 mRNA fit the consensus sequences closely (Lamb and Lai, 1980). Thus, NS1 mRNA has the sequence CAG/GUAGAG at its 5' splice site, compared with the consensus sequence $\frac{C}{A}$AG/GU$\frac{A}{G}$AGU (the underlined GU is found at all 5' splice sites) and has the sequence $\frac{U}{C}_9$CCAG/G at its 3' splice site, compared with the consensus sequence $\frac{U}{C}_n$N$\frac{C}{U}$AG/G (the underlined AG is found at all 3' splice sites). Nonetheless, using uninfected HeLa cell nuclear extracts, no splicing of NS1 mRNA to form NS2 mRNA was detected in vitro (Plotch and Krug, 1986). A very small amount of lariat formation was observed, and the branch point was mapped to an A residue 20 nucleotides upstream from the 3' splice junction. The sequence around this A residue (AUCUCAC) fits the loose consensus for a mammalian branch-point sequence ($\frac{U}{C}$NCUGAG) (Plotch and Krug, 1986; Reed and Maniatis, 1988). Varying the reaction conditions in many different ways did not lead to detectable splicing (Plotch and Krug, 1986; Agris et al., 1989). Nor did nuclear extracts from influenza virus-infected cells catalyze splicing of NS1 mRNA in vitro, but these extracts also failed to splice other pre-mRNAs efficiently (Plotch and Krug, 1986). This poor splicing was presumably caused by the presence of large amounts of the viral cap-dependent endonuclease, which most likely clipped off the 5' ends of most of the pre-mRNAs added to the extracts. Consequently, it was not clear how to interpret the results obtained with infected cell nuclear extracts.

To determine which features of NS1 mRNA rendered it a poor substrate for splicing, chimeric precursors containing both NS1 and β-globin sequences were tested for their ability to be spliced *in vitro* (Plotch and Krug, 1986). A chimeric precursor that contained the 5' exon and 5' splice site of NS1 mRNA was efficiently spliced, indicating that the NS1 5' splice site is capable of functioning in splicing. By contrast, a chimeric precursor containing NS1 sequences starting from 23 nucleotides downstream from the NS1 5' splice site was not spliced. This indicated that the sequence element(s) of NS1 mRNA that blocked its splicing was in the intron and/or 3' exon of NS1. Although the branch point and 3' splice site of NS1 mRNA fit the consensus for these sequences, it was nonetheless conceivable that these NS1 sequences were defective. Consequently, NS1 mRNA precursors were constructed in which the branch point and/or 3' splice site of NS1 mRNA were replaced by the comparable regions from the β-globin precursor. These replacements did not correct the splicing defect of NS1 mRNA, suggesting that another sequence element(s) causes this defect (Plotch and Krug, 1986; C. H. Agris and R. M. Krug, unpublished experiments).

One possibility was that these sequence element(s) blocked the binding of snRNPs and other splicing factors to the branch point and/or 3' splice site of NS1 mRNA, so that the formation of 50–60S ribonucleoprotein complexes (spliceosomes) would not occur. The virtual absence of splicing intermediates strongly argued for this possibility. Surprisingly, this was not the case. NS1 mRNA very efficiently formed ATP-dependent 55S complexes during *in vitro* splicing reactions (Agris *et al.*, 1989). Under the same conditions, an adenovirus pre-mRNA also formed 55S complexes. The adenovirus 55S complexes were active in producing spliced RNA and as a result dissociated into 20–40S complexes containing the spliced RNA product. In order to compare 55S complex formation with these two pre-mRNAs directly, it was necessary to employ conditions under which the dissociation of the 55S complexes formed with the adenovirus pre-mRNA was minimized, if not eliminated. These conditions were obtained by using partially purified splicing fractions that restricted the splicing of the adenovirus pre-mRNA to the production of splicing intermediates (Furneaux *et al.*, 1985; Perkins *et al.*, 1986). Under these conditions, both the NS1 mRNA and the adenovirus pre-mRNA formed ATP-dependent 55S complexes (Agris *et al.*, 1989) (Fig. 17). The adenovirus 55S complexes contained splicing intermediates, 5' exon and lariat structures (see Fig. 17), whereas the NS1 55S complexes contained only the original unspliced NS1 mRNA.

To determine whether the 55S complexes formed with NS1 mRNA contained the same *sn*RNPs as those formed with the adenovirus premRNA, biotinylated NS1 mRNA and adenovirus pre-mRNA were prepared and used to form 55S complexes (Argis *et al.*, 1989). These complexes were then affinity purified on streptavidin–agarose beads, and the identity of the snRNPs in the affinity-purified 55S complexes was determined (Fig. 18). The affinity-purified 55S complexes formed with biotiny-

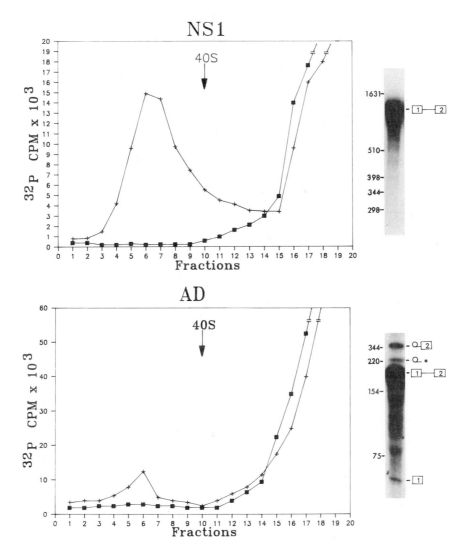

FIGURE 17. Formation of 55S complexes with NS1 mRNA and an adenovirus pre-mRNA using partially purified splicing fractions that restricted the splicing of the adenovirus pre-mRNA to the production of splicing intermediates. NS1 mRNA $(1.5 \times 10^5$ cpm) and the adenovirus pre-mRNA $(6.5 \times 10^5$ cpm) were each incubated for 2 hr at 30°C with the partially purified splicing fractions in the presence $(+)$ and absence $(\blacksquare)$ of ATP and the ATP-generating system. The reaction mixtures were analyzed on sucrose gradients. The position of the 40S ribosomal marker is shown. The RNAs in fractions 6 and 7 of the NS1 mRNA (with ATP) and in fraction 6 of the adenovirus pre-mRNA (with ATP) gradients were isolated and analyzed by gel electrophoresis. The identity of the RNA species is indicated on the right. In addition to the lariat-3' exon, the adenovirus product contained a second lariat species marked with an asterisk (*). This most likely represents the intron-3' exon lariat that has been digested by a 3' exonuclease present in one of the partially purified splicing fractions (Perkins et al., 1986). The 101-nucleotide 5' exon of NS1 was not detected with shorter times of gel electrophresis. From Agris et al. (1989).

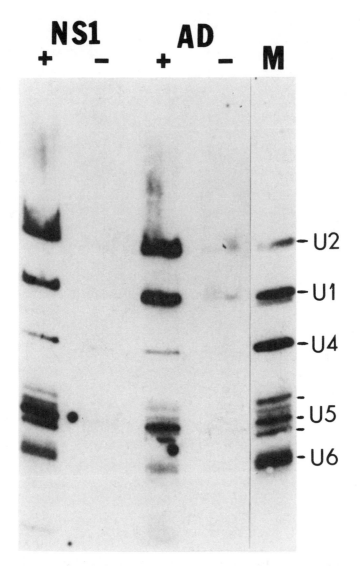

FIGURE 18. The *sn*RNP composition of NS1 and adenovirus (AD) 55S complexes. NS1 mRNA and the adenovirus pre-mRNA were synthesized in the presence (+) or absence (−) of biotin-11-UTP, using [α-$^{35}$S]UTP as the labeled precursor. After incubation of NS1 mRNA (1.2 × 10$^6$ cpm) and of the adenovirus pre-mRNA (2.5 × 10$^6$ cpm) with the partially purified splicing fractions for 2 hr at 30°C, the 55S complexes were isolated by sucrose density-gradient centrifugation (e.g., fractions 6 and 7 from the NS1 sucrose gradient and fractions 5 and 6 from the AD sucrose gradient of Fig. 3) and affinity-purified on strep-tavidin–agarose beads. The RNAs eluted from the beads were electrophoresed on an 8% polyacrylamide gel. The gel was electroblotted onto a nylon membrane, which was then hybridized with $^{32}$P-labeled riboprobes specific for the U1, U2, U4, U5, and U6 *sn*RNAs (50 × 10$_6$ cpm of each riboprobe). After being washed, the membranes were autoradiographed. For the marker (M) lane, an aliquot of one of the partially purified splicing fractions was electrophoresed on the same gel. After electroblotting, this lane was hybridized with a

lated NS1 mRNA contained the U1, U2, U4, U5, and U6 snRNAs (the NS1 + lane); only extremely low amounts of these RNAs were found using the control, nonbiotinylated NS1 mRNA (the NS1 − lane). This indicates that the RNPs containing these snRNAs were specifically associated with NS1 mRNA in the 55S complexes. It was estimated that the molar amounts of U1/U2/U4/U5/U6 in the NS1 55S complexes were approximately 1/1/1/1/1 (see legend of Fig. 18). The same five snRNPs were associated with the adenovirus pre-mRNA (AD + lane). However, the adenovirus 55S complexes contained somewhat less (25–50%) of the U4, U5, and U6 snRNPs than did the NS1 55S complexes. Two studies that estimated snRNP composition of the spliceosomes formed with a β-globin pre-mRNA found approximately equimolar amounts of the U2, U4, U5, and U6 snRNPs, but differed in their estimate of the amount of U1 snRNPs (Bindereif and Green, 1987; Reed et al., 1988).

Consequently, in in vitro splicing reactions, influenza virus NS1 mRNA formed ATP-dependent 55S complexes containing the U1, U2, U4, U5, and U6 snRNPs, yet essentially no catalysis of the first step of splicing, 5′ cleavage and lariat formation, occurred in these complexes (Agris et al., 1989). Several conclusions can be derived from these results. First, although binding of these five snRNPs to a splicing precursor is apparently required for splicing (Grabowski and Sharp, 1986; Konarska and Sharp, 1986, 1987; Bindereif and Green, 1987; Frendeway et al., 1987; Zillman et al., 1988), this binding is not sufficient for the catalysis of splicing. Some additional step(s) must occur subsequent to this binding. The necessity for one or more steps after snRNP binding has already been documented in yeast: with the yeast rna 2 mutant, binding of these snRNPs has been shown to occur in the absence of subsequent catalysis of splicing (Cheng and Abelson, 1987). Recent data indicate that catalysis actually occurs concomitantly with the dissociation of snRNP-containing complexes. It has been reported that production of splicing intermediates may be coupled with the release of U4 snRNPs from splicing complexes (Lamond et al., 1988). One or more steps in addition to U4 snRNP release is also probably required, as release of the U4 snRNPs from splicing complexes has been shown to occur with the yeast rna 2 mutant in the absence of subsequent catalysis (Cheng and Abelson, 1987). The observation that the 55S complexes formed with the adenovirus pre-mRNA

---

mixture of the snRNA riboprobes that yielded approximately equal autoradiographic signals for the five snRNAs. Because this mixture of riboprobes was quite different from that used for the NS1 and AD lanes, the relative intensities of labeling of the different snRNPs in lane M cannot be compared with the relative intensities observed in the NS1 and AD lanes. The relative amounts of the snRNPs in 55S complexes were estimated by comparing the autoradiographic signals of the riboprobes hybridized to the snRNAs and by correcting for the efficiencies of hybridization of the different riboprobes to the snRNAs. The efficiencies of hybridization were determined by comparing the ethidium bromide-staining intensities of the snRNAs present in the partially purified splicing fraction with the autoradiographic signals obtained after hybridization with the labeled riboprobes. From Agris et al. (1989).

contained somewhat less U4, U5, and U6 snRNPs than did the NS1 55S complexes (Agris et al., 1989) might indicate that some of the adenovirus 55S complexes were undergoing dissociation coupled with catalysis of the synthesis of splicing intermediates.

In addition, the ability of NS1 mRNA to form 55S complexes containing the U1, U2, U4, U5, and U6 snRNPs indicates that the 5' splice site, 3' splice site, and branch point of NS1 mRNA, all of which fit the consensus for these sequences, are capable of interacting with the splicing machinery. The U1 snRNP has been shown to bind to the 5' splice site of mammalian pre-mRNAs, almost certainly via base pairing (Black et al., 1985; Zhuang and Weiner, 1986; Maniatis and Reed, 1987). The U2 snRNP binds to branch-point sequences (Black et al., 1985; Maniatis and Reed, 1987); this interaction apparently requires a factor designated U2AF that binds to the 3' splice site (Ruskin et al., 1988). Finally, it is believed that the U5 snRNP binds to the 3' splice site (Chabot et al., 1985). Consequently, it is almost certain that these snRNPs bind to the 5' splice site, 3' splice site, and branch point of NS1 mRNA, but that at least one other sequence element in NS1 mRNA blocks the resolution of the 55S complexes that leads to the catalysis of splicing. On the basis of the results obtained with the NS1-β-globin chimeric precursors (see earlier in this section), these "blocking" sequence elements could be anywhere in the intron and/or 3' exon of NS1 mRNA.

The in vitro splicing results enabled one to restate in more concrete terms the possible cis and trans mechanisms for the control of the extent of splicing of influenza virus NS1 mRNA (Agris et al., 1989). In the trans mechanism, a virus-specific protein(s) would interact with the NS1 sequence element(s) that blocks splicing after 55S complex formation, obviating this block. Some evidence has been presented that influenza virus RNA splicing in infected cells requires the synthesis of virus-specific protein(s) (Inglis and Brown, 1984; D. B. Smith and Inglis, 1985). The question of the role of virus-specific proteins in the splicing of NS1 mRNA can be resolved using DNA vectors. Because splicing of NS1 mRNA occurs during infection with a SV40 recombinant containing the NS1 gene (Lamb and Lai, 1984), the trans model would predict that the NS1 protein, or possibly the NS2 protein, is the virus-specific protein that alleviates the block in NS1 mRNA splicing. This possibility can be tested by introducing amber mutations into the NS1 and NS2 reading frames of an NS1 gene carried in a DNA vector.

In the cis mechanism, the sequence element(s) in NS1 mRNA that blocks splicing in vitro after 55S complex formation itself regulates the extent of splicing in vivo. The blocking in splicing in the in vitro system has been found to be essentially complete, with only very small amounts of lariat formation detected (Plotch and Krug, 1986; Agris et al., 1989). However, it is quite possible that the block in vivo is partial, rather than complete. Nuclear splicing extracts—both the unfractionated extract and the purified fractions obtained from it—have been developed using pre-mRNAs that are readily spliced *in vitro*, i.e., that presumably lack sequences analogous to the putative blocking sequence(s) in NS1 mRNA. As a consequence, these extracts may not contain optimal amounts of the factors required for the catalysis of splicing after 55S complex formation. These factors would not be expected to be limiting *in vivo*, so that *in vivo* the blocking sequence(s) in NS1 mRNA might decrease the rate of splicing, rather than block splicing completely. Coupled with the rapid transport of influenza virus mRNAs from the nucleus to the cytoplasm (Herz *et al.*, 1981; Shapiro *et al.*, 1987), this decreased rate of splicing could lead to the extent of splicing of NS1 mRNA observed *in vivo*. If the *cis* mechanism is operating, the challenge will be to identify the sequence element(s) in NS1 mRNA that regulate its splicing and to determine the mechanism by which this element acts. It is quite possible that the mechanism of regulation of the extent of splicing of influenza virus NS1 mRNA is similar to that of retrovirus mRNAs. Recent experiments indicate that the extent of splicing of full-length avian sarcoma virus is regulated by *cis*-acting sequences (Arrigo and Beemon, 1988; Katz *et al.*, 1988).

As noted previously, the extent of splicing of the M1 mRNA (of influenza A virus) is regulated as well. The splicing of the M1 mRNA is complicated by the use of alternative 5' splice sites. When the 5' splice site at position 11 (the first virus-coded nucleotide at the 5' end of the mRNA is assigned position number 1) is used, mRNA$_3$ is generated. This RNA has only a 9-amino acid open reading frame; it is not likely that this RNA is a functional mRNA. Perhaps it has some other function during infection. The mRNA$_3$ 5' splice site, CAG/GUAGAU, fits the consensus sequence closely. By contrast, the 5' splice site for the M2 mRNA (at position 51), AAC/GUAUGU, does not fit the consensus as well in that there is a C rather than a G at the 3' end of the 5' exon. The M2 mRNA encodes an integral membrane protein of 97 amino acids that is expressed on the surface of infected cells and is also found in virions in lower amounts (Lamb *et al.*, 1985; Zebedee and Lamb, 1988). The common 3' splice site, C$_7$G(A)$_4$(U)$_3$GCAG/G does not contain an immediately adjacent pyrimidine-rich tract. When the M1 gene was inserted into an SV40 virus, two of the three splice sites were used: the common 3' splice site and the mRNA$_3$ 5' splice site, but not the M2 5' splice site (Lamb and Lai, 1982). This raises the question of how the M2 mRNA as well as the mRNA$_3$ 5' splice site is used in influenza virus-infected cells.

Splicing also occurs with the B and C strains of influenza virus. As with influenza A virus, the NS1 mRNA of influenza B virus is spliced to form a smaller NS2 mRNA (Briedis and Lamb, 1982). Both the open reading frame and the intron of the NS1 mRNA of the B strain (B/Lee/40) are larger than that of the NS1 mRNA of A/Udorn/72; the sizes of the NS2 mRNAs of the two virus strains are similar. The 5' splice site, GAG/GUGGGU, of the NS1 mRNA of the B strain fits the consensus sequence closely, whereas the 3' splice site, GAUCGGACAG/U, deviates from the consensus in two ways: (1) absence of a pyrmidine tract immediately upstream of the 3'-terminal AG of the intron; and (2) the presence of a U rather than a G at the 5' end of the 3' exon. The extent of splicing of the B NS1 mRNA is regulated similar to that of the A NS1 mRNA; the steady-state level of the spliced B NS2 mRNA is only about 5–10% of that of the unspliced B NS1 mRNA (Briedis and Lamb, 1982).

In contrast to the documented splicing of the NS1 mRNA of the B virus, it is not clear that the M1 mRNA of the B virus is spliced (Briedis *et al.*, 1982). The M gene of B/Lee/40 and of B/Singapore/222/79 contains two open reading frames (Briedis *et al.*, 1982; Hiebert *et al.*, 1986). The open reading frame at the end of the gene corresponding to the 5' end of the M1 mRNA encodes a protein that has clear homology to the M1 protein of influenza A virus. A second open reading frame, which overlaps that of the M1 protein by 86 amino acids, could code for 195 amino acids in the +2 reading frame. The putative protein, which shows no obvious homology to the M2 protein of influenza A virus, has not yet been identified, nor has a spliced RNA containing the M2 open reading frame been identified. It remains to be determined whether a protein encoded by the second reading frame is synthesized and, if so, whether this synthesis is directed by a spliced RNA.

Evidence has been obtained that a spliced mRNA encodes the M1 (matrix) protein of influenza C virus (Yamashita *et al.*, 1988). The M genomic RNA of the C/JJ/50 virus strain has a single open reading frame that could code for a protein of 374 amino acids (predicted molecular weight of 41,700). However, the M1 protein has a molecular weight of about 27,000 (Compans *et al.*, 1977; Sugawara *et al.*, 1983; Yokota *et al.*, 19823), and M-specific mRNA isolated from infected cells directed the synthesis of a 27,000-molecular-weight protein and not a 41,700-molecular-weight protein (Yamashita *et al.*, 1988). Analysis of the structure of this M1 mRNA indicated that it lacked 228 internal nucleotides encoded in the M genomic RNA (Yamashita *et al.*, 1988), suggesting that it arose via a splicing event. The apparent 5' splice site (AUG/GUUAGU) deviates from the consensus most notably by the presence of a U rather than an A at the 3' penultimate position of the 5' exon, and the apparent 3' splice site (GCUCGGCAG/A) deviates substantially from the consensus sequence. Consequently, although splicing by host-cell nuclear enzymes is a likely mechanism for the generation of this interrupted M1 mRNA, it will be important to verify this by expressing the M segment of influenza virus C in a DNA vector. Influenza C virus differs from both A and B virus in that its M1 protein is encoded not in an unspliced mRNA that is colinear with the genomic RNA but rather in an interrupted mRNA that probably arises by splicing of the colinear mRNA. In addition, the M gene of influenza C virus does not contain a second reading frame. Finally, the extent of production of the interrupted M1 mRNA of influenza C virus is apparently not regulated like the splicing of the NS1 and M1 mRNAs of influenza A virus and of the NS1 mRNA of influenza B virus. In cells infected by influenza C virus, only a small amount of a colinear M transcript was detected; this transcript might actually represent M template RNA (Yamashita *et al.*, 1988).

## VIII. MECHANISMS FOR THE SELECTIVE AND EFFICIENT TRANSLATION OF INFLUENZA VIRUS mRNAs

During influenza virus infection, a dramatic switch from cellular to viral protein synthesis occurred (Lazarowitz *et al.*, 1971; Skehel, 1972; Katze and Krug, 1984) (see Fig. 12; Fig. 19a). This switch resulted both from nuclear and from cytoplasmic processes: (1) newly synthesized cellular polymerase II transcripts were degraded in the nucleus, preventing the appearance of newly synthesized cellular mRNAs in the cytoplasm (Katze and Krug, 1984); and (2) the translation of cytoplasmic cellular mRNAs, which were synthesized predominantly prior to infection, was blocked at both initiation and elongation (Katze *et al.*, 1986b).

To determine the effects of influenza virus infection on the metabolism of newly synthesized polymerase II transcripts, two representative transcripts in chicken embryo fibroblasts (CEFs), those coding for β-actin and for avian leukosis virus (ALV) proteins, were examined (Katze and Krug, 1984). Proviral ALV DNA was integrated into host cell DNA by prior infection with ALV. Within 1 hr after influenza virus infection, the apparent rate of transcription of β-actin and ALV sequences decreased 40–60%, as determined by labeling the cells for 5 min with [³H]uridine and by *in vitro* runoff assays with isolated nuclei. No further decrease in transcription rates occurred at later times of infection. The transcripts that continued to be synthesized did not appear in the cytoplasm as mature mRNAs, and the kinetics of labeling of these transcripts indicated that they were degraded in the nucleus. By S1 endonuclease assay, it was confirmed that nuclear ALV transcripts disappeared very early after infection, already decreasing approximately 80% by 1 hr p.i. A plausible explanation for this nuclear degradation is that it is initiated by the cleavage of the 5′ ends of polymerase II transcripts by the viral cap-dependent endonuclease. The resulting decapped RNAs would likely be more susceptible to degradation by cellular nucleases, as it has been shown that the 5′ cap structure stabilizes RNAs against nucleolytic degradation both *in vivo* and in cell extracts (Banerjee, 1980). In fact, the apparent reduction in transcription rate of the β-actin and ALV transcripts might actually reflect rapid degradation of nascent chains rather than inhibition of transcription per se. A 5-min [³H]uridine pulse in infected cells might be measuring the balance between the rate of transcription of ALV and β-actin sequences and the rate of degradation of nascent transcripts. Similarly, the *in vitro* nuclear runoff assay can be considered to measure the rate of transcription only if nascent polymerase II transcripts are not being degraded. It is therefore likely that the same virus-specific reaction, i.e., endonucleolytic cleavage of polymerase II tran-

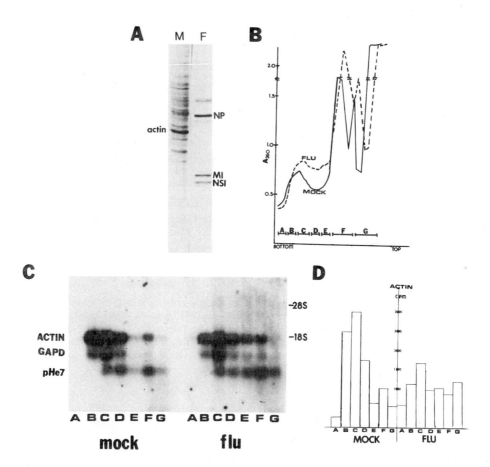

FIGURE 19. Shutoff of cellular protein synthesis and association of host-cell mRNAs with polysomes after influenza virus infection. (a) HeLa cells were either mock infected (M) or infected with influenza virus (F) at 50 plaque-forming units (PFU) per cell. At 4 hr p.i., the cells were labeled with [$^{35}$S]methionine for 30 min. The labeled proteins were analyzed on a 14% gel. The positions of representative influenza virus-specific proteins are shown on the right, and the position of actin is shown on the left. (b) Sedimentation profile of poly-ribosomes from mock-infected and influenza virus-infected (FLU) cells. A cytoplasmic extract from approximately $2 \times 10^8$ cells was subjected to centrifugation on a 10–50% sucrose gradient. Fractions were collected, and the absorbance at 160 nm (A$_{260}$) was determined. The gradient fractions were pooled as shown, yielding samples A–G. (c) Northern blot analysis of cell-equivalent amounts of poly(A)$^+$ RNAs in samples A–G prepared from mock-infected and influenza virus-infected cells (FLU), using as probes $^{32}$P-labeled actin, glyceraldehyde 3-phosphate dehydrogenase (GAPD), and pHe7 DNA clones. The positions of 18S and 28S ribosomal RNA markers are shown on the right. (d) Dot-blot analysis of distribution of actin mRNA sequences. Poly(A)$^+$ RNA from each gradient sample was denatured and dotted onto nitrocellulose hybridized with $^{32}$P-labeled actin DNA. The labeled dots were cut out and counted in a liquid scintillation counter. The total counts of actin mRNA sequences in the mock-infected and influenza virus-infected (FLU) samples were 9606 and 7433 cpm, respectively. From Katze *et al.* (1986b).

scripts, that allowed influenza virus to synthesize its own mRNAs also caused the degradation of new host mRNAs in the nucleus.

The degradation of newly synthesized cellular mRNAs in the nucleus is not sufficient to explain the shutoff of host-cell mRNA translation, because high levels of functional host-cell mRNAs remained in the cytoplasm (Katze and Krug, 1984). Thus, in contrast to the nuclear transcripts, the cytoplasmic ALV β-actin mRNAs in CEFs, measured by Northern blot and S1 nuclease analyses, were stable and were not significantly reduced in amount until some time after 3 hr p.i. The shutoff of the synthesis of host-cell proteins, including actin and the ALV p27 protein, preceded the decrease in the amounts of the encoding mRNAs by about 1 hr. In another cell line, HeLa cells, the shutoff of host-cell protein synthesis was also complete by 3 hr p.i., while the cytoplasmic levels of three representative mRNAs remained at high levels for at least 4.5 hr p.i. (Katze and Krug, 1984). Moreover, these cellular mRNAs isolated from infected cells were efficiently translated *in vitro* in reticulocyte extracts, indicating that influenza virus did not significantly inactivate cytoplasmic host mRNAs. Consequently, the shutoff of host-cell protein synthesis, which was complete by about 3 hr in both CEFs and HeLa cells, did not result from the degradation or modification of cytoplasmic host-cell mRNAs.

Despite the presence of high levels of functional cellular mRNAs in the cytoplasm of infected cells, host-cell protein synthesis was effectively shut off. The step at which the block in the translation of host-cell mRNAs occurred was determined by examining the polysome association of several representative cellular mRNAs in uninfected and infected HeLa cells (Katze *et al.*, 1986b). Average polysome size, i.e., the average number of ribosomes per given mRNA molecule, should be proportional to both the length of the mRNA coding region and the rate at which it initiates translation and should be inversely proportional to the rate at which the encoded protein is elongated (Lodish, 1976; Lodish and Froshauer, 1977). If the block in host-cell mRNA translation was at the initiation step alone, cellular mRNAs would be displaced from polysomes after infection. By contrast, if the block in translation were at the elongation step alone, the polysomes containing host-cell mRNAs would increase in size after infection. A combined initiation–elongation block would give an intermediate result: a significant proportion of the host-cell mRNAs would remain associated with polysomes, which would probably not increase in size and, in fact, some cellular mRNAs might be on smaller polysomes or even displaced from polysomes. The polysomal pattern of three mRNAs were examined, those encoding β-actin, glyceraldehyde 3-phosphate dehydrogenase (GAPD), and the pHe 7 protein (Katze *et al.*, 1986b) (Fig. 19). The polysome pattern was that predicted for a combined initiation–elongation block in translation. Most of these three cellular mRNAs remained polysome associated after virus infection, indicating that the elongation of the proteins encoded by these cellular mRNAs must have been inhibited. For example, in mock-infected cells, 84% of the actin mRNA was in polysomes (predominantly in sucrose gradient samples B, C, and D; Fig. 19c,d). In infected cells, 72% of this mRNA was on polysomes, but the relative amount in the faster sedimenting polysomes in sucrose-gradient samples B, C, and D decreased, with a concomitant increase in the relative amount in the smallest polysomes (sucrose-gradient sample E) and in the material remaining at the top of the sucrose gradient (sample G). Thus, the polysomes containing actin mRNA (and also the two other cellular mRNAs) did not increase in size, but either remained the same size or decreased in size. This indicated that the initiation step as well as the elongation step in cellular protein synthesis was defective. Several control experiments established that the cellular mRNAs sedimenting in the polysome region of sucrose gradients were in fact associated with polysomes (Katze *et al.*, 1986b). Most definitively, puromycin treatment of infected cells caused the dissociation of polysomes and the release of cellular, as well as viral, mRNAs from the polysomes, indicating that the cellular mRNAs were associated with polysomes capable of forming at least a single peptide bond.

The molecular mechanisms by which influenza virus imposes a selective block against the initiation and elongation of cellular, but not viral, proteins have not been determined. It can be argued that the block in the translation of cellular mRNAs is tied in with the block in the appearance of newly synthesized cellular mRNAs in the cytoplasm. Perhaps influenza virus has a mechanism for inhibiting the elongation of protein chains on polysomes encoded by preexisting mRNAs, whereas newly synthesized mRNAs, which are viral mRNAs only, could form polysomes that are not inhibited in elongation. This hypothesis implies that newly synthesized mRNAs have access to a different pool of ribosomes than do old mRNAs. A similar hypothesis has been proposed for the combined initiation–elongation translation block that occurs after cells are stressed by heat shock or amino acid analogues (Thomas and Mathews, 1984). One possibility was that cellular mRNAs in influenza virus-infected cells were on polysomes that were not associated with the cytoskeleton, as it has been proposed that only actively translated mRNAs are on polysomes associated with the cytoskeleton (Lenk *et al.*, 1977; Lenk and Penman, 1979). However, both cellular and viral mRNAs were found to be predominantly on cytoskeleton-associated polysomes in influenza virus-infected cells (Katze *et al.*, 1989).

The selectivity at the initiation step might be easier to understand than the selectivity at the elongation step. Selectivity at initiation might involve, at least in part, competition between the viral and cellular mRNAs for limiting components of the translation machinery, presumably mainly initiation factors (Lodish and Porter, 1980; Brendler *et al.*, 1981; Walden *et al.*, 1981; Jen and Thach, 1982; Kozak, 1986). Viral mRNAs might have a competitive advantage at the level of initiation because they were (1) essentially the only newly synthesized mRNAs in the cytoplasm of infected cells; (2) intrinsically better initiators of translation (discussed below); or (3) present in much larger amounts than the cellular mRNAs. It is unlikely that viral mRNA mass per se contributes greatly to selective translation because the host shutoff preceded maximal accumulation of viral mRNAs (Katze et al., 1984). Furthermore, after infection at the nonpermissive temperature by certain influenza virus ts mutants (i.e., NP mutants) that synthesized only 10–20% of the wild-type levels of the viral mRNAs, host protein synthesis was still inhibited (G. Shapiro and R. M. Krug, unpublished experiments). In addition to these various types of competition, influenza virus might induce changes in initiation factors that also might have a role in the selective initiation by viral mRNAs.

Convincing evidence that influenza viral mRNAs are intrinsically very efficient initiators of translation came from experiments in which adenovirus-infected cells at late times of infection were superinfected with influenza virus (Katze et al., 1984, 1986b). These experiments also demonstrated that influenza virus has a mechanism for maintaining overall protein synthesis in infected cells at a high level (discussed later in this section). Adenovirus at late times of infection has been shown to block the expression of host-cell mRNAs: the transport of newly synthesized host cell mRNAs from the nucleus to the cytoplasm is greatly reduced or essentially eliminated (Beltz and Flint, 1979; Babich et al., 1983; Flint et al., 1983), and the translation of preexisting host cell mRNAs is drastically inhibited (Anderson et al., 1973; Bello and Ginsberg, 1976; Beltz and Flint, 1979; Babich et al., 1983). By contrast, influenza viral mRNAs in doubly infected cells overcame these two blocks and were efficiently transported from the nucleus and translated (Katze et al., 1984). Although a significant proportion of the influenza viral mRNAs possessed at their 5' ends 10 or 11 nucleotides from the 5' ends of the major late adenovirus transcripts, both the transport of influenza viral mRNAs from the nucleus and their association with polysomes were independent of these adenovirus 5' ends. Both influenza viral and adenoviral protein synthesis proceeded at levels comparable to those in singly infected cells (Katze et al., 1984), indicating that a sufficient number of functional ribosomes existed in the doubly infected cells to accommodate most of both viral mRNAs. Evidence was obtained that the influenza viral mRNAs were better initiators of translation than the late adenovirus mRNAs: the influenza virus NP mRNA was on larger polysomes than were several late adenovirus mRNAs with comparably sized coding regions (Katze et al., 1986b). Strong support for this conclusion came from the observation that in cells infected with influenza virus and the adenovirus mutant d133, influenza viral mRNA translation occurred at high levels, whereas adenovirus mRNA translation was severely depressed (Katze et al., 1984, 1986a,b). These doubly infected cells contained a limiting amount of functional eIF-2, a translation-initiation factor. This limitation led to a general, nonspecific reduction in the rate of initiation of translation. The model for protein synthesis described by Lodish (1976) predicted that any nonspecific "reduction in the rate of polypeptide chain initiation steps at or before binding of mRNA will result in preferential inhibition of translation of mRNAs with lower rate constants for polypeptide chain initiation (the poorer mRNAs)." The limitation in functional eIF-2 in cells infected by both influenza virus and the adenovirus dl331 mutant would constitute such a nonspecific reduction in the rate of initiation that resulted in the preferential translation of the better mRNAs (influenza virus mRNAs) at the expense of the poorer mRNAs (adenovirus mRNAs). The influenza viral NP and other influenza viral mRNAs were on polysomes in these doubly infected cells, whereas late adenovirus mRNAs were not (Katze et al., 1986b).

Thus, these results strongly suggested that the influenza viral mRNAs were intrinsically good mRNAs and as a consequence in doubly infected cells could outcompete the late adenovirus mRNAs in translational initiation. It was conceivable, however, that the efficient translation of influenza viral mRNAs in doubly infected cells was mediated by an influenza virus gene product. To discriminate between these two possibilities, the influenza virus NP mRNA was expressed in late adenovirus-infected cells in the absence of other influenza virus gene products (Alonso-Caplen et al., 1988). The NP gene was inserted into the adenovirus genome under the control of the adenovirus major late promoter (MLP); this promoter replaced part of the early (E1A) region of the adenovirus genome. The resulting NP mRNAs contained either none or various portions of the common 5' untranslated region found on most adenovirus mRNAs—the tripartite leader, which is generated by splicing of sequences from three distinct regions of the adenovirus genome (Berget et al., 1977; Zain et al., 1979). Previous results concerning the effect of the tripartite leader on the translation of marker mRNAs (i.e., mRNAs not made at late times of adenovirus infection) synthesized under the control of the MLP gave conflicting results. With three marker mRNAs, adenovirus E1A mRNAs (Logan and Shenk, 1984), hepatitis B surface antigen (HBSAg) mRNA (Davis et al., 1985), and mouse dihydrofolate reductase mRNA (Berkner and Sharp, 1985), the tripartite leader was reported to enhance translation. Specifically, the marker mRNA lacking any tripartite leader sequences or containing only the first segment of the leader at its 5' end was translated with much less efficiency than the same mRNA containing the entire, or almost entire, tripartite leader. By contrast, with the SV40 T-antigen mRNA, it was found that the tripartite leader did not enhance translation (Mansour et al., 1986). Little or no difference in the efficiency of translation of this mRNA was seen whether it contained only a portion or the entirety of the tripartite leader.

Two lines of evidence indicated that the translation of the influenza viral NP mRNA inserted into the adenovirus genome was independent of the presence of tripartite leader sequences. First, the relative amounts of the NP protein synthesized in cells infected by the recombinant adenoviruses were directly proportional to the amounts of the NP mRNA made (Alonso-Caplen et al., 1988), indicating that the presence of 5' tripartite leader sequences did not enhance the translation of NP mRNA. Different amounts of NP mRNA were made because the presence of tripartite leader sequences immediately downstream of the MLP enhanced the transcription of the NP gene, as shown by Northern blot analysis of in vivo NP mRNA levels and by in vitro runoff assays with isolated nuclei. Other studies have also shown that tripartite leader sequences enhance transcription from the MLP (Lewis and Manley, 1985; Mansour et al., 1986). The second line of evidence was that the sizes of the polysomes containing NP mRNA were not increased by the presence of tripartite leader sequences (Alonso-Caplen et al., 1988), indicating that the initiation of translation was not enhanced by these sequences. The sizes of the polysomes containing the NP mRNA with no tripartite leader sequences were similar to those found for the NP mRNA synthesized in cells infected with both influenza virus and wild-type adenovirus. Thus, it can be concluded that the structure of influenza virus NP mRNA per se allowed it to initiate translation efficiently in competition with late adenovirus mRNAs in the absence of other influenza virus gene products. Furthermore, the observations that two marker mRNAs, influenza virus NP mRNA (Alonso-Caplen et al., 1988) and SV40 T-antigen mRNA (Mansour et al., 1986), and at least one adenovirus mRNA, the pIX mRNA (Lawrence and Jackson, 1982), did not require the tripartite leader for efficient translation in adenovirus-infected cells at late times of infection clearly established that a 5' tripartite leader is not a general requirement for efficient translation of mRNAs in these cells. On the basis of these results, it was predicted that those mRNAs that are intrinsically good initiators of translation will not need 5' tripartite leader sequences to be translated efficiently at late times of adenovirus infection (Alonso-Caplen et al., 1988).

As mentioned above, the experiments with cells infected by both adenovirus and influenza virus indicated that influenza virus has a mechanism for maintaining overall protein synthesis in infected cells at a high level (Katze, et al., 1984, 1986a,b). In particular, this was shown by the experiments using the adenovirus mutant dl331, which does not synthesize VA1 RNA, an adenovirus-encoded polymerase III product (Thimmappaya et al., 1982). In cells infected by dl331 alone, both adenovirus and host-cell protein synthesis was severely inhibited (Thimmappaya et al., 1982). The translation defect resulted from a deficiency in functional eIF-2 (Reichel et al., 1985; Siekierka et al., 1985; Schneider et al., 1985). This initiation factor has been shown to form the ternary complex (eIF-2)-GTP-(Met-tRNA$_i$) that binds to the initiating 40S ribosomal subunit before mRNA is bound (Jagus et al., 1981). Inactivation of eIF-2 in dl331-infected cells resulted from the phosphorylation of its α-subunit by a cellular protein kinase (Siekierka et al., 1985; Schneider et al., 1985;

O'Malley et al., 1986; Kitajewski et al., 1986; Katze et al., 1987). This phosphorylation prevented the recycling of eIF-2–GDP to form the functional form of eIF-2, eIF-2–GTP. Recycling has been shown to be catalyzed by the factor eIF-2B, which is trapped in an inactive complex with eIF-2–GDP when the α-subunit of eIF-2 is phosphorylated (Konieczny and Safer, 1983; Panniers and Henshaw, 1983; Safer, 1983). Without this recycling of eIF-2 by eIF-2B, protein synthesis initiation was effectively blocked. The cellular protein kinase that was activated during dl331 infection has been shown to be identical to the double-stranded RNA-activated kinase induced by IFN treatment (referred to as P68 based on its molecular weight of 68,000, but also referred to by several groups as DAI or P1/eIF-2 α kinase) (Lengyel, 1982; O'Malley et al., 1986; Kitajewski et al., 1986; Katze et al., 1987). During wild-type adenovirus infection, VA1 RNA blocks the activation of the protein kinase by forming a complex with P68 (Katze et al., 1987). Most likely, the kinase was activated by adenovirus-specific RNA (presumably double-stranded) (Siekierka et al., 1985; Schneider et al., 1985; O'Malley et al., 1986), so that in the absence of VA1 RNA, as is the case in dl331 infections, the P68 kinase was activated, thereby phosphorylating eIF-2 and shutting down protein synthesis.

Consequently, at 16 hr p.i. with the adenovirus dl331 mutant, the time at which the superinfection with influenza virus was carried out, the cells contained both activated P68 kinase and phosphorylated eIF-2. Nonetheless, after influenza virus superinfection, efficient synthesis of influenza virus proteins occurred (Katze et al., 1984, 1986a,b). Several experimental approaches showed that influenza virus superinfection suppressed the activity of the P68 kinase. First, using crude cytoplasmic extracts, it was found that kinase activity specific for the α-subunit of eIF-2 was suppressed about 5- to 10-fold after influenza virus superinfection (Katze et al., 1986a). Subsequently, the P68 kinase was immunopurified on a Sepharose column containing a monoclonal antibody specific for P68, and the activity of the purified P68 was measured by phosphorylation of exogenously added eIF-2 (Katze et al., 1988). P68 activity in the doubly infected cells was about fourfold lower than that in cells infected by dl331 alone. The reduction in P68 activity resulted in a reduction in the level of phosphorylation of the endogenous eIF-2 in doubly infected cells compared with that in cells infected by dl331 alone. These results indicated that influenza virus most likely encodes a gene product that, analogous to the adenovirus VA1 RNA, inhibited P68 activation, thereby preventing the phosphorylation of eIF-2 and the consequent shutdown of protein synthesis. In the doubly infected cells, the reduction in eIF-2 phosphorylation was not complete (Katze et al., 1986a, 1988). As a result, some of the eIF-2B was sequestered in an inactive complex with phosphorylated eIF-2–GDP, resulting in a limitation in functional eIF-2—GTP complexes and 25–50% inhibition of protein synthesis. Under these conditions, influenza viral mRNAs (the more effec-

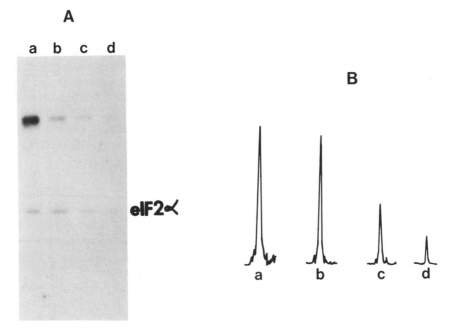

FIGURE 20. Repression of P68 activity in cells infected by influenza virus alone. (A) Analysis of P68 activity. The P68 analyzed for functional activity was immunopurified from extracts prepared from uninfected cells (lane a) and from cells infected with influenza virus for 1 hr (lane b), 2 hr (lane c), or 3 hr (lane d). Purified eIF-2 (0.5 µg) was added to the immunoprecipitate; the mixture was incubated in the presence of [$\gamma$-$^{32}$P]-ATP for 30 min at 30°C. The reaction was stopped by blocking, and the labeled proteins were analyzed by gel electrophoresis. The labeled protein that migrated with lower mobility than eIF-2 represents autophosphorylated P68. (B) Densitometric analysis of the eIF-2 $\alpha$-bands shown in (a). The relative areas under the peaks were as follows: lane a, 3,393,600; lane b, 3,005,600; lane c, 1,267,600; lane d, 400,040. From Katze *et al.* (1986a).

tive initiators of translation) were preferentially translated over adenoviral mRNAs (the less effective initiators of translation) (Katze *et al.*, 1984, 1986a,b).

The putative influenza virus gene product(s) suppressing the P68 kinase would be expected to operate in cells infected by influenza virus alone. This has been directly demonstrated (Katze *et al.*, 1988). P68 kinase was immunopurified from infected (and, as a control, uninfected) cell extracts, and kinase activity was measured with purified eIF-2 as an exogenous substrate (Fig. 20). The activity of P68 was suppressed three-fold at 2 hr p.i. and eightfold at 3 hr p.i. In addition, the level of phosphorylation of the endogenous eIF-2 in the infected cells was reduced. The suppression of P68 activity required viral gene expression, specifically steps after primary transcription. When virus gene expression was restricted to primary transcription by carrying out infection in the presence of anisomycin, the P68 kinase was not suppressed (Katze *et al.*,

1988). In fact, P68 activity was increased over the control (uninfected, anisomycin-treated) level. This suggested that influenza virus mRNAs (and probably other virus-specific RNAs), presumably by forming double-stranded RNA structures, activated the P68 kinase, and that the virus had to encode a P68 suppressor in order to counteract this activation of P68 by virus-specific RNAs. An *in vitro* assay has been developed that detected a P68-suppressing activity in infected, but not in uninfected cell extracts (Katze *et al.*, 1988). It is hoped that the use of this assay will lead to the purification and identification of the influenza virus gene product(s) suppressing the P68 kinase.

ACKNOWLEDGMENTS. We are grateful to Dr. Martin Nemeroff and Barbara Broni for their comments on this review and to Eveyon Maldonado for her careful typing of this manuscript. We have attempted to quote all papers relevant to this review and apologize to any authors whose papers we may have inadvertently omitted. Research in the authors' laboratories were supported by U.S. Public Health Service grants AI-11772, AI-22646, and CA-08747 from the National Institutes of Health and by grant MV-260 from the American Cancer Society. F. V. Alonso-Caplen was supported by fellowships from Exxon Corporation and from the American Heart Association, New York City Affiliate, and I. Julkunen was supported by a Fogarty International Fellowship F05-TW03/36 and by a Clinical Scholars Fellowship provided by Memorial Sloan-Kettering Cancer Center.

# REFERENCES

Agris, C. H., Nemeroff, M. E., and Krug, R. M., 1989, A block in mammalian splicing occurring after formation of large complexes containing U1, U2, U4, U5 and U6 small nuclear ribonucleoproteins, *Mol. Cell. Biol.* **9:**259–267.

Akkina, R. K., Chambers, T. M., Londo, D. R., and Nayak, D. P., 1987, Intracellular localization of the viral polymerase proteins in cells infected with influenza virus and cells expressing PB$_1$ protein from cloned cDNA, *J. Virol.* **61:**2217–2224.

Aloni, Y., Dhar, R., and Khoury, G., 1979, Methylation of nuclear simian virus 40 RNAs, *J. Virol.* **32:**52–60.

Alonso-Caplen, F. V., Katze, M. G., and Krug, R. M., 1988, Efficient transcription, not translation, is dependent on adenovirus tripartite leader sequences at late times of infection, *J. Virol.* **62:**1606–1616.

Anderson, C. W., Baum, P. R., and Gesteland, R. F., 1973, Processing of adenovirus 2-induced proteins, *J. Virol.* **12:**241–252.

Arnheiter, H., and Haller, O., 1988, Antiviral state against influenza virus neutralized by microinjection of antibodies to interferon induced Mx proteins, *EMBO J.* **5:**1315–1320.

Arnheiter, H., and Staehli, P., 1983, Expression of interferon dependent resistance to influenza virus in mouse embryo cells, *Arch. Virol.* **76:**127–137.

Arnheiter, H., Haller, O., and Lindenmann, J., 1980, Host gene influence on interferon action in adult mouse hepatocytes: Specificity for influenza virus, *Virology* **103:**11–20.

Arrigo, S., and Beemon, K., 1988, Regulation of Rous sarcoma virus RNA splicing and stability, *Mol. Cell. Biol.* **8:**4858–4867.

Babich, A., Feldman, L., Nevins, J., Darnell, J., and Weinberger, C., 1983, Effects of adenovirus on metabolism of specific host mRNAs: Transport control and specific translation discrimination, *Mol. Cell. Biol.* **3**:1212–1221.

Banerjee, A. K., 1980, 5'-terminal cap structure in eukaryotic messenger ribonucleic acids, *Microbiol. Rev.* **44**:175–205.

Barrett, T., Wolstenholme, A. J., and Mahy B. W. J., 1979, Transcription and replication of influenza virus RNA, *Virology* **98**:211–225.

Beaton, A. R., and Krug, R. M., 1984, Synthesis of the templates for influenza virion RNA replication *in vitro, Proc. Natl. Acad. Sci. USA* **81**:4682–4686.

Beaton, A. R., and Krug, R. M., 1986, Transcription antitermination during influenza viral template RNA synthesis requires the nucleocapsid protein and the absence of a 5' capped end, *Proc. Natl. Acad. Sci. USA* **83**:6282–6286.

Beemon, K., and Keith, J., 1977, Localization of $N^6$-methyladenosine in the Rous sarcoma virus genome, *J. Mol. Biol.* **113**:165–179.

Bello, J., and Ginsberg, H. S., 1976, Inhibition of host protein synthesis in type 5 adenovirus-infected cells, *J. Virol.* **1**:843–850.

Beltz, G. A., and Flint, S. J., 1979, Inhibition of HeLa cell protein synthesis in type 5 adenovirus infection, *J. Mol. Biol.* **131**:353–373.

Berget, S. M., Moore, C., and Sharp, P. A., 1977, Spliced segments at the 5' terminus of adenovirus 2 late mRNA, *Proc. Natl. Acad. Sci. USA* **74**:3171–3175.

Berkner, K. L., and Sharp, P. A., 1985, Effect of the tripartite leader on synthesis of a nonviral protein in an adenovirus 5 recombinant, *Nucl. Acids Res.* **13**:841–857.

Bindereif, A., and Green, M. R., 1986, Ribonucleoprotein complex formation during pre-mRNA splicing *in vitro, Mol. Cell. Biol.* **6**:2582–2592.

Bindereif, A., and Green, M. R., 1987, An ordered pathway of snRNP binding during mammalian pre-mRNA splicing complex assembly, *EMBO J.* **6**:2415–2424.

Blaas, D., Patzelt, E., and Keuchler, E., 1982, Identification of the cap binding protein of influenza virus, *Nucl. Acids Res.* **10**:4803–4812.

Black, L. D., Chabot, B., and Steitz, J. A., 1985, U2 as well as U1 small nuclear ribonucleoproteins are involved in pre-messenger RNA splicing, *Cell* **42**:737–750.

Bouloy, M., Plotch, S. J., and Krug, R. M., 1978, Globin mRNAs are primers for the transcription of influenza viral RNA *in vitro, Proc. Natl. Acad. Sci. USA* **75**:4886–4890.

Bouloy, M., Morgan, M. A., Shatkin, A. J., and Krug, R. M., 1979, Cap and internal nucleotides of reovirus mRNA primers are incorporated into influenza viral complementary RNA during transcription *in vitro, J. Virol.* **32**:895–904.

Bouloy, M., Plotch, S. J., and Krug, R. M., 1980, Both the 7-methyl and 2'-O-methyl groups in the cap of an mRNA strongly influence its ability to act as a primer for influenza viral RNA transcription, *Proc. Natl. Acad. Sci. USA* **77**:3952–3956.

Braam, J., Ulmanen, I., and Krug, R. M., 1983, Molecular model of a eukaryotic transcription complex: Functions and movements of influenza P proteins during capped RNA-primed transcription, *Cell* **34**:609–618.

Brendler, T., Godefroy-Colburn, J., Yu, S., and Thach, R. E., 1981, The role of mRNA competition in regulating translation, *J. Biol. Chem.* **256**:11755–11761.

Briedis, D. J., and Lamb, R. A., 1982, Influenza B virus genome: sequences and structural organization of RNA segment 8 and the mRNAs coding for the $NS_1$ and $NS_2$ proteins, *J. Virol.* **42**:186–193.

Briedis, D. J., Lamb, R. A., and Choppin, P. W., 1982, Sequence of RNA segment 7 of the influenza B virus genome: Partial amino acid homology between the membrane proteins ($M_1$) of influenza A and B viruses and conservation of a second open reading frame, *Virology* **116**:581–588.

Brody, E., and Abelson, J., 1985, The spliceosome: Yeast pre-messenger RNA associates with a 40S complex in a splicing dependent reaction, *Science* **228**:963–967.

Canaani, D., Kahana, C., Lavi, C. S., and Groner, Y., 1979, Identification and mapping of $N^6$-methyladenosine containing sequences in Simian Virus 40 RNA, *Nucl. Acids Res.* **6**: 2879–2899.

Carroll, A. R., and Wagner, R. R., 1979, Role of the membrane (M) protein in endogenous inhibition of *in vitro* transcription by vesicular stomatitis virus, *J. Virol.* **29:**134–142.

Chabot, B., Black, L. D., LeMaster, D. M., and Steitz, J. A., 1985, The 3′ splice site of pre-messenger RNA is recognized by a small nuclear ribonucleoprotein, *Science* **230:**1344–1349.

Chen-Kiang, S., Nevins, J. R., and Darnell, J. E., 1973, *N*-6-methyladenosine in adenovirus type 2 nuclear RNA is conserved in the formation of messenger RNA, *J. Mol. Biol.* **135:**733–752.

Cheng, S.-C., and Abelson, J., 1987, Spliceosome assembly in yeast, *Genes Dev.* **1:**1014–1027.

Clinton, G. M., Little, S. P., Hagen, F. S., and Huang, A. S., 1978, The matrix (M) protein of vesicular stomatitis protein regulates transcription, *Cell* **15:**1455–1462.

Compans, R. W., Content, J., and Duesburg, P. H., 1972, Structure of the ribonucleoprotein of influenza virus, *J. Virol.* **10:**795–800.

Compans, R. W., Bishop, D. H., and Meier-Ewert, H., 1977, Structural components of influenza C virions, *J. Virol.* **21:**658–665.

Davey, J., Colman, A., and Dimmock, N. J., 1985a, Location of influenza virus M, NP and NS1 proteins in microinjected cells, *J. Gen. Virol.* **66:**2319–2334.

Davey, J., Dimmock, N. J., and Colman, A., 1985b, Identification of the sequence responsible for nuclear accumulation of the influenza virus nucleoprotein in *Xenopus* oocytes, *Cell* **40:**657–667.

Davis, A. R., Kostek, B., Mason, B. B., Hsiao, C. L., Morin, J., Dheer, S. K., and Hung, P. P., 1985, Expression of hepatitis B surface antigen with a recombinant adenovirus, *Proc. Natl. Acad. Sci. USA* **82:**7560–7564.

De, B. P., Thornton, G. B., Luk, D., and Banerjee, A. K., 1982, Purified matrix protein in vesicular stomatitis virus blocks viral transcription *in vitro, Proc. Natl. Acad. Sci. USA* **79:**7137–7141.

Desselberger, U., Racaniello, V. R., Zazra, J. J., and Palese, P., 1980, The 3′ and 5′ end terminal sequences of influenza A, B, and C virus RNA segments are highly conserved and show partial inverted complementarity, *Gene* **8:**315–328.

Detjen, B. M., St. Angelo, C., Katze, M. G., and Krug, R. M., 1987, The three influenza virus polymerase (P) proteins not associated with viral nucleocapsids in the infected cell are in the form of a complex, *J. Virol.* **61:**16–22.

Dimock, K., and Stoltzfus, C. M., 1977, Sequence specificity of internal methylation in B77 avian sarcoma virus RNA subunits, *Biochemistry* **16:**471–478.

Dreiding, P., Staehli, P., and Haller, O., 1985, Interferon-induced protein Mx accumulates in nuclei of mouse cells expressing resistance to influenza viruses, *Virology* **140:**192–196.

Emerson, S. U., and Yu, Y.-H., 1975, Both NS and L proteins are required for *in vitro* RNA synthesis by vesicular stomatitis virus, *J. Virol.* **15:**1348–1356.

Flint, S. J., Beltz, G. A., and Linzer, D. I. H., 1983, Synthesis and processing of simian virus 40-specific RNA in adenovirus-infected simian virus 40-transformed human cells, *J. Mol. Biol.* **167:**335–359.

Frendeway, D., and Keller, W., 1985, Stepwise assembly of a pre-mRNA splicing complex requires U-snRNPs and specific intron sequences, *Cell* **42:**355–367.

Frendeway, D., Kramer, A., and Keller, W., 1987, Different small nuclear ribonucleoprotein particles are involved in different steps of splicing complex formation, *Cold Spring Harbor Symp. Quant. Biol.* **52:**287–298.

Furneaux, H. M., Perkins, K. K., Freyer, G. A., Arenas, J., and Hurwitz, J., 1985, Isolation and characterization of two fractions from HeLa cells required for mRNA splicing *in vitro, Proc. Natl. Acad. Sci. USA* **82:**4351–4355.

Grabowski, P. J., and Sharp, P. A., 1986, Affinity chromatography of splicing complexes: U2, U5, and U4 + U6 small nuclear ribonucleoprotein particles in the spliceosome, *Science* **233:**1294–1299.

Grabowski, P. J., Seller, S. R., and Sharp, P. A., 1985, A multicomponent complex is involved in the splicing of messenger RNA precursors, *Cell* **42:**345–353.

Greenspan, D., Krystal, M., Nakada, S., Arnheiter, H., Lyles, D. S., and Palese, P., 1985, Expression of influenza virus NS2 nonstructural protein in bacteria and localization of NS2 in infected eucaryotic cells, *J. Virol.* **54:**833–843.

Greenspan, D., Palese, P., and Krystal, M., 1988, Two nuclear location signals in the influenza virus $NS_1$ nonstructural protein, *J. Virol.* **62:**3020–3026.

Gregoriades, A., 1977, Influenza virus-induced proteins in nuclei and cytoplasm of infected cells, *Virology* **79:**449–454.

Gupta, K. C., and Kingsbury, D. W., 1982, Conserved polyadenylation signals in two negative-strand RNA virus families, *Virology* **12:**951–961.

Haller, O., 1981, Inborn resistance of mice to orthomyxoviruses, *Curr. Top. Microbiol. Immunol.* **92:**25–52.

Haller, O., Arnheiter, H., Gresser, I., and Lindenmann, J., 1979, Genetically determined, interferon-dependent resistance to influenza virus in mice, *J. Exp. Med.* **149:**601–612.

Haller, O., Arnheiter, H., Lindenmann, J., and Gresser, I., 1980, Host gene influences sensitivity to interferon action selectively for influenza virus, *Nature (Lond.)***283:**660–662.

Haller, O., Arnheiter, H., Gresser, I., and Lindenmann, J., 1981, Virus-specific interferon action: Protection of newborn Mx carriers against lethal-infection with influenza virus, *J. Exp. Med.* **154:**199–203.

Haller, O., Acklin, M., and Staehli, P., 1987, Influenza virus resistance of wild mice: Wild-type and mutant Mx alleles occur at comparable frequencies, *J. Interferon Res.* **7:**647–656.

Hay, A. J., and Skehel, J. J., 1975, Studies on the synthesis of influenza virus proteins, in: *Negative Strand Viruses*, Vol. 2 (B. W. J. Mahy and R. D. Barry, eds.), pp. 635–655, Academic, London.

Hay, A. J., Lomnicizi, B., Bellamy, A. R., and Skehel, J. J., 1977a, Transcription of the influenza virus genome, *Virology* **83:**337–355.

Hay, A. J., Abraham, G., Skehel, J. J., Smith, J. C., and Fellner, P., 1977b, Influenza virus messenger RNAs are incomplete transcripts of the genome RNA, *Nucl. Acids Res.* **4:**4197–4209.

Hay, A. J., Skehel, J. J., and McCauley, J., 1982, Characterization of influenza virus RNA complete transcripts, *Virology* **116:**517–522.

Herz, C., Stavnezer, E., Krug, R. M., and Gurney, T., Jr., 1981, Influenza virus, an RNA virus, synthesizes its messenger RNA in the nucleus of infected cells, *Cell* **26:**391–400.

Hiebert, S. W., Williams, M. A., and Lamb, R. A., 1986, Nucleotide sequence of RNA segment 7 of influenza B/Singapore/222/79: Maintenance of a second large open reading frame, *Virology* **155:**747–751.

Horisberger, M. A., 1988, The action of recombinant bovine interferons on influenza virus replication correlates with the induction of two Mx-related proteins in bovine cells, *Virology* **162:**181–186.

Horisberger, M. A., Haller, O., and Arnheiter, H., 1980, Interferon-dependent genetic resistance to influenza virus in mice: Virus replication in macrophages is inhibited at an early step, *J. Gen. Virol.* **50:**205–210.

Horisberger, M. A., Staehli, P., and Haller, O., 1983, Interferon induces a unique protein in mouse cells bearing a gene for resistance to influenza virus, *Proc. Natl. Acad. Sci. USA* **80:**1910–1914.

Horisberger, M. A., and Hochkeppel, H. K., 1987, IFN-α induced human 78 kD protein: Purification and homologies with the mouse Mx protein, production of monoclonal antibodies and potentiation effect of IFN-γ, *J. Interferon Res.* **7:**331–343.

Horowitz, S., Horowitz, A., Nilsen, T. W., Munns, T. W., and Rottman, F. M., 1984, Mapping of $N^6$-methyladenosine residues in bovine prolactin mRNA, *Proc. Natl. Acad. Sci. USA* **81:**5667–5671.

Hsu, M.-T., Parvin, J. D., Gupta, S., Krystal, M., and Palese, P., 1987, Genomic RNAs of influenza viruses are held in a circular conformation in virions and in infected cells by a terminal panhandle, *Proc. Natl. Acad. Sci. USA* **84:**8140–8144.

Hug, H., Costas, M., Staehli, P., Aebi, M., and Weissmann, C., 1988, Organization of the murine Mx gene and characterization of its interferon- and virus-inducible promoter, *Mol. Cell. Biol.* **8:**3065–3079.

Inglis, S. C., and Brown, C. M., 1981, Spliced and unspliced RNAs encoded by virion RNA segment 7 of influenza virus, *Nucl. Acids Res.* **9:**2727–2740.

Inglis, S. C., and Brown, C. M., 1984, Differences in the control of virus mRNA splicing during permissive or abortive infection with influenza A (Fowl Plague) virus, *J. Gen. Virol.* **65:**153–164.

Inglis, S. C., Carroll, A. R., Lamb, R. A., and Mahy, B. W. J., 1976, Polypeptides specified by the influenza virus genome. I. Evidence for eight distinct gene products specified by fowl plague virus, *Virology* **74:**489–503.

Jackson, D. A., Caton, A. J., McCready, S. J., and Cook, P. R., 1982, Influenza virus RNA is synthesized at a fixed site in the nucleus, *Nature (Lond.)* **296:**366–368.

Jagus, R., Anderson, W. F., and Safer, B., 1981, The regulation of initiation of mammalian protein synthesis, *Prog. Nucl. Acid Res.* **25:**127–185.

Jen, G., and Thach, R. E., 1982, Inhibition of host translation in encephalomyocarditis virus infected L cells: A novel mechanism, *J. Virol.* **43:**250–261.

Jones, I. M., Reay, P. A., and Philpott, K. L., 1986, Nuclear location of all three influenza polymerase proteins and a nuclear signal in polymerase PB2, *EMBO J.* **5:**2371–2376.

Kane, S. E., and Beemon, K., 1985, Precise localization of m6A in Rous sarcoma virus RNA reveals clustering of methylation sites: Implications for RNA processing, *Mol. Cell. Biol.* **5:**2298–2306.

Kato, A., Mixumoto, K., and Ishihama, A., 1985, Purification and enzymatic properties of an RNA-polymerase–RNA complex from influenza virus, *Virus Res.* **3:**115–127.

Katz, R. A., Kotler, M., and Skalka, A. M., 1988, *cis*-Acting intron mutations that affect the efficiency of avian retroviral RNA splicing: Implication for mechanisms of control, *J. Virol.* **62:**2686–2695.

Katze, M. G., and Krug, R. M., 1984, Metabolism and expression of RNA polymerase II transcripts in influenza virus-infected cells, *Mol. Cell. Biol.* **4:**2198–2206.

Katze, M. G., Chen, Y. T., and Krug, R. M., 1984, Nuclear-cytoplasmic transport and VAI RNA independent translation of influenza viral messenger RNAs in late adenovirus-infected cells, *Cell* **37:**483–490.

Katze, M. G., Detjen, B. M., Safer, B., and Krug, R. M., 1986a, Translational control of influenza virus: Suppression of the kinase that phosphorylates the alpha subunit of initiation factor eIF-2 and selective translation of influenza viral mRNAs, *Mol. Cell. Biol.* **6:**1741–1750.

Katze, M. G., DeCorato, D., and Krug, R. M., 1986b, Cellular mRNA translation is blocked at both initiation and elongation after infection by influenza virus or adenovirus, *J. Virol.* **60:**1027–1039.

Katze, M. G., DeCorato, D., Safer, B., Galabru, J., and Hovanessian, A. G., 1987, The adenovirus VA1 RNA complexes with the P68 protein kinase to regulate its autophosphorylation and activity, *EMBO J.* **6:**689–697.

Katze, M. G., Tomita, J., Black, T., Krug, R. M., Safer, B., and Hovanessian, A., 1988, Influenza virus regulates protein synthesis during infection by repressing the autophosphorylation and activity of the cellular 68,000 $M_r$ protein kinase, *J. Virol.* **62:**3710–3717.

Katze, M. G., Lara, J., and Wamback, M., 1989, Nontranslated cellular mRNAs are associated with the cytoskeletal framework in influenza virus or adenovirus infected cells, *Virology* (in press).

Kawakami, K., and Ishihama, A., 1983, RNA polymerase of influenza virus. III. Isolation of RNA polymerase–RNA complexes from influenza virus PR8, *J. Biochem.* **93:**989–996.

Kingsbury, D. W., Jones, I. M., and Murti, K. G., 1987, Assembly of influenza ribonucleoprotein *in vitro* using recombinant nucleoprotein, *Virology* **156:**396–403.

Kitajewski, J., Schneider, R. J., Safer, B., Muenmitsu, M., Samuel, C. E., Thimmappaya, B., and Shenk, T., 1986, Adenovirus VA1 RNA antagonizes the antiviral action of interferon by preventing activation of the interferon-induced eIF-2 kinase, *Cell* **45:**195–200.

Konarska, M. M., and Sharp, P. A., 1986, Electrophoretic separation of complexes involved in the splicing of precursors of mRNAs, *Cell* **46:**845–855.

Konarska, M. M., and Sharp, P. A., 1987, Interactions between small nuclear ribonucleoprotein particles in the formation of spliceosomes, *Cell* **49:**763–774.

Konieczny, A., and Safer, B., 1983, Purification of the eukaryotic initiation factor 2-eukaryotic initiation factor 2B complex and characterization of its guanine nucleotide exchange activity during protein synthesis, *J. Biol. Chem.* **256:**3402–3408.

Kozak, M., 1986, Regulation of protein synthesis in virus-infected animal cells, *Adv. Virus Res.* **31:**229–292.

Krug, R. M., 1972, Cytoplasmic and nucleoplasmic viral RNPs in influenza virus-infected MDCK cells, *Virology* **50:**103–136.

Krug, R. M., 1981, Priming of influenza viral RNA transcription by capped heterologous RNAs, *Curr. Top. Microbiol. Immunol.* **93:**125–150.

Krug, R. M., 1983, Transcription and replication of influenza virus, in: *Genetics of Influenza Virus* (P. Palese and D. Kingsbury, eds.), pp. 70–98, Springer-Verlag, New York.

Krug, R. M., and Etkind, P., 1973, Cytoplasmic and nuclear virus-specific proteins in influenza virus-infected MDCK cells, *Virology* **56:**334–348.

Krug, R. M., and Soeiro, R., 1975, Studies on the intranuclear localization of influenza virus-specific proteins, *Virology* **64:**378–387.

Krug, R. M., Ueda, M., and Palese, P., 1975, Temperature-sensitive mutants of influenza WSN virus defective in virus-specific RNA synthesis, *J. Virol.* **16:**790–796.

Krug, R. M., Broni, B. B., and Bouloy, M., 1979, Are the 5′ ends of influenza viral mRNAs synthesized *in vivo* donated by host mRNAs?, *Cell* **18:**329–334.

Krug, R. M., Broni, B. A., Lafiandra, A. J., Morgan, M. A., and Shatkin, A. J., 1980, Priming and inhibitory activities of RNAs for the influenza viral transcriptase do not require base pairing with the virion template RNA, *Proc. Natl. Acad. Sci. USA* **77:**5874–5878.

Krug, R. M., Shaw, M., Broni, B., Shapiro, G., and Haller, O., 1985, Inhibition of influenza viral messenger RNA synthesis in cells expressing the interferon-induced Mx gene product, *J. Virol.* **56:**201–206.

Krug, R. M., St. Angelo, C., Broni, B., and Shapiro, G., 1987, Transcription and replication of influenza virion RNA in the nucleus of infected cells, *Cold Spring Harbor Symp. Quant. Biol.* **LII:**353–358.

Lamb, R. A., and Lai, C.-J., 1980, Sequence of interrupted and uninterrupted mRNAs and cloned DNA coding for the two overlapping nonstructural proteins of influenza virus, *Cell* **21:**475–485.

Lamb, R. A., and Lai, C.-J., 1982, Spliced and unspliced messenger RNAs synthesized from cloned influenza virus M DNA in an SV40 vector: Expression of the influenza virus membrane protein, *Virology* **123:**237–256.

Lamb, R. A., and Lai, C.-J., 1984, Expression of unspliced NS1 mRNA, spliced NS2 mRNA, and a spliced chimera mRNA from cloned influenza virus NS DNA in an SV40 vector, *Virology* **135:**139–147.

Lamb, R. A., Choppin, P. W., Channock, R. M., and Lai, C.-J., 1980, Mapping of the two overlapping genes for polypeptides NS1 and NS2 on RNA segment 8 of influenza virus genome, *Proc. Natl. Acad. Sci. USA* **77:**1857–1861.

Lamb, R. A., Lai, C.-J., and Choppin, P. W., 1981, Sequences of mRNAs derived from genome RNA segment 7 of influenza virus: Colinear and interrupted mRNAs code for overlapping proteins, *Proc. Natl. Acad. Sci. USA* **78:**4170–4174.

Lamb, R. A., Zebedee, S. L., and Richardson, C. D., 1985, Influenza virus $M_2$ protein is an integral membrane protein expressed on the infected-cell surface, *Cell* **40:**627–633.

Lamond, A. I., Konarska, M. M., Grabowski, P. J., and Sharp, P. A., 1988, Spliceosome assembly involves the binding and release of U4 small nuclear ribonucleoprotein, *Proc. Natl. Acad. Sci. USA* **85:**411–415.

Lawrence, C. B., and Jackson, K. J., 1982, Translation of adenovirus serotype 2 late mRNAs, *J. Mol. Biol.* **162:**317–334.

Lazarowitz, S. G., Compans, R. W., and Choppin, P. W., 1971, Influenza virus structural and nonstructural proteins in infected cells and their plasma membranes, *Virology* **46**:830–843.

Lengyel, P., 1982, Biochemistry of interferons and their actions, *Annu. Rev. Biochem.* **51**:251–282.

Lenk, R., Ransom, L., Kaufmann, Y., and Penman, S., 1977, A cytoskeletal structure with associated polyribosomes obtained from HeLa cells, *Cell* **10**:67–78.

Lenk, R., and Penman, S., 1979, The cytoskeleton framework and poliovirus metabolism, *Cell* **16**:289–301.

Lewis, E. D., and Manley, J. L., 1985, Control of adenovirus late promoter expression in two human cell lines, *Mol. Cell. Biol.* **5**:2433–2442.

Lindenmann, J., 1962, Resistance of mice to mouse adapted influenza A virus, *Virology* **16**:203–204.

Lindenmann, J., 1964, Inheritance of resistance to influenza in mice, *Proc. Soc. Exp. Biol. Med.* **116**:505–509.

Lindenmann, J., and Klein, P. A., 1966, Further studies on the resistance of mice to myxoviruses, *Arch. Ges. Virusforsch* **19**:1–12.

Lindenmann, J., Lance, C. A., and Hobson, D., 1963, The resistance of A2G mice to myxoviruses, *J. Immunol.* **90**:942–951.

Lodish, H. F., 1976, Translational control of protein synthesis, *Annu. Rev. Biochem.* **45**:39–72.

Lodish, H., and Froshauer, S., 1977, Rates of initiation of protein synthesis by two purified species of vesicular stomatitis virus messenger RNA, *J. Biol. Chem.* **52**:8804–8811.

Lodish, H. F., and Porter, M., 1980, Translational control of protein synthesis after infection by vesicular stomatitis virus, *J. Virol.* **36**:719–733.

Logan, J., and Shenk, T., 1984, Adenovirus tripartite leader sequence enhances translation of mRNAs late after infection, *Proc. Natl. Acad. Sci. USA* **81**:3655–3659.

Mahy, B. W. J., Barrett, T., Nichol. S. T., Penn, C. R., and Wolstenholme, A. J., 1981, Analysis of the functions of influenza virus genome RNA segments by use of temperature-sensitive mutants of fowl plague virus, in: *The Replication of Negative-Stranded Viruses* (D. H. L. Bishop and R. W. Compans, eds.), pp. 379–387, Elsevier/North-Holland, New York.

Mandler, J., and Scholtissek, C., 1989, Determination of the mutations responsible for the temperature sensitive defects in the nucleoproteins (NP) of two influenza A virus mutants with different phenotypes, in: *Genetics and Pathogenecity of Negative Strand Viruses* (B. W. J. Mahy, D. Kolakofsky, and A. Flammand, eds.) (in press).

Maniatis, T., and Reed, R., 1987, The role of small nuclear ribonucleoprotein particles in pre-mRNA splicing, *Nature (Lond.)* **325**:673–678.

Mansour, S. L., Grodzicker, T., and Tjian, R., 1986, Downstream sequences affect transcription initiation from the adenovirus major late promoter, *Mol. Cell. Biol.* **6**:2684–2694.

Mark, G. E., Taylor, J. M., Broni, B., and Krug, R. M., 1979, Nuclear accumulation of influenza viral RNA and the effects of cyclohexamide actinomycin D and alpha amanitin, *J. Virol.* **29**:744–752.

McGeoch, D., and Kitron, N., 1974, Influenza virion RNA-dependent RNA polymerase: Stimulation by guanosine and related compounds, *J. Virol.* **15**:686–695.

Meier, E., Fah, J., Grob, M. S., End, R., Staehli, P., and Haller, O., 1988, A family of interferon-induced Mx-related mRNAs encodes cytoplasmic and nuclear proteins in rat cells, *J. Virol.* **62**:2386–2393.

Meyer, T., and Horisberger, M. A., 1984, Combined action of mouse α and β interferons in influenza virus-infected macrophages carrying the resistance gene Mx, *J. Virol.* **49**:709–716.

Mortier, C., and Haller, O., 1987, Homologs to mouse Mx protein induced by interferon in various species, in: *The Biology of the Interferon System* (K. Cantell and H. Schellekens, eds.), pp. 79–84, Martinus Nijhoff, Dordrecht.

Mowshowitz, S. L., 1981, RNA synthesis of temperature-sensitive mutants of WSN influenza virus, in: *The Replication of Negative-Strand Viruses* (D. H. L. Bishop and R. W. Compans, eds.), pp. 317–323, Elsevier/North-Holland, New York.

Murti, K. G., Webster, R. G., and Jones, I. M., 1988, Localization of RNA polymerase on influenza viral ribonucleoproteins by immunogold labeling, *Virology* **164**:562–566.

Narayan, P., Ayers, D. F., Rottman, F. M., Maroney, P. A., and Nilsen, T. W., 1987, Unequal distribution of $N^6$-methyladenosine in influenza virus mRNAs, *Mol. Cell. Biol.* **7**: 1572–1575.

Noteborn, M., Arnheiter, H., Richter-Mann, L., Browing, H., and Weissmann, C., 1987, Transport of the murine Mx protein into the nucleus is dependent on a basic carboxy terminal sequence, *J. Interferon Res.* **7**:657–669.

O'Malley, R. P., Mariano, T. M., Siekierka, J., and Mathews, M. B., 1986, A mechanism for the control of protein synthesis by adenovirus VA RNA₁, *Cell* **44**:391–400.

Oxford, J. S., and Schild, G. C., 1975, Immunological studies with influenza virus matrix protein, in: *Negative Strand Viruses* (B. W. J. Mahy and R. D. Barry, eds.), pp. 611–620, Academic, Orlando, Florida.

Padgett, R. A., Konarska, M. M., Grabowski, P. J., and Sharp, P. A., 1984, Lariat RNAs as intermediates and products in the splicing of messenger RNA precursors, *Science* **225**: 898–903.

Pal, R., Grinnell, B. W., Snyder, R. M., and Wagner, R. R., 1985, Regulation of viral transcription by the matrix protein of vesicular stomatitis virus probed by monoclonal antibodies and temperature-sensitive mutants, *J. Virol.* **56**:386–394.

Panniers, R., and Henshaw, E. C., 1983, A GDP/GTP exchange factor essential for eukaryotic initiation factor 2 cycling in Ehrlich ascites tumor cells and its regulation by eukaryotic initiation factor 2 phosphorylation, *J. Biol. Chem.* **258**:7982–7934.

Patton, J. T., Davis, N. L., and Wertz, G., 1983, Cell-free synthesis and assembly of vesicular stomatitis virus nucleocapsids, *J. Virol.* **45**:155–164.

Patton, J. T., Davis, N. L., and Wertz, G., 1984, N protein alone satisfies the requirement for protein synthesis during RNA replication of vesicular stomatitis virus, *J. Virol.* **49**:303–309.

Peluso, R. W., and Moyer, S. A., 1983, Initiation and replication of vesicular stomatitis virus genome RNA in a cell-free system, *Proc. Natl. Acad. Sci. USA* **80**:3198–3202.

Peluso, R. W., and Moyer, S. A., 1984, Vesicular stomatitis virus proteins required for the *in vitro* replication of defective interfering particle genome RNA, in: *Nonsegmented Negative Strand Viruses* (D. H. L. Bishop and R. Compans, eds.), pp. 153–160, Academic, Orlando, Florida.

Perkins, K. K., Furneaux, H. M., and Hurwitz, J., 1986, RNA splicing products formed with isolated fractions from HeLa cells are associated with fast-sedimenting complexes, *Proc. Natl. Acad. Sci. USA* **83**:887–981.

Plotch, S. J., and Krug, R. M., 1977, Influenza virion transcriptase: Synthesis *in vitro* of large, polyadenylic acid-containing complementary RNA, *J. Virol.* **21**:24–34.

Plotch, S. J., and Krug, R. M., 1978, Segments of influenza virus complementary RNA synthesized *in vitro*, *J. Virol.* **25**:579–586.

Plotch, S. J., and Krug, R. M., 1986, *In vitro* splicing of influenza viral NS1 mRNA and NS1-β-globin chimeras: Possible mechanisms for the control of viral mRNA splicing, *Proc. Natl. Acad. Sci. USA* **83**:5444–5448.

Plotch, S., Bouloy, M., and Krug, R. M., 1979, Transfer of 5' terminal cap of globin mRNA to influenza viral complementary RNA during transcription *in vitro*, *Proc. Natl. Acad. Sci. USA* **76**:1618–1622.

Plotch, S. J., Bouloy, M., Ulmanen, I., and Krug, R. M., 1981, A unique cap($m^7$GpppX$^m$)-dependent influenza virion endonuclease cleaves capped RNAs to generate the primers that initiate viral RNA transcription, *Cell* **23**:847–858.

Pons, M. W., 1971, Isolation of influenza virus ribonucleoprotein from infected cells: Demonstration of the presence of negative-stranded RNA in viral RNP, *Virology* **46**:149–160.

Portela, A., Meleor, J. A., Martinez, C., Domingo, E., and Oritin, J., 1985, Oriented synthesis and cloning of influenza virus nucleoprotein cDNA that leads to its expression in mammalian cells, *Virus Res.* **4**:69–82.

Reed, R., and Maniatis, T., 1988, The role of the mammalian branchpoint sequence in pre-mRNA splicing, *Genes Dev.* **2**:1268–1276.

Reed, R., Griffith, J., and Maniatis, T., 1988, Purification and visualization of native spliceosomes, *Cell* **53**:949–961.

Reeves, R. H., O'Hara, B. F., Pavan, W. J., Gearhart, J. D., and Haller, O., 1988, Genetic mapping of the Mx influenza virus resistance gene within the region of mouse chromosome 16 that is homologous to human chromosome 21, *J. Virol.* **62**:4372–4375.

Reichel, P. A., Merrick, W. C., Siekierka, J., and Mathews, M. B., 1985, Regulation of a protein synthesis initiation factor by adenovirus virus-associated RNA₁, *Nature (Lond.)* **313**:196–200.

Robertson, J. S., 1979, 5′ and 3′ terminal nucleotide sequences of the RNA genome segments of influenza virus, *Nucl. Acids Res.* **6**:3745–3757.

Robertson, J. S., Schubert, M., and Lazzarini, R. A., 1981, Polyadenylation sites for influenza virus mRNA, *J. Virol.* **38**:157–163.

Ruskin, B., Krainer, A. R., Maniatis, T., and Green, M. R., 1984, Excision of an intact intron as a novel lariat structure during pre-mRNA splicing *in vitro*, *Cell* **38**:317–331.

Ruskin, B., Zamore, P. D., and Green, M. R., 1988, A factor U2AF is required for U2 snRNP binding and splicing complex assembly, *Cell* **52**:207–219.

Rychlik, W., Domier, L. L., Gardner, P. R., Hellman, G. M., and Rhoads, R. E., 1987, Amino acid sequence of the mRNA cap-binding protein from human tissues, *Proc. Natl. Acad. Sci. USA* **84**:945–949.

Safer, B., 1983, 2B or not 2B: Regulation of the catalytic utilization of eIF-2, *Cell* **33**:7–8.

Schibler, U., Kelley, D. E., and Perry, R. P., 1977, Comparison of methylated sequences in messenger RNA and heterogeneous nuclear RNA from mouse L cells, *J. Mol. Biol.* **115**:696–714.

Schneider, R. J., Safer, B., Munemitsu, S., Samuel, C., and Shenk, T., 1985, Adenovirus VA1 RNA prevents phosphorylation of the eukaryotic initiation factor 2 α subunit subsequent to infection, *Proc. Natl. Acad. Sci. USA* **82**:4321–4325.

Scholtissek, C., 1978, The genome of influenza virus, *Curr. Top. Microbiol. Immunol.* **80**:139–169.

Scholtissek, C., and Bowles, A. L., 1975, Isolation and characterization of temperature-sensitive mutants of fowl plague virus, *Virology* **67**:576–587.

Schubert, M., Keene, J. D., Herman, R. C., and Lazzarini, R. A., 1980, Site on the vesicular stomatitis virus genome specifying polyadenylation and the end of the L gene mRNA, *J. Virol.* **34**:550–559.

Shapiro, G. I., and Krug, R. M., 1988, Influenza virus RNA replication *in vitro*: Synthesis of viral template RNAs and virion RNAs in the absence of an added primer, *J. Virol.* **62**:2285–2290.

Shapiro, G. I., Gurney, T., Jr., and Krug, R. M., 1987, Influenza virus gene expression: Control mechanisms at early and late times of infection and nuclear cytoplasmic transport of virus specific RNAs, *J. Virol.* **61**:764–773.

Siekierka, J., Mariano, T. M., Reichel, P. A., and Mathews, M. B., 1985, Translational control by adenovirus: Lack of virus-associated RNA₁ during adenovirus infection results in phosphorylation of initiation factor eIF-2 and inhibition of protein synthesis, *Proc. Natl. Acad. Sci. USA* **82**:1959–1963.

Skehel, J. J., 1972, Polypeptide synthesis in influenza virus-infected cells, *Virology* **49**:23–36.

Skehel, J. J., and Hay, A. J., 1978, Nucleotide sequence of the 5′ termini of influenza virus RNAs and their transcripts, *Nucl. Acids Res.* **5**:1207–1219.

Smith, D. B., and Inglis, S. C., 1985, Regulated production of an influenza virus spliced mRNA mediated by virus-specific products, *EMBO J.* **4**:2313–2319.

Smith, G. L., and Hay, A. J., 1982, Replication of the influenza virus genome, *Virology* **118**:96–108.

Smith, G. L., Levin, J. Z., Palese, P., and Moss, B., 1987, Synthesis and cellular location of the ten influenza polypeptides individually expressed by recombinant vaccinia viruses, *Virology* **160:**336–345.

St. Angelo, C., 1988, Expression of the influenza viral polymerase and nucleocapsid proteins and reconstitution studies with the expressed proteins. Ph.D. thesis, Cornell University Graduate School of Medical Sciences, New York, New York.

St. Angelo, C., Smith, G. E., Summers, M. D., and Krug, R. M., 1987, Two of the three influenza viral polymerase proteins expressed by using baculovirus vectors form a complex in insect cells, *J. Virol.* **61:**361–365.

Staehli, P., and Haller, O., 1985, Interferon-induced human protein with homology to protein Mx of influenza virus-resistant mice, *Mol. Cell. Biol.* **5:**2150–2153.

Staehli, P., and Haller, O., 1987, Interferon-induced Mx protein: A mediator of cellular resistance to influenza virus, *Interferon* **8:**1–23.

Staehli, P., Horisberger, M. A., and Haller, O., 1984, Mx-dependent resistance to influenza virus is induced by mouse interferons alpha and beta but not gamma, *Virology* **132:**456–461.

Staehli, P., Dreiding, P., Haller, O., and Lindenmann, J., 1985, Polyclonal and monoclonal antibodies to the interferon-inducible protein Mx of influenza virus-resistant mice, *J. Biol. Chem.* **260:**1823–1825.

Staehli, P., Haller, O., Boll, W., Lindenmann, J., and Weissmann, C., 1986a, Mx protein: Constitutive expression in 3T3 cells transformed with cloned Mx cDNA confers selective resistance to influenza virus, *Cell* **44:**147–158.

Staehli, P., Pravtcheva, D., Lundin, L.-G., Acklin, M., Ruddle, F., Lindenmann, J., and Haller, O., 1986b, Interferon-related influenza virus resistance gene Mx is localized on mouse chromosome 16, *J. Virol.* **58:**967–969.

Staehli, P., Grob, R., Meier, E., Sutcliffe, J. G., and Haller, O., 1988, Influenza virus-susceptible mice carry Mx genes with a large deletion or a nonsense mutation, *Mol. Cell. Biol.* **8:**4518–4523.

Stoltzfus, C. M., and Dane, R. W., 1982, Accumulation of spliced avian retrovirus mRNA is inhibited in *S*-adenosylmethionine-depleted chicken embryo fibroblasts, *J. Virol.* **42:** 918–931.

Sugawara, K., Nakamura, K., and Homma, M., 1983, Analyses of structural polypeptides of seven different isolates of influenza C virus, *J. Gen. Virol.* **64:**579–587.

Sugiura, A., Ueda, M., Tobita, K., and Enomoto, C., 1975, Further isolation and characterization of temperature-sensitive mutants of influenza virus, *Virology* **65:**363–373.

Szewczyk, B., Laver, W. G., and Summers, D. F., 1988, Purification, thioredoxin renaturation, and reconstituted activity of the three subunits of the influenza A virus RNA polymerase, *Proc. Natl. Acad. Sci. USA* **85:**7907–7911.

Thimmappaya, B., Weinberger, C., Schneider, R. J., and Shenk, T., 1982, Adenovirus VAl RNA is required for efficient translation of viral mRNAs at late times after infection, *Cell* **31:**543–551.

Thomas, G. P., and Mathews, M. B., 1984, Alterations of transcription and translation in HeLa cells exposed to amino acid analogs, *Mol. Cell. Biol.* **4:**1063–1072.

Ulmanen, I., Broni, B. A., and Krug, R. M., 1981, The role of two of the influenza virus core P proteins in recognizing cap 1 structures ($m^7$GpppNm) on RNAs and in initiating viral RNA transcription, *Proc. Natl. Acad. Sci. USA* **78:**7355–7359.

Ulmanen, I., Broni, B. A., and Krug, R. M., 1983, Influenza virus temperature-sensitive cap($m^7$GpppNm)-dependent endonuclease, *J. Virol.* **45:**27–35.

Varmus, H., and Swanstrom, R., 1983, Replication of retroviruses, in: *RNA Tumor Viruses* (R. Weiss, N. Teich, H. Varmus, and J. Coffin, eds.), pp. 369–512, Cold Spring Harbor Laboratory, Cold Spring Harbor, New York.

Walden, W. E., Godefroy-Colburn, T., and Thach, R. E., 1981, The role of mRNA competition in regulating translation, *J. Biol. Chem.* **256:**11739–11746.

Wei, C.-M., and Moss, B., 1977, Nucleotide sequence at the $N^6$-methyladenosine sites of HeLa cell messenger ribonucleic acid, *Biochemistry* **16:**1672–1676.

Wolstenholme, A. J., Barrett, T., Nichol, S. T., Mahy, B. W. J., 1980, Influenza virus-specific RNA and protein synthesis in cells infected with temperature-sensitive mutants defective in the genome segment encoding nonstructural protein, *J. Virol.* **35**:1–7.

Yamashita, M., Krystal, M., and Palese, P., 1988, Evidence that the matrix protein of influenza C virus is coded for by a spliced mRNA, *J. Virol.* **62**:3348–3355.

Yokota, M., Nakamura, K., Sugawara, K., and Homma, M., 1983, The synthesis of polypeptides in influenza C virus-infected cells, *Virology* **130**:105–117.

Young, R. J., and Content, J., 1971, 5' terminus of the influenza virus RNA, *Nature (Lond.)* **230**:140–142.

Zain, S., Sambrook, J., Roberts, R., Keller, W., Fried, M., and Dunn, A. R., 1979, Nucleotide sequence analysis of the leader segments in a cloned copy of adenovirus 2 fiber mRNA, *Cell* **16**:851–861.

Zebedee, S. L., and Lamb, R. A., 1988, Influenza A virus $M_2$ protein: Monoclonal antibody restriction of virus growth and detection of $M_2$ in virions, *J. Virol.* **62**:2762–2772.

Zeitlin, S., and Efstratiadis, A., 1984, *In vivo* splicing products of the rabbit β-globin pre-mRNA, *Cell* **39**:589–602.

Zhuang, Y., and Weiner, A. M., 1986, A compensatory base change in U1 snRNA suppresses 5' splice site mutations, *Cell* **46**:827–835.

Zillman, M., Zapp, L. M., and Berget, S. M., 1988, Gel electrophoretic isolation of splicing complexes containing U1 small nuclear ribonucleoprotein particles, *Mol. Cell. Biol.* **8**:814–821.

Zvonarjev, A. Y., and Ghendon, Y. Z., 1980, Influence of a membrane (M) protein on influenza A virus virion transcriptase *in vitro* and its susceptibility to rimantadine, *J. Virol.* **33**:583–586.

CHAPTER 3

# Structure, Function, and Antigenicity of the Hemagglutinin of Influenza Virus

S. A. WHARTON, W. WEIS, J. J. SKEHEL, AND
D. C. WILEY

## I. INTRODUCTION

The major surface glycoprotein of influenza virus is hemagglutinin (HA). This chapter reviews the two major functions of HA: (1) its involvement in binding to receptors on cells before their infection, and (2) its role in the fusion of viral and endosomal membranes, necessary for the release of the viral genome into the cell. In addition, HA is the viral antigen that interacts with infectivity-neutralizing antibodies; alterations in the molecule enable the virus to escape immune surveillance and cause epidemics of disease. The nature of these changes in antigenicity is discussed.

## II. STRUCTURE OF HEMAGGLUTININ

Hemagglutinin is synthesized in the rough endoplasmic reticulum (RER) of infected cells as an uncleaved precursor, $HA_0$. During passage to

S. A. WHARTON and J. J. SKEHEL • Division of Virology, National Institute for Medical Research, London, NW7 1AA, England.    W. WEIS and D. C. WILEY • Department of Biochemistry and Molecular Biology, Harvard University, Cambridge, Massachusetts 02138.

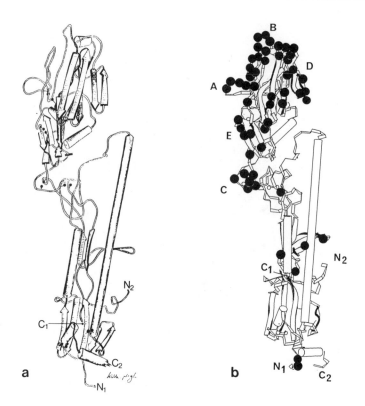

FIGURE 1. (a) Diagrammatic representation of one subunit of BHA showing the carboxyl-
and amino-termini of HA$_1$ and HA$_2$ (C1 C2, N1 N2) and regions of $\alpha$-helix (cylinders) and $\beta$-
sheet (arrows). The trimeric interface is to the right, and the viral membrane is at the base of
the molecule. (b) Diagrammatic representation of one subunit of BHA showing the sites of
amino acid substitutions in HA$_1$ that have occurred in natural isolates of H$_3$ viruses since
1968 as a result of antigenic drift. The five main antigenic regions are labeled, as are the
carboxyl- and amino-termini of HA$_1$ and HA$_2$ (C1C2, N1N2). The diagram was generated
using the computer program designed by Lesk and Hardman (1982).

the plasma membrane, the molecule is glycosylated in several positions
(seven in the case of the Hong Kong HA) and is cleaved by unidentified
host cell protease(s). Cleavage is essential for infectivity (Klenk *et al.*,
1975) and results in two polypeptides, HA$_1$ (328 residues) and HA$_2$ (221
residues), which are disulfide linked to form subunits associated non-
covalently as trimers. The carboxy-terminus of HA$_2$ anchors each monomer in the virus membranes with the 15 C-terminal residues on the
cytoplasmic side. It is unclear whether this region is important in con-
trolling intracellular transport (reviewed by Matlin, 1986) or virus assembly. Bromelain digestion cleaves HA on virus membranes at HA$_2$ 175 and
results in the release of the entire ectodomain (BHA) as a soluble trimer
(Brand and Skehel, 1972). The three-dimensional structure of X31 BHA
has been determined to 3 Å by X-ray crystallography (Wilson *et al.*, 1981).

A schematic diagram of one monomer is shown in Fig. 1a. The monomer is 13.4 nm long. Basically, each monomer consists of a globular membrane distal region consisting only of $HA_1$ on top of an elongated stem consisting of all of $HA_2$ and parts of $HA_1$. Proceeding through the molecule from the amino-terminus of $HA_1$ (N1 in Fig. 1a), close to the viral membrane, the first 63 amino aids are in an extended conformation, stretching 9.6 nm from the viral membrane. Residues 63–305 form the distal globular domain, which contains an eight-stranded antiparallel β-sheet and two short α-helices. The receptor-binding site is located in this region (see Section IV). Residues 305–328 return down the stem of the molecule running approximately antiparallel to the amino-terminal extended chain of $HA_1$. The amino-terminus of $HA_2$ (N2) is 2.2 nm from the carboxy-terminus of $HA_1$ (C1), showing that a change in conformation in this region occurred following the proteolytic cleavage of the precursor $HA_0$. The amino-terminus of $HA_2$ is located at the trimeric interface of the molecule, 3 nm from the viral membrane. The conserved hydrophobic amino-terminal region involved in membrane fusion activity (see Section V), which is rich in glycine residues, forms an unusual helical structure extending away from the trimeric interface. Residues 22–38 form an antiparallel β-sheet near the viral membrane. Residues 39–58 form an α-helix extending up the outside of the stem of the molecule, residues 59–75 are in an extended configuration at the top of the stem region, linking this short helix to the major structural component of $HA_2$, an 8-nm-long α-helix composed of residues 76–129. This α-helix forms a coiled coil with the other two equivalent helices in the trimer which separate at the membrane proximal end so as to accommodate the amino-terminus of $HA_2$. The rest of the ectodomain consists of a short antiparallel β-sheet and two short α-helices, all near the viral membrane. The bromelain cleavage site is at residue 175, beyond which $HA_2$ enters the membrane, with 15 carboxy-terminal residues being enclosed by the virus. The major points of contact between $HA_1$ and $HA_2$ are in the stem of the molecule, and the single disulfide bond between $HA_1$ residue 14 and $HA_2$ residue 137 is at the membrane proximal end of the molecule. The points of contact between trimers are primarily the coiled coil of the long α-helices, but there are also hydrophobic interactions between the amino-termini of $HA_2$, and also interactions at the globular distal end of the molecule between $HA_1$ subunits.

All oligosaccharides are N-linked to asparagine residues. In the X31 (H3N2) molecule, there are six oligosaccharides on $HA_1$ at residues, 8, 22, 38, 81, 165, and 285 and one on $HA_2$ at residue 154; the number and position of sugar chains varies from strain to strain and subtype to subtype. All the oligosaccharides are complex and processed except those at $HA_1$ residues 81 and 165. All the oligosaccharides, except that at $HA_1$ 165, are on lateral surfaces of the molecule. The carbohydrate side chain at $HA_1$ 165 stretches across $HA_1$ of an adjacent monomer; it could have a role in stabilizing the structure of the trimer.

Throughout this chapter, amino acids are numbered from the amino-terminus of $HA_1$ or $HA_2$. Amino acid substitutions are shown in the form Leu226Gln, where the leucine at position 226 has been replaced by a glutamine residue.

## III. ANTIGENIC VARIATION

Antigenic variation of influenza virus involves primarily the HA and NA glycoproteins. The internal proteins are not invariant, varying by up to 8% (Winter and Fields, 1981; Huddleston and Brownlee, 1982), compared with the membrane glycoproteins, which differ by up to 80% and are placed in different subtypes on the basis of their reaction with hyperimmune sera (WHO Memorandum, 1980). There are 13 subtypes of HA. All 13 have been found in avian species, whereas more restricted ranges are seen in horses, swine, seals, and humans. Since the first isolation of influenza virus in 1933 (Smith *et al.*, 1933), three subtypes of HA and two of neuraminidase (NA) have been found in humans. A sudden appearance of a different HA subtype in the population gives rise to pandemics, since the new virus is initially not neutralized by the host immune system. This phenomenon is known as antigenic shift. In addition, because anti-HA antibodies neutralize virus infectivity, amino acid substitutions that arise by mutation in the genes for HA over the years cause new epidemics, as again the molecules escape immune surveillance. This process is called antigenic drift.

### A. Antigenic Shift

Antigenic shift is known to have occurred in the human population at least three times: in 1918, when H1N1 viruses appeared; in 1957, with the appearance of H2N2 or Asian viruses; and in 1968, with H3N2 or Hong Kong influenza. From direct observation and indirectly from serological studies, changes probably occurred during the nineteenth century (Mulder and Masurel, 1958), and viruses appear to be of the H2 or H3 subtypes. In addition, viruses of the H1 subtype have recurred since 1977, as the younger members of the population could not have developed immunity to this subtype before 1957.

Antigenic analysis of avian and equine viruses isolated before 1968 showed that the HAs of A/duck/Ukraine/63 (Coleman *et al.*, 1968) and A/equine/Miami/63 (Waddell *et al.*, 1963) are closely related to the human H3 subtype in 1968 isolates. These, and other studies, involving protein (Laver and Webster, 1973; Ward and Dopheide, 1981) and gene sequencing (Fang *et al.*, 1981; Daniels *et al.*, 1985a) strongly suggest that antigenic shift results from the formation of recombinant viruses during mixed infections of a host with human and animal or avian viruses. The site of reassortment is unknown, but evidence suggests that swine are involved as intermediates in the process (Scholtissek et al., 1985). Reassortments occur readily in vivo and in vitro because of the segmented nature of the viral genome (Tumova and Pereira, 1965; Webster et al., 1971). In the case of the 1968 shift, RNA hybridization analyses showed that only the HA gene was changed (Scholtissek et al., 1978a). The shift in 1957 is thought also to have been due to reassortments. However, the H1 subtype that reappeared in 1977 seems, from RNA hybridization (Scholtissek et al., 1978b) and sequencing analyses (Young et al., 1979) to be closely related to the H1 viruses circulating in 1950; thus, it appears that this antigenic shift is attributable to the emergence of a dormant virus from an unknown reservoir.

)

## B. Antigenic Drift

Evidence for antigenic drift has come chiefly from the study of human viruses, although there is evidence that it also occurs at least in horses (Daniels et al., 1985a). The most studied human subtype is the H3 Hong Kong subtype. Hemagglutination inhibition studies using convalescent ferret antisera that are highly strain specific (Pereira, 1982), as well as gene sequence analyses (Both et al., 1983; Skehel et al., 1983; Daniels et al., 1985b; Raymond et al., 1983, 1986) have indicated that antigenic drift results from the accumulation of amino acid substitutions primarily in the $HA_1$ chain. Since 1968 73 of the 328 residues of $HA_1$ have changed compared with 12 of 221 in $HA_2$. The changes in $HA_1$ are clustered in five surface regions, A–E (Wiley et al., 1981) in the distal domain of $HA_1$ (Fig. 1b). Some changes involve the addition of new sites for glycosylation that could mask antigenically significant regions. Site A is centered around a protruding loop of residues 140–146. Residues 138, 139, 147, and 148 are conserved, thus preserving the structural foundation of the site. Site B, at the extreme membrane distal end of the molecule, is centered on a loop of residues 155–160 as well as the α-helix of residues 188 to 198 situated at the edge of the receptor-binding site. Site C is situated at the base of the globular domain in the antiparallel strands of $HA_1$. Site D is situated near the trimeric interface of the globular domains of $HA_1$ on the top of the molecule. Site E is near the bottom of the globular distal domain between sites C and A. The antigenic significance of the five sites is confirmed by the observation that mouse monoclonal antibody selected mutants of H3 viruses have amino acid substitutions at the same sites (Laver et al., 1979, 1981; Daniels et al., 1983a; Newton et al., 1983; Webster et al., 1983). Crystallographic studies on two such mutants with substitutions at residues $HA_1$Gly146Asp (Knossow et al., 1984) and $HA_1$Asn188Asp (Weis, 1987) have given evidence that the relevant antibody binds to the region of the molecule where the amino acid substitution occurs as structural changes in these mutants are localized to the vicinity of the amino acid substitution.

Studies of H1 subtype viruses isolated between 1934–1957 and 1977 to the present show that the antigenically significant regions appear to be in similar positions to those of the H3 subtype (Gerhard et al., 1981; Caton et al., 1982; Daniels et al,. 1985b; Raymond et al., 1986). The rate of change of amino acids in H3 viruses is quite constant at 1.1%/year for 1968–1979 and at 1.2% for 1979–1986 (Both et al., 1983; Stevens et al., 1987). The H1 subtype changed at 1.4% for 1950–1957, but at a lower frequency, 0.8%/year, for 1977–1983. The former higher value may reflect the higher level of immunity in the population in the fourth decade of H1 prevalence.

The information obtained from the above studies indicate that antigenic drift results from amino acid substitutions in the membrane distal region of the HA that prevent binding of antibodies induced by a previous infection; thus, the viruses have the ability to reinfect the host. However, the precise mechanism of their selection is unknown. The frequency of selection of antigenic variants by monoclonal antibodies in vitro is $10^{-4}$–$10^{-5}$ (Yewdell et al., 1979) and, as any antibody that recognizes any of the five antigenic regions is sufficient for neutralization, selection of a mutant with the ability to reinfect most of the population would be expected to occur spontaneously at a very low frequency. By contrast, analyses of the variety of antibody specificities in postinfection human sera (Wang et al., 1986) and the restricted ability of human sera to neutralize monoclonal antibody-selected mutants (Natali et al., 1981) have suggested that sequential changes may occur during reinfection of a partially immune host.

The reason for the characteristic high frequency of antigenic change is unknown, but it seems likely that the existence of virus reservoirs in other species and their transfer to humans at infrequent intervals during antigenic shift are major factors in disturbing the approach to equilibrium that similar but less variable viruses appear to achieve.

## C. The Importance of Receptor Specificity in Antigenic Change

An essential function of HA is its ability to bind the appropriate receptors of cells. The position of the receptor binding site is shown in Fig. 2a. As the receptor-binding specificity differs among humans, birds, and horses, the number of HA molecules that can transfer successfully from one species to another during antigenic shift may be limited. In addition, changes in the receptor-binding site may limit the viability of certain antigenic variants. The sites of substitution that occur during antigenic drift encircle this pocket (Fig. 2b) and do not intrude into it. Relationships between receptor binding and antigenicity may be an important consideration when assessing the significance of natural antigenic variants detected in influenza surveillance studies, as these are routinely passaged in hens eggs, which may impose selection pressure in favor of certain receptors (Schild et al., 1983). The next section discusses

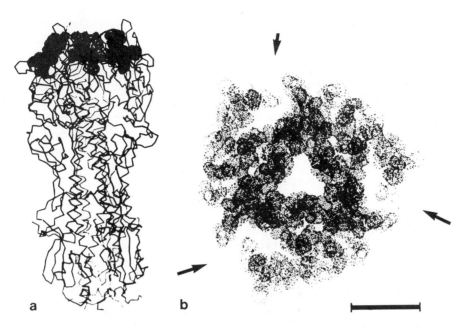

FIGURE 2. (a) Side view of the hemagglutinin trimer with the position of the receptor-binding site highlighted at the top of the molecule. (b) Top view of the hemagglutinin trimer. The dotted spheres illustrate sites involved in antigenic variation around the three receptor-binding sites (arrows). The bar represents 25 Å, the approximate size of an antibody "footprint." From Amit *et al.* (1986).

the structure of the receptor pocket and the information that antigenic variants have given to receptor-binding specificity.

## IV. RECEPTOR BINDING

The receptor-binding site of HA is located at the membrane distal tip of the molecule (Fig. 2a). Direct evidence of this location was obtained by isolating viruses that could grow in the presence of equine $\alpha_2$-macroglobulin. These viruses exhibited altered receptor-binding specificities and were found to have amino acid substitutions at $HA_1$ 226 (Rogers *et al.*, 1983).

Sialic acid (NeuAc) is an essential component of cellular receptors of influenza A virus, since treating erythrocytes with neuraminidase destroys hemagglutination. A common feature of viruses of the H3 subtype isolated from humans is the preference of NeuAc being linked to galactose by an $\alpha2,6$ bond. Wild-type X31 HA has leucine at $HA_1$ 226, and the receptor-binding mutant with glutamine at $HA_1$ 226 (Leu226Gln) bound more efficiently to NeuAc$\alpha2,3$-Gal linkages. Most receptor-binding specificities were determined using erythrocytes derivatized with NeuAc in one or other specific linkage to the penultimate saccharide of the carbohydrate side chains (Rogers et al., 1985). Variations in receptor-binding affinities have been observed in natural strains of virus, e.g., A/duck/Ukraine/63 and A/equine/Miami/63, and the importance in adaptation to different hosts has been discussed. Certain antibody-selected mutants (Fazekas de St. Groth, 1977; Underwood et al., 1987; Yewdell et al., 1986; Daniels et al., 1984, 1987) have been shown to have altered binding characteristics. The amino acid substitutions they contain are located near the receptor-binding site. Yewdell et al. (1986) isolated A/PR/8/34 (H1N1) mutants under conditions of partial neutralization that showed increased avidity for host cell receptors. These variants were not antigenic variants, as shown by enzyme-linked immunosorbent assays (ELISA). Underwood et al. (1987) selected mutants of X31 by monoclonal antibodies that had altered receptor-binding characteristics, as determined by affinity for periodate-treated erythrocytes. Daniels et al. (1987) isolated a variety of mutants, again not antigenic, which differed in their binding specificities. Substitutions Gly218Arg, Gly218Glu, Leu226Pro and deletion of $HA_1$ residues 224–230 caused reductions in affinity of X31 HA for α2,6 linkages and increased affinity for NeuAcα2,3 linkages. In addition, an antibody-selected mutant of the α2,3-binding variant of X31 had increased binding α2,6 linkages and decreased binding to α2,3 linkages. Although the above studies shed light on the nature of HA-receptor binding, it is only recently that the crystal structure of HA bound to a receptor analogue sialyllactose has been obtained (Weis et al., 1988).

## A. Structure of the Receptor-Binding Site

The crystal structure of sialyllactose (NeuAcα2,3 or α2,6 Galβ1,4-glucose) complexed with HA was obtained by comparison of the structure of HA alone and HA soaked in sialyllactose, using either the α2,6-sialyllactose with X31 HA or α2,3-sialyllactose with the receptor-binding mutant Leu226Gln. In each case, the NeuAc binds in a similar configuration. With the α2,6-sialyllactose, only signals from the C5 and C6 galactose atoms can be observed, and even these are absent from the α2,3-sialyllactose–mutant Leu226Gln complex. The absence of these signals means that in both cases the galactose and glucose residues are spatially disordered.

The binding site of HA is a depression at the bottom of which are Tyr98 and Trp153 (Fig. 3). Glu190 and Leu194 project down from the short α-helix to define the rear of the site with His183 and Thr155. Residues 134–138 form the right side of the site and residues 224–228 form the left side. A chain of hydrogen bonds orientates several residues for binding to ligand. Thus, Trp153 donates a proton to the hydroxyl group of Tyr195, which donates to the δ-nitrogen of His183, and the ε-nitrogen of His183 donates a proton to the Tyr98 hydroxyl group, which may be linked to the carboxylate of Glu190 via bound water.

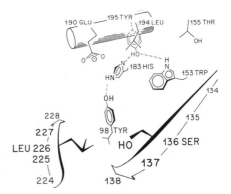

FIGURE 3. Schematic diagram of the wild-type receptor-binding site. Dashed lines represent hydrogen bonds. See text for details.

When NeuAc is bound in the pocket, each of the three hexose ring substitutents—the acetamido, the glycerol, and the carboxylate—fit tightly into, and interact specifically with, the protein (Fig. 4a). However, the hydroxyl group at position 4 does not. The numbering system for NeuAc is also shown in Fig. 4a. The methyl group of the acetamido is in Van der Waals contact with the six-membered ring of Trp153 and with the carbon of Gly134, both of which are conserved residues. The carbonyl group of the acetamido makes Van der Waals contact with the conserved Leu194 and the amide proton hydrogen bonds with the peptide carbonyl oxygen of residue Gly135. In the glycerol side chain, hydroxyl 09 hydrogen bonds to Glu190 and Ser228, and hydroxyl 08 hydrogen bonds to Tyr98. Carbon 7 is in Van der Waals contact with Trp153. There are no positive charges in the vicinity of the NeuAc carboxylate group. One carboxylate oxygen is in a position to hydrogen-bond to the hydroxyl group of Ser136 and the peptide amide proton of Asn137. Overall, the binding of NeuAc to the HA is very polar with hydrogen bonds to conserved polar atoms of residues 98, 136, 137, and 190, with several polar ligands displacing bound water molecules. However, a nonpolar face is formed by the acetamido methyl group, and glycerol C7 and C9, which fit against the conserved Trp153, His183, and Gly134 residues.

The mode of NeuAc binding observed in crystals is consistent with the relative binding affinities of several sialosides (Pritchett et al., 1987) α-anomers of NeuAc, in which the carbohydrate group is axial and projects into the binding pocket, are 30-fold better inhibitors of hemagglutination than β-anomers (T. J. Pritchett, R. Brossmer, and J. C. Paulson, unpublished observations), which have a different configuration around C-2, so that the carboxylate group is equatorial and is orientated away from the molecule toward solution without making contact with the receptor pocket. Most HA molecules have very low affinity for $N$-glycolyl NeuAc, in which the acetamidomethyl group of NeuAc is replaced by a hydroxymethyl group. However, HA molecules that have a Tyr residue at position 155 of $HA_1$, instead of a Thr residue, have increased affinity for $N$-glycolyl-Neu, so that they bind this molecule as well as NeuAc (Anders et al., 1986). The orientation of NeuAc deduced

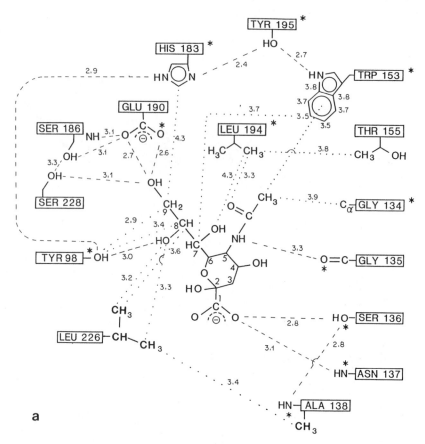

FIGURE 4. (a) Interaction of NeuAc with wild-type hemagglutinin. Potential hydrogen bonds are dashed lines and Van der Waals contacts dotted lines. The distances shown are in angstroms (Å). Some atoms are shown with more hydrogen bonds than they can participate in, because of uncertainties in the accuracy of the coordinates (~0.3 Å). The stars denote conserved atoms and conserved residues (a star next to a box). The numbering of the carbon atoms on NeuAc is also shown. (b) Equivalent diagram of the Leu226Gln mutant hemagglutinin.

by crystallography places the acetamidomethyl group facing toward residue 155.

## B. The Binding Site of a Receptor-Binding Mutant

The crystal structure of the binding mutant of X31, which has a Leu226Gln substitution (Weis *et al.*, 1988), indicates that there are small adjustments (<1 Å) to fill the void created by the loss of the δ1-carbon of Leu226 and to accommodate the new hydrogen-bonding potential of Gln226. The Gln226 carboxyl oxygen and amino nitrogen atoms project

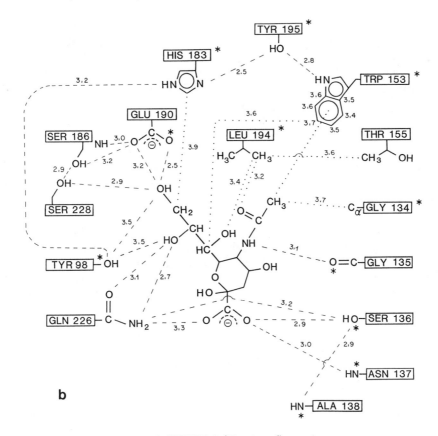

FIGURE 4. (*Continued*)

into the site above the position previously occupied by Leu226. This adjustment involves Tyr98, Trp153, and Phe147, and His183. The loss of a contact between the Leu226 δ1-carbon and the Ala 138 β-carbon allows the right side of the site to be closer to the left, and a new hydrogen bond is formed between Gln226 and Ser136 (Fig. 4b). Small but significant alterations are also seen extending out of the binding site as a result of the adjustments described above.

The molecular basis of receptor specificity differences of α2,6 and α2,3 recognition is unclear. There might be a small difference in the configurations of α2,6- and α2,3-sialyllactose below the current level of detection of 3 Å, which can nevertheless be distinguished by the Leu226Gln mutant. Alternatively, the lactose moieties, whose flexibilities preclude observation in the crystal structures, may form weak contacts with the protein or its oligosaccharide at Asn165 that are distinguished by the mutation, since the orientation of the galactose and glucose residues of sialyllactose will be different in α2,3-linkage from that in α2,6-linkage.

No antibodies have been found to be directed against the residues involved in binding NeuAc. However, some monoclonal antibody selected HA variants have amino acid substitutions close to the receptor-binding pocket; these have been found to have different receptor-binding specificities, as determined using chemically or enzymatically modified erythrocytes. These variants have substitutions in $HA_1$ at residues 144, 145, 158, 193, and 198 (Underwood et al., 1987) as well as deletion of residues 224–230 that constitute the left side of the binding pocket (Daniels et al., 1987). Variants with substitutions quite remote from the receptor-binding site (1–2 nm) also have altered binding specificities, e.g., at residues 188, 189 (Underwood et al., 1987), 185, 231, 244 (Yewdell et al., 1986), and 218 (Daniels et al., 1987), indicating, in agreement with the conclusions from the structural analysis of the receptor-binding mutant, that differences in structure extending out of the receptor-binding pocket influence receptor specificity. Other evidence for this was obtained by Daniels et al. (1984), who found that monoclonal antibodies directed against residues close to the receptor pocket (193, 199, 219, and 229) bound to wild-type HA, but not to the α2,3-NeuAc-binding variant (Leu226Gln).

## V. FUSION PROPERTIES OF HEMAGGLUTININ

In order for the genome of enveloped viruses to be transferred into cells, their lipid membranes fuse with a membrane of the host cell. In common with a number of other viruses, such as alpha and rhabdo viruses, influenza viruses use the endocytic pathway of the cell (reviewed by White et al., 1983; Wharton, 1987). The virus attaches to receptors on the cell surface (respiratory epithelial cells in the case of influenza virus infections), and the virus is taken into endosomes. These are subsequently acidified by the action of proton pumps and at a specific pH level of 5–6, the virus is triggered to fuse with the endosomal membrane. The viral genome is thus released into the cell, where transcription commences.

The first in vitro observation of the fusogenic activity of HA was that at low pH levels, influenza virus was responsible for the lysis of erythrocytes, a result of membrane leakiness to ions after erythrocyte–virus fusion (Maeda and Ohnishi, 1980; Huang et al., 1981; White et al., 1981). More recently, a direct method of monitoring fusion of virus with artificial lipid vesicles (liposomes) has been used. This method involves measuring the decrease of energy transfer between two fluorescent lipids embedded in the liposomal membrane that occurs after a fusion event as a result of dilution of the liposomal lipid by the viral lipid (Stegmann et al., 1985, 1986; Wharton et al., 1986). There is no absolute requirement for any particular lipid molecules to be in the liposomal membrane, although varying the lipid composition alters the rate of fusion to a small extent. Fusion is essentially complete within 3 min at 37°C pH 5.0, this rate can be slowed either by increasing the pH or by decreasing the temperature. Liposomes and virus must both be present when the pH is lowered and virus which has previously been incubated at low pH cannot fuse subsequently when added to liposomes at any pH (Wharton *et al.*, 1986).

There is a great deal of evidence to demonstrate that HA is the protein responsible for the fusogenic activity of the virus. Cells expressing HA and no other viral proteins fuse when the pH is lowered (White *et al.*, 1982). Rosettes of detergent-isolated HA hemolyze (Sato *et al.*, 1983) and fuse liposomes (Wharton *et al.*, 1986) at low pH, and antihemagglutinin antibodies inhibit fusion (Wharton *et al.*, 1986).

## VI. ACID-INDUCED CONFORMATIONAL CHANGE OF HEMAGGLUTININ

*In vitro*, at pH 5.6 37°C, X31 Bromelain released HA, BHA, undergoes a conformational change that involves an increase in hydrophobicity as determined by detergent and liposome-binding studies (Skehel *et al.*, 1982; Doms *et al.*, 1985) and, in the absence of detergent or lipid, the molecule aggregates into rosettes of approximately eight molecules (Skehel *et al.*, 1982; Ruigrok *et al.*, 1986a). In these rosettes, $HA_1$ is arranged distally, as rosettes are capable of hemagglutination. The region of the molecule responsible for the aggregation is the amino-terminus of $HA_2$ because removal of this region by the protease thermolysin results in solubilization of the low pH-induced aggregate (Daniels *et al.*, 1983b; Ruigrok *et al.*, 1988). Unlike the native molecule, low pH BHA is susceptible to trypsin cleavage at $HA_1$ Lys27 and $HA_1$ Arg224 (Skehel *et al.*, 1982). Circular dichroism (CD) studies show that there is no difference in secondary structure between native and low pH BHA; thus, the conformational change involves the movement of domains rather than a general denaturation (Skehel *et al.*, 1982; Wharton *et al.*, 1988b). The ability of low pH BHA to react with most anti-native BHA antibodies (Daniels *et al.*, 1983a) and the retained hemagglutination ability of low pH BHA agree with this conclusion. Without the direct determination of the structure of low-pH BHA by X-ray diffraction (XRD), investigators have had to use other methods to obtain information on its tertiary and quaternary structures. Electron microscopy has been used to study the morphological changes in HA (Ruigrok *et al.*, 1986a). A thinning and elongation of the molecules was observed after low-pH treatment involving possible changes in the tertiary structure of the stem region and in the globular distal region.

Assays of fusion, protease sensitivity, and lipid binding do not, however, determine whether any localized regions of HA are important in triggering the conformational change. Toward this end, Daniels *et al.*

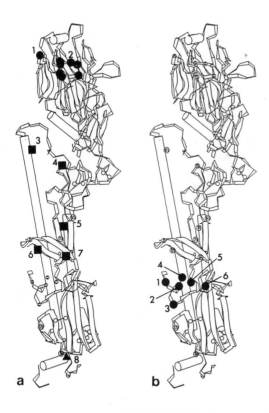

**a**    **b**

FIGURE 5. Positions of substitutions of amino acids in hemagglutinin mutants that fuse at a higher pH than is possible with the wild type. (a) Substitutions at interfaces between $HA_1HA_1$ (●) (Daniels *et al.*, 1987), $HA_1HA_2$, or $HA_2/HA_2$ subunits (■) (Daniels *et al.*, 1985). The substitution of the variant isolated by Doms *et al.* (1986) is also shown (▲). (b) Substitutions around the amino-terminus of $HA_2$ (●).

|   | Position | Substitution | pH |
|---|----------|-------------|-----|
| 1 | $HA_1$ 218 | Gly→Glu | 0.5 |
|   |          | Gly→Arg | 0.5 |
| 2 | $HA_1$ 224–230 | Deleted | 0.2 |
| 3 | $HA_2$ 81 | Glu→Gly | 0.4 |
| 4 | $HA_1$ 300 | Arg→Ser | 0.3 |
| 5 | $HA_2$ 54 | Arg→Lys | 0.3 |
| 6 | $HA_2$ 47 | Gln→Arg | 0.5 |
| 7 | $HA_2$ 105 | Gln→Lys | 0.3 |
| 8 | $HA_2$ 132 | Asp→Asn | 0.2 |

|   | Position | Substitution | pH |
|---|----------|-------------|-----|
| 1 | $HA_2$ 3 | Phe→Leu | 0.3 |
| 2 | $HA_2$ 6 | Ile→Met | 0.4 |
| 3 | $HA_2$ 9 | Phe→Leu | 0.4 |
| 4 | $HA_2$ 112 | Asp→Glu | 0.4 |
|   |          | Asp→Gly | 0.5 |
| 5 | $HA_2$ 114 | Glu→Lys | 0.5 |
| 6 | $HA_1$ 17 | His→Gln | 0.9 |
|   |          | His→Arg | 0.8 |

(1985c) selected variants that had the ability of fusing membranes at an elevated pH level. This was achieved by growing the virus in cells in the presence of amantadine, a weak base, which elevates endosomal pH and selects viruses that can fuse at a higher pH than can the wild type. Base substitutions were only detected in HA genes and, when the positions of the resulting amino acid changes were mapped on the three-dimensional structure of BHA, the positions of the substitutions fell into two main groups: at interfaces between subunits and in the vicinity of the amino-terminus of $HA_2$. The positions that fall into the first category are shown in Fig. 5a. All substitutions involve a change in charge that would modify salt bridges. Two of the substitutions are considered in detail. $HA_2$ Glu81 forms a salt link with $HA_2$ Arg76 at the top of the long helix of $HA_2$, which is repeated by symmetry in the trimer. A Glu81Gly substitution would therefore modify $HA_2/HA_2$ contacts in this region. $HA_2$ Arg54 forms a hydrogen bonded salt bridge with $HA_2$ Glu97 of an adjacent subunit and a hydrogen bond with $HA_1$ Thr28 of an adjacent subunit. Therefore, the $HA_2$ Arg54Lys substitution would alter $HA_2/HA_2$ and $HA_2/HA_1$ contacts. The effects of other substitutions are discussed by Daniels *et al.* (1985c). Mutants with an elevated pH of fusion have arisen naturally (Doms *et al.*, 1986) and by antibody selection (Daniels *et al.*, 1987). In the latter case, the alterations were at the distal end of the molecule; they also altered receptor-binding specificity (see Section IV). Other evidence has led to the conclusion that the conformational change involves changes in intersubunit contacts. $HA_1$ can be separated from $HA_2$ in the low pH form as monomers by the action of reducing agents (Graves *et al.*, 1983) or trypsin (Skehel *et al.*, 1982), epitopes at the interface of $HA_1/HA_1$ (site D) are destroyed after low pH treatment (Daniels *et al.*, 1983a) and $HA_1$ and $HA_2$ thermally denature at different temperatures in the low pH form, whereas in the native form they co-denature (Ruigrok *et al.*, 1986b). The substitutions of high pH-fusing mutants clustered around the amino-terminus of $HA_2$ are shown in Fig. 5b. A closer view of the region is shown in Fig. 6.

A change of $HA_1$ His17Gln that elevated the pH of fusion 0.9 units would influence the hydrogen bonds involving the peptide carboxyl groups of residues $HA_2$ 6 and $HA_2$ 10. Altering residue $HA_2$ Asp 112 would affect hydrogen bonds with the peptide amide nitrogens of residues 3–6 of $HA_2$. These substitutions could locally destabilize the structure and therefore implicate a role of the amino-terminus of $HA_2$ in the conformational change. The amino-terminal region is the most likely candidate for interacting with a membrane during the fusion process for several reasons: the amino-terminus is a region of highly conserved hydrophobic residues as found in other fusogenic viruses (White *et al.*, 1983), it is the site of hydrophobic aggregation of low pH BHA (Daniels *et al.*, 1983b) and mutagenized variants of HA with alterations in this region exhibited different fusion characteristics from wild-type HA (Gething *et al.*, 1986). For example, a substitution of glutamic acid for glycine at $HA_2$

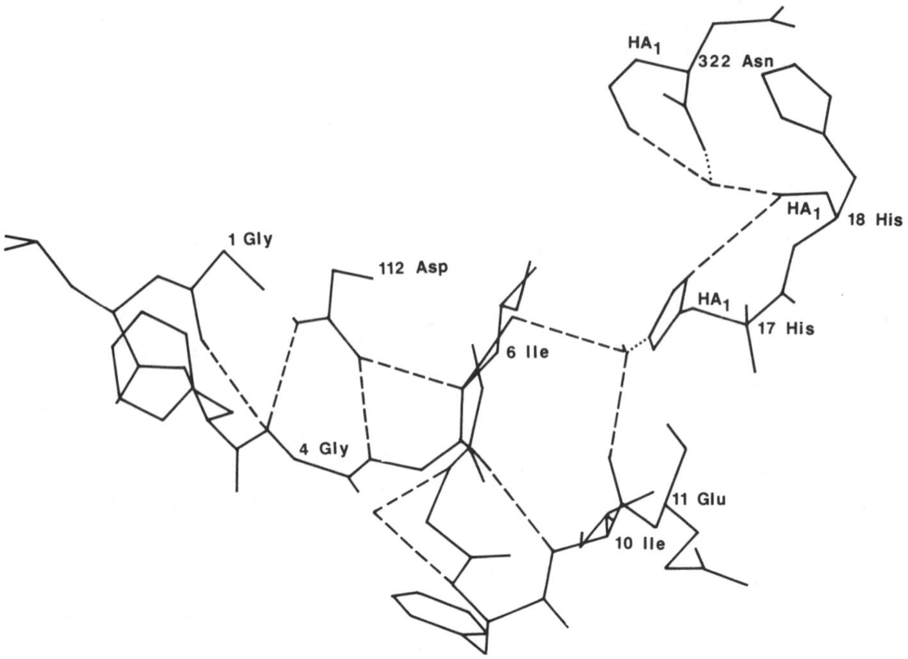

FIGURE 6. Structure of the amino-terminus of HA$_2$. Possible hydrogen bonds are shown as dashed lines. X is a water molecule.

residue 1 abolished fusion activity. Recently, synthetic peptides corresponding to the amino-terminus of HA$_2$ were found to have the ability to fuse liposomes at low pH (Lear and de Grado, 1987; Murata et al., 1987; Wharton et al., 1988a); under certain conditions, this could occur at neutral pH. Furthermore, the peptides corresponding to the sequences of the mutagenized variants of HA of Gething et al. (1986) exhibited different fusion characteristics (Wharton et al., 1988a); notably, the peptide with a Glu at the amino-terminus instead of Gly, i.e., HA$_2$ Gly1Glu, was capable of interacting with liposomes but not fusing them. Also, Duzgunes and Gambale (1988) reported that a peptide with a similar substitution did not cause conductance changes in planar membranes, whereas the wild-type peptide did.

The molecular mechanism of HA-mediated fusion is unknown. The conformational change in the HA could bring membranes close enough together to overcome the repulsive forces arising from their shell of hydration (Rand, 1981). The amino-terminus could then interact with one or other of the membranes, causing a local destabilization and subsequent fusion. It is still unclear with which membrane the amino-terminus of HA$_2$ interacts—the viral or the endosomal membrane—but, in the absence of exogenous membranes, the amino-terminus appears to insert into the viral membrane (Ruigrok et al., 1986a). If this is the case, when endosomal membranes are present, alternative regions of the HA molecule—possibly residues at the viral membrane distal end of the coiled coil of α-helices in $HA_2$—may interact with the endosomal membranes. Alternatively, the amino-terminus of $HA_2$ could insert into the endosomal membrane and some other event could occur at the viral membrane, possibly involving in some way the acyl chain covalently attached to $HA_2$ (Lambrecht and Schmidt, 1986), to destabilize its structure. Since at a low pH level, HA remains as a stable trimeric 11-nm-long rod protruding from the viral membrane (Ruigrok *et al.*, 1988), it is unclear how membranes are brought close enough for fusion to occur. Many future experiments are aimed at elucidating the molecular basis of membrane fusion. The destabilization could involve a transition of lipid packing from a lamellar Lα to a hexagonal $H_{11}$ state (Gruner *et al.*, 1985) that has been reported to occur in $Ca^{2+}$-mediated fusion of negatively charged liposomes (Ellens *et al.*, 1986). In this respect, the amino-terminus of $HA_2$ may interact with membranes in similar ways to some toxins, e.g., diphtheria or tetanus toxins, which also interact with membranes in a pH-dependent manner (Neville and Hudson, 1986).

## VII. CONCLUSIONS

We have described antigenic variation and the receptor-binding and fusion properties of influenza virus HA. Although vaccines may be effective, the phenomenon of antigenic drift means that the duration of the clinical value of vaccines is limited; constant monitoring of viruses in circulation is essential to ensure the availability of the most efficient vaccines. Armed with an understanding of the nature of HA–receptor interactions, it is hoped that agents might be developed that are capable of binding very tightly to HA and that competitively prevent virus–cell binding and subsequent infection. Such molecules may have clinical value as antiviral agents. Similarly, knowledge of the fusion mechanism may lead to the identification of molecules that block the required changes in HA conformation, (e.g., by the binding of a multivalent molecule between adjacent $HA_1$ subunits, perhaps by binding to the receptor pocket), thereby preventing the conformational change of HA, as this seems to require the distal regions of $HA_1$ to move apart. Since both membrane fusion and receptor-binding activities involve highly conserved regions of the molecule and alterations in sequence of these regions have considerable effects on their functions, it is less likely that viruses resistant to such agents arise by spontaneous mutation. Thus, the duration of their efficacy will be prolonged.

## REFERENCES

Amit, A. G., Marizza, R. A., Phillips, S. E. V., and Poljak, R. J., 1986, Three-dimensional structure of an antigen-antibody complex at 2.8 Å resolution, *Science* **233**:747–753.
Anders, E. M., Scalzo, A. A., Rogers, G. N., and White, D. O., 1986, Relationship between

mitogenic activity of influenza viruses and the receptor-binding specificity of their haemagglutinin molecules, *J. Virol.* **60:**476–482.

Both, G. W., Sleigh, M. J., Cox, N. J., and Kendal, A. P., 1983, Antigenic drift in influenza virus H3 haemagglutinin from 1968 to 1980. Multiple evolutionary pathways and sequential amino acid changes at key antigenic sites, *J. Virol.* **48:**52–60.

Brand, C. M., and Skehel, J. J., 1972, Crystalline antigen from the influenza virus envelope, *Nature New Biol.* **238:**145–147.

Caton, A. J., Brownlee, G. G., Yewdell, J. W., and Gerhard, W., 1982, The antigenic structure of the influenza virus A/PR/8/34 haemagglutinin (H1 subtype), *Cell* **31:**417–427.

Coleman, M. T., Dowdle, W. R., Pereira, H. G., Schild, G. C., and Chang, W. K., 1968, The Hong Kong/68 influenza A2 variant, *Lancet* **2:**1384–1413.

Daniels, R. S., Douglas, A. R., Skehel, J. J., and Wiley, D. C., 1983a, Analyses of the antigenicity of influenza haemagglutinin at the pH optimum for virus-mediated membrane fusion, *J. Gen. Virol.* **64:**1657–1661.

Daniels, R. S., Douglas, A. R., Skehel, J. J., Waterfield, M. D., Wilson, I. A., and Wiley, D. C., 1983b, Studies of the influenza virus haemagglutinin in the pH5 conformation, in: *The Origin of Pandemic Influenza Viruses* (W. G. Laver, ed.), pp. 1–7, Elsevier, New York.

Daniels, R. S., Douglas, A. R., Skehel, J. J., Wiley, D. C., Naeve, C. W., Webster, R. G., Rogers, G. N., and Paulson, J. C., 1984, Antigenic analyses of influenza virus haemagglutinins with different receptor-binding specificities, *Virology* **138:**174–177.

Daniels, R. S., Skehel, J. J., and Wiley, D. C., 1985a, Amino acid sequences of haemagglutinins of influenza viruses of the H3 subtype isolated from horses, *J. Gen. Virol.* **66:** 457–464.

Daniels, R. S., Douglas, A. R., Skehel, J. J., and Wiley, D. C., 1985b, Antigenic and amino acid sequence analyses of influenza viruses of the H1N1 subtype isolated between 1982 and 1984, *Bull. WHO* **63:**273–277.

Daniels, R. S., Downie, J. C., Knossow, M., Skehel, J. J., Wang, M.-L., and Wiley, D. C., 1985c, Fusion mutants of influenza virus haemagglutinin glycoprotein, *Cell* **40:**431–439.

Daniels, R. S., Jeffries, S., Yates, P., Schild, G. C., Rogers, G. N., Paulson, J. C., Wharton, S. A., Douglas, A. R., Skehel, J. J., and Wiley, D. C., 1987, The receptor binding and membrane fusion properties of influenza virus variants selected using anti-haemagglutinin monoclonal antibodies, *EMBO J.* **6:**1459–1465.

Doms, R. W., Helenius, A. H., and White, J., 1985, Membrane fusion activity of the influenza virus haemagglutinin, *J. Biol. Chem.* **260:**2973–2981.

Doms, R. W., Gething, M.-J., Henneberry, J., White, J., and Helenius, A., 1986, Variant influenza virus haemagglutinin that induces fusion at elevated pH, *J. Virol.* **57:**603–613.

Duzgunes, N., and Gambale, F., 1988, Membrane action of synthetic N-terminal peptides of influenza virus haemagglutinin and its mutants, *FEBS Lett.* **227:**110–114.

Ellens, H., Bentz, J., and Szoka, F. C., 1986, Fusion of phosphatidylethanolamine-containing liposomes and the mechanism of the $L_\alpha$-$H_{II}$ phase transition, *Biochemistry* **25:**4141–4147.

Fang, R., Min Jou, W., Huylebroeck, D., Devos, R., and Fiers, W., 1981, Complete structure of A/duck/Ukraine/63 influenza haemagglutinin gene: Animal virus as progenitor of human H3 Hong Kong 1968 influenza haemagglutinin, *Cell* **25:**315–323.

Fazekas, de St. Groth, S., 1977, Antigenic, adaptive and adsorptive variants of the influenza A haemagglutinin, in: *Topics in Infectious Diseases*, Vol. 3 (R. G. Laver, H. Bachmayer, and R. Weil, eds.), pp. 25–48, Springer-Verlag, Vienna.

Gerhard, W., Yewdell, J., Frankel, M. E., and Webster, R. G., 1981, Antigenic structure of influenza virus haemagglutinin defined by hybridoma antibodies, *Nature (Lond.)* **290:** 713–717.

Gething, M.-J., Doms, R. W., York, D., and White, J., 1986, Studies on the mechanism of membrane fusion: Site-specific mutagenesis of the haemagglutinin of influenza virus, *J. Cell Biol.* **102:**11–23.

Graves, P. N., Schulman, J. F., Young, J. F., and Palese, P., 1983, Preparation of influenza virus subviral particles lacking the HA₁ subunit of haemagglutinin: Unmasking of cross reactive HA₂ determinants, *Virology* **126**:106–116.

Gruner, S. M., Cullis, P. R., Hope, M. J., and Tilcock, C. P. S., 1985, Lipid polymorphism: The molecule basis of nonbilayer phases, *Annu. Rev. Biophys. Biophys. Chem.* **14**:211–238.

Huang, R. T. C., Rott, R., and Klenk, H.-D., 1981, Influenza viruses cause haemolysis and fusion of cells, *Virology* **110**:243–247.

Huddleston, J. A., and Brownlee, G. G., 1982, The sequence of the nucleoprotein gene of human influenza A virus, strain A/NT/60/68, *Nucl. Acid. Res.* **10**:1029–1038.

Klenk, H.-D., Rott, R., Orlich, M., and Blodorn, J., 1975, Activation of influenza A viruses by trypsin treatment, *Virology* **68**:426–439.

Knossow, M., Daniels, R. S., Douglas, A. R., Skehel, J. J., and Wiley, D. C., 1984, Three-dimensional structure of an antigenic mutant of the influenza virus haemagglutinin, *Nature (Lond.)* **311**:678–680.

Lambrecht, B., and Schmidt, M. F. G., 1986, Membrane fusion induced by influenza virus haemagglutinin requires protein bound fatty acids, *FEBS Lett.* **202**:127–132.

Laver, W. G., and Webster, R. G., 1973, Studies on the origin of pandemic influenza. III. Evidence implicating duck and equine influenza viruses as possible progenitors of the Hong Kong strain of human influenza, *Virology* **51**:383–391.

Laver, W. G., Air, G. M., Webster, R. G., Gerhard, W., Ward, C. W., and Dopheide, T. A., 1979, Antigenic drift in type A influenza virus: Sequence differences in the haemagglutinin of Hong Kong (H3N2) variants selected with monoclonal hybridoma antibodies, *Virology* **98**:226–237.

Laver, W. G., Air, G. M., and Webster, R. G., 1981, Mechanism of antigenic drift in influenza virus. Amino and sequence changes in an antigenically active region of Hong Kong (H3N2) influenza virus haemagglutinin, *J. Mol. Biol.* **145**:339–361.

Lear, J. D., and de Grado, W. F., 1987, Membrane binding and conformational properties of a peptide representing the amino terminus of influenza virus HA₂, *J. Biol. Chem.* **262**:6500–6505.

Lesk, A. M., and Hardman, K. D., 1982, Computer-generated schematic diagrams of protein structures, *Science* **216**:539–540.

Maeda, T., and Ohnishi, S., 1980, Activation of influenza virus by acidic media causes haemolysis and fusion of erythrocytes, *FEBS Lett.* **122**:283–287.

Matlin, K. S., 1986, The sorting of proteins to the plasma membrane in epithelial cells, *J. Cell Biol.* **103**:2565–2568.

Mulder, J., and Masurel, N., 1958, Pre-epidemic antibody against the 1957 strain of Asiatic influenza in the serum of older persons living in the Netherlands, *Lancet* **1**:810.

Murata, M., Sugahara, Y., Takahashi, S., and Ohnishi, S.-I., 1987, pH-dependent membrane fusion activity of a synthetic twenty amino acid peptide with the same sequence as that of the hydrophobic segment of influenza virus haemagglutinin, *J. Biochem.* **102**:957–962.

Natali, A., Oxford, J. A., and Schild, G. C., 1981, Frequency of naturally occurring antibody to influenza virus antigenic variants selected with monoclonal antibody, *J. Hyg. Camb.* **87**:185–190.

Neville, D. M., and Hudson, T. H., 1986, Transmembrane transport of diphtheria toxin, related toxins, and colicins, *Annu. Rev. Biochem.* **55**:195–224.

Newton, S. E., Air, G. M., Webster, R. G., and Laver, W. G., 1983, Sequence of the haemagglutinin gene of influenza virus A/Memphis/1/71 and previously uncharacterized monoclonal antibody derived variants, *Virology* **128**:495–501.

Pereira, M. S., 1982, Persistence of influenza in a population, in: *Virus Persistence*, Thirty-third Symposium of the Society for General Microbiology (B. W. J. Mahy, A. C. Minson, and G. K. Darby, eds.), pp. 15–37, Cambridge University Press, Cambridge.

Pritchett, T. J., Brossmer, R., Rose, R., and Paulson, J. C., 1987, Recognition of monovalent sialosides by influenza virus H3 haemagglutinin, *Virology* **160**:502–506.

Rand, R. P., 1981, Interacting phospholipid bilayers: Measured forces and induced structural changes, *Annu. Rev. Biophys. Bioeng.* **10**:277–314.

Raymond, F. L., Caton, A. J., Cox, N. J., Kendal, A. P., and Brownlee, G. G., 1983, Antigenicity and evolution amongst recent influenza viruses of the H1N1 subtypes, *Nucl. Acid Res.* **11**:7191–7203.

Raymond, F. L., Caton, A. J., Cox, N. J., Kendal, A. P., and Brownlee, G. G., 1986, The antigenicity and evolution of influenza H1 haemagglutinin from 1950–1957 and 1977–1983: Two pathways from one gene. *Virology* **148**:275–287.

Rogers, G. N., Paulson, J. C., Daniels, R. S., Skehel, J. J., Wilson, I. A., and Wiley, D. C., 1983, Single amino acid substitutions in influenza haemagglutinin change receptor binding specificity, *Nature (Lond.)* **304**:76–79.

Rogers, G. N., Daniels, R. S., Skehel, J. J., Wiley, D. C., Wang, X.-F., Higa, H. H., and Paulson, J. C., 1985, Host-mediated selection of influenza virus receptor variants, *J. Biol. Chem.* **260**:7362–7367.

Ruigrok, R. W. H., Wrigley, N. G., Calder, L. J., Cusack, S., Wharton, S. A., Brown, E. B., and Skehel, J. J., 1986a, Electron microscopy of the low pH structure of influenza virus haemagglutinin, *EMBO J.* **5**: 41–49.

Ruigrok, R. W. H., Martin, S. R., Wharton, S. A., Skehel, J. J., Bayley, P. M., and Wiley, D. C., 1986b, Conformational changes in the haemagglutinin of influenza virus which accompany heat-induced fusion with liposomes, *Virology* **155**:484–497.

Ruigrok, R. W. H., Aitken, A., Calder, L. J., Martin, S. R., Skehel, J. J., Wharton, S. A., Weis, W., and Wiley, D. C., 1988, Studies on the structure of the influenza virus haemagglutinin at the pH of membrane fusion, *J. Gen. Virol.* **69**:2785–2795.

Sato, S. B., Kawasaki, K., and Ohnishi, S.-I., 1983, Haemolytic activity of influenza virus haemagglutinin glycoproteins activated in mildly acidic environments, *Proc. Natl. Acad. Sci. USA* **80**:3153–3157.

Schild, G. C., Oxford, J. S., de Jong, J. C., and Webster, R. G., 1983, Evidence for host-cell selection of influenza virus antigenic variants, *Nature (Lond.)* **303**:706–709.

Scholtissek, C., Rohde, W., Van Hoyningen, V., and Rott, R., 1978a, On the origin of the human influenza virus subtypes H2N2 and H3N2, *Virology* **87**:13–20.

Scholtissek, C., Van Hoyningen, V., and Rott, R., 1978b, Genetic relatedness between the new 1977 epidemic strains (H1N1) of influenza and human influenza strains isolated between 1947 and 1957, *Virology* **89**:613–617.

Scholtissek, C., Burger, H., Kistner, O., and Shortridge, K. F., 1985, The nucleoprotein as a possible major factor in determining host specificity of influenza H3N2 viruses, *Virology* **147**:287–294.

Skehel, J. J., Bayley, P. M., Brown, E. M., Martin, S. R., Waterfield, M. D., White, J. M., Wilson, I. A., and Wiley, D. C., 1982, Changes in the conformation of influenza virus haemagglutinin at the pH optimum of virus-mediated membrane fusion, *Proc. Natl. Acad. Sci. USA* **79**:968–972.

Skehel, J. J., Daniels, R. S., Douglas, A. R., and Wiley, D. C., 1983, Antigenic and amino acid sequence variations in the haemagglutinins of type A influenza viruses recently isolated from human subjects, *Bull. WHO* **61**:671–676.

Smith, W. C., Andrewes, C. H., and Laidlaw, P. P., 1933, A virus obtained from influenza patients, *Lancet* **2**:66–68.

Stegmann, T., Hoekstra, D., Scherphof, G., and Wilschut, J., 1985, Kinetics of pH-dependent fusion between influenza virus and liposomes, *Biochemistry* **24**:3107–3113.

Stegmann, T., Hoekstra, D., and Wilschut, J., 1986, Fusion activity of influenza virus: A comparison between biological and artificial target membrane vesicles, *J. Biol. Chem.* **261**:10966–10969.

Stevens, D. J., Douglas, A. R., Skehel, J. J., and Wiley, D. C., 1987, Antigenic and amino acid sequence analysis of the variants of H1N1 influenza virus in 1986, *Bull. WHO* **65**:177–180.

Tumova, B., and Pereira, H. G., 1965, Genetic interaction between influenza A viruses of human and animal origin, *Virology* **27**:253–261.

Underwood, P. A., Skehel, J. J., and Wiley, D. C., 1987, Receptor binding characteristics of monoclonal antibody-selected antigenic variants of influenza virus, *J. Virol.* **61**:206–208.

Waddell, G. H., Tiegland, M. B., and Sigel, M. M., 1963, A new influenza virus associated with equine respiratory disease, *J. Am. Vet. Med. Assoc.* **143**:587–590.

Wang, M.-L., Skehel, J. J., and Wiley, D. C., 1986, Comparative analyses of the specificities of anti-influenza haemagglutinin antibodies in human sera, *J. Virol.* **57**:124–128.

Ward, C. W., and Dopheide, T. A., 1981, Evolution of the Hong Kong influenza A subtype. Structural relationship between the haemagglutinin from A/duck/Ukraine/63 (Hav7) and the Hong Kong (H3) haemagglutinins, *Biochem. J.* **195**:337–340.

Webster, R. G., Campbell, C. H., and Granoff, A., 1971, The *in vivo* production of "new" influenza A viruses. 1. Genetic recombination between avian and mammalian influenza viruses, *Virology* **44**:317–328.

Webster, R. G., Brown, L. E., and Jackson, D. C., 1983, Changes in the antigenicity of the haemagglutinin molecule of H3 influenza virus at acidic pH, *Virology* **126**:587–599.

Weis, W., 1987, Receptor binding to the influenza virus hemagglutinin, Doctoral thesis, Harvard University, Cambridge, Massachusetts.

Weis, W., Brown, J. H., Cusack, S., Paulson, J. C., Skehel, J. J., and Wiley, D. C., 1988, Structure of the influenza virus haemagglutinin complexed with its receptor, sialic acid, *Nature (Lond.)* **333**:426–431.

Wharton, S. A., 1987, The role of influenza virus haemagglutinin in membrane fusion, *Microbiol. Sci.* **4**:119–124.

Wharton, S. A., Skehel, J. J., and Wiley, D. C., 1986, Studies of influenza virus haemagglutinin-mediated membrane fusion, *Virology* **149**:27–35.

Wharton, S. A., Martin, S. R., Ruigrok, R. W. H., Skehel, J. J., and Wiley, D. C., 1988a, Membrane fusion by peptide analogues of influenza virus haemagglutinin, *J. Gen. Virol.* **69**:1847–1857.

Wharton, S. A., Ruigrok, R. W. H., Martin, S. R., Skehel, J. J., Bayley, P. M., Weis, W., and Wiley, D. C., 1988b, Conformational aspects of the acid-induced fusion mechanism of influenza virus haemagglutinin: Circular dichroism and fluorescence studies, *J. Biol. Chem.* **263**:4474–4480.

White, J., Matlin, K., and Helenius, A., 1981, Cell fusion by Semliki forest, influenza and vesicular stomatitis virus, *J. Cell Biol.* **89**:674–679.

White, J., Helenius, A., and Gething, M.-J., 1982, Haemagglutinin of influenza virus expressed from a cloned gene promotes membrane fusion, *Nature (Lond.)* **300**:658–659.

White, J., Kielian, M., and Helenius, A., 1983, Membrane fusion proteins of enveloped animal viruses, *Q. Rev. Biophys.* **16**:151–195.

WHO Memorandum, 1980, A revision of the system of nomenclature for influenza viruses: A WHO memorandum, *Bull. WHO* **58**:585–591.

Wiley, D. C., and Skehel, J. J., 1987, The structure and function of the haemagglutinin membrane glycoprotein of influenza virus, *Annu. Rev. Biochem.* **56**:365–394.

Wiley, D. C., Wilson, I. A., and Skehel, J. J., 1981, Structural identification of the antibody-binding sites of Hong Kong influenza haemagglutinin and their involvement in antigenic variation, *Nature (Lond.)* **289**:373–378.

Wilson, I. A., Skehel, J. J., and Wiley, D. C., 1981, Structure of the haemagglutinin membrane glycoprotein of influenza virus at 3 Å resolution, *Nature (Lond.)* **289**:366–373.

Winter, G., and Fields, S., 1981, The structure of the gene encoding the nucleoprotein of human influenza virus A/PR/8/34, *Virology* **114**:423–428.

Yewdell, J. W., Webster, R. G., and Gerhard, W., 1979, Antigenic variation in three distinct determinants of an influenza type A haemagglutinin molecule, *Nature (Lond.)* **279**:246–248.

Yewdell, J. W., Caton, A. J., and Gerhard, W., 1986, Selection of influenza A virus adsorptive mutants by growth in the presence of a mixture of monoclonal antihaemagglutinin antibodies, *J. Virol.* **57**:623–628.

Young, J. F., Desselberger, U., and Palese, P., 1979, Evolution of human influenza A viruses in nature: Sequential mutations in the genomes of new H1N1 isolates, *Cell* **18**:73–83.

CHAPTER 4

# Neuraminidase
## Enzyme and Antigen

P. M. COLMAN

## I. INTRODUCTION

The discovery of enzymatic activity on the surface of influenza virus is attributable to Hirst (1942), who observed that it was not possible to reagglutinate red blood cells (RBC) after they had once been agglutinated by the virus. The modification to the cells resulting from the first contact with virus is now known to be the specific removal of sialic acid (N-acetylneuraminic acid) from glycoconjugates on the RBC surface by the viral neuraminidase. The enzyme action of the neuraminidase destroys the receptors for the viral hemagglutinin. Bacterial neuraminidases, as found for example in culture filtrates of *Vibrio cholerae* also render RBC nonagglutinable by influenza virus (Burnet and Stone, 1947).

Hirst originally ascribed the properties of agglutination and receptor destruction to a single protein on the virus. For the parainfluenza viruses, this is indeed the case (Scheid *et al.*, 1972). The neuraminidase and hemagglutinin activities of influenza A and B viruses, however, are associated with separate polypeptide chains. This chapter discusses current knowledge of the gene and protein structure of the neuraminidase, particularly in relationship to its enzymatic and antigenic properties. Recent reviews of neuraminidase include those by Colman and Ward, (1985), Colman (1984), and Bucher and Palese (1975).

P. M. COLMAN • CSIRO Division of Biotechnology, Parkville 3052, Australia.

## A. Biological Properties

### 1. Receptor Removal

Neuraminidase catalyzes the removal of terminal sialic acid residues from oligosaccharide chains. N-acetylneuraminic acid is but one member of a family of neuraminic acid derivatives known collectively as sialic acids. The enzyme from influenza and other ortho- and paramyxoviruses, preferably cleaves the $\alpha2-3$ linkage between terminal N-acetylneuraminic acid and the penultimate sugar (Corfield et al., 1982, 1983; Paulson et al., 1982). With some exceptions, bacterial neuraminidases show a broad range of specificity for $\alpha2-3$, $\alpha2-6$, and, less frequently, $\alpha2-8$ linkages (Drzenick, 1972; Davis et al., 1979; Uchida et al., 1979; Milligan et al., 1980).

The biological consequences of this activity are not well understood. Cells of the upper respiratory tract are a common target for infection by influenza virus. These cells are bathed in mucosal secretions rich in sialic acid-containing macromolecules (Gottschalk 1958, 1972; Allen, 1983). Entrapment of the virus in these secretions by the action of the hemagglutinin is likely (Burnet et al., 1947; Burnet, 1948) in the absence of neuraminidase activity. The virus can also be immobilized at the surface of infected cells. Newly synthesized virus particles bud from the plasma membrane of infected cells. The hemagglutinin and neuraminidase molecules are both glycosylated by host-cell processes; in particular, terminal sialic acid is added to these oligosaccharide chains (Basak et al., 1985). Cleavage of these sugars is a prerequisite for elution of virus from infected cells (Palese et al., 1974; Palese and Compans, 1976; Griffin and Compans, 1979). Release of virus from cells is also controlled by cell-surface sialic acid. Griffin et al. (1983) observed aggregation of progeny virus particles at the cell surface under conditions of hexose starvation, which blocks the addition of sialic acid to the virus glycoproteins and results in virions with viable hemagglutinin molecules but no neuraminidase activity.

While the specificity of the hemagglutinin and the neuraminidase of influenza A and B viruses is for N-acetylneuraminic acid, influenza C viruses are directed to N-acetyl-9-O-acetylneuraminic acid receptors on cell surfaces. The presence of receptor-destroying activity on influenza C virions was first observed by Hirst (1950), but only recently has the enzymatic activity been characterized. Herrler et al. (1985a) demonstrated the involvement of sialic acid in the influenza C virus receptor by showing that, under certain conditions, bacterial neuraminidases could remove the receptor from chick RBC. Furthermore, influenza C virus converts N-acetyl-9-O-acetylneuraminic acid into N-acetylneuraminic acid (Herrler et al., 1985b). The influenza C virus enzyme is therefore not a neuraminidase, but a neuraminate-O-acetylesterase. Influenza C virus is unique among the myxoviruses in that all three functions of the membrane glycoproteins (i.e., attachment, receptor destruction, and fusion) are apparently associated with a single polypeptide.

## 2. Membrane Fusion

Debate over a role for neuraminidase in fusion of viral and host-cell membranes was initiated by Huang et al. (1980). They observed that liposomes bearing viral glycoproteins were fused with chick embryo cells only when neuraminidase was present, either bound to the liposomes or in a soluble form. Studies with cloned hemagglutinin (White et al., 1982a) indicate no requirement for neuraminidase, however, at least for low pH-dependent fusion of cells; there is also no requirement for cell-surface sialic acid for fusion (White and Helenius, 1980; White et al., 1982b). Conformational changes in the hemagglutinin molecule following exposure to acid pH (Skehel et al., 1982) and the recent characterization of hemagglutinin mutants modulating the pH optimum for fusion (Daniels et al., 1985) provide direct structural explanations for the role of the hemagglutinin in low pH membrane fusion. Huang et al. (1985) have now extended their earlier studies to include the observation that anti-neuraminidase antisera can inhibit fowl plague virus-induced fusion of erythrocytes and that fusion can be restored upon addition of Vibrio cholerae neuraminidase. Partial inhibition of hemolysis was also caused by the same antisera. Huang et al. (1985) went on to show that the surface properties of erythrocytes were the determining factors in low pH-dependent fusion and hemolysis by influenza virus. The influence of target membrane composition on fusion has also been demonstrated by Haywood and Boyer (1985). It therefore appears to be important to consider target membrane variability in assessing factors influencing virus-mediated membrane fusion.

## 3. Antigenicity

Influenza virus neuraminidase is both enzyme and antigen. Anti-neuraminidase antibodies appear incapable of neutralising viral infectivity, except at very high concentrations (Kilbourne et al., 1968). They do, however, modify the disease in favor of the host by reducing both the levels of virus in lungs and the extent of lung lesions (Schulman, 1975). Antigenic variation of the neuraminidase polypeptides is as extensive as that of the hemagglutinin, and neuraminidase variants can be selected by monoclonal antibodies (Webster et al., 1982). It is therefore likely that antibodies to neuraminidase have an important role in the epidemiology of influenza.

Nine serologically distinct subtypes of neuraminidase are now recognized (Schild et al., 1980), only two of which, N1 and N2, have been found so far on human influenza viruses. N1 neuraminidase was associated with viruses circulating between 1933 and 1957 and reappeared in

1977 on the Russian influenza strains. The N2 enzyme was first characterized on human influenza virus on the Asian strains which arose in 1957. Although the hemagglutinin subtype changed to H3 in 1968 with the appearance of the Hong Kong influenza virus, N2 persists on these strains, which are still in circulation.

## 4. Sialic Acid and Biological Regulation

Little is known about the secondary effects of the viral neuraminidase on the host during an infection. Levels of neuraminidase activity in the lungs of mice decline precipitously 5 days after initial infection despite the persistence of viral antigens at elevated levels for a further 8 weeks (Astry et al., 1984). Sialic acid on the surface of cells and soluble glycoproteins plays a variety of control functions. These include masking penultimate sugar residues on oligosaccharide chains from specific receptors on cell surfaces (Ashwell and Morell, 1974; Jancik et al., 1975; Schauer, 1982, 1985), modulation of complement activation (Wedgewood et al., 1956; Kazatchkine et al., 1979; McSharry et al., 1981; Hirsch et al., 1980, 1981), and modification of immune responses (Woodruff and Gesner, 1969; Kolb-Bachofen and Kolb, 1979). Desialylation of glycoconjugates encountered by the virus may therefore disturb the regulatory processes controlled by sialic acid.

## B. Physical Characteristics

### 1. Membrane Association, Size, and Shape

Neuraminidase is an integral membrane glycoprotein. It is fixed in the viral membrane by a hydrophobic anchor sequence stretching from amino acid residues 7–35 (Fields et al., 1981; Ward et al., 1982). This orientation is opposite that of the viral hemagglutinin (Skehel and Waterfield, 1975; Dopheide and Ward, 1981).

Neuraminidase has been solubilized from viral membranes by treatment with detergents and imaged in the electron microscope. Images of negatively stained protein show a mushroom morphology, with a box-shaped head measuring $80 \times 80 \times 40$ Å attached to a slender stalk of diameter 15 Å and length up to 100 Å (Laver and Valentine, 1969; Wrigley et al., 1973). In the absence of detergent on the microscope grid, aggregation of these protomers is observed to occur via the end of the stalk distal to the globular head, a region of the protein formerly in contact with the lipid of the viral envelope. The molecular weight of the detergent-solubilized polypetide is 60 kDa, or 240 kDa for the tetramer (Bucher and Kilbourne, 1972; Wrigley et al., 1973). Neuraminidase solubilized by proteolysis from the virus has lost the stalklike structure (Wrigley et al., 1973) but retains the full enzymatic and antigenic capacity of virus-associated neuraminidase (Drzeniek *et al.*, 1968; Wrigley *et al.*, 1977; Laver, 1978). Electron microscopic images of the globular head regions provide direct evidence for the tetrameric nature of the enzyme and further suggest that the four subunits are in a coplanar arrangement (Wrigley *et al.*, 1973, 1977). Protease-solubilized heads have a tetramer molecular weight of only 200 kDa (Wrigley *et al.*, 1973; Blok *et al.*, 1982). The four polypeptide chains in the tetramer are identical; X-ray diffraction studies have shown that they are related by a fourfold rotation symmetry axis (Varghese *et al.*, 1983).

### 2. Surface Distribution

Bucher and Palese (1975) estimated that there are approximately 50 neuraminidase tetramers on the surface of a virion. The calculation is based on estimates of neuraminidase being 7% of total viral protein. Estimates of the surface area of influenza virions were recently confirmed by electron microscopic examination of ice-embedded particles (Booy *et al.*, 1985). The radius of the virus is 635 Å (±35 Å), and the resulting surface area is sufficient for some 500 neuraminidase tetramers to cover the viral surface completely (here using the X-ray-determined dimensions of the heads as $100 \times 100$ Å (see Section II), or alternately 1000 hemagglutinin trimers of lateral surface area 5000 Å$^2$ at their distal ends (Wilson *et al.*, 1981). However, apparently the packing density of viral glycoproteins on the surface is only about 50%. Booy *et al.* (1985) estimate that there are 530 peplomers on the surface of a typical virus particle, in good agreement with earlier estimates of 550 (Tiffany and Blough, 1970).

Experiments with immunoglobulin–gold staining of fixed influenza virus particles imply that the neuraminidase is not uniformly distributed over the viral envelope but is capped (Murti and Webster, 1986). Studies with Fab–gold would be a more convincing demonstration of this claim.

## C. Gene Structure

### 1. Types A and B

The neuraminidase polypeptide is coded for by a single segment of RNA, which normally runs as the sixth least mobile band of total viral RNA extract (reviewed by Scholtissek, 1978). Gene sequences from representative strains of the N1, N2, N7, N8, and N9 subtypes of influenza A virus and from type B influenza have been determined. Partial sequence data at the 3' end of all the A subtypes barring N9 are also published. The information available is summarized in Table I. The nucleotide sequence homology between subtypes is approximately 0.5 (Elleman *et al.*, 1982). The homology between the subtype sequences is higher in the noncoding regions as predicted (Robertson, 1979). Two different size classes of gene

TABLE I. Gene-Sequence Data Base for Influenza
Neuraminidases

| Subtype | Strain | References |
|---------|--------|------------|
| N1 | A/PR8/34 | Fields et al. (1981) |
| N1 | A/WSN/33 | Hiti and Nayak (1982) |
| N1 | A/NJ/8/76 | Miki et al. (1983) |
| N1 | A/USSR/90/77 | Concannon et al. (1984) |
| N1 | A/Parrot/Ulster/73 | Steuler et al. (1984) |
| N2 | A/RI/5-/37 | Elleman et al. (1982) |
| N2 | A/Tokyo/3/67 | Lentz et al. (1984) |
| N2 | A/NT/60/68 | Bentley and Brownlee (1982) |
| N2 | A/Udorn/307/72 | Markoff and Lai (1982) |
| N2 | A/Vic/3/75 | Van Rompuy et al. (1982) |
| N2 | A/Bangkok/1/79 | Martinez et al. (1983) |
| N2 | A/Chicken/Penn/83 | Desphande et al. (1985) |
| N7 | A/Equ/Cor/16/74 | Dale et al. (1986) |
| N8 | A/Equ/Ken/1/81 | Dale et al. (1986) |
| N9 | A/Tern/Australia/G70C/75 | Air et al. (1985a) |
| N9 | A/Whale/Maine/1/84 | Air et al. (1987) |
| N1-N8 | 3' terminal sequences | Blok and Air (1980; 1982a;c) |
| B | B/Lee/40 | Shaw et al. (1982) |

are observed as a result of a deletion in some strains from that region of
the gene coding for the protein stalk. These block deletions (Blok and Air,
1982a) are not a characteristic of subtype or species specificity and have
been observed in N1 and N2 genes, in both human and avian viruses.
Apart from occasional insertions and deletions of base triplets, the
lengths of the gene region coding for the head of the protein are the same
across subtypes. Only one of these genes contains a second open reading
frame and that is B/Lee/40 (Shaw et al., 1982). This alternate frame
overlaps the main reading frame for most of its length and could code for
a 100-amino acid protein. Evidence for the existence of this protein in
infected cells has now been found (Lamb et al., 1983), although its function is unknown. The B-type neuraminidase gene is further distinguished
from the A sequences by the presence of an elongated flanking region at
both the 3' and 5' ends. In the complementary DNA (cDNA) sense, approximately 20 bases precede the initiating AUG codon and some 30
bases follow the termination codon of A-subtype neuraminidases, while
54 and 100 bases, respectively, flank the coding region of the B neuraminidase gene. It is not known whether the failure of A and B influenza genes
to form mixed recombinants is due to peculiarities in their gene structures or protein structures. The translated protein sequences from these
genes are discussed below.

## 2. Type C

The glycoprotein of influenza type C virus is related to the hemagglutinin of type A and B viruses (Pfeifer and Compans, 1984). Neverthe-

less, it is also a receptor-destroying enzyme and is briefly included here on that basis. The gene codes for a protein with a hydrophobic leader sequence, an arginine residue at the site of cleavage into a two-chain structure, a hydrophobic sequence at the second N-terminal segment, and a hydrophobic C-terminal segment believed to be the membrane anchor (Pfeifer and Compans, 1984). Low-level (10–15%) sequence homology exists between the influenza type C glycoprotein and the hemagglutinin of types A and B. The enzymatic activity of this protein is not a neuraminidase, but an esterase, and the absence of sequence homology to the type A and B neuraminidases is therefore not surprising.

## 3. HN Gene of Paramyxoviruses

The hemagglutinin–neuraminidase (HN) genes from four paramyxoviruses have been sequenced. The gene from Simian virus 5 (SV5) encodes a protein of 565 amino acids with a single hydrophobic domain sufficient for membrane anchoring at residues 18–36 (Hiebert et al., 1985). Other data support the orientation of the C-terminal on the exterior of the viral envelope (Schuy et al., 1984). Hiebert et al. (1985) detected no significant level of sequence homology between HN and either hemagglutinin or neuraminidase of influenza, although the membrane orientation is the same as for neuraminidase.

Blumberg et al., (1985) sequenced the HN gene from Sendai virus. The primary translation product is 576 amino acids in length, and residues 36–59 comprise the putative trans-membrane sequence. Blumberg and co-workers claim that low levels of sequence homology between Sendai virus HN and both influenza hemagglutinin and neuraminidase speak for a common ancestral gene. With respect to their alignment of neuraminidase with HN, it should be noted that only five of the active site residues in neuraminidase (see Section II.B.4) are matched in the Sendai HN sequence and that three of these are not conserved between Sendai and Simian virus 5 HN.

HN sequences from two different strains of Newcastle disease virus (NDV) have also been reported (Miller et al., 1986; Jorgensen et al., 1987), along with a sequence from parainfluenza type 3 (Elango et al., 1986). Comparison of sequences from all four virus types—SV5, Sendai, NDV, and parainfluenza (Jorgensen et al., 1987)—shows that only 65 of some 650 amino acids are strictly conserved. Although secondary structure predictions show alternating β-strand/loop topology for the HN proteins, reminiscent of the influenza neuraminidase structure (see Section II.B.1), the sequence homology between HN and neuraminidase is too low to support the claim that the structures are folded similarly. Such a possibility should not be ruled out, however, given the observation of similar structures in two sequence-unrelated variable surface glycoproteins from trypanosomes (Metcalf et al., 1987). There is, then, no convincing evidence yet for a structural relationship between HN and influenza neuraminidase.

## II. PROTEIN STRUCTURE

### A. Chemical Aspects

The globular head region of the neuraminidase from A/Tokyo/3/67 has been sequenced entirely by protein sequencing methods (Ward *et al.*, 1982). Other data come from translated gene-sequence analyses discussed in Section I.C.1. Direct studies of the protein have been important (1) for identifying the site at which pronase cleaves the polypeptide to liberate neuraminidase heads from the virus, (2) to determine whether post-translational cleavages have occurred, (3) to locate which of the Asn-X-Ser/Thr sequences are glycosylated, and (4) to fix the pattern of intermolecular disulfide bonds.

The sequence comparisons discussed below suggest a partitioning of the structure into four sections: a cytoplasmic domain of 6 amino acids, a hydrophobic membrane anchor of 29 amino acids, a stalk of approximately 45 amino acids (or as few as 25 amino acids in strains with stalk deletions), and a head of approximately 390 amino acids. Occasionally, the three N-terminal sections are referred to collectively as the stalk.

### 1. Sequence Alignment across Types and Subtypes

Representative sequences from the five A subtypes, N2, N1, N7, N8, N9, and type B neuraminidase, are shown aligned in Table II. N2, N1, and B were aligned by a three-way matching algorithm (Murata *et al.*, 1985), and the other three sequences were grafted manually to this alignment. There are some important differences to previously published alignments (Colman, 1984; Colman and Ward, 1985). Most significant is the improved matching over the C-terminal 100 residues, where only very low levels of homology had been detected previously between A- and B-type neuraminidases. In particular, the alignment now demonstrates that Arg 371 (N2 numbering) and Tyr 406 are strictly conserved across the subtypes, an observation consistent with the location of these groups within the catalytic cavity (see Section II.B.4). Thr 365 is also now seen to be invariant.

The homology between subtypes is highest for comparisons within the A-type neuraminidases, as shown in the similarity matrix in Table III. In the head region of the sequence, 74 of 390 residues are matched in all the sequences, with the highest levels of homology occurring in the region 118–375 (N2 numbers, homology 0.25) and lower levels at the N- and C-terminal ends. In structural terms, these latter sequences comprise the sixth β-sheet and the outer half of the fifth sheet (see Section II.B.1) and the sequence homology there is only 0.10 over 120 amino acids. The five A strain sequences remain as homologous in this section of the structure as elsewhere in the heads.

TABLE II. Alignment of Six Subtype Sequences of Influenza Virus
Neuraminidase[a]

```
              5                    10                   15                   20                   25
N2  Met Asn Pro Asn Gln  -  Lys Thr Ile Thr Ile Gly Ser Val Ser Leu Thr Ile Ala Thr Val Cys Phe Leu Met Gln
N1      Asn     Asn Gln  -  Lys Ile Ile Thr Ile Gly Ser Ile Cys Met Ala Ile Gly  -   -   "   "  Ile Ile Ser
N7      Asn     Asn Gln  -  Lys Leu Phe Ala Ser Ser Gly Ile Ala Ile Val Leu Gly Ile Ile Asn Leu Leu Ile Gly
N8      Asn     Asn Gln  -  Lys Ile Ile Ala Ile Gly Ser Ala Ser Leu Gly Ile Leu Ile Leu Asn Val Ile Leu His
N9      Asn     Asn Gln  "  Lys Ile Leu Cys Thr Ser Ala Thr Ala Leu Val Ile Gly Thr Ile Ala Val Leu Ile Gly
B       Leu     Ser Thr Val Gln Gln Thr Leu Thr Leu Leu Leu Thr Ser Gly Gly Val  "   "   "   "   "  Leu Leu Ser

             30                   35                   40                   45                   50
N2  Ile Ala Ile Leu Ala Thr Thr Val Thr Leu His Phe Lys Gln His  -  Glu Cys Asp Ser Pro Ala Ser Asn Gln Val
N1  Leu Ile Leu Gln Ile Gly Asn Ile Ile Ser Ile Trp Val Ser His  -  Ser Ile Gln Thr Gly Ser Gln Asn His Thr
N7  Ile Ser Asn Met Ser Leu Asn Ile Ser Leu Tyr Ser Lys Gly Glu Ser His Lys Asn Asn Asn Leu Thr Cys Thr Asn
N8  Val Val Ser Ile Ile Val Thr Val Leu Asn Gly  -  Thr Gly Leu Asn Cys Asn Cys Ser Asn Gly Thr Ile Ile
N9  Ile Thr Asn Leu Gly Leu Asn Ile Gly Leu His Leu Lys Pro Ser Cys Asn Cys Ser His Ser Gln Pro Glu Ala Thr
B   Leu Tyr Val Ser Ala Ser Leu Ser Tyr Leu Leu Tyr Ser Asp Val  "  Leu Leu Lys Phe Ser Ser Thr Lys Thr Thr

             55                   60                   65                   70                   75
N2  Met Pro Cys Glu Pro Ile Ile Ile Glu Arg  -  Asn Ile Thr  "   -  Glu Ile Val Tyr Leu Asn Asn Thr Thr Ile Glu Lys
N1  Gly Ile Cys Asn Gln Arg Ile Ile Thr Tyr Glu Asn Ser Thr Trp Val Asn Gln Thr Tyr Val Asn Ile Ser Asn Thr Asn Val
N7  Ile Asn Gln Asn Asp Thr Thr Met Val Asn  "   "   "   "   "   -   -  Thr Tyr Ile Asn Asn Ala Thr Ile Ile Asp Lys
N8  Arg Glu Tyr Asn Glu Thr Val Arg Val  "   "  Glu Arg  "   -   "  Ile Thr Gln Trp Tyr Asn Thr Asn Thr Ile Glu Tyr
N9  Asn Ala Ser Gln Thr Ile Ile Asn Asn Tyr Tyr Asn Asp Thr  -   "  Asn Ile Thr Gln Ile Ser Asn Thr Asn Ile Gln Val
B   Ala Pro Thr Met Ser Leu Glu Cys Thr  "   -  Asn Ala Ser  "   "  Asn Ala Gln Thr Val Asn His Ser Ala Thr Lys Glu

             80                   85                   90                   95                   100
N2  Glu Ile Cys Pro Glu Val Val Glu Tyr Arg Asn Trp Ser Lys Pro Gln  "  Cys Gln Ile Thr Gly Phe Ala  -   -   -  Pro Phe
N1  Val Ala Gly Lys Asp Thr Thr Ser Met Thr Leu Ala Gly Asn Ser Ser Leu     Pro Ile Arg Gly Trp Ala  -   "   "  Ile Tyr
N7  Ser Thr  -  Lys Glu Asn Pro Gly Tyr Leu Leu Leu Asn Lys Ser Leu  -     Asn Val Glu Gly Trp Val  "   "   "  Val Ile
N8  Ile Glu Arg Pro Ser Asn Glu Tyr Tyr Met Asn Asn Thr Glu Pro Leu  -     Glu Ala Gln Gly Phe Ala  -   "   "  Pro Phe
N9  Glu Glu  -  Arg Ala Ile Arg Asp Phe Asn Asn Leu Thr Lys Gly Leu  "     Thr Ile Asn Ser Trp His  -   -   "  Ile Tyr
B   Met Thr Phe Pro Pro Pro Glu Pro Glu Trp Thr Tyr Pro Arg Leu Ser  "     Gln Gly Ser Thr Pro Gln Lys Ala Leu Leu Ile

             105                  110                  115                  120                  125
N2  Ser Lys Asp Asn Ser Ile Arg Leu Ser Ala Gly Gly Asp Ile Trp Val Thr Arg Glu Pro Tyr Val Ser Cys Asp
N1  Ser Lys Asp Asn Ser Ile Arg Ile Gly Ser Lys Gly Asp Val Phe Val Ile             Phe Ile Ser         Ser
N7  Ala Lys Asp Asn Ala Ile Phe Gly Glu Ile Lys Gly Gln Ser Ile Leu Val Thr         Tyr Val Ser         Asp
N8  Ser Lys Asp Asn Gly Ile Arg Ile Gly Ser Arg Gly His Val Phe Val Ile             Phe Val Ser         Ser
N9  Gly Lys Asp Asn Ala Val Ala Val Gly Glu Val Gln Asp Asn Thr Leu Val Thr         Tyr Val Ser         Asp
B   Ser Pro His Arg Phe Gly Glu Ile Lys Gly Asn Ser Ala Pro Leu Ile Ile             Phe Val Ala         Gly

             130                  135                  140                  145                  150
N2  Pro Gly Lys Cys Tyr Gln Phe Ala Leu Gly Gln Gly Thr Thr Leu Asp Asn Lys His Ser Asn Gly Thr Ile His
N1  His Leu Glu     Arg Thr Phe Phe     Thr Gln Gly Ala Leu Leu Asn Asp Lys His Ser         Gly     Val Lys
N7  Pro Leu Ser     Lys Met Tyr Ala     His Gln Gly Thr Phe Ile Leu Asn Asp Lys His Ser     Thr     Thr His
N8  Pro Leu Glu     Arg Thr Phe Phe     Thr Gln Gly Ser Leu Leu Asn Asp Lys His Ser         Gly     Thr His
N9  Pro Asp Glu     Arg Phe Tyr Ala     Ser Gln Gly Thr Thr Ile Arg Gly Lys His Ser         Gly     Ile His
B   Pro Lys Glu     Arg Gln His Phe Ala     Thr His Tyr Ala Ala Gln Pro Gly Gly Tyr Tyr     Gly     Arg Lys

             155                  160                  165                  170                  175
N2  Asp Arg Ile Pro His Arg Thr Leu Leu Met Asn Glu Leu Gly Val Pro Phe His Leu  -  Gly Thr Lys Gln Val Cys
N1      Ser Pro Tyr     Ala     Met Ser Cys Pro Ile Gly Glu Ala Pro Ser Pro Tyr Asn Ser Arg Phe Glu Ser
N7      Thr Ala Phe     Gly     Ile Ser Thr Pro Leu Gly Ser Pro Pro Thr Val Ser Asn Ser Glu Phe Ile Cys
N8      Ser Pro Tyr     Thr     Met Ser Val Lys Val Gly Ser Asn Pro Asn Val Tyr Gln Ala Ala Phe Glu Ser
N9      Ser Gln Tyr     Ala     Ile Ser Trp Pro Leu Ser Ser Pro Pro Thr Val Tyr Asn Ser Arg Val Glu Cys
B       Asn Lys Leu     His     Val Ser Val Leu Gly Lys Ile Pro Thr Val Glu Asn Ser Ile Phe His Met

             180                  185                  190                  195                  200
N2  Val Ala Trp Ser Ser Ser Ser Cys His Asp Gly Lys Ala Trp Leu His Val Cys Val Thr Gly Asp Asp Arg Asn
N1  Val Ala     Ala Ser Ala     Tyr         Met Gly Thr Val Lys Asp Ile Ile Ser     Pro Arg Gly Ala Gly
N7  Val Gly     Ser Thr Ser     His         Val Asn Arg Met Thr Ile Cys Val Gln     Asp Asn Glu Asn
N8  Val Ala     Ala Thr Ala     His         Lys Lys Val Met Ser Ile Cys Ile Ser     Pro Asn Ala Gln
N9  Ile Gly     Ser Thr Ser     His         Lys Thr Arg Met Ser Ile Cys Ile Ser     Pro Asn Asn Asn
B   Ala Ala     Gly Ser Ala     His         Arg Glu Thr Thr Ile Gly Val Asp     Pro Asp Asn Asp

             205                  210                  215                  220                  225
N2  Ala Thr Ala Ser Phe Ile Tyr Asp Gly Arg Leu Val Asp Ser Ile Gly Ser Trp Ser Gln Asn Ile Leu Arg Thr
N1      Val     Val Leu Lys     Asn Gly Ile Ile Thr Glu Thr Ile Lys Ser Trp Arg Lys Gln
N7      Thr     Thr Val Tyr     Asn Lys Asn Leu Thr Thr Thr Gly Ser Asp Thr Arg Ala Asp
N8      Val     Val Val Asn     Gly Gly Val Pro Val Asp Ile Ile Asn Ser Trp Gly Arg Asp
N9      Ser     Val Ile Trp     Asn Arg Arg Thr Tyr Asn Ser Thr Asn Thr Lys Thr Arg Ala Asp
B       Leu     Lys Ile Lys     Gly Glu Ala Tyr Thr Asp Thr Tyr His Ser Tyr Ala His Asn

             230                  235                  240                  245                  250
N2  Gln Glu Ser Glu Cys Val Cys Ile Asn Gly Thr Cys Thr Val Val Met Thr Asp Gly Ser Ala Ser Gly Arg Ala
N1      Glu     Val         Val Asn Gly Ser     Phe Thr Ile Met     Pro Ser Asp Gly Pro Ala
N7      Glu     Val         His Asn Ser Thr     Val Val Met     Pro Ala Asn Asn Gln Ala
N8      Ser     Thr         Ile Lys Gly Asp     Tyr Trp Val Met     Pro Ala Asn Arg Gln Ala
N9      Glu     Val         His Ser Gln Val     Pro Val Phe     Ser Ala Thr Gly Pro Ala
B       Ala     Asn         Ile Gly Gly Asp     Tyr Leu Met Ile     Ser Ala Ser Gly Ile Ser

             255                  260                  265                  270                  275
N2  Asp Thr Arg Ile Leu Phe Ile Lys Glu Gly Lys Ile Val His Ile Ser Pro Leu Ser Gly Ser Ala Gln His Ile
N1  Ser Tyr Arg Ile Phe Lys Ile Glu Lys         Thr Lys Ser Ile Glu Leu Asp Ala Pro Asn Ser         Ile
N7  Phe Tyr Lys Val Ile Tyr Phe His Lys         Met Ile Lys Glu Glu Ser Leu Lys Gly Ser Ala Lys     Ile
N8  Lys Tyr Arg Ile Phe Lys Ala Lys Asp         Arg Ile Lys Gln Thr His Asp Ser Ile Ser Phe Asn Gly Ile Ile
N9  Glu Thr Arg Ile Tyr Tyr Phe Lys Glu         Lys     Leu Lys Trp Glu Pro Leu Ala Gly Thr Ala Lys Ile
B   Lys Cys Arg Phe Leu Lys Ile Arg Arg         Lys     Ile Lys Glu Ile Leu Pro Thr Gly Arg Val Glu Tyr
```

(continued)

## TABLE II. (Continued)

```
                    280                     285                     290                     295                     300
N2  Glu Glu Cys Ser Cys  -  Tyr Pro Arg Tyr Pro Asp Val Arg Cys Ile Cys Arg Asp Asn Trp Lys Gly Ser Asn Arg
N1                  Ser  -  Tyr Pro Asp Thr Gly Thr Val Met     Val             Trp His Gly Ser Asn
N7                  Ser  -  Tyr Gly His Asn Gln Arg Val Thr     Val             Trp Gln Gly Ala Asn
N8                  Ser  -  Tyr Pro Asn Glu Gly Lys Glu         Val             Trp Thr Gly Thr Asn
N9                  Ser  -  Tyr Gly Glu Arg Ala Glu Ile Thr     Thr             Trp Gln Gly Ser Asn
B                   Thr     Gly Phe Ala Ser Asn Lys Thr Ile Glu Ala             Ser Tyr Thr Ala Lys

                    305                     310                     315                     320                     325
N2  Pro Val Ile Asp Ile Asn Met Glu Asp Tyr Ser Ile Asp Ser Ser Tyr Val Cys Ser Gly Leu Val Gly Asp Thr
N1      Trp Val Ser Phe Asn Gln Asn Leu Asp Tyr  -  Gln Ile Gly Tyr Ile     Ser Gly Val Phe Gly     Asn
N7      Ile Ile Glu Ile Asp Met Asn Lys Leu Glu His Thr Ser Arg Tyr Ile     Thr Gly Val Leu Thr     Thr
N8      Ile Leu Val Ile Ser Pro Asp Leu Ser Tyr  -  Thr Val Gly Tyr Leu     Ala Gly Ile Pro Thr     Thr
N9      Val Ile Arg Ile Asp Pro Val Ala Met Thr His Thr Gln Tyr Ile         Ser Pro Val Leu Thr     Thr
B       Phe Val Lys Leu Asn Val Glu Thr Asp Thr Ala Glu Ile Arg Leu Met     Thr Lys Thr Tyr Leu     Thr

                    330                     335                     340                     345                     350
N2  Pro Arg Asn Asp Asp Ser Ser Ser Asn Ser Asn Cys Arg Asp Pro Asn Asn Gln Arg Gly Asn Pro Gly Val Lys
N1      Pro Pro Lys     Gly Lys  -   -  Gly Arg  -  Asp Pro Val Asn Val Asp Gly Ala Asp     Val
N7  Ser     Pro Lys     Lys Thr Ile  -  Gly Glu Phe Asn Pro Ile Thr Gly Ser Gly Ala Pro     Ile
N8  Pro     Gly Glu     Ser Gln Phe Thr Gly Ser     Thr Ser Pro Leu Gly Asn Lys Gly Tyr  -  Val
N9      Pro Pro Asn     Pro Thr Val  -  Gly Lys     Asn Asp Pro Tyr Pro Gly Asn Asn Asn Asn Val
B       Pro Pro Asp     Gly Ser Ile Ala Gly Pro  -  Glu Ser Asn Gly Asp Lys Trp Leu Gly     Ile

                    355                     360                     365                     370                     375
N2  Gly  -  Trp Ala Phe Asp Asn Gly Asp  -   -   -  Asp Val Trp Met Gly Arg Thr Ile Asn Lys Glu Ser Arg Ser Gly Tyr Glu
N1       -  Phe Ser Tyr Arg Tyr Gly Asn  -   -  Gly Gly     Ile Gly     Lys Ser Asn Ser     Ser         Lys     Phe
N7       -  Phe Gly Phe Leu Asn Glu Asp  -   -  Asn Thr     Leu Gly     Ile Ser Pro Arg Leu             Ser     Phe
N8       -  Phe Gly Phe Arg Gln Gly Asn  -   -  Asp Val     Ala Gly     Ile Ser Arg Thr Ser             Ser     Phe
N9       -  Phe Ser Tyr Leu Asp Gly Val  -   -  Asn Thr     Leu Gly     Ile Ser Ile Ala Ser             Ser     Tyr
B   Gly Phe Val His Gln Arg Met Ala Ser Lys Ile Gly     Arg Tyr Ser     Met Ser Lys Thr Asn             Met     Met

                    380                     385                     390                     395                     400
N2  Thr Phe Lys Val Ile Gly Gly Trp Ser Thr Pro Asn Ser Lys Ser Gln Ile Asn Arg Gln Val Ile Val Asp Asn
N1  Met Ile Trp Asp Pro Asn Gly Trp Thr Asp Pro Asp     Asn Phe Leu Val  -  Lys     Asp Ile Val Ala Met
N7  Met Leu Lys Ile Pro Asn Ala Gly Thr Asp Pro Glu     Lys Ile Lys Glu  -  Arg     Glu Ile Val Ser Glu
N8  Ile Ile Lys Ile Ala Gly Trp Thr Gln  -  Asn Lys     Asp Gln Ile Arg Leu Arg     Val Ile Val Asp Asn
N9  Met Leu Lys Val Pro Asn Ala Leu Thr Asp Asp Lys     Lys Pro Thr Gln  -  Gly     Thr Ile Val Leu Asn
B   Leu Tyr Val Lys Ser Gly Asp Pro Trp Thr Pro Asp     Asp Ala Leu Thr Leu Ser     Val Met Val Ser Ile

                    405                     410                     415                     420                     425
N2  Asn Asn Trp Ser Gly Tyr Ser Gly Ile Phe  -   -   -   -  Ser Val Glu Gly Lys Ser Cys Ile Asn Arg Cys Phe Tyr Val Glu
N1  Thr Asp Trp ser Gly             Gly Arg     Val Gln His Pro Glu Thr Glu Gly Leu Asp     Met Arg Pro     Phe Trp Val
N7  Asp Asn Trp Ser Gly             Gly Ser      -  Ile Asp Tyr Trp Asn Asp Asn Ser Glu     Tyr Asn Pro     Phe Tyr Val
N8  Leu Asn Trp Ser Gly             Gly Ser     Thr Leu Pro Val Glu Leu Thr Lys Lys Gly     Leu Val Pro     Phe Trp Val
N9  Thr Asp Trp Ser Gly             Gly Ser      -  Met Asp Tyr Ala Tyr Ala Leu Gly Gly Glu Tyr Arg Ala     Phe Tyr Val
B   Glu Glu Pro Gly Trp             Phe Gly      -   -   -   -  Glu Lys Glu Asp Lys Lys     Asp Val Pro     Ile Gly Ile

                    430                     435                     440                     445                     450
N2  Leu Ile Arg Gly Arg Pro Gln Glu Thr Arg  -  Val Trp Trp Thr Ser Asn Ser Ile Val Ala Phe Cys Gly Thr Ser
N1  Leu Ile Arg Gly Arg Pro Arg Glu Lys Thr  -  Thr Ile     Thr         Gly Ser Ser Ile Ser Phe         Gly Val Asn
N7  Leu Ile Arg Gly Arg Glu Glu Lys Ala Lys Tyr Val Glu     Thr         Asn Ser Leu Ile Ala Leu         Gly Ser Pro
N8  Met Ile Arg Gly Lys Pro Glu Asp Thr Thr  -   -  Ile     Thr         Ser Ser Ser Ile Val Met         Gly Val Asp
N9  Leu Ile Arg Gly Arg Pro Lys Glu Asp Lys  -  Val Trp     Thr         Asn Ser Asn Ser Ile Val Ser Met Ser Ser Thr
B   Met Val His Asp Gly Gly Lys Asp Thr  -   -   -   -  Arg         Ala Ala Thr Ala Ile Tyr             Leu Met Gly

                    455                     460                     465                     470
N2  Gly Thr Tyr Gly Thr Gly Ser Trp Pro Asp Gly Ala Asn Ile Asn Phe Met Pro Ile  -   -
N1  Ser Asp Thr Val Asn Trp Ser Trp Pro Asp Gly Ala Glu Leu Pro Phe Thr Ile Pro Lys  -
N7  Ile Ser Val Gly Ser Gly Ser Phe Asp Gly Ala Gln Ile Lys Tyr Phe Ser  -   -   -
N8  His Lys Ile Ala Ser Ser Trp Asp Gly Ala Glu Leu Pro Phe His Ile Asp Lys Ala Lys Ile
N9  Glu Phe Leu Gly Gln Trp Asp Trp Pro Asp Gly Ala Lys Ile Glu Tyr Phe Leu  -   -   -
B   Ser Gly Gln Leu Leu Trp Asp Thr Val Thr Gly Val Asp Met Ala Leu  -   -   -   -
```

[a] N2 (A/RI/5-/57); N1 (A/USSR/90/77); N7 (A/Equ/Cor/16/74); N8 (A/Equ/Ken/1/81); N9 (A/Tern/Australia/G70c/75) B (/Lee/40).
[b] See Table I for sources.

## TABLE III. Similarity Matrix for the Five A-Subtype and B-Type Neuraminidase Sequences in Table II[a]

|    | N2  | N1  | N7  | N8  | N9  |
|----|-----|-----|-----|-----|-----|
| N1 | 168 | —   | —   | —   | —   |
| N7 | 190 | 166 | —   | —   | —   |
| N8 | 188 | 206 | 174 | —   | —   |
| N9 | 188 | 183 | 222 | 169 | —   |
| B  | 119 | 129 | 107 | 126 | 114 |

[a] The stalk peptide (1–80) has been omitted for the purpose of this comparison. Entries show the number of identical residues in the head regions for any given pair of sequences out of a total of approximately 390 residues.

## 2. Processing

The membrane orientation was first demonstrated directly by sequencing of the pronase-released heads of A/Tokyo/3/67 neuraminidase (Ward et al., 1982). Pronase cuts the N2 neuraminidase in at least two positions to give a ragged N-terminal starting at position 74 or 77. The protein sequence then continues uninterrupted to residue 469, indicating that there is neither additional processing by pronase nor posttranslational cleavage in the head regions by host or viral enzymes. Automated amino acid sequencing of detergent-solubilized RI/5+/57 neuraminidase by Blok et al. (1982) showed that the N-terminal residue is indeed the initiating methionine and that no signal peptide was removed during biosynthesis and membrane translocation. This is in marked contrast to the proteolytic processing that accompanies hemagglutinin biosynthesis (Air, 1979; McCauley et al., 1979) where a signal peptide is removed from the N-terminus, and the chain is later cut in two to produce HA1 and HA2 chains as required for virions to be infectious.

## 3. Carbohydrate

Potential glycosylation sites in the representative sequences of Table II are summarized in Table IV. The sequence alignment is not particularly reliable in the stalk region, where the homology is low, but in the heads it is seen that only one of the glycosylation triplets is conserved across the four subtypes, and that is the sequence at Asn 146 (N2 numbering). Only one neuraminidase sequence is known in which this sequence is not found. The neuraminidase from WSN/33 (Hiti and Nayak, 1982), a neurovirulent influenza strain, has a double base change in the relevant Asn codon with respect to the PR/8/34 sequence, giving rise to Arg at this position. The unique growth properties of the WSN/33 virus in MDBK cells (Choppin, 1969) maps to the neuraminidase gene (Schulman and Palese, 1977; Sugiura and Ueda, 1980; Nakajima and Sugiura, 1980).

TABLE IV. N-Glycosyl Attachment Sequences in Different Neuraminidase Types and Subtypes[a]

| Subtype | | | Stalk | | ; | | Heads | | | | | |
|---------|---|---|---|---|---|---|---|---|---|---|---|---|
| | | | | | | | Asn-X-Ser/Thr sequences | | | | | |
| N2 | | | 61 | 69 | 70; | 86 | 146 | 200 | 234 | | 402 | |
| N1 | 44 | | 58 | | ; | 88 | 146 | | 235 | | 365 | 455 |
| N7 | 46 | 55 | | 66 | ; | 85 | 144 | 198 | 232 | | 399 | |
| N8 | 46 | 54 | | | ; | 84 | 144 | | | 293 | 398 | |
| N9 | 42 | 52 | 63 | 66 | ; | 87 | 147 | 202 | | | | |
| B | | | 56 | 64 | | | 144 | | 284 | | | |

[a]Note that Asn 402 in N2 is not glycosylated and that the homologous Asn in N7 and N8 might also be free of sugar.

One outstanding characteristic of the protein sequence of this neuraminidase is the loss of this carbohydrate site.

Protein sequencing of the Tokyo/3/67 N2 strain has shown that the Asn residue at position 402 is not glycosylated and that simple sugars are associated with residues 86 and 200, while complex sugars are attached at residues 146 and 234 (Ward *et al.*, 1983b). N-Acetyl galactosamine is found on the oligosaccharide at residue 146 but in none of the other complex carbohydrate residues of influenza neuraminidase or hemagglutinin (Basak *et al.*, 1981; Ward and Dopheide, 1981; Brown *et al.*, 1982; Ward *et al.*, 1983b; Keil *et al.*, 1985). Indeed this sugar is rare in N-glycosidic carbohydrate units (Wagh and Bahl, 1981).

## 4. Disulfide Bonds

An alignment of Cys residues extracted from the amino acid sequence alignment in Table II is given in Table V. For the stalk regions of the molecule, little is known about the pattern of disulfide bonding. Cys 54 of the B enzyme has been demonstrated to be involved in inter subunit bridging (Allen *et al.*, 1977), and at the C-terminal end of the stalk, Cys 78 in the N2 enzyme is also involved in linking pairs of polypeptide chains (Blok *et al.*, 1982; Varghese *et al.*, 1983). The intrachain pairing shown in Table V comes from chemical (Ward *et al.*, 1983a) and crystallographic (Varghese *et al.*, 1983) studies. Disulfide bridging assignments in N1, N7, N8, N9 and B strains are only by analogy with N2.

There are eight conserved disulfide bridges and one additional bridge in the N2, N8, and N9 strains. The unpaired cysteine in N1 sequences, residue 161, could link the chains in the tetramer into dimers by virtue of its proximity to the central fourfold axis of the tetramer (see Section II.B.2). The same does not hold for the unpaired cysteine in the B neuraminidase sequence.

## 5. Membrane Anchor and Stalk

Amino acid sequence homology across subtypes is lowest in the N-terminal region encompassing the transmembrane anchor and in the stalk that connects the anchor to the head. Since neuraminidase heads, solubilised from the virus with enzymes, are antigenically indistinguishable from neuraminidase on virus (Laver, 1978), variation in the structure of the N-terminal region does not appear to be a consequence of host immune pressure.

Blok and Air (1982b,c) sequenced the gene corresponding to the N-terminal region of 8 influenza A neuraminidase subtypes, and observed that all sequences have at least one Cys residue in the first 75 amino acids. Highly significant homology can be detected in the first 12 or so amino acids, the first six of which are believed to be on the cytoplasmic side of the viral membrane. Weaker homology extends through to residue

TABLE V. Cysteine Residues and Disulfide Bonds in Different Neuraminidase Subtypes[a]

| N2 | N1 | N7 | N8 | N9 | B |
|---|---|---|---|---|---|
| | | Membrane anchor region | | | |
| | | | | 9 | |
| | 14 | | | | |
| 21 | | | | | |
| | | Stalk region | | | |
| | | | | 41 | |
| 42 | | | 45 | 43 | |
| | | 49 | | | |
| 53 | 49 | | | | 54 |
| | | Head region | | | |
| 78 | | | | | |
| 92–417 | 92–417 | 91–417 | 90–417 | 93–419 | 87–420 |
| 124–129 | 124–129 | 121–126 | 122–127 | 125–130 | 122–127 |
| | 161 | | | | |
| 175–193 | | | 173–181 | 177–195 | |
| 183–230 | 184–231 | 181–228 | 182–229 | 185–232 | 182–229 |
| 232–237 | 233–238 | 230–235 | 231–236 | 234–239 | 231–236 |
| | | | | | 251 |
| 278–291 | 279–292 | 276–289 | 277–290 | 280–293 | 277–291 |
| 280–289 | 281–290 | 278–287 | 279–288 | 282–291 | 279–289 |
| 318–337 | 318–335 | 316–334 | 316–335 | 320–338 | 318–337 |
| 421–447 | 421–447 | 421–447 | 421–445 | 423–449 | 424–447 |

[a]Numbers relate directly to the particular sequence. Horizontal alignments follow Table II.

40, i.e., just beyond the membrane spanning region (Colman and Ward, 1985), but the stalk peptides show no homology from one subtype to another.

The function of the stalk is unclear. Indeed its length varies from 25 to 57 amino acids. Els et al. (1985) examined the rate of elution of influenza virus from red cells for parental A/RI/5+/57 virus (43 amino acid stalk) and a mutant of that virus known as "stubby" (25-amino acid stalk). Stubby virus eluted more slowly and indeed failed to elute completely over a 7-hr period. Electron micrographs of negatively stained SDS solubilized rosettes of stubby and full-length neuraminidase showed a reduction in the diameter of stubby rosettes of 50 Å. This corresponds to a reduction in the length of the stalk of 1.5 Å for each of the 18 amino acids that are deleted in stubby with respect to the wild type (Els et al., 1985). Although this figure is reminiscent of the rise per residue of an α-helix, there is no direct evidence for helical structure in the stalk. If the stalk functioned simply as a spacer between the membrane and the enzyme activity, there would be little pressure for conserved stalk sequences.

Estimates of the stalk length based on these figures lead to a value of 60 Å which, when added to the thickness of the neuraminidase head, places the upper surface of the head some 120 Å from the membrane. The hemagglutinin molecule protrudes a similar distance (135 Å) from the membrane (Wilson *et al.*, 1981), while stubby neuraminidase will presumably display its enzymatic activity only some 95 Å from the membrane. The extent to which this argument provides a rationale for the elution kinetics of stubby virus from red cells is dependent in part on the extent to which neuraminidase is capped on the virus (see Section I.B.2).

## B. Three-Dimensional Structure of Head

Few examples of crystalline integral membrane protein crystals have been reported to date (Michel, 1983). The difficulties derive from the insoluble nature of these proteins and the requirement for detergent in the crystallization buffer. To determine the three-dimensional structures of both influenza antigens, solubilization from the viral envelope has been effected by proteases (Wilson *et al.*, 1981; Varghese *et al.*, 1983). Crystallization of whole molecules of neuraminidase is also expected to be hampered by the supposed flexibility of the stalk peptide, which might be a source of conformational impurity.

Three different neuraminidases have now been crystallized in a form suitable for high-resolution X-ray diffraction studies. Two of these are from human influenza viruses of the N2 subtype (Laver, 1978; Colman and Laver, 1981; Varghese *et al.*, 1983) and one from an avian influenza virus of the N9 subtype (Laver *et al.*, 1984). The structures of the N2 enzymes have been determined (Varghese *et al.*, 1983), and the N9 study has recently demonstrated that it has an essentially identical fold to the N2 proteins (Baker *et al.*, 1987). The following discussion relates specifically to the N2 structure, although much of it is equally valid for N9.

### 1. Polypeptide Folding

The polypeptide backbone of each of the four subunits of the neuraminidase head is folded into a large single domain composed of six β-sheets. All sheets are topologically identical, constructed of four antiparallel strands connected by reverse turns (Fig. 1). The N-terminus of the head, residue 77, is seen to be near the molecular fourfold symmetry axis, which relates the four subunits in the tetrameric enzyme. From there, the chain extends across the bottom of the subunit, as viewed from outside the virus, in a semicircular arc before building the outermost (fourth) strand of the sixth sheet. A loop on the upper surface then connects to the first strand of sheet one, which like all subsequent sheets has the topology of a W, or +1, +1, +1 in Richardson's (1981) notation. The first strand of each sheet is nearest the subunit center, where all six such strands run parallel to each other. The last strand of each sheet is at the subunit surface. Each sheet is twisted in the same way as the blades of a propeller, with central strands almost parallel to the molecular symmetry axis and the outer strands more nearly in the plane of the tetramer. In

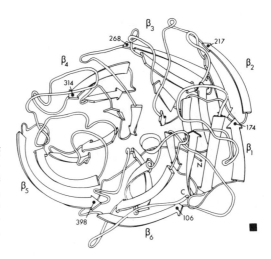

FIGURE 1. Schematic diagram of polypeptide fold in the neuraminidase subunit. The view is down the molecular fourfold symmetry axis, shown at bottom right. The six β-sheets are labeled as are the N- and C-terminals. From Varghese *et al.* (1983).

each sheet, two of the loops connecting strands are on the upper surface (L01 connecting adjacent sheets and L23 connecting strands 2 and 3 within a sheet) and two are on the bottom surface (L12 and L34). If the *j*th strand of the *i*th sheet is described as β*i*S*j*, then the folding pattern reads as

. . N terminal arm, β6S4
β1L01, β1S1, β1L12, β1S2, β1L23, β1S3, β1L34, β1S4, β2L01, β2S2, β2L12 . . . . .
                        . . . β6S3,
C-terminal arm at subunit interface

The loop structures connecting the strands of the sheets are variable in length, and carry amino acids implicated in both the enzyme activity and antigenic variation (Colman *et al.*, 1983).

There is a loose topological similarity here with the so-called α/β-proteins in which an eight-stranded parallel β-barrel is surrounded by eight α-helices. The parallelism of central strands in the structure is the common feature with the propeller fold, although in the latter case there is not the full capacity for hydrogen bonding of those strands. Furthermore, the active sites in the different structural classes are at opposite ends of the parallel strand structure (Branden, 1980; see Section II.B.4).

Apart from the catalytic site, amino acids conserved across subtypes are found in key places in reverse turns as well as in the interfaces between the six sheets (Colman *et al.*, 1983; Colman and Ward, 1985).

2. Disulfide Bonds

A schematic of the nine disulfide bonds in the N2 head overlaid on a diagram of the propeller fold is shown in Fig. 2. On this sketch, the

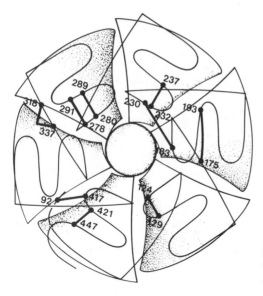

FIGURE 2. Disulfide bonds in N2 neuraminidase overlaid on the sheets of the propeller fold. From Colman and Ward (1985).

molecular fourfold axis is in the lower right-hand corner, perpendicular to the page. The additional cysteine at position 161 in N1 sequences is in β1L34 near the symmetry axis, from whence it might link subunits in dimers as discussed in Section II.A.4. The cysteine residues at positions 175 and 193 are not present in the N1, N7, and B sequences.

The only bridge between linearly distant residues is from 92 to 417. All others can be regarded as local. Four of them are found between the first two strands of sheets (sheets 1, 3, and 4), adjacent to residues implicated in catalysis.

## 3. Carbohydrate

Two of the glycosylation sites on the N2 proteins are on the upper surface (Asn 146 and Asn 200), and two are on the underneath surface (Asn 86 and Asn 234). The unglycosylated Asn-X-Ser sequence at residue 402 is found to have a conformation in which the X amino acid is oriented away from the surface in a manner that could prevent simultaneous recognition of the amino acids either side of it by glycosyl transferases. This finding suggests that the structure here might have folded before the protein was exposed to the sugar attachment machinery of the host cell, and is consistent with a distance lag of 45–80 amino acids between protein synthesis and sugar transfer (Rothman and Lodish, 1977; Hubbard and Ivatt, 1981).

The sugar moiety most clearly seen in the present electron-density map of the N2 structure is that associated with Asn 200. The oligosaccharide attached there makes contact with the surface of a neighboring subunit in the neuraminidase tetramer and in this way will contribute to the stabilization of the four subunits within the head.

There are several interesting aspects to the sugar moiety at residue 146. With the exception of the neurovirulent influenza virus, WSN/33, this is the only conserved sugar attachment site across neuraminidase types and subtypes. Neurovirulence of the WSN/33 virus is believed to be a property of the neuraminidase gene (Schulman and Palese, 1977; Sugiura and Ueda, 1980; Nakajima and Sugiura, 1980), but sequence comparisons show no obvious unique properties for this protein, apart from the glycosylation anomaly. Factors determining neurovirulence may well be complex, and might not even map to coding regions of the gene. The chemical composition of this oligosaccharide is also unusual (see Section II.A.3). Finally, carbohydrate has been implicated, directly or otherwise, in enzymatic activity (Griffin et al., 1983), and Asn 146 is near the active site.

### 4. The Enzyme Active Center

The active site of neuraminidase has been directly identified by soaking sialic acid into crystals of Tokyo/3/67 neuraminidase and using difference Fourier methods to determine the binding site (Colman et al., 1983). When the subunit is viewed into the top face down the molecular symmetry axis, the sialic acid binding site is almost directly above the first strands of sheets 3 and 4 ($\beta$3S1, $\beta$4S1) and is located in a large depression or pocket in the surface of the molecule (Fig. 3). Unlike the $\alpha/\beta$-proteins referred to above (see Section II.B.1), here the active site is at the N-terminal end of the central parallel strands.

Assignment of particular amino acids to a substrate-binding or catalytic function is dependent on their satisfying two criteria. They should first be located in the sialic acid-binding cavity and then be conserved across all neuraminidase sequences. The amino acids listed in Table VI meet both conditions, with one exception. Asp 198 (N2 numbers) was originally identified as a catalytic residue by Colman et al. (1983). Subsequent sequence data for N9 and N7 subtype neuraminidase have shown that Asn replaces Asp at this position in these two strains.

The most remarkable aspect of the sialic acid-binding site is the large number of charged residues. This stands in marked contrast to the receptor-binding pocket of hemagglutinin in which a single glutamic acid residue is found (Wilson et al., 1981).

Site-directed mutants of neuraminidase are now being constructed with a view to mapping the catalytic site in more detail. In the first of these experiments, Lentz and Air (1985) showed that Trp 178 to Leu is a mutation that abolishes the activity of the enzyme. Other such mutants are Asn 146 to Ser, Arg 152 to Ile or Lys, Asp 198 to Asn, Glu 277 to Asp, and Tyr 406 to Thr (Lentz et al., 1987). Arg 371 to Lys reduces enzymatic activity and His 274 to Tyr or Asn produces an enzyme in which the pH optimum is shifted from 6 to 5 (Lentz et al., 1987).

Preservation of the enzymatic function in the face of host immune pressure is presumably a result of the inability of antibodies to bind

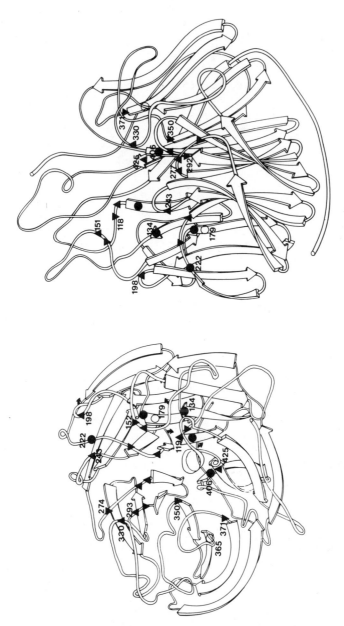

FIGURE 3. Conserved amino acids in the sialic acid binding pocket (see Table VI). (▲) acidic groups, (●) hydrophobic groups, (○) neutral hydrophilic groups, (▼) basic groups. Left-hand view as in Fig. 1. Arrows show glycosylation sites in N2 strains. Right-hand view is perpendicular to the molecular fourfold symmetry axis, which is vertical and at the left rear (not labeled).

TABLE VI. Amino Acids Implicated in
Substrate Binding or Catalysis

| Acids | Bases | Hydrophobics | Hydrophilics |
|-------|-------|--------------|--------------|
| Glu 119 | Arg 118 | Leu 134 | Ser 179 |
| Asp 151 | Arg 152 | Trp 178 | Thr 365 |
| (Asp 198) | Arg 224 | Ile 222 | |
| Glu 227 | His 274 | Tyr 406 | |
| Asp 243 | Arg 292 | | |
| Glu 276 | Arg 371 | | |
| Glu 277 | | | |
| Asp 293 | | | |
| Asp 330 | | | |
| Glu 425 | | | |

exclusively to catalytically essential residues (Colman *et al.*, 1983). The same argument has recently been presented as the canyon hypothesis for picornaviruses (Luo *et al.*, 1987), where strain variation is also common.

## 5. Subtype Variation

The only three-dimensional structure data for comparing influenza viral antigens of different subtypes are the neuraminidase N2 and N9 structures (Baker *et al.*, 1987). The results there are not yet complete but permit some preliminary conclusions.

The polypeptide chain fold in the two subtype structures is identical. The failure of antisera to cross-react with two antigens can therefore not be taken to imply anything at all about the similarity of the structure of the two antigens. All that is important for absence of cross-reaction is the absence of common epitopes, which is easily achievable with sequence homologies as high as 50%.

The nature of the intersubunit contacts in N2 and N9 is different. Amino acids located in the subunit interface that are not conserved in the two structures are more polar in character in N9 than are their hydrophobic counterparts in N2.

For other subtype comparisons in which no three-dimensional structure information is available, insertions and deletions of amino acids in the alignment shown in Table II are for the most part easily accommodated on external regions of the reference N2 structure. The insertion of three amino acids in the B sequence at residue 98 of N2 is one exception; the homology suffers little if the insert is moved forward or backward a few residues to place it at either end of the fourth strand of sheet 6. The insertion of a glycine in B neuraminidase after residue 351 of N2 is likely to cause minor adjustment in sheet 5.

The lower homology between A and B neuraminidases in sheets 5 and 6 (see Section II.A.1) is not easily rationalised. Of the 24 conserved residues listed in Table VI and implicated in catalytic activity, only two derive from the C-terminal 100-amino acids in which the intersubtype homology is lowest. It may be that A- and B-type structures are less constrained by functional dictates in this region of the molecule.

### 6. Hemagglutinin Activity of N9

The capacity of N9 neuraminidase to agglutinate red blood cells (RBC) was described by Laver *et al.* (1984) and, as a result of the study of a number of laboratory variants of N9 (Webster *et al.*, 1987) and their hemagglutinin properties, there is now a firm indication as to the location of the site on the three-dimensional structure responsible for this activity. Single amino acid substitutions at positions (in N2 numbering) 369, 370, 372, or 400 abolish the activity and at 368 or 432 reduce the activity (Webster *et al.*, 1987). These residues form, or are adjacent to, a shallow pocket that is separated from the enzyme-active center by the antigenic surface loop 367–370. This observation is consistent with the failure of enzyme inhibitors to block the hemagglutinin activity (Laver *et al.*, 1984). It is not known why the N9 neuraminidase carries the extra activity, but it is conserved in another N9 neuraminidase isolated from a whale virus (Air *et al.*, 1987) (see Section III.A.3). The HN protein of paramyxoviruses, which exhibits both hemagglutinin and neuraminidase activities, also appears to have the functional centers separated (Portner, 1981).

## III. ANTIGENIC STRUCTURE

Antigenic analysis of neuraminidase is based on two major classes of study. The first is the study of amino acid sequence changes within subtypes and the location of strain-variable residues within the three-dimensional structure of neuraminidase. The second is the analysis of neuraminidase variants selected with monoclonal antibodies and of competition assays of antibody binding to the antigen.

### A. Amino Acid Sequence Variation within Subtypes

#### 1. N2

An amino acid sequence alignment of the N2 strains listed in Table I is presented in Table VII. The data show that antigenic drift in the N2 neuraminidase results from an accumulation of point mutations at a rate similar to that found for the H3 hemagglutinin, i.e., 0.68 residues/year per 100 amino acid residues. A matrix of amino acid sequence homologies is given in Table VIII.

## TABLE VII. Alignment of N2 Subtype Sequences[a]

```
                            5              10             15             20             25
RI/5⁻/57       Met Asn Pro Asn Gln Lys Thr Ile Thr Ile Gly Ser Val Ser Leu Thr Ile Ala Thr Val Cys Phe Leu Met Gln
Tokyo/3/67                         Ile                 Val                         Val
NT/60/68                           Ile                 Val                         Val
Udorn/307/72                       Ile                 Val                         Ile
Vic/3/75                           Ile                 Val                         Ile
Bangkok/1/79                       Ile                 Val                         Ile
Chick/Penn/83                      Ile                 Ile                         Val
                                                                                                               50
RI/5⁻/57       Ile Ala Ile Leu Ala Thr Thr Val Thr Leu His Phe Lys Gln His Glu Cys Asp Ser Pro Ala Ser Asn Gln Val
Tokyo/3/67             Leu Val         Thr             Lys     His     Cys Asp Ser     Ala Ser             Val
NT/60/68              Leu  al         Thr             Lys     Tyr     Cys Asp Ser     Ala Ser             Val
Udorn/307/72          Gln Val         Thr             Lys     Tyr     Cys Asp Ser     Ala Asn             Val
Vic/3/75              Leu Val         Thr             Lys     Tyr     Cys Asp Ser     Ala Asn             Val
Bangkok/1/79          Leu Val         Thr             Lys     Tyr     Cys Ser Ser     Pro Asn             Val
Chick/Penn/83         Leu Ala         Asn             Arg     Asn     His Ser Ile     Ala Tyr             Thr
                                                                                                               75
RI/5⁻/57       Met Pro Cys Glu Pro Ile Ile Ile Glu Arg Asn Ile Thr Glu Ile Val Tyr Leu Asn Asn Thr Thr Ile Glu Lys
Tokyo/3/67     Met         Glu         Ile                             Asn
NT/60/68       Met         Glu         Ile                             Asn
Udorn/307/72   Met         Glu         Ile                             Thr
Vic/3/75       Met         Glu         Ile                             Thr
Bangkok/1/79   Met         Glu         Ser                             Thr
Chick/Penn/83  Thr         Lys         Ile        ̄   ̄   ̄   ̄   ̄   ̄   ̄   ̄   ̄   ̄   ̄   ̄   ̄   ̄   ̄   ̄
                                                                                                              100
RI/5⁻/57       Glu Ile Cys Pro Glu Val Val Glu Tyr Arg Asn Trp Ser Lys Pro Gln Cys Gln Ile Thr Gly Phe Ala Pro Phe
Tokyo/3/67     Glu         Lys Val                                         Gln
NT/60/68       Glu         Lys Val                                         Gln
Udorn/307/72   Gly         Lys Leu                                         Lys
Vic/3/75       Gly         Lys Leu                                         Lys
Bangkok/1/79   Glu         Lys Leu                                         Lys
Chick/Penn/83   ̄   ̄   ̄   ̄   ̄   ̄   ̄   ̄                                   Gln
                                                                                                              125
RI/5⁻/57       Ser Lys Asp Asn Ser Ile Arg Leu Ser Ala Gly Gly Asp Ile Trp Val Thr Arg Glu Pro Tyr Val Ser Cys Asp
Tokyo/3/67                                             Asp
NT/60/68                                               Asp
Udorn/307/72                                           Asp
Vic/3/75                                               Asp
Bangkok/1/79                                           Gly
Chick/Penn/83                                          Gly
                                                                                                              150
RI/5⁻/57       Pro Gly Lys Cys Tyr Gln Phe Ala Leu Gly Gln Gly Thr Thr Leu Asp Asn Lys His Ser Asn Gly Thr Ile His
Tokyo/3/67     Pro Val                                     Asp Asn Lys         Asp     Val
NT/60/68       His Gly                                     Asp Asn Lys         Asp     Ile
Udorn/307/72   Pro Gly                                     Asp Asn Lys         Asp     Ile
Vic/3/75       Pro Arg                                     Glu Asn Lys         Asp     Ile
Bangkok/1/79   Pro Gly                                     Asp Asn Lys         Asp     Ile
Chick/Penn/83  Pro Ser                                     Asp His Asn         Gly     Ile
                                                                                                              175
RI/5⁻/57       Asp Arg Ile Pro His Arg Thr Leu Leu Met Asn Glu Leu Gly Val Pro Phe His Leu Gly Thr Lys Gln Val Cys
Tokyo/3/67             Ile     His                                                         Arg
NT/60/68               Ile     His                                                         Arg
Udorn/307/72           Ile     His                                                         Arg
Vic/3/75               Thr     His                                                         Arg
Bangkok/1/79           Thr     Tyr                                                         Arg
Chick/Penn/83          Thr     His                                                         Arg
RI/5⁻/57       Val Ala Trp Ser Ser Ser Ser Cys His Asp Gly Lys Ala Trp Leu His Val Cys Val Thr Gly Asp Asp Arg Asn
Tokyo/3/67     Ile Ala         Ser                                             Ile     Asp         Lys
NT/60/68       Ile Ala         Ser                                             Ile     Asp         Lys
Udorn/307/72   Ile Gly         Ser                                             Val     Tyr         Lys
Vic/3/75       Ile Ala         Ser                                             Val     Tyr         Lys
Bangkok/1/79   Ile Ala         Ser                                             Val     Tyr         Lys
Chick/Penn/83  Ile Ala         Arg                                             Val     Asp         Arg
                                                                                                              225
RI/5⁻/57       Ala Thr Ala Ser Phe Ile Tyr Asp Gly Arg Leu Val Asp Ser Ile Gly Ser Trp Ser Gln Asn Ile Leu Arg Thr
Tokyo/3/67                         Asp     Arg                             Gln
NT/60/68                           Asp     Arg                             Gln
Udorn/307/72                       Asp     Arg                             Gln
Vic/3/75                           Asp     Arg                             Gln
Bangkok/1/79                       Asp     Arg                             Lys
Chick/Penn/83                      Asn     Met                             Gln
                                                                                                              250
RI/5⁻/57       Gln Glu Ser Glu Cys Val Cys Ile Asn Gly Thr Cys Thr Val Val Met Thr Asp Gly Ser Ala Ser Gly Arg Ala
Tokyo/3/67                                                                             Gly Arg
NT/60/68                                                                               Gly Arg
Udorn/307/72                                                                           Gly Arg
Vic/3/75                                                                               Gly Arg
Bangkok/1/79                                                                           Glu Arg
Chick/Penn/83                                                                          Gly Lys
                                                                                                              275
RI/5⁻/57       Asp Thr Arg Ile Leu Phe Ile Lys Glu Gly Lys Ile Val His Ile Ser Pro Leu Ser Gly Ser Ala Gln His Ile
Tokyo/3/67             Arg             Glu                             Ser                         Val
NT/60/68               Arg             Glu                             Ala                         Val
Udorn/307/72           Lys             Glu                             Ser                         Val
Vic/3/75               Lys             Glu                             Ser                         Val
Bangkok/1/79           Lys             Glu                             Ser                         Val
Chick/Penn/83          Arg             Arg                             Ser                         Ile
```

(continued)

## TABLE VII. (Continued)

```
                                                                                                        300
RI/5"/57       Glu Glu Cys Ser Cys Tyr Pro Arg  -  Tyr Pro Asp Val Arg Cys Ile Cys Arg Asp Asn Trp Lys Gly Ser Asn Arg
Tokyo/3/67                             Arg  -            Gly             Ile
NT/60/68                               Arg  -            Gly             Ile
Udorn/307/72                           Arg  -            Gly             Ile
Vic/3/75                               Arg  -            Gly             Ile
Bangkok/1/79                           Arg  -            Gly             Val
Chick/Penn/83                          Lys Ser           Asn             Val
                                                                                                        325
RI/5"/57       Pro Val  -  Ile Asp Ile Asn Met Glu Asp Tyr Ser Ile Asp Ser Ser Tyr Val Cys Ser Gly Leu Val Gly Asp Thr
Tokyo/3/67              -  Val         Met Glu             Asp
NT/60/68                -  Val         Met Glu             Asp
Udorn/307/72            -  Val         Val Lys             Asp
Vic/3/75                -  Val         Val Lys             Asp
Bangkok/1/79            -  Val         Val Lys             Val
Chick/Penn/83      Lys Leu             Met Ala             Asp
                                                                                                        350
RI/5"/57       Pro Arg Asn Asp Asp Ser Ser Ser Asn Ser Asn Cys Ser Arg Asp Pro Asn Asn Glu Arg Gly Asn Pro Gly Val Lys
Tokyo/3/67         Asn Asp     Arg         Asn             Asn         Asn             Arg     Thr Gln
NT/60/68           Asn Asp     Arg         Asn             Asn         Asn             Arg     Asn Gln
Udorn/307/72       Asn Asn     Arg         Asn             Tyr         Asn             Lys     Asn His
Vic/3/75           Lys Asn     Arg         Ser             Tyr         Asn             Lys     Ile His
Bangkok/1/79       Lys Asn     Arg         Ser             Tyr         Asn             Lys     Asn His
Chick/Penn/83      Asn Asp     Ser         Ser             Asn         Asp             Arg     Asn Pro
                                                                                                        375
RI/5"/57       Gly Trp Ala Phe Asp Asn Gly Asp Asp Val Trp Met Gly Arg Thr Ile Asn Lys Glu Ser Arg Ser Gly Tyr Glu
Tokyo/3/67             Asn         Asn         Leu                         Ser Lys Asp Leu
NT/60/68               Asn         Asp         Val                         Ser Lys Asp Leu
Udorn/307/72           Asp         Asn         Val                         Ser Glu Asp Ser
Vic/3/75               Asp         Asn         Val                         Ser Glu Asp Ser
Bangkok/1/79           Asp         Asn         Val                         Ser Glu Glu Ser
Chick/Penn/83          Ile         Asp         Val                         Ser Lys Asp Ser
                                                                                                        400
RI/5"/57       Thr Phe Lys Val Ile Gly Gly Trp Ser Thr Pro Asn Ser Lys Ser Gln Val Asn Arg Gln Val Ile Val Asp Asn
Tokyo/3/67             Lys             Gly         Ser         Pro             Ser Ile                             Ser
NT/60/68               Lys             Gly         Ser         Pro             Ser Ile                             Ser
Udorn/307/72           Lys             Gly         Ser         Pro             Leu Ile                             Ser
Vic/3/75               Lys             Gly         Ser         Pro             Leu Ile                             Ser
Bangkok/1/79           Lys             Gly         Ser         Pro             Leu Ile                             Ser
Chick/Penn/83          Arg                         Ala         Ala             Ser Thr                             Asn
                                                                                                        425
RI/5"/57       Asn Asn Trp Ser Gly Tyr Ser Gly Ile Phe Ser Val Glu Gly Lys Ser Cys Ile Asn Arg Cys Phe Tyr Val Glu
Tokyo/3/67     Asp     Arg                                 Gly
NT/60/68       Asp     Arg                                 Gly
Udorn/307/72   Asp     Arg                                 Gly
Vic/3/75       Ala     Arg                                 Gly
Bangkok/1/79   Asp     Arg                                 Gly
Chick/Penn/83  Asn     Trp                                 Ser
                                                                                                        450
RI/5"/57       Leu Ile Arg Gly Arg Pro Gln Glu Thr Arg Val Trp Trp Thr Ser Asn Ser Ile Val Val Phe Cys Gly Thr Ser
Tokyo/3/67                     Lys             Thr                 Ser
NT/60/68                       Lys             Ala                 Ser
Udorn/307/72                   Glu             Thr                 Ser
Vic/3/75                       Glu             Thr                 Ser
Bangkok/1/79                   Glu             Thr                 Ser
Chick/Penn/83                  Pro             Asn                 Thr

RI/5"/57       Gly Thr Tyr Gly Thr Gly Ser Trp Pro Asp Gly Ala Asn Ile Asn Phe Met Pro Ile
Tokyo/3/67                                             Asn             Phe             Ile
NT/60/68                                               Asn             Phe             Ile
Udorn/307/72                                           Asp             Leu             Ile
Vic/3/75                                               Asp             Leu             Ile
Bangkok/1/79                                           Asp             Leu             Ile
Chick/Penn/83                                          Asn             Phe             Leu
```

[a]See Table I for sources.

## TABLE VIII. Similarity Matrix for N2 Sequences[a]

|              | RI/5+/57 | Tok/3/67 | NT/60/68 | Udn/307/72 | Vic/3/75 | Bang/1/79 |
|--------------|----------|----------|----------|------------|----------|-----------|
| Tok/3/67     | 440      |          |          |            |          |           |
| NT/60/68     | 443      | 461      |          |            |          |           |
| Udn/307/72   | 429      | 442      | 443      |            |          |           |
| Vic/3/75     | 424      | 437      | 437      | 459        |          |           |
| Bang/1/79    | 420      | 432      | 432      | 453        | 454      |           |
| Chk/Penn/83  | 403      | 392      | 394      | 385        | 385      | 385       |

[a]Numbers of identical residues out of a total of approximately 470 residues.

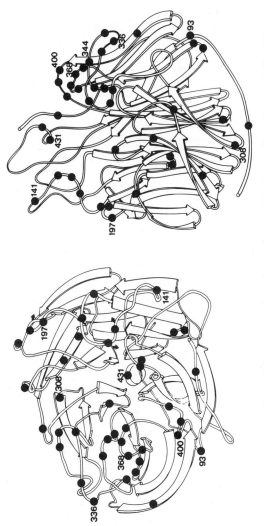

FIGURE 4. Strain variable surface amino acid residues in human N2 viruses. Two views as in Fig. 3. From Colman and Ward (1985).

Within the neuraminidase head, where the antigenic sites are confined, the N2 molecules from human influenza viruses show variation at 54 positions. Not all these sites are on the surface of the molecule. Those judged to be surface accessible in the present neuraminidase model are flagged in Fig. 4 (Colman *et al.*, 1983; Colman and Ward, 1985). These variable residues are distributed around the protein surface. Most are on the upper surface (with respect to the viral membrane) encircling the catalytic cavity. Some are on the side and underneath surfaces, sites at which antibody binding is not likely to influence the enzymatic activity directly.

On the upper surface, the variation is found in the loop structures, L01 and L23, connecting the strands of the β-sheets. It should be recalled that amino acids strictly conserved across subtypes and implicated in enzymatic activity are also found on these same loops. The most variable of these structures are the loops L01 and L23 on sheet 5, involving amino acid residues 327–347 and 367–370. All these loops are within 15 Å of the sialic acid binding site (Colman *et al.*, 1983), and antibodies bound to any of them could readily block access of macromolecular substrates to the active site of the enzyme.

There is no easy way to subdivide the variable regions around the active site into nonoverlapping clusters of variation. Any one of the loop structures in question is within 17 Å of at least one other. Although antigen-combining sites of antibodies show extreme variation from one molecule to another, the six complementarity-determining loops of an antibody describe a surface having a radius of at least 15 Å. Thus, if the sites of variation on the upper surface of neuraminidase are indeed indicative of antigenic sites, the antigenic structure is best described as a spatially continuous epitope encircling the sialic acid-binding cavity. By contrast, antigenic sites on the hemagglutinin appear to be segregated from each other (Wiley *et al.*, 1981; Daniels *et al.*, 1983).

The significance of the amino acid sequence variation occurring on the bottom surface of the neuraminidase head is unknown. Nor is it known whether antibodies have access to this surface on influenza virions. It should be noted that antibodies neutralize infectivity or modify disease by processes that do not necessarily involve direct inhibition of some viral function (Hirsch, 1982; Dimmock, 1987).

The avian N2 sequence in the alignment in Table VII is the neuraminidase from the avirulent strain of Chick/Penn/83. Four amino acid sequence changes distinguish the virulent and avirulent forms, two of which are in the head region (Desphande *et al.*, 1985) at residues 286 and 434 (human N2 numbering). Both positions are in strain-variable regions of the human influenza N2 structures, the first on the underneath surface and the second on the upper surface. The virulence characteristics of these two viruses are believed to map to the hemagglutinin gene (Kawaoka *et al.*, 1984).

There are a number of unique aspects to the avian N2 sequences, in

## TABLE IX. Alignment of N1 Subtypes[a]

```
                          5                  10                 15                 20                 25
WSN/33          Met Asn Pro Asn Gln Lys Ile Ile Thr Ile Gly Ser Ile Cys Met Val Val Gly Ile Ile Ser Leu Ile Leu Gln
PR/8/34                     Pro         Lys                 Ser         Leu Val Val     Leu             Ile
NJ/8/76                     Thr         Arg                 Thr         Leu Ile Val     Ile             Leu
USSR/90/77                  Pro         Lys                 Ser         Met Ala Ile     Ile             Ile
Parrot/Ulster/73            Pro         Lys                 Ser         Met Val Val     Ile             Ile
                                                                                                              50
WSN/33          Ile Gly Asn Ile Ile Ser Ile Trp Ile Ser His Ser Ile Gln Thr Gly Asn Gln Asn His Thr Gly Ile Cys Asn
PR/8/34                             Ser Ile     Ile                     Ser Gln Asn His Thr Gly Ile
NJ/8/76                             Leu Leu     Met                     Glu Lys Ser His Pro Lys Val
USSR/9077                           Ser Ile     Val                     Ser Gln Asn His Thr Gly Ile
Parrot/Ulster/PU                    Ser Ile     Val                     Asn Gln Asn Gln Pro Glu Thr
                                                                                                              75
WSN/33          Gln Gly Ser Ile Thr Tyr Lys  -   -   -   -   -   -   -   -   -   -   -   -   -   -   -   Val Val
PR/8/34             Asn Ile         Lys Asn Ser Thr Trp Val  -   -   -   -   -   -
NJ/8/76             Ser Val         Glu Asn Ser Thr Trp Val Asn Gln Thr Tyr Val Asn Ile Ser Asn Thr Asn Ile Ala
USSR/90/77          Arg Ile         Glu Asn Ser Thr Trp Val Asn Gln Thr Tyr Val Asn Ile Ser Asn Thr Asn Val Val
Parrot/Ulster/PU    Ser Ile         Glu Asn Asn Thr Trp Val Asn Gln Thr Tyr Val Asn Ile Ser Asn Thr Asn Phe Val
                                                                                                              100
WSN/33          Ala Gly Gln Asp Ser Thr Thr Ser Val Ile Leu Thr Gly Asn Ser Ser Leu Cys Pro Ile Arg Gly Trp Ala Ile His
PR/8/34          "   "  Lys Asp Thr Thr Ser Val Ile Leu Thr                             Arg                         Tyr
NJ/8/76         Ala Gly Gln Gly Val Thr Pro Ile Ile Leu Ala                             Ser                         Tyr
USSR/90/77      Ala Gly Lys Asp Thr Thr Ser Met Thr Leu Ala                             Arg                         Tyr
Parrot/Ulster/PU Ala Glu Gln Ala Val Ala Pro Val Ala Leu Ala                            Ser                         Tyr
                                                                                                              125
WSN/33          Ser Lys Asp Asn Gly Ile Arg Ile Gly Ser Lys Gly Asp Val Phe Val Ile Arg Glu Pro Phe Ile Ser Cys Ser
PR/8/34                 Ser                     Lys             Val     Ile
NJ/8/76                 Ser                     Lys             Ile     Met
USSR/90/77              Ser                     Lys             Val     Ile
Parrot/Ulster/PU        Gly                     Arg             Val     Ile
                                                                                                              150
WSN/33          His Leu Glu Cys Arg Thr Phe Phe Leu Thr Gln Gly Ala Leu Leu Asn Asp Lys His Ser Arg Gly Thr Phe Lys
PR/8/34                         Phe                                     Arg         Asn         Val
NJ/8/76                         Phe                                     Arg         Asn         Val
USSR/90/77                      Phe                                     Lys         Asn         Val
Parrot/Ulster/PU                Leu                                     Lys         Asn         Val
                                                                                                              175
WSN/33          Asp Arg Ser Pro Tyr Arg Ala Leu Met Ser Cys Pro Val Gly Glu Ala Pro Ser Pro Tyr Asn Ser Arg Phe Glu
PR/8/34                         Ala             Val
NJ/8/76                         Thr             Ile
USSR/90/77                      Ala             Ile
Parrot/Ulster/PU                Thr             Val
                                                                                                              200
WSN/33          Ser Val Ala Trp Ser Ala Ser Ala Cys His Asp Gly Val Gly Trp Leu Thr Ile Gly Ile Ser Gly Pro Asp Asp
PR/8/34                             His         Met                     Ile                         Asn
NJ/8/76                             His         Met                     Ile                         Asn
USSR/90/77                          Tyr         Met                     Ile                         Asp
Parrot/Ulster/PU                    His         Ile                     Val                         Asn
                                                                                                              225
WSN/33          Gly Ala Val Ala Val Leu Lys Tyr Asn Gly Ile Ile Thr Glu Thr Ile Lys Ser Trp Arg Asn Lys Ile Leu Arg
PR/8/34                                         Glu                 Lys Lys
NJ/8/76                                         Asp                 Asn Lys
USSR/90/77                                      Glu                 Lys Gln
Parrot/Ulster/PU                                Asp                 Asn Asn
                                                                                                              250
WSN/33          Thr Gln Glu Ser Glu Cys Thr Cys Val Asn Gly Ser Cys Phe Thr Ile Met Thr Asp Gly Pro Ser Asp Gly Leu
PR/8/34                         Ala     Val                         Ile             Asp         Leu
NJ/8/76                         Val     Ile                         Ile             Asn         Gln
USSR/90/77                      Val     Val                         Ile             Asp         Pro
Parrot/Ulster/PU                Ala     Val                         Val             Asn         Gln
                                                                                                              275
WSN/33          Ala Ser Tyr Lys Ile Phe Lys Ile Glu Lys Gly Lys Val Thr Lys Ser Ile Glu Leu Asn Ala Pro Asn Ser His
PR/8/34                 Lys Ile     Ile             Val     Ile             Asn                         Ser
NJ/8/76                 Lys Leu     Met             Ile Ile         Ile             Asp                 Ser
USSR/90/77              Arg Ile     Ile             Ile Thr         Ile             Asp                 Ser
Parrot/Ulster/PU        Lys Ile     Ile             Val Val         Val             Asn                 Tyr
                                                                                                              300
WSN/33          Tyr Glu Glu Cys Ser Cys Tyr Pro Asp Thr Gly Lys Val Met Cys Val Cys Arg Asp Asn Trp His Gly Ser Asn
PR/8/34                                 Thr     Lys Val Met                                         Gly
NJ/8/76                                 Thr     Lys Val Val                                         Ala
USSR/90/77                              Thr     Thr Val Met                                         Gly
Parrot/Ulster/PU                        Ala     Glu Ile Thr                                         Gly
                                                                                                              325
WSN/33          Arg Pro Trp Val Ser Phe Asp Gln Asn Leu Asp Tyr Lys Ile Gly Tyr Ile Cys Ser Gly Val Phe Gly Asp Asn
PR/8/34                             Asp         Asp     Gln
NJ/8/76                             Asp         Asp     Gln
USSR/90/77                          Asn         Asp     Gln
Parrot/Ulster/PU                    Asn         Glu     Gln
                                                                                                              350
WSN/33          Pro Arg Pro Lys Asp Gly Thr Gly Ser Cys Gly Pro Val Ser Ala Asp Gly Ala Asn Gly Val Lys Gly Phe Ser
PR/8/34                 Pro Lys     Thr     Ser     Gly         Tyr Val Asp     Asn     Val
NJ/8/76                 Ser Asn     Lys     Asn     Gly         Leu Ser Asn     Asn     Tyr
USSR/90/77              Pro Lys     Lys     Arg     Asp         Asn Val Asp     Asp     Val
Parrot/Ulster/PU        Pro Asn     Thr     Ser     Gly         Ser Ser Asn     Tyr     Val
                                                                                                              375
WSN/33          Tyr Lys Tyr Gly Asn Gly Val Trp Ile Gly Arg Thr Lys Ser Asp Ser Ser Arg His Gly Phe Glu Met Ile Trp
PR/8/34         Tyr Arg         Val Trp                         His             His                     Ile
NJ/8/76         Phe Arg         Val Met                         Ile             Arg                     Ile
USSR/90/77      Tyr Arg         Gly Trp                         Asn             Lys                     Ile
Parrot/Ulster/PU Phe Lys        Val Trp                         Thr             Ser                     Val
```

(continued)

TABLE IX. (Continued)

```
                                                                                      400
WSN/33            Asp Pro Asn Gly Trp Thr Glu Thr Asp Ser Arg Phe Ser Met Arg Gln Asp Val Val Ala Ile Thr Asn Arg Ser
PR/8/34                                   Glu Thr         Lys     Ser Val Arg         Val Val Ala Met     Asp Trp
NJ/8/76                                   Glu Thr         Ser     Ser Met Lys         Ile Ile Ala Leu     Asp Trp
USSR/90/77                                Asp Pro         Asn     Leu Val Lys         Ile Val Ala Met     Asp Trp
Parrot/Ulster/PU                          Glu Thr         Ser     Ser Val Lys         Val Ala Ile Thr     Asp Trp
                                                                                      425
WSN/33            Gly Tyr Ser Gly Ser Phe Val Gln His Pro Glu Leu Thr Gly Leu Asp Cys Met Arg Pro Cys Phe Trp Val Glu
PR/8/34                   Ser                             Leu Asp     Ile
NJ/8/76                   Ser                             Met Asn     Ile
USSR/90/77                Arg                             Leu Asp     Met
Parrot/Ulster/PU          Ser                             Leu Asp     Met
                                                                                      450
WSN/33            Leu Ile Arg Gly Leu Pro Glu Glu Asp Ala  -  Ile Trp Thr Ser Gly Ser Ile Ile Ser Phe Cys Gly Val Asn Ser
PR/8/34                   Arg         Lys     Lys Thr  -                 Ala     Ser
NJ/8/76                   Gln         Lys     Ser Thr  -                 Gly     Ser
USSR/90/77                Arg         Arg     Lys Thr Thr                Gly     Ser
Parrot/Ulster/PU          Arg         Lys     Asn Thr  -                 Gly     Ser

WSN/33            Asp Thr Val Asp Trp Ser Trp Pro Asp Gly Ala Glu Leu Pro Phe Thr Ile Asp Lys
PR/8/34           Asp     Val Asp             Gly     Glu         Phe         Asp
NJ/8/76           Gly     Ala Ser             Gly     Asp         Phe         Asp
USSR/90/77        Asp     Val Asn             Asp     Glu         Leu         Pro
Parrot/Ulster/PU  Asp     Val Gly             Gly     Glu         Phe         Asp
```

[a]See Table I for sources.

particular the additional glycosylation site at position 143 (human virus N2 numbers), the large side chain (Arg) at residue 180 and the single residue insertions at positions 283 and 302. The homology between the avian and human viruses is highest for the early human N2 strains and lowest for the later strains (Desphande et al., 1985). Chicken/Penn/83 and Bangkok/1/79 are less alike than any other pair of N2 sequences characterized to date with 68 amino acid differences (see Tables VII and VIII).

For the most part, amino acid differences between the avian N2 and the human N2 are found at locations in which variation occurs within the human N2 sequences. New clusters (i.e., a pair or more) of changes are found at positions 142 and 143, 208 and 210, and 378, 381, 384, and 386. The 210 and 380 regions are both on the underside of the head.

## 2. N1

An amino acid sequence alignment of the N1 strains listed in Table I is given in Table IX. The available data do not permit a chronological analysis of antigenic drift as for N2, but sequence differences nevertheless highlight potentially antigenic regions of the molecule.

Concannon et al. (1984) reported that amino acid sequence variation among the N1 proteins is consistent with antigenic sites in locations homologous to those defined on the same basis for N2 (Colman et al., 1983). One important exception was the observation that N1 sequences show an additional hypervariable site, between residues 385 and 400 (N2 numbering) (see Table II). Varghese et al. (1983) observed that carbohydrate, covalently attached to Asn 200 on a neighboring subunit in the neuraminidase tetramer, covers at least a part of this surface in the N2 structure. Concannon et al. (1984) conclude that this N2 oligosaccharide, which is not present in N1 molecules, shields that surface from antibodies, rendering it nonantigenic in N2 structures. Their observation of variation in this region of the N1 structure is substantiated and extended by the two additional N1 sequences (NJ/8/76 and Parrot/Ulster/73) included here in Table IX, and is reminiscent of a parallel observation in the H1 and H3 hemagglutinins (Caton *et al.*, 1982).

The pattern of glycosylation triplets is not conserved in these N1 sequences. WSN/33 neuraminidase has lost the potential site at residue 146 (N2 numbering) (see Tables II and IX), a site otherwise conserved across and within all subtypes characterized to date. The WSN/33 sequence also displays the triplet sequence at residue 402 found in N2 sequences but not in other N1 sequences. Note here that in N2, the only subtype for which direct protein sequence data are available, Asn 402 is not glycosylated.

In USSR/90/77 neuraminidase there are two additional potential carbohydrate sites at residues 368 and 455 (N2 numbering) (see Table II). It must be emphasized here that no data yet address the question of whether these asparagine residues are indeed glycosylated.

The first of these two additional sites could mask the variable region on β5L23, and possibly adjacent sites of variation on β5L01 (Colman *et al.*, 1983). The second potential site is particularly interesting in light of the discussion above on the extra variable region in N1 proteins and its possible relationship to the lack of carbohydrate at position 200 (N2 numbering) in N1 structures. Residue 455 of N2 is within about 6 Å of residue 200 on a neighboring subunit of the tetramer. Thus, oligosac-

TABLE X. Amino Acid Sequence Differences
between N9 Neuraminidases Isolated from Avian
(A/Tern/Australia/G70c/75) and Mammalian
(A/Whale/Maine/1/84) Influenza Viruses[a]

| N9 Residue number (homologous N2 number) | | Whale | Tern |
|---|---|---|---|
| 27 | (27) | Val | Thr |
| 45 | (44) | Arg | His |
| 64 | (62) | Glu | Asp |
| 82 | (81) | Ser | Ile |
| 84 | (83) | Glu | Asp |
| 158 | (157) | Asp | Ala |
| 189 | (187) | Arg | Lys |
| 190 | (188) | Ala | Thr |
| 235 | (233) | Gln | His |
| 271 | (269) | Thr | Ala |
| 286 | (284) | Gln | Arg |
| 288 | (286) | Gly | Glu |
| 289 | (287) | Val | Ile |
| 306 | (304) | Gln | Arg |
| 359 | (358) | Gly | Val |
| 388 | (388) | Arg | Lys |
| 459 | (457) | Asn | Asp |

[a]From Air *et al.* (1987).

charide attached there in N1 could easily cover the homologous surface to that covered by sugars attached at Asn 200 in N2. If this were indeed the case, the new variable region on N1 would be expected to revert to a more constant state after 1977.

### 3. N9

Two N9 sequences have been characterized, one from an avian (tern) influenza virus and the other from a mammalian (whale) virus (Air *et al.*, 1987) (see Table X). Seventeen amino acid sequence changes differentiate the two molecules, three changes in the stalk region of the molecule and 14 in the head region. These substitutions are documented in Table X.

Unlike the sequence variation seen within N1 and N2 subtypes, which shows about one half the variable amino acids located on the upper surface of the neuraminidase head, the pattern of N9 variation, admittedly only in two sequences, shows 13 of 14 changes occurring on the bottom surface of the head. It is not clear what, if any, selection pressures have given rise to this result. One possibility could be the preservation of the hemagglutinating activity of N9 (see Section II.B.6) first observed for the tern virus (Laver *et al.*, 1984) and also found for the whale virus (Air *et al.*, 1987).

## B. Antigenic Analysis with Antibodies

Monoclonal antibodies have been used in a number of studies to probe the antigenic structure of neuraminidase. These studies have as their goal a mapping of the surface into well-defined antigenic areas and the identification of particular amino acids with particular epitopes. The first part derives from competition and cross-reactivity analyses and the second from the characterization of neuraminidase variants selected with different monoclonal antibodies. The three-dimensional structure of a monoclonal variant of hemagglutinin has shown only local structural changes around the mutated site, confirming that, at least in that case, the antibody binds directly to that amino acid as part of the epitope (Knossow *et al.*, 1984). The structure of a monoclonal variant of N2 neuraminidase is described in Section III.B.3.

### 1. N2

Monoclonal antibodies raised against different members of the N2 subtype were tested for their capacity to inhibit neuraminidase activity across the N2 subtype (Webster *et al.*, 1982). In most cases, these antibodies had neuraminidase inhibitory activity restricted to virus strains isolated within 3 or 4 years from the strain used for raising that particular antibody. By contrast, very little antigenic variation in avian N2 molecules was observed, all avian N2 strains being closely related to 1957 human influenza N2 isolates (Webster *et al.*, 1982). Results obtained with enzyme-linked immunosorbent assay (ELISA) in place of neuraminidase inhibition were similar but not always identical.

Competitive radioimmunoassays (RIA) with a panel of 40 monoclonal antibodies against RI/5+/57 neuraminidase have shown that there are four antigenic regions on this neuraminidase and that in most cases there is overlap between them. Most of these antibodies belong to the so-called group 2, and some of them compete with antibodies in group 1, 3, or 4 (Webster *et al.*, 1984). Group 2 antibodies can be further divided into overlapping subgroups A, B, C, and D.

Group 2 and 3 antibodies inhibit enzymatic activity on a macromolecular substrate (fetuin) and occasionally on the trisaccharide neuraminyllactose (Webster *et al.*, 1984). Whereas all antibodies tested inhibit the release of virus from Madin–Darby canine kidney (MDCK) cells when incorporated in an agar overlay, only the group 2 and 3 antibodies permit selection of antigenic variants, either in eggs or in MDCK cells.

Characterization of amino acid sequence changes in 12 monoclonal variants of RI/5+/57 neuraminidase has now permitted an implicit structural identification of regions of the antigen involved in the binding of group 2A, 2B, and 2D antibodies (Air *et al.*, 1985b), as summarized in Table XI. Antibodies in other groups have failed to select variants to date. The placement of monoclonal variants showing amino acid sequence changes extending from position 329 around to 370 within the same antigenic group, 2B, is consistent with the three-dimensional proximity of these residues (20 Å) and earlier interpretations of the overlapping nature of the antigenic surface (Colman *et al.*, 1983).

Monoclonal variants of a second N2 neuraminidase, Tokyo/3/67, have been characterized and those data are also shown in Table XI. Differences in reactivity of these variants which led to their classification as group Ia or Ib are for the most part substantiated by the sequence data; one exception is the 344 Arg to Ile variant (Webster *et al.*, 1982; Lentz *et al.*, 1984). Antibody 16/8 selects two different variants at position 344, both of which produce a new glycosylation triplet (Lentz *et al.*, 1984), although it is not known whether sugar is indeed attached at that site in the variants.

With one exception, residue 253, all these altered amino acids are on the upper surface of the neuraminidase head, in close proximity to the rim of the catalytic cavity (Colman *et al.*, 1983). The effect of the substitution at position 253 of Arg to Ser in Tokyo/3/67 neuraminidase on the binding of anti-Jap113/2 neuraminidase antibody is not understood. It is not clear whether an antibody making contact with this residue on the side surface of the head could inhibit the enzyme without inducing a structural change at the active site. The remaining upper surface sites of substitution all occur in loops of the structure showing variation between different field strains of influenza virus. Monoclonal variants of N2 are shown in Fig. 5.

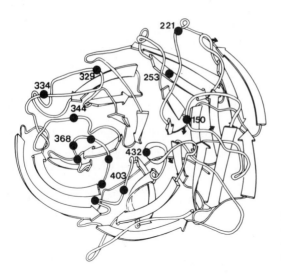

FIGURE 5. Location of substitutions observed in monoclonal variants of N2 and N9 neuraminidases (see Table XI).

### 2. N9

Monoclonal variants of the N9 subtype neuraminidase have now been characterized (Tulloch *et al.*, 1986; Webster, *et al.*, 1987), and are compiled in Table XI. With two exceptions, the same surface loops mapped in the N2 variants are scored in the N9 experiments. No N9 variants have yet been observed in the vicinity of residue 344, but, on the other hand, a variant at 432 has further implicated that region in antibody binding. Residues 431 and 434 are variable in N2 field strains.

One of the antibodies used in the study by Webster *et al.* (1987) has been crystallized in complex with N9 (Laver *et al.*, 1987) and the three-dimensional structure determined (Colman *et al.*, 1987) (see Section III.C.2). The variants selected by NC41, at residues 367 and 400 map to surface loops that are clearly part of the epitope, as shown by that structure, although precise details of their involvement or otherwise as contact residues to the antibody are not yet available.

### 3. Structure of an N2 Monoclonal Variant

The variant of A/Tokyo/3/67 neuraminidase selected by the monoclonal antibody S10/1 (see Table XI) has now been studied by X-ray diffraction (Varghese *et al.*, 1987). The substitution of glutamic acid for lysine at residue 368 has no consequences other than relocation of the substituted side chain. In the wild-type structure, Lys 368 is salt-linked to Asp 369, and in the variant Glu 368, it is bent away from residue 369 as if by simple charge repulsion.

This result is very similar to that reported by Knossow *et al.* (1984) for a monoclonal variant of the hemagglutinin; i.e., local structural changes suffice to abolish the binding of an antibody to an antigen effec-

TABLE XI. Monoclonal Variants
of N2 and N9

| Antibody group | | Selected variant | | |
|---|---|---|---|---|
| RI/5+/57(N2) | | | | |
| 81/4 | 2A | 403 | W to R | |
| 514/1 | 2B | 370 | S to L | |
| m220/2 | 2B | 370 | S to L | |
| 664/1 | 2B | 344 | R to K | |
| 561/3 | 2B | 344 | R to K | |
| 415/1 | 2B | 344 | R to G | |
| 112/2 | 2B | 334 | N to S; 368 K to E | |
| m474/1 | 2B | 334 | N to S; 368 K to E | |
| m490/3 | 2B | 334 | N to S; 368 K to E | |
| m193/2 | 2B | 329 | D to N | |
| 145/1 | 2D | 150 | H to N | |
| 480/2 | 2D | 150 | H to Q | |
| Tokyo/3/67(N2) | | | | |
| S25/3 | Ia | 344 | R to I | |
| 25/4 | Ia | 344 | R to I | |
| 32/3 | Ia | 344 | R to I | |
| 23/9 | Ib | 344 | R to K | |
| 23/9 | Ib | 344 | R to I | |
| 23/9 | Ib | 344 | R to G | |
| 16/8 | Ib | 344 | R to T | |
| 16/8 | Ib | 344 | R to S | |
| S10/1 | Ia | 368 | K to E | |
| Jap113/2[b] | II | 253 | R to S | |
| Tx18/1[b] | III | 221 | N to H | |
| A/Tern/Australia/G70c/75(N9) | | | | |
| NC10 | | 331 | (N2 329) N to D | |
| NC11 | | 370 | (N2 369) A to D | |
| NC17 | | 370 | (N2 369) A to D | |
| NC20 | | 370 | (N2 369) A to D | |
| NC24 | | 369 | (N2 368) I to R | |
| NC31 | | 370 | (N2 369) A to D | |
| NC34 | | 434 | (N2 432) K to N | |
| NC35 | | 331 | (N2 329) N to D | |
| NC41 | | 368 | (N2 367) S to N | |
| NC41 | | 400 | N to K | |
| NC42 | | 400 | N to K | |
| NC44 | | 373 | (N2 372) S to Y | |
| NC45 | | 370 | (N2 369) A to D | |
| NC47 | | 368 | (N2 367) S to N | |
| NC66 | | 370 | (N2 369) A to D | |
| 32/3 | | 331 | (N2 329) N to D | |
| 32/3 | | 371 | (N2 370) S to L | |

[a]From Laver et al. (1982), Webster et al. (1984), Lentz et al. (1984), Air et al. (1985b), and Webster et al. (1987).
[b]Jap antibody is raised against Japan/305/57 neuraminidase and Tx antibody against Texas/77.

tively. Although it is not possible to rule out formally the possibility of some monoclonal variants exerting their influence on antibody binding to an epitope distant from the site of the substitution, it is likely that the substituted residue in many of these variants will reside within the epitope for the selecting antibody.

## C. Structure of Immune Complexes

### 1. Structure of an Antineuraminidase Fab Fragment

Following earlier studies of the three-dimensional structures of Fab fragments of myeloma proteins (reviewed by Davies and Metzger, 1983), attention is now turning to the structural details of monoclonal antibodies of defined specificity, with a view to studying stereochemical aspects of antibody–antigen binding. The first such Fab fragment to be crystallized was from an antineuraminidase antibody (Colman et al., 1981), and the structure has been recently reported (Colman and Webster, 1987).

The details of the amino acid sequence of this protein, the S10/1 antibody referred to in Table XI, are not yet available, and the cell line has died. The structure shows a quaternary arrangement of variable module and constant module similar to that of most other isolated Fab fragments, namely, a bent elbow structure with the pseudo-diads of the two modules subtending an angle of some 130°. Furthermore, the pairing of VL and VH domains in the V module, and of CL and CH1 domains in the C module, conforms to that observed in a number of free Fab fragment structures (Davies and Metzger, 1983).

The characterization of the monoclonal variant selected by this antibody (Lys 368 to Glu) (see Section III.B.3) permits implicit identification of the epitope it recognizes. In the absence of amino acid sequence data for the complementarity-determining residues and the unavailability of the cell line, it is unlikely that this particular study will be able to proceed further.

### 2. Three-Dimensional Structure of Neuraminidase–Antibody Complexes

The first crystals of an antibody–antigen complex were reported by Colman et al. (1981). The antigen was neuraminidase and the antibody was S10/1 (see Table XI). No progress has been made on the study of those thin crystals, but electron microscopic evaluation of a number of different antibody–neuraminidase complexes has recently been reported (Tulloch et al., 1986). Low-resolution lattice image analysis of these crystals and of isolated protomers from dissolved crystals have demonstrated self-consistent images of Fab fragments and whole immunoglobulin attached to neuraminidase heads (Tulloch et al., 1986). The sites on the neuraminidase to which the antibodies bind are, to a low-

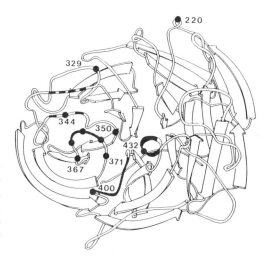

FIGURE 6. View, as in Fig. 1, of schematic of the neuraminidase structure, showing the epitope for the NC41 antibody as solid shading on the relevant chain segments. The assignment of segments around 344 and 333 to the epitope is tentative. From Colman *et al.* (1987).

resolution approximation, consistent with the notion that single amino acid changes selected by the antibodies are directly involved in binding epitopes for the antibodies used for their selection. A similar observation has been made for hemagglutinin–antibody complexes (Wrigley *et al.*, 1983).

Single crystals of an Fab fragment of the NC41 antibody complexed with N9 neuraminidase have been grown (Laver *et al.*, 1987) and the three-dimensional structure described (Colman *et al.*, 1987). The epitope on the neuraminidase comprises four discontinuous segments of the structure (Fig. 6), namely, the loops 367–370, 400–403, 430–434, and parts of 325–350. These data are consistent with selection of variants by the NC41 antibody at residues 367 and 400 and by the failure of other variants at residues 369, 370, and 432 to bind this antibody. Two other mutants, at positions 329 and 368, show partial binding to the antibody, while a third, at residue 220, behaves as wild-type N9 (Colman *et al.*, 1987). Thus, in this case at least, the behavior of mutants with the antibody provides an accurate map of the epitope.

Crystals of the N9 variant Asn 329 to Asp complexed with NC41 Fab are isomorphous with wild-type N9–NC41 Fab complex, indicating that, despite reduced binding in this complex, there is no rearrangement of the antibody on the antigen.

It has been suggested that epitopes on protein molecules are less well-ordered regions of the protein structure (Westhof *et al.*, 1984; Tainer *et al.*, 1984, 1985). Binding to antibody is thus facilitated by these flexible structures, accommodating the combining site of the antibody. There remains some debate over whether epitopes are not simply exposed regions (Novotny *et al.*, 1986), which are typically the less rigid structures of a protein molecule, but, at least in some cases, flexibility appears to be more important in determining antigenicity than does mere exposure (Geysen *et al.*, 1987). For the N9–NC41 complex, the epitope includes some of the more flexible regions of the neuraminidase (J. N. Varghese and P. M. Colman, unpublished data). In particular, the 325–350 loop is the last element of the structure remaining to be described accurately, testimony to its disorder with respect to the rest of the molecule.

Parts of the antigen have changed structure compared with that found in uncomplexed N9 (Colman et al., 1987). Although the structural alterations are small, they are likely to be energetically highly relevant to the interaction with antibody.

The NC41 antibody reduces enzymatic activity toward even a trisaccharide substrate, although there appears to be no steric impedence to access to the active site (Fig. 7) as a result of NC41 binding (Colman et al., 1987). Several factors might explain this observation:

1. The position of the aglycon moiety during catalysis is unknown, and there might be some hindrance there, especially if it is located in the region of the 370 loop.
2. Electrostatic, or other long-range forces, might perturb the diffusion of substrates into, or out of, the active site.
3. A small displacement of the end of Arg 371 might distort the geometry of the catalytic residues sufficiently to effect catalysis.
4. The position of the 370 loop might become frozen on binding the antibody, damping some catalytically important breathing motion of the structure.

Which, if any, of these mechanisms is operating should become clearer as the structure is refined and more is learned about the catalytic mechanism.

Another finding from the study of the N9–NC41 complex is that the pairing of the VL and VH domains is unlike that observed in other Fab

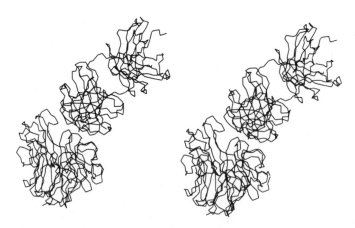

FIGURE 7. Stereo drawing of alpha-carbon atoms in one subunit of neuraminidase (view as in Fig. 3, right) and the attached Fab fragment. The active site of the enzyme is below and in front of the epitope. From Colman et al. (1987).

fragments (Colman *et al.*, 1987), including Fab in complex with lysozyme as antigen (Amit *et al.*, 1986). Differences of up to 6° in the way these domains associate in uncomplexed Fab structures have been attributed to the role of the complementarity-determining regions (CDRs) in the VL–VH interface (Davies *et al.*, 1975), where they constitute about 25% of the buried surface area. The implication is that the bulk of the buried surface in that interface can make small adjustments depending on the antigen-specific sequences of the CDRs.

The NC41 antibody complexed to N9 has an VL–VH pairing that differs by 8–12° from other Fab fragments (Colman *et al.*, 1987). Two rationales can be advanced to account for this observation. The first argues that the CDR sequences of NC41 have not been observed before in an Fab structure, and that they determine the pairing pattern by extension, and enlargement, of the arguments proposed to account for the smaller variation seen among Fab structures in general. There is no obvious correlation between the NC41 CDR sequences and an altered quaternary state of the V module, but this is not to say that it does not exist. The second allows for antigen to utilise the demonstrated slipperiness of the VL–VH interface to induce a change in the pairing for optimal binding of antigen to antibody; i.e., if a few amino acids within the CDRs can modulate the geometry of the VL–VH interface, there are at least no conceptual problems in envisaging how the presentation of a macromolecular surface to the 6 CDRs might do likewise. The acid test of this hypothesis would be the determination of the structure of the free Fab fragment, but crystals of that material have not yet been grown.

The suggested picture of antigen–antibody interactions that emerges from the induced-sliding hypothesis is that, not of a binary complex, but rather a ternary interaction among VL, VH, and antigen. Three contact surfaces are involved—VL–VH, VL–antigen, and VH–antigen—and modulation of the first might facilitate the simultaneous formation of the second and third.

The conclusions and hypotheses from this structure stand in contrast to those from the study of an antibody–lysozyme complex (Amit *et al.*, 1985, 1986). The size of the epitopes in the two cases is likely to be similar (16 amino acids in lysozyme), but no conformational changes are reported in either lysozyme or in the Fab fragment in that complex.

## IV. SUMMARY

A considerable body of data has now been assembled on both the hemagglutinin and neuraminidase membrane glycoproteins of influenza. Protein and gene sequences, together with three-dimensional structure data, provide a firm basis for understanding both the nature of antigenic variation in influenza virus and, to a lesser degree, the biological properties of the two surface antigens.

Although the enzymatic active center of neuraminidase is well characterized, the mechanism of action is not understood, and crystallographic studies with substrate analogues are required to address this question. Characterization of new site-directed mutants with respect to substrate binding and catalysis will further clarify the issue. Whether these studies uncover new inhibitors of the enzyme with antiviral activity remains to be seen.

The first data on the structure of a viral antigen–antibody complex have demonstrated that antibodies can, at least locally, deform viral antigens. Mutant antigens that show modified binding to the antibody are within the epitope, and one that does not is outside the epitope. The possibility that the antibody has changed its shape on binding antigens has raised many questions, while offering a rationale for further diversification of the immune system beyond the genetically accountable 10 or 100 million antibody specificities.

The active site of the enzyme is immunologically privileged. Its shape and location suggest that antibodies will not be able to bind to it at the exclusion of nearby, catalytically nonessential, surface amino acids. This is not just a result of the concave nature of the active site, but also of the small surface area of that site, with concomitant limited capacity to form stable complexes with antibodies on its own. In this way, enzymatically viable antigenic variants continue to emerge in response to host immune selection pressure.

## NOTE ADDED IN PROOF

Since this manuscript was completed in 1987, several relevant developments have occurred. The sequence of a neuraminidase of subtype N5 has been determined by Harley et al. (1989). The pattern of conserved amino acids in the enzyme active site is consistent with that sequence. Refinement of the three-dimensional structure of N2 (J.N. Varghese and P.M. Colman, unpublished data) has indicated that the chain folding around residue 330 is not as shown here in various places. The conserved aspartic acid at that position now appears not to be involved in substrate binding or catalysis. Further studies of 3-dimensional structures of antigen–antibody complexes have been published (Sheriff et al., 1987; Colman et al., 1989) and compared with other examples of protein–protein interactions (Colman, 1988).

## REFERENCES

Air, G. M., 1979, Nucleotide sequence for the signal peptide and N-terminus of the haemagglutinin from an Asian (H2N2) strain of influenza virus, Virology 97:468–472.
Air, G. M., Ritchie, L. R., Laver, W. G., and Colman, P. M., 1985a, Gene and protein sequence of an influenza neuraminidase with haemagglutinin activity, Virology 145: 117–122.

Air, G. M., Els, M. C., Brown, L. E., Laver, W. G., and Webster, R. G., 1985b, Location of antigenic sites on the three-dimensional structure of the influenza N2 virus neuraminidase, *Virology* **145**:237–248.

Air, G. M., Webster, R. G., Colman, P. M., and Laver, W. G., 1987, Distribution of sequence differences in influenza N9 neuraminidase of tern and whale viruses and crystallisation of the whale neuraminidase complexed with antibodies, *Virology* **160**:346–354.

Allen, A., 1983, Mucus—A protective secretion of complexity, *TIBS* **8**:169–173.

Allen, A. K., Skehel, J. J., and Yuferof, V. J., 1977, The amino acid and carbohydrate composition of the neuraminidase of B/Lee/40 influenza virus, *J. Gen. Virol.* **37**:625–628.

Amit, A. G., Mariuzza, R. A., Phillips, S. E. V., and Poljak, R. J., 1985, Three dimensional structure of an antigen–antibody complex at 6A resolution, *Nature (Lond.)* **313**:156–158.

Amit, A. G., Mariuzza, R. A., Phillips, S. E. V., and Poljak, R. J., 1986, Three dimensional structure of an antigen–antibody complex at 2.8 A resolution, *Science* **233**:747–753.

Ashwell, G., and Morell, A. G., 1974, The role of surface carbohydrates in the hepatic recognition and transport circulating glycoproteins, *Adv. Enzymol.* **41**:99–128.

Astry, C. L., Yolken, R. H., and Jakab, G. J., 1984, Dynamics of viral growth, viral enzymatic activity, and antigenicity in murine lungs during the course of influenza pneumonia, *J. Med. Virol.* **14**:81–90.

Baker, A. T., Varghese, J. N., Laver, W. G., Air, G. M., and Colman, P. M., 1987, The three-dimensional structure of neuraminidase of subtype N9 from an avian influenza virus, *Proteins: Structure, Function and Genetics* **2**:111–117.

Basak, S., Pritchard, D. G., Bhown, A. S., and Compans, R. W., 1981, Glycosylation sites of influenza viral glycoproteins—Characterisation of tryptic glycopeptides from the A/USSR (H1N1) haemagglutinin glycoprotein, *J. Virol.* **37**:549–558.

Basak, S., Tomana, M., and Compans, R. W., 1985, Sialic acid is incorporated into influenza haemagglutinin glycoproteins in the absence of viral neuraminidase, *Virus Res.* **2**:61–68.

Bentley, D. R., and Brownlee, G. G., 1982, Sequence of the N2 neuraminidase from influenza virus A/NT/60/68, *Nucleic Acids Res.* **10**:5033–5042.

Blok, J., and Air, G. M., 1980, Comparative nucleotide sequences at the 3′ end of the neuraminidase gene from eleven influenza type A viruses, *Virology* **107**:50–60.

Blok, J., and Air, G. M., 1982a, Block deletions in the neuraminidase genes from some influenza A viruses of the N1 subtype, *Virology* **118**:229–234.

Blok, J., and Air, G. M., 1982b, Variation in the membrane insertion and "stalk" sequences in eight subtypes of influenza type A virus neuraminidase, *Biochemistry* **21**:4001–4007.

Blok, J., and Air, G. M., 1982c, Sequence variation at the 3′ end of the neuraminidase gene from 39 influenza type A viruses, *Virology* **121**:211–229.

Blok, J., Air, G. M., Laver, W. G., Ward, C. W., Lilley, G. G., Woods, E. F., Roxburgh, C. M., and Inglis, A. S., 1982, Studies on the size, chemical composition and partial sequence of the neuraminidase (NA) from type A influenza virus show that the N-terminal region of the NA is not processed and serves to anchor the NA in the viral membrane, *Virology* **119**:109–121.

Blumberg, B., Giorgi, C., Roux, L., Raju, R., Dowling, P., Chollet, A., and Kolakofsky, D., 1985, Sequence determination of the Sendai virus HN gene and its comparison to the influenza glycoproteins, *Cell* **41**:269–278.

Booy, F. P., Ruigrok, R. W. H., and van Bruggen, E. F. J., 1985, Electron microscopy of influenza virus. A comparison of negatively stained and ice-embedded particles, *J. Mol. Biol.* **184**:667–676.

Branden, C.-I., 1980, Relation between structure and function of alpha/beta proteins, *Q. Rev. Biophys.* **13**:317–338.

Brown, L. E., Ward, C. W., and Jackson, D. C., 1982, Antigenic determinants of influenza virus haemagglutinin. IX. The carbohydrate side chains from an Asian strain, *Mol. Immunol.* **19**:329–338.

Bucher, D. J., and Kilbourne, E. D., 1972, A2 (N2) neuraminidase of the X-7 influenza virus recombinant: Determination of molecular size and subunit composition of the active unit, *J. Virol.* **10**:60–66.

Bucher, D. J., and Palese, P., 1975, The biologically active proteins of influenza virus. Neuraminidase, in: *Influenza Virus and Influenza* (E. D. Kilbourne, ed.), pp. 83–123, Academic, New York.

Burnet, F. M., 1948, Mucins and mucoids in relation to influenza virus action. IV. Inhibition by purified mucoid of infection and haemagglutinin with the virus strain WSE, *Aust. J. Exp. Biol. Med. Sci.* **26**:381–387.

Burnet, F. M., and Stone, J. D., 1947, The receptor destroying enzyme of *V. cholerae, Aust. J. Exp. Biol. Med. Sci.* **25**:227–233.

Burnet, F. M., McCrea, J. F., and Anderson, S. G., 1947, Mucin as a substrate of enzyme action by viruses of the mumps influenza group, *Nature (Lond.)* **160**:404–405.

Caton, A. J., Brownlee, G. G., Yewdell, J. W., and Gerhard, W., 1982, The antigenic structure of the influenza virus A/PR/8/34 haemagglutinin (H1 subtype), *Cell* **31**:417–427.

Choppin, P. W., 1969, Replication of influenza virus in a continuous cell line: High yield of infective virus from cells infected at high multiplicity, *Virology* **38**:130–134.

Colman, P. M., 1984, The structure and function of neuraminidase, *Peptide Protein Rev.* **4**: 215–255.

Colman, P. M., 1988, Structure of antibody–antigen complexes, *Adv. Immunol.* **43**:99–132.

Colman, P. M., and Laver, W. G., 1981, The structure of influenza virus neuraminidase heads at 5 Å resolution, in: *Structural Aspects of Recognition and Assembly in Biological Macromolecules* (M. Balaban, ed.), pp. 869–872, I.S.S., Rehovot.

Colman, P. M., and Ward, C. W., 1985, Structure and diversity of influenza virus neuraminidase, *Curr. Top. Microbiol. Immunol.* **114**:177–255.

Colman, P. M., and Webster, R. G., 1987, The structure of an antineuraminidase monoclonal Fab fragment and its interaction with the antigen, in: P. and S. Biomedical Sciences Symposium, *Biological Organisation and Macromolecular Interactions at High Resolution* (R. M. Burnet and H. J. Vogel, eds.), pp. 125–133, Academic, New York.

Colman, P. M., Gough, K. H., Lilley, G. G., Blagrove, R. J., Webster, R. G., and Laver, W. G., 1981, Crystalline monoclonal Fab fragment with specificity towards an influenza virus neuraminidase, *J. Mol. Biol.* **152**:609–614.

Colman, P. M., Varghese, J. N., and Laver, W. G., 1983, Structure of the catalytic and antigenic sites in influenza virus neuraminidase, *Nature (Lond.)* **303**:41–44.

Colman, P. M., Laver, W. G., Varghese, J. N., Baker, A. T., Tulloch, P. A., Air, G. M., and Webster, R. G., 1987, Three-dimensional structure of a complex of antibody with influenza virus neuraminidase, *Nature (Lond.)* **326**:358–363.

Colman, P. M., Tulip, W. R., Varghese, J. N., Tulloch, P. A., Baker, A. T., Laver, W. G., Air, G. M., and Webster, R. G., 1989, Three-dimensional structures of influenza virus neuraminidase–antibody complexes, *Phil. Trans. Roy Soc. (Lond.)* (in press).

Concannon, P., Kwolek, C. J., and Salser, W. A., 1984, Nucleotide sequence of the influenza virus A/USSR/90/77 neuraminidase gene, *J. Virol.* **50**:654–656.

Corfield, A. P., Wember, M., Schauer, R., and Rott, R., 1982, The specificity of viral sialidases. The use of oligosaccharide substrates to probe enzyme characteristics and strain specific differences, *Eur. J. Biochem.* **124**:521–525.

Corfield, A. P., Higa, H., Paulson, J. C., and Schauer, R., 1983, The specificity of viral and bacterial sialidases for $\alpha$2-3 and $\alpha$2-6 linked sialic acids in glycoproteins, *Biochim. Biophys. Acta* **744**:121–126.

Dale, B., Brown, R., Miller, J., White, R. T., Air, G. M., and Cordell, B., 1986, Nucleotide and deduced amino acid sequence of the influenza neuraminidase genes of two equine serotypes, *Virology* **155**:460–468.

Daniels, R. S., Douglas, A. R., Gonsalves-Scarano, F., Palu, G., Skehel, J. J., Brown, E., Knossow, M., Wilson, I. A., and Wiley, D. C., 1983, Antigenic structure of influenza virus haemagglutinin, in: *Origin of Pandemic Influenza Viruses* (W. G. Laver, ed.), pp. 9–18, Elsevier, New York.

Daniels, R. S., Downie, J. C., Hay, A. J., Knossow, M., Skehel, J. J., Wang, M. L., and Wiley, D. C., 1985, Fusion mutants of the influenza virus haemagglutinin glycoprotein, *Cell* **40**:431–439.

Davies, D. R., Padlan, E. A., and Segal, D. M., 1975, Three-dimensional structure of immunoglobulins, *Annu. Rev. Biochem.* **44**:639–667.

Davies, D. R., and Metzger, H., 1983, Structural basis of antibody function, *Annu. Rev. Immunol.* **1**:87–117.

Davis, L., Baig, M. M., and Ayoub, E. M., 1979, Properties of extracellular neuraminidase produced by group A *Streptococcus*, *Infect. Immunol.* **24**:780–786.

Desphande, K. L., Naeve, C. W., and Webster, R. G., 1985, The neuraminidases of the virulent and avirulent A/Chicken/Penn/83 (H5N2) influenza A viruses. Sequence and antigenic analysis, *Virology* **147**:49–60.

Dimmock, N. J., 1987, Multiple mechanisms of neutralisation of animal viruses, *TIBS* **12**: 70–75.

Dopheide, T. A. A., and Ward, C. W., 1981, The location of the bromelain cleavage site in a Hong Kong influenza virus haemagglutinin, *J. Gen. Virol.* **52**:367–370.

Drzeniek, R., 1972, Viral and bacterial neuraminidases, *Curr. Top. Microbiol. Immunol.* **59**: 35–74.

Drzenick, R., Frank, H., and Rott, R., 1968, Electron microscopy of purified influenza virus neuraminidase, *Virology* **36**:703–707.

Elango, N., Coligan, J. E., Jambou, R. C., and Venkatesan, S., 1986, Human parainfluenza type 3 virus haemagglutinin-neuraminidase glycoprotein: Nucleotide sequence of mRNA and limited amino acid sequence of CNBr peptides of the purified protein, *J. Virol.* **57**:481–489.

Elleman, T. C., Azad, A. A., and Ward, C. W., 1982, Neuraminidase gene from the early Asian strain of human influenza virus A/RI/5-/57 (H2N2), *Nucleic Acids Res.* **10**: 7005–7015.

Els, M. C., Air, G. M., Murti, K. G., Webster, R. G., and Laver, W. G., 1985, An 18 amino acid deletion in an influenza neuraminidase, *Virology* **142**:241–247.

Fields, S., Winter, G., and Brownlee, G. G., 1981, Structure of the neuraminidase in human influenza virus A/PR/8/34, *Nature (Lond.)* **290**:213–217.

Geysen, H. M., Tainer, J. A., Rodda, S. J., Mason, T. J., Alexander, H. Getzoff, E. D., and Lerner, R. A., 1987, Chemistry of antibody binding to a protein, *Science* **235**:1184–1190.

Gottschalk, A., 1958, Neuraminidase: Its substrate and mode of action, *Adv. Enzymol.* **20**: 135–145.

Gottschalk, A., 1972, Historical introduction, in: *Glycoproteins, Their Composition, Structure and Function* (A. Gottschalk, ed.), pp. 2–3, Elsevier, Amsterdam.

Griffin, J. A., and Compans, R. W., 1979, Effect of cytochalasin B on the maturation of enveloped viruses, *J. Exp. Med.* **150**:379–391.

Griffin, J. A., Basak, S., and Compans, R. W., 1983, Effects of hexose starvation and the role of sialic acid in influenza virus release, *Virology* **125**:324–334.

Harley, V. R., Ward, C. W., and Hudson, P. J., 1989, Molecular cloning and analysis of the NS neuraminidase subtype from an avian influenza virus, *Virology* **169**:239–243.

Haywood, A. M., and Boyer, B. P., 1985, Fusion of influenza virus membranes with liposomes at pH 7.5, *Proc. Natl. Acad. Sci. USA* **82**:4611–4615.

Herrler, G., Rott, R., and Klenk, H.-D., 1985a, Neuraminic acid is involved in the binding of influenza C virus to erythrocytes, *Virology* **141**:144–147.

Herrler, G., Rott, R., and Klenk, H.-D., Mueller, H.-P., Shukla, H. K., and Schauer, R., 1985b, The receptor destroying enzyme of influenza C virus is neuraminate-O-acetylesterase, *EMBO J.* **4**:1503–1506.

Hiebert, S. W., Paterson, R. G., and Lamb, R. A., 1985, Haemagglutinin-neuraminidase protein of the paramyxovirus Simian virus 5: Nucleotide sequence of the mRNA predicts an N-terminal membrane anchor, *J. Virol.* **54**:1–6.

Hirsch, R. L., 1982, The complement system: Its importance in the host response to viral infection, *Microbiol. Rev.* **46**:71–85.

Hirsch, R. L., Winkelstein, J. A., and Griffin, D. E., 1980, The role of complement in viral infections. III. Activation of the classical and alternate pathways by Sindbis virus, *J. Immunol.* **124:**2507–2510.

Hirsch, R. L., Griffin, D. E., and Winkelstein, J. A., 1981, Host modification of Sindbis virus sialic acid content influences alternative complement pathway activation and virus clearance, *J. Immunol.* **127:**1740–1743.

Hirst, G. K., 1942, Adsorption of influenza haemagglutinins and virus by red blood cells, *J. Exp. Med.* **76:**195–209.

Hirst, G. K., 1950, The relationship of the receptors of a new strain of virus to those of the mumps–NDV–influenza group, *J. Exp. Med.* **91:**177–184.

Hiti, A. L., and Nayak, D. P., 1982, Complete nucleotide sequence of the neuraminidase gene of human influenza virus A/WSN/33, *J. Virol.* **41:**730–734.

Huang, R. T. C., Rott, R., Wahn, K., Klenk, H.-D., and Kohama, T., 1980, The function of the neuraminidase in membrane fusion induced by myxoviruses, *Virology* **107:**313–319.

Huang, R. T. C., Dietsch, E., and Rott, R., 1985, Further studies on the role of neuraminidase and the mechanism of low pH dependence in influenza virus-induced membrane fusion, *J. Gen. Virol.* **66:**295–301.

Hubbard, S. C., and Ivatt, R. J., 1981, Synthesis and processing of asparagine-linked oligosaccharides, *Annu. Rev. Biochem.* **50:**555–583.

Jancik, J. M., Schauer, R., and Streicher, H.-J., 1975, Influence of membrane bound N-acetylneuraminic acid on the survival of erythrocytes in man, *Z. Physiol. Chem.* **356:**1329–1331.

Jorgensen, E. D., Collins, P. L., and Lomedico, P., 1987, Cloning and nucleotide sequence of Newcastle Disease Virus haemagglutinin-neuraminidase mRNA: Identification of a putative sialic acid binding site, *Virology* **156:**12–24.

Kawaoka, Y., Naeve, C., and Webster, R. G., 1984, Is virulence of H5N2 influenza viruses in chickens associated with the loss of carbohydrate from the haemagglutinin?, *Virology* **139:**303–316.

Kazatchkine, M. D., Fearon, D. T., and Austen, K. F., 1979, Human alternative complement pathway: Membrane associated sialic acid regulates the competition between β and β1H for cell bound C3b, *J. Immunol.* **122:**75–81.

Keil, W., Geyer, R., Dabrowski, J., Dabrowski, U., Niemann, H., Stirm, S., and Klenk, H.-D., 1985, Carbohydrates of influenza virus. Structural elucidation of the individual glycans of the FPV haemagglutinin by two dimensional H n.m.r. and methylation analysis, *EMBO J.* **4:**2711–2720.

Kilbourne, E. D., Laver, W. G., Schulman, J. L., and Webster, R. G., 1968, Antiviral activity of antiserum specific for an influenza virus neuraminidase, *J. Virol.* **2:**281–288.

Knossow, M., Daniels, R. S., Douglas, A. R., Skehel, J. J., and Wiley, D. C., 1984, Three dimensional structure of an antigenic mutant of the influenza virus haemagglutinin, *Nature (Lond.)* **311:**678–680.

Kolb-Bachofen, V., and Kolb, H., 1979, Autoimmune reactions against liver cells by syngeneic neuraminidase-treated lymphocytes, *J. Immunol.* **123:**2830–2834.

Lamb, R. A., Shaw, M. W., Breidis, D. J., and Choppin, P. W., 1983, The nucleotide sequence of the neuraminidase gene of influenza B virus reveals two overlapping reading frames, in: *The Origin of Pandemic Influenza Viruses* (W. G. Laver, ed.), pp. 77–86, Elsevier, New York.

Laver, W. G., 1978, Crystallisation and peptide maps of neuraminidase heads from H2N2 and H3N2 influenza virus strains, *Virology* **86:**78–87.

Laver, W. G., and Valentine, R. C., 1969, Morphology of the isolated haemagglutinin and neuraminidase subunits of influenza virus, *Virology* **38:**105–119.

Laver, W. G., Air, G. M., Webster, R. G., and Markoff, L. J., 1982, Amino acid sequence changes in antigenic variants of type A influenza virus N2 neuraminidase, *Virology* **122:**450–460.

Laver, W. G., Colman, P. M., Webster, R. G., Hinshaw, V. S., and Air, G. M., 1984, Influenza virus neuraminidase with haemagglutinin activity, *Virology* **137:**314–323.

Laver, W. G., Webster, R. G., and Colman, P. M., 1987, Crystals of antibodies complexed with influenza virus neuraminidase show isosteric binding of antibody to wild-type and variant antigens, *Virology* **156**:181–184.

Lentz, M. R., and Air, G. M., 1985, Loss of enzyme activity in a site directed mutant of influenza neuraminidase compared to expressed wild-type protein, *Virology* **148**:74–83.

Lentz, M. R., Air, G. M., Laver, W. G., and Webster, R. G., 1984, Sequence of the neuraminidase gene from influenza virus A/Tokyo/3/67 and previously uncharacterised monoclonal variants, *Virology* **135**:257–265.

Lentz, M. R., Air, G. M., and Webster, R. G., 1987, Site-directed mutation of the active site of influenza neuraminidase and implications for the catalytic mechanism, *Biochemistry* **26**:5351–5358.

Luo, M., Vriend, G., Kamer, G., Minor, I., Arnold, E., Rossman, M. G., Boege, U., Scraba, D. G., Duke, G. M., and Palmenberg, A. C., 1987, The atomic structure of Mengo Virus at 3 Å resolution, *Science* **235**:182–191.

Markoff, L., and Lai, C. J., 1982, Sequence of the influenza A/Udorn/72 (H3N2) virus neuraminidase gene as determined from cloned full-length DNA, *Virology* **119**: 288–297.

Martinez, C., Del Rio, L., Portelo, A., Domingo, E., and Ortin, J., 1983, Evolution of the influenza virus neuraminidase gene during drift of the N2 subtype, *Virology* **130**:539–545.

McCauley, J. W., Bye, J., Elder, K., Gething, M.-J., Skehel, J. J., Smith, A., and Waterfield, M. D., 1979, Influenza virus haemagglutinin signal sequence, *FEBS Lett.* **108**:422–428.

McSharry, J. J., Pickering, R. J., and Caliguiri, L. A., 1981, Activation of the alternate complement pathway by enveloped viruses containing limited amounts of sialic acid, *Virology* **114**:507–515.

Metcalf, P., Blum, M., Freyman, D., Turner, M., and Wiley, D. C., 1987, Two variant surface glycoproteins of *Trypanosoma brucei* of different sequence classes have similar 6 A resolution X-ray structures, *Nature (Lond.)* **325**:84–86.

Michel, H., 1983, Crystallisation of membrane proteins, *TIBS* **8**:56–59.

Miki, T., Nishida, Y., Hisajima, H., Takashi, M., Kumahara, Y., Nerome, K., Oya, A., Fukui, T., Ohtsuka, E., Ikehara, M., and Honjo, T., 1983, The complete nucleotide sequence of the influenza virus neuraminidase gene of A/NJ/8/76 strain and its evolution by segmental duplication and deletion, *Mol. Biol. Med.* **1**:401–413.

Millar, N. S., Chambers, P., and Emmerson, P. T., 1986, Nucleotide sequence analysis of the haemagglutinin-neuraminidase gene of Newcastle Disease Virus, *J. Gen. Virol.* **67**: 1917–1927.

Milligan, T. W., Mattingly, S. J., and Straus, D. C., 1980, Purification and partial characterisation of neuraminidase from Type III group B *Streptococci*, *J. Bacteriol.* **144**:164–172.

Murata, M., Richardson, J. S., and Sussman, J. L., 1985, Simultaneous comparison of three protein sequences, *Proc. Natl. Acad. Sci. USA* **82**:3073–3077.

Murti, K. G., and Webster, R. G., 1986, Distribution of haemagglutinin and neuraminidase on influenza virus revealed by immune electron microscopy, *Virology* **149**:36–43.

Nakajima, S., and Sugiura, A., 1980, Neurovirulence of influenza in mice. II. Mechanism of virulence as studied in a neuroblastoma cell line, *Virology* **101**:450–457.

Novotny, J., Handschumacher, M., Haber, E., Bruccoleri, R. E., Carlson, W. B., Fanning, D. W., Smith, J. A., and Rose, G. D., 1986, Antigenic determinants in proteins coincide with surface regions accessible to large probes (antibody domains), *Proc. Natl. Acad. Sci. USA* **83**:226–230.

Palese, P., and Compans, R. W., 1976, Inhibition of influenza virus replication in tissue culture by 2-deoxy-2,3-dehydro-N-trifluoro-acetyl-neuraminic acid (FANA): Mechanism of action, *J. Gen. Virol.* **33**:159–163.

Palese, P., Tobita, K., Ueda, M., and Compans, R. W., 1974, Characterisation of temperature sensitive influenza virus mutants defective in neuraminidase, *Virology* **61**:397–410.

Paulson, J. C., Weinstein, J., Dorland, L., Van Halbeek, H., and Vliegenthart, J. F. G., 1982,

Newcastle disease virus contains a linkage-specific glycoprotein sialidase, *J. Biol. Chem.* **257**:12734–12738.

Pfeifer, J. B., and Compans, R. W., 1984, Structure of the influenza C glycoprotein gene as determined from cloned DNA, *Virus Res.* **1**:281–296.

Portner, A., 1981, The HN glycoprotein of Sendai virus: Analysis of site(s) in haemagglutinin and neuraminidase activities, *Virology* **115**:375–384.

Richardson, J. S., 1981, The anatomy and taxonomy of protein structure, *Adv. Protein Chem.* **34**:167–339.

Robertson, J. S., 1979, 5′ and 3′ terminal nucleotide sequences of the RNA genome segments of influenza virus, *Nucleic Acids Res.* **6**:3745–3757.

Rothman, J. E., and Lodish, H. F., 1977, Synchronised transmembrane insertion and glycosylation of a nascent membrane protein, *Nature (Lond.)* **269**:775–780.

Schauer, R., 1982, Sialic acids, *Adv. Carbohydr. Chem. Biochem.* **40**:131–234.

Schauer, R., 1985, Sialic acids and their role as biological masks, *TIBS* **10**:357–360.

Scheid, A., Caliguiri, L. A., Compans, R. W., and Choppin, P. W., 1972, Isolation of paramyxovirus glycoproteins. Association of both haemagglutinin and neuraminidase activities with the larger SV5 glycoprotein, *Virology* **50**:640–652.

Schild, G. C., Newman, R. W., Webster, R. G., Major, D., and Hinshaw, V. S., 1980, Antigenic analysis of the haemagglutinin, neuraminidase and nucleoprotein antigens of influenza A viruses, in: *Structure and Variation in Influenza Viruses* (W. G. Laver and G. M. Air, eds.), pp. 373–384, Elsevier/North-Holland, New York.

Scholtissek, C., 1978, The genome of influenza virus, *Curr. Top. Microbiol. Immunol.* **80**: 139–169.

Schulman, J. L., 1975, Immunology of influenza, in: *The Influenza Viruses and Influenza* (E. D. Kilbourne, ed.), pp. 373–393, Academic, New York.

Schulman, J. L., and Palese, P., 1977, Virulence factors of influenza A virus: WSN virus neuraminidase required for plaque production in MDBK cells, *J. Virol.* **24**: 170–176.

Schuy, W., Garten, W., Linder, D., and Klenk, H.-D., 1984, The carboxy-terminus of the haemagglutinin-neuraminidase of Newcastle disease virus is exposed at the surface of the viral envelope, *Virus Res.* **1**:415–426.

Shaw, M. W., Lamb, R. A., Erikson, B. W., Breidis, D. J., and Choppin, P. W., 1982, Complete nucleotide sequence of the neuraminidase gene of influenza B virus, *Proc. Natl. Acad. Sci. USA* **79**:6817–6821.

Sheriff, S., Silverton, E. W., Padlan, E. A., Cohen, G. H., Smith-Giu, S. J., Finzel, B. C., and Davies, D. R., 1987, Three-dimensional structure of an antibody–antigen complex, *Proc. Natl. Acad. Sci. USA* **84**:8075–8079.

Skehel, J. J., and Waterfield, M. D., 1975, Studies on the primary structure of the influenza virus haemagglutinin, *Proc. Natl. Acad. Sci. USA* **72**:93–97.

Skehel, J. J., Bayley, P. M., Brown, E. B., Martis, S. R., Waterfield, M. D., White, J. M., Wilson, I. A., and Wiley, D. C., 1982, Changes in the conformation of influenza virus haemagglutinin at the pH optimum of virus-mediated membrane fusion, *Proc. Natl. Acad. Sci. USA* **79**:968–972.

Steuler, H., Rohde, W., and Scholtissek, C., 1984, Sequence of the neuraminidase gene of an avian influenza A virus (A/Parrot/Ulster/73, H7N1), *Virology* **135**:118–124.

Sugiura, A., and Ueda, M., 1980, Neurovirulence of influenza in mice. I. Neurovirulence of recombinants between virulent and avirulent viral strains, *Virology* **101**:440–449.

Tainer, J. A., Getzoff, E. D., Alexander, H., Houghten, R. A., Olson, A. J., and Lerner, R. A., 1984, The reactivity of antipeptide antibodies is a function of the atomic mobility of sites in a protein, *Nature (Lond.)* **312**:127–133.

Tainer, J. A., Getzoff, E. D., Paterson, Y., Olson, A. J., and Lerner, R. A., 1985, The atomic mobility component of protein antigenicity, *Annu. Rev. Immunol.* **3**:501–535.

Tiffany, J. M., and Blough, H. A., 1970, Estimation of the number of surface projections on myxo- and paramyxoviruses, *Virology* **41**:392–394.

Tulloch, P. A., Colman, P. M., Davis, P. C., Laver, W. G., Webster, R. G., and Air, G. M.,

1986, Electron and X-ray diffraction studies of influenza neuraminidase complexed with monoclonal antibodies, *J. Mol. Biol.* **190**:215–225.

Uchida, Y., Tsukada, Y., and Sugimori, T., 1979, Enzymatic properties of neuraminidases from *Arthrobacter ureafaciens, J. Biochem.* **86**:1573–1585.

Van Rompuy, L., Min Jou, W., Huylebroeck, D., and Fiers, W., 1982, Complete nucleotide sequence of a human influenza neuraminidase gene of subtype N2 (A/Vic/3/75), *J. Mol. Biol.* **161**:1–11.

Varghese, J. N., Laver, W. G., and Colman, P. M., 1983, Structure of the influenza virus glycoprotein antigen neuraminidase at 2.9 Å resolution, *Nature (Lond.)* **303**:35–40.

Varghese, J. N., Webster, R. G., Laver, W. G., and Colman, P. M., 1988, Structure of an escape mutant of glycoprotein N2 neuraminidase of influenza virus A/Tokyo/3/67 at 3 Å resolution, *J. Mol. Biol.* **200**:201–203.

Wagh, P. V., and Bahl, O. P., 1981, Sugar residues on proteins, *Crit. Rev. Biochem.* **10**:307–377.

Ward, C. W., and Dopheide, T. A. A., 1981, Amino acid sequence and oligosaccharide distribution of the haemagglutinin from an early Hong Kong variant A/Aichi/2/68(X-31), *Biochem J.* **193**:953–962.

Ward, C. W., Elleman, T. C., and Azad, A. A., 1982, Amino acid sequence of the pronase-released heads of neuraminidase subtype N2 from the Asian strain A/Tokyo/3/67 of influenza virus, *Biochem. J.,* **207**:91–95.

Ward, C. W., Colman, P. M., and Laver, W. G., 1983a, The disulphide bonds of an Asian influenza virus neuraminidase, *FEBS Lett.* **153**:29–33.

Ward, C. W., Murray, J. M., Roxburgh, C. M., and Jackson, D. C., 1983b, Chemical and antigenic characterisation of the carbohydrate side chains of an Asian (N2) influenza virus neuraminidase, *Virology* **126**:370–375.

Webster, R. G., Hinshaw, V. S., and Laver, W. G., 1982, Selection and analysis of antigenic variants of the neuraminidase of N2 influenza virus with monoclonal antibodies, *Virology* **117**:93–104.

Webster, R. G., Brown, L. E., and Laver, W. G., 1984, Antigenic and biological characterisation of influenza virus neuraminidase (N2) with monoclonal antibodies, *Virology* **135**:30–42.

Webster, R. G., Air, G. M., Metzger, D. W., Colman, P. M., Varghese, J. N., Baker, A. T., and Laver, W. G., 1987, Antigenic structure and variation in influenza N9 neuraminidase, *J. Virol.* **61**:2910–2916.

Wedgewood, R. J., Ginsberg, H. S., and Pillemer, L., 1956, The properdin system and immunity. VI. The inactivation of Newcastle disease virus by the properdin system, *J. Exp. Med.* **104**:707–725.

Westhof, E., Altschuh, D., Moras, D., Bloomer, A. C., Mondragon, A., Klug, A., and Van Regenmortel, M. H. V., 1984, Correlation between segmental mobility and the location of antigenic determinants in proteins, *Nature (Lond.)* **311**:123–126.

White, J., and Helenius, A., 1980, pH dependent fusion between Semliki forest virus membrane and liposomes, *Proc. Natl. Acad. Sci. USA* **77**:3273–3277.

White, J., Helenius, A., and Gething, M.-J., 1982a, Haemagglutinin of influenza virus expressed from a cloned gene promotes membrane fusion, *Nature (Lond.)* **300**:658–659.

White, J., Kartenbeck, J., and Helenius, A., 1982b, Membrane fusion activity of influenza virus, *EMBO J.* **1**:217–222.

Wiley, D. C., Wilson, I. A., and Skehel, J. J., 1981, Structural identification of the antibody binding sites of the Hong Kong influenza haemagglutinin and their involvement in antigenic variation, *Nature (Lond.)* **298**:373–378.

Wilson, I. A., Skehel, J. J., and Wiley, D. C., 1981, Structure of the haemagglutinin membrane glycoprotein of influenza virus at 3 Å resolution, *Nature (Lond.)* **289**:366–373.

Woodruff, J. J., and Gesner, B. M., 1969, The effect of neuraminidase on the fate of transfused lymphocytes, *J. Exp. Med.* **129**:551–567.

Wrigley, N. G., Skehel, J. J., Charlwood, P. A., and Brand, C. M., 1973, The size and shape of influenza virus neuraminidase, *Virology* **51**:525–529.

Wrigley, N. G., Laver, W. G., and Downie, J. C., 1977, Binding of antibodies to isolated haemagglutinin and neuraminidase molecules of influenza virus observed in the electron microscope, *J. Mol. Biol.* **109**:405–421.

Wrigley, N. G., Brown, E. B., Daniels, R. S., Douglas, A. R., Skehel, J. J., and Wiley, D. C., 1983, Electron microscopy of influenza haemagglutinin–monoclonal antibody complexes, *Virology* **131**:308–314.

## CHAPTER 5

# Membrane Insertion and Intracellular Transport of Influenza Virus Glycoproteins

MICHAEL G. ROTH, MARY-JANE GETHING, AND JOE SAMBROOK

## I. INTRODUCTION

While the major features of the intracellular route traveled by the hemagglutinin (HA) and neuraminidase (NA) glycoproteins of influenza virus were established more than a decade ago by a synthesis of studies of virus morphogenesis with investigations of the secretory pathway in mammalian cells, our understanding of the biosynthesis of these viral envelope glycoproteins has expanded dramatically during the past 5 years. This progress has depended in part on the availability of detailed information on the structure of the HA and NA glycoproteins, and on the ability to express cloned genes encoding these polypeptides. These advances rest on a foundation of more than half a century of investigation of the nature of the surface antigens of a virus that remains one of the uncontrolled pathogens of man.

The envelope glycoproteins of influenza viruses were first identified as hemagglutinating and neuramidase activities associated with the vir-

MICHAEL G. ROTH and JOE SAMBROOK • Department of Biochemistry, University of Texas Southwestern Medical Center, Dallas, Texas 75235-9038.    MARY-JANE GETHING • Department of Biochemistry and Howard Hughes Medical Institute, University of Texas Southwestern Medical Center, Dallas, Texas 75235-9038.

ion surface (Hirst, 1941, 1942, 1943; McClelland and Hare, 1941; Gottschalk, 1957). In some of the earliest applications of electron microscopy to the study of virus-infected cells, it was discovered that influenza viruses matured by the then novel process of budding from the cell surface (Murphy and Bang, 1952)and that influenza virions were covered with a fuzzy coat (Morgan et al., 1956). As preservation techniques improved, this outer fuzz was seen to be composed of spikelike projections (Horne et al., 1960), which were later shown to be embedded in a lipid bilayer derived from areas of the plasma membrane modified by the virus (Bachi et al., 1969; Compans and Dimmock, 1969). That the NA and HA activities resided on separate proteins was established by analyzing viruses with reassorted genomes that express the HA of one antigentically distinct parent and the NA of the other (Tumova and Pereira, 1965; Laver and Kilbourne, 1966). Through the use of such viruses, antisera specific for each spike protein were prepared, and HA was identified as the antigen responsible for eliciting neutralizing antibodies (Seto and Rott, 1966; Webster and Laver, 1967).

The molecular analysis of HA and NA began with the discovery that influenza viruses could be dissociated with sodium dodecyl sulfate (SDS) into subunits that retained HA or NA activity. These subunits could be purified from certain virus strains by electrophoresis (Laver, 1963, 1964; Laver and Webster, 1968). Using electron microscopy, Laver and Valentine (1969) were able to show correspondence between these isolated subunits and two kinds of functionally and morphologically distinct spikes, each anchored to the virus envelope by a terminal hydrophobic domain. Soon afterward, experiments that employed controlled proteolysis of the virion surface established that HA was attached to the viral membrane by its carboxy-terminus (Compans et al., 1970; Brand and Skehel, 1972). Several groups used electrophoretic and immunochemical techniques to show that the HA and NA polypeptides were glycosylated as well as to determine their molecular weights (Dimmock, 1969; Schultz, 1975; Compans et al., 1970; Haslam et al., 1970; Webster, 1970; Skehel and Schild, 1971). Comparison of the molecular weights of the HA and NA polypeptides with the size and morphology of the intact spikes observed by electron microscopy led to the idea that HA was most likely a trimer (Laver, 1973; Griffith, 1975; Schultz, 1975) and NA a tetramer (Wrigley et al., 1973). As the details of the structure of the influenza virus glycoproteins became known, awareness increased that these proteins were interesting not only for the role they played in viral infection, but also as model cell-surface proteins that were easy to isolate and characterize biochemically.

Four key observations led to the understanding that influenza virus morphogenesis, and in particular the biosynthetic pathway of the envelope proteins, reflected the traffic patterns of normal membrane proteins within animal cells. First, the viral spike proteins were observed at the plasma membrane before budding of virus particles began, and thus were presumably transported to the cell surface independently of the internal components of the virus (Compans and Dimmock, 1969; Compans and Choppin, 1971). Second, it was discovered that in some, but not all, cell types, HA polypeptides were cleaved into two disulfide-linked subunits (Laver, 1971; Stanley and Haslam, 1971; Lazarowitz et al., 1971). Third, influenza virus glycoproteins isolated from viruses grown in different cell types differed in their electrophoretic mobilities although nonglycosylated influenza proteins did not (Compans et al., 1970). These observations indicated that HA and NA were processed post-translationally by cellular mechanisms. Fourth, the size of the RNA genome of influenza viruses suggested that the virus encoded few, if any, gene products in addition to the known influenza structural proteins (Skehel, 1971; Bishop et al., 1971; Lewandowski et al., 1971). Thus it seemed likely that influenza viruses used host mechanisms in order to synthesize their envelope components. Confirmation of this idea came from experiments using pulse-chase protocols and cell fractionation techniques. Several groups (Compans, 1973a, b; Stanley et al., 1973; Hay, 1974; Klenk et al., 1974) demonstrated that influenza virus glycoproteins were glycosylated and transported to the plasma membrane along a path that in major respects resembled that previously established for cellular secretory proteins (reviewed by Palade, 1975). Experiments have since confirmed that transport of the envelope glycoproteins of simple RNA viruses, including influenza virus HA and NA, does not differ in any significant respect from that of cellular exocytic proteins (Knipe, 1977; Bergmann et al., 1981; Green et al., 1981; Wehland et al., 1982; Bergmann and Singer, 1983; Saraste and Hedman, 1983; Rodriguez-Boulan et al., 1984; Balch et al., 1984; Rindler et al., 1985; Griffiths et al., 1985). Thus, some 40 years after they were first described as hemagglutinating and receptor-destroying activities in influenza virus preparations, the influenza virus envelope glycoproteins have become appreciated as a superb model system for study of the biosynthesis and transport of cell-surface glycoproteins.

Efforts to produce a universal vaccine to influenza viruses have generated a large body of information relating the structure, the biological function, and the antigenicity of influenza virus glycoproteins (reviewed by Ward, 1981; Webster et al., 1983; Colman and Ward, 1985; see also Chapter 4, this volume). This work has focused primarily on the type A influenza virus proteins, and the three-dimensional structures of wild-type and mutant influenza A virus HAs of an H3 subtype (Wilson et al., 1981; Daniels et al., 1985) and of two NAs of the N2 subtype (Varghese et al., 1983) have been solved to a resolution of 0.3 nm. These structural data assume an increasingly important role in the genetic analysis of the biosynthetic pathway of influenza virus glycoproteins. In particular, they permit an understanding of the structural consequences of mutations in HA or NA that is not presently possible for less characterized glycoproteins.

In recent years, our knowledge of the biosynthesis and transport of the influenza virus glycoproteins has been expanded enormously by the ability to clone and express complementary DNA (cDNA) copies of wild-type and mutant forms of the viral polypeptides. The genes encoding HA or NA glycoproteins from a large number of human, animal, and avian type A viruses have been cloned and sequenced, as have lesser numbers of genes from the type B viruses and the gene that specifies the single gp88 glycoprotein of type C influenza virus (for review, see Air and Compans, 1983). The majority of expression studies have employed eukaryotic viral vectors based on simian virus 40 (SV40) (for review, see Gething and Sambrook, 1983) or vaccinia virus (Smith *et al.*, 1983; Panicali *et al.*, 1984). More recently, continuous cell lines that express high levels of HA have been established (Sambrook *et al.*, 1985), so that for the first time it has been possible to study the biosynthesis of HA in an uninfected cell. This opens the possibility of selecting cells that are defective in their ability to process HA correctly and consequently of elucidating the cellular components required for the synthesis and transport of influenza virus glycoproteins.

This chapter reviews what is known of the biosynthesis of influenza virus glycoproteins: where the proteins travel, how they are recognized and sorted to their proper destination, and how the proteins are processed by the cell to acquire the mature structures ultimately incorporated into virions. In the following sections, the exocytic pathway has been dissected into its successive stages. In each case, we have presented the available data for HA and NA in the broader context of our knowledge of the synthesis, processing, and transport of eukaryotic membrane proteins in general. We have relied heavily on data obtained for the biosynthesis of the glycoproteins of type A influenza viruses; however, comparisons of the amino acid sequences of type A and B HAs and of type A and B NAs indicate a strong conservation of residues known to be important in maintaining protein structures. It is therefore likely that the three-dimensional structures of the glycoproteins of type A and type B influenza viruses are quite similar and that their synthesis and transport do not differ significantly. By contrast, the single glycoprotein of type C viruses is only distantly related to the HAs of type A and B viruses (Pfeiffer and Compans, 1984; Nakada *et al.*, 1984); the degree to which its biosynthesis resembles that of the glycoproteins of influenza types A and B is not yet apparent.

It is now clear that a common pathway exists for transport of proteins to the cell surface (for review, see Palade, 1975; Farquhar and Palade, 1981; Sabatini *et al.*, 1982; Kornfeld and Kornfeld, 1985; Farquhar, 1985). This pathway is organized both spacially and temporally as a sequence of membrane bound compartments each of which contains its own complement of processing enzymes (Fig. 1). Movement through the exocytic pathway is vectorial; there is no evidence for a backflow of proteins from a later organelle of the pathway to an earlier one, at least

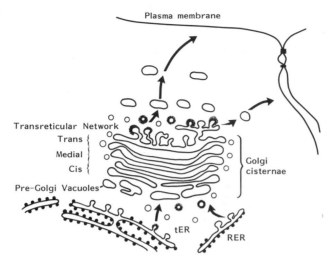

FIGURE 1. Membrane-bound compartments of the exocytic pathway. The arrows show the direction of vectorial protein transport. Nascent polypeptides synthesized on ribosomes bound to the cytoplasmic face of the endoplasmic reticulum (RER) enter the exocytic pathway. In the transitional or smooth endoplasmic reticulum (tER), proteins to be transported to the Golgi complex are collected into transport vesicles. Transport vesicles from the tER appear to fuse with pre-Golgi vacuoles at the *cis*-face of the Golgi complex. Proteins pass from the cis to the medial to the *trans*-Golgi cisternae. Sorting of proteins to separate post-Golgi destinations appears to occur in the *trans*-reticular network associated with the trans Golgi cisternae. Proteins leave the *trans*-reticular network in small vesicles and are next observed in larger, noncoated vesicles that in some cases have been seen fusing with the cytoplasmic face of the plasma membrane.

not before proteins pass beyond the Golgi apparatus (Rodriguez-Boulan *et al.*, 1978; Amar-Costesec *et al.*, 1984; Lewis *et al.*, 1985; Brands *et al.*, 1985; Yamamoto *et al.*, 1985). Although the reason for the high degree of compartmentalization of exocytosis in eukaryotic cells is not well understood, it is possible that this organization plays a role in regulating protein traffic by allowing some proteins to be retained in the endoplasmic reticulum (ER) or in specific cisternae of the Golgi complex through which other proteins freely move. Furthermore, compartmentalization might provide a mechanism for sorting proteins into the separate pathways that lead to the plasma membrane, lysosomes, endosomes, or other destinations. The ability of the cell to segregate proteins bound for different destinations implies that proteins in transit are recognized by elements of the exocytic apparatus. The features of proteins that serve as addresses specifying the extent and direction of movement within the pathway have been very difficult to identify and have been convincingly demonstrated for only a single class of proteins—the lysosomal enzymes. In marked contrast, structural elements that direct the membrane insertion, transmembrane anchoring, and primary glycosylation of nascent

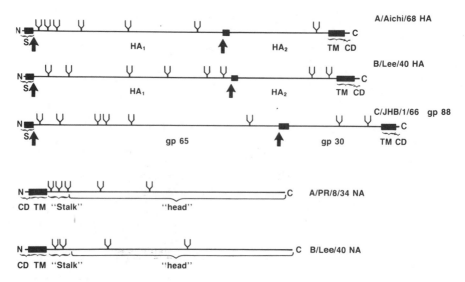

FIGURE 2. Linear maps of the polypeptides of influenza virus glycoproteins. The maps are drawn to scale. The boxed areas indicate the location of hydrophobic amino acid sequences that interact with cell membranes. The positions of oligosaccharides are indicated by the branched figures above each line. Arrows below the HAs indicate the position of proteolytic cleavage sites on the HAs and gp88. S, signal peptide; TM, transmembrane domain; CD, cytoplasmic domain; N, amino-terminus; C, carboxy-terminus.

polypeptides have been relatively easy to recognize. A likely explanation for this disparity is that recognition events occurring early during biosynthesis involve relatively simple features of unfolded polypeptides and do not require higher order structure. Later during exocytosis, proteins not only assume the three-dimensional shape required for their biological functions, but also pass through organelles that differ in their internal chemical environment. This raises the possibility that the signals recognized during exocytosis may depend on conformations assumed only transiently and points to a possible relationship between protein folding and the control of movement through the exocytic pathway.

The biosynthesis of influenza virus envelope glycoproteins can be divided into four stages: (1) translation and membrane insertion, (2) folding of the polypeptides and oligomerization, (3) sorting and processing of the assembled proteins during transport to the cell surface, and (4) incorporation into virions. During the events of the first stage in their biosynthesis, specific regions of HA, NA, and the influenza C gp88 polypeptides are recognized by host cell mechanisms responsible for membrane insertion, primary glycosylation, and anchoring in the membrane of the ER. The locations of these regions on linear maps of HA, NA, and gp88 polypeptides are shown in Fig. 2.

## II. ENTRY OF INFLUENZA GLYCOPROTEINS INTO THE EXOCYTIC PATHWAY

### A. Co-translational Events

The steps by which a nascent polypeptide is directed to and across the membrane of the ER occur quite rapidly and have therefore been difficult to analyze. Most of the information about this process has been derived from experiments performed *in vitro* in which translation is at least an order of magnitude slower than *in vivo* (reviewed by Walter *et al.*, 1984); the precise relationship of these *in vitro* results to the situation *in vivo* is unclear. However, for such polypeptides as HA, NA, and gp88 which have a single N-terminal signal sequence, results from a number of laboratories suggest that the process of insertion of the nascent polypeptide into the ER membrane has the general outline described below.

The synthesis of proteins destined to be exported or sequestered within organelles of the exocytic pathway begins on free ribosomes (Blobel and Doberstein, 1975a, b; Szcesma and Boime, 1976). As the nascent polypeptide chain emerges from the ribosome, the hydrophobic signal sequence (Blobel and Sabbatini, 1970, 1971; Milstein *et al.*, 1972) is recognized by a ribonucleoprotein complex called the signal-recognition particle (SRP) (Walter and Blobel, 1980; Walter *et al.*, 1981). The SRP–ribosome complex then interacts with the SRP receptor or docking protein, an integral membrane protein of the ER membrane (Meyer *et al.*, 1982; Gilmore *et al.*, 1982a, b); a functional ribosome–membrane junction is established, permitting translocation of the nascent chain across the ER membrane. SRP and SRP receptor are not constituents of the ribosome–membrane junction and are likely to be recycled once they have accomplished the initial targeting of the ribosome/nascent chain (Gilmore and Blobel, 1983). Attachment of the ribosome to the membrane occurs through ionic interactions between the large ribosomal subunit and integral membrane proteins of the ER, which remain unidentified (Adelman *et al.*, 1973; Borgese *et al.*, 1974). Possible candidates for this function are a secondary signal receptor (Gilmore and Blobel, 1985) and ribophorins I and II (Kreibich *et al.*, 1978a, b), two glycoproteins known to be limited to the rough endoplasmic reticulum (RER). In microsomes, the ribophorins are present in amounts equal to the number of ribosomes capable of being bound by those membranes (Amar-Costesec *et al.*, 1984; Marcantonio *et al.*, 1984).

The mechanism by which proteins are translocated across the membrane of the ER is not understood. It has been hypothesized that the movement of largely hydrophilic polypeptides through the lipid bilayer requires a protein pore (Blobel and Dobberstein, 1975a; Blobel, 1980). Alternatively, it has been argued on thermodynamic grounds that a loop of polypeptide might pass through the membrane without the aid of accessory proteins (Engelman and Steitz, 1981). Although it is clear that many proteins are translocated co-translationally, some are capable of posttranslational translocation, at least *in vitro* (reviewed by Sabatini *et al.*, 1982; Walter *et al.*, 1984; Wickner and Lodish, 1985; Schatz, 1986). As the nascent polypeptide chain enters the lumen of the ER, several processes occur rapidly, in some cases while the polypeptide is still attached to its ribosome. These include glycosylation, disulfide bond formation, development of secondary structure, and signal peptide cleavage.

For many proteins with N-terminal signal peptides, the signal is cleaved co-translationally (Blobel and Dobberstein, 1975b; Palmiter *et al.*, 1977; Lingappa *et al.*, 1978) by a cisternal peptidase (Jackson and Blobel, 1977; Walter *et al.*, 1979). The mammalian signal peptidase has recently been purified and consists of a complex of at least four and as many as six polypeptides, two of which are glycoproteins (Evans *et al.*, 1986). The amount of this protein complex that is present in microsomes roughly equals the amount of bound ribosomes; thus, the signal peptidase complex may also function in ribosome binding or in another aspect of translocation, such as the formation of a pore through the membrane (Evans *et al.*, 1986). If this were the case, one would expect this complex to be found to interact not only with proteins, such as HA, from which the signal is removed, but also with those like NA for which the signal remains uncleaved and serves as a membrane anchor.

Oligosaccharides are added to a polypeptide quite rapidly after specific asparagine residues are exposed in the lumen of the ER. Glycosylation occurs by the *en bloc* transfer of a branched oligosaccharide of the structure $GlcNAc_2Man_9Glc_3$ from a dolichol-lipid carrier to an asparagine that is part of the tripeptide sequence Asn-X-(Thr or Ser) (reviewed by Struck and Lennarz, 1980; Hubbard and Ivatt, 1981; Kornfeld and Kornfeld, 1985). Secondary structure develops in at least some polypeptides during their synthesis; disulfide bonds may form (Bergman and Kuehl, 1979), and folding may cause potential glycosylation sites on the protein to become inaccessible to the glycosylating enzymes at the luminal face of the ER membrane (Pless and Lennarz, 1977; Glabe *et al.*, 1980).

Translocation of the nascent polypeptide chain through the membrane can be halted by the presence of a hydrophobic halt-transfer sequence that anchors the protein within the lipid bilayer (Blobel, 1980; Gething and Sambrook, 1982; Sveda *et al.*, 1982; Yost *et al.*, 1983; Guan and Rose; 1984). Normally, such sequences are always of sufficient length (20–25 residues) to form an α-helix that crosses the lipid bilayer, usually followed by one or more basic amino acids (Sabatini *et al.*, 1982). Recent experiments with artificially shortened hydrophobic halt-transfer sequences have demonstrated that for a number of proteins the minimum number of contiguous hydrophobic residues required for anchoring alone is less than that required to form an α-helix spanning the ER membrane (Adams and Rose, 1985; Davis and Model, 1985; Doyle *et al.*, 1986; Davis and Hunter, 1986). These observations are curious, as some glycoproteins contain internal hydrophobic sequences that have the length and overall hydrophobicity of minimal membrane anchor sequences, yet these internal sequences fail to halt translocation (White *et al.*, 1983; Davis and Model, 1985).

## 1. Translocation of Influenza Virus Glycoproteins across the ER Membrane

*In vitro* translation of the HA from the A/Japan/305/57 strain of influenza virus in the presence or absence of added microsomal membranes has established that the insertion of the native polypeptide chain across the microsomal membrane is co-translational (Elder *et al.*, 1979). In the absence of membranes, HA contains an N-terminal extension that is not observed when HA is synthesized in their presence. (McCauley *et al.*, 1979, 1980).

Construction and expression of proteins with mutant or heterologous N-terminal sequences (Table I) has provided information on the role of the signal peptide. Expression of HA cDNAs from which sequences encoding the signal peptide have been deleted results in the synthesis of HA molecules that remain in the cytoplasm, demonstrating that the HA signal peptide is necessary for translocation (Gething and Sambrook, 1982; Sekikawa and Lai, 1983). That the HA signal peptide is also sufficient for translocation has been shown in experiments in which sequences encoding that region were fused to sequences encoding the N-termini of proteins that normally do not cross the ER membrane. When preceded by the HA signal peptide, both SV40 large T Ag and a modified aminoglycoside 3′ phosphotransferase II, proteins found normally in the nucleus and in the cytoplasm, are translocated into the ER and glycosylated (Sharma *et al.*, 1985; P. Bird, unpublished results). Furthermore, it has been shown that the HA signal can be replaced by heterologous signal sequences, a finding consistent with the earlier observation that there is little or no amino acid conservation within the hydrophobic core of signal peptides of HAs from different virus strains (Blok and Air, 1982a). Correctly translocated and processed HA is expressed from a gene in which sequences encoding the HA signal peptide are replaced either with those encoding the N-terminal 64-amino acid leader and signal peptide of the Rous sarcoma virus glycoprotein (M. J. Gething and E. Hunter, unpublished results) or with those encoding the signal of bovine preprochymosin (S. Sharma, unpublished results).

On the basis of an analysis of the nucleic acid sequence of the cloned NA cDNA that showed that the NA protein does not contain a C-terminal hydrophobic sequence, Fields *et al.* (1981) proposed that NA was attached to cell membranes by an N-terminal hydrophobic region extensive enough to span a cell bilayer as an α-helix, a hypothesis that was subsequently shown to be correct by protein sequencing (Blok *et al.*,

TABLE I. Influenza Glycoproteins with Mutations in the Signal Sequence[a]

| Mutant | Mutation | Surface expression | Transport defect | Reference |
|---|---|---|---|---|
| **A/Japan/305/57 HA** | | | | |
| HA7-S⁻ | 14 aa signal replaced by Met-Glu-Leu | − | Not translocated | Gething and Sambrook (1982) |
| env$_s$HA-12 | HA signal and 1st two aa's replaced by signal and 1st six aa's of RSV Pr95$^{env}$ | − | Retained in ER | Gething et al (1986b); M. J. Gething and E. Hunter (unpublished results) |
| env$_s$HA-16 | HA signal replaced by signal from RSV Pr95$^{env}$ | + + | No defect | M. J. Gething and E. Hunter (unpublished results) |
| S1 + S2 | Preprochymosin signal N-terminal to HA signal | + + | No defect | S. Sharma, M. J. Gething, and J. Sambrook (unpublished results) |
| S1 | 17 N-terminal aa's of HA replaced by preprochymosin signal + 12 aa's | − | Retained in ER | S. Sharma, M. J. Gething, and J. Sambrook (unpublished results) |
| X-HA | HA signal replaced by preprochymosin signal | + + | No defect | S. Sharma, M. J. Gething, and J. Sambrook (unpublished results) |
| **A/Udorn/68 HA** | | | | |
| dl HA | 11aa hydrophobic core of signal removed | − | Not translocated | Sekikawa and Lai (1983) |
| 28 | Multiple aa substitutions in signal | − | Not translocated | Lai et al. (1984) |
| 7 | Multiple aa substitutions in signal | − | Retained in ER | Lai et al. (1984) |
| **A/WSN/33 HA** | | | | |
| GHA | 48 N-terminal aa's of G replace 31 N-terminal aa's of HA | − | Retained in ER | McQueen et al. (1984) |
| N4OH | 40 N-terminal aa's of NA replace HA signal | − | Retained in ER | Bos et al. (1984) |
| N4OH482 | 40 N-terminal aa's of NA replace HA signal, C-terminal 66 aa's of HA deleted | − | Retained in ER | Bos et al. (1984) |
| H548 | Signal-minus HA | − | Not translocated | Bos et al. (1984) |
| **A/Udorn/72 NA** | | | | |
| dlK | Deletion of aa's 29–86 | + | ? | Markoff et al. (1984) |
| dlI | Deletion of aa's 16–78 | − | Not translocated | Markoff et al. (1984) |
| dlZ | Deletion of aa's 13–105 | − | Not translocated | Markoff et al. (1984) |
| **A/WSN/33 NA** | | | | |
| SN26 | Deletion of 26 N-terminal aa's | − | Not translocated | Davis et al. (1984) |
| SN10 | Deletion of 10 N-terminal aa's | + | ? | Davis et al. (1984) |

[a]ER, endoplasmic reticulum.

1982c; Ward *et al.*, 1982). Direct evidence that this region is necessary for membrane insertion as well as membrane anchoring of NA has been obtained by removing the sequences encoding the N-terminal hydrophobic residues from two different NA cDNAs, resulting in the expression of proteins that remained in the cytoplasm (Markoff *et al.*, 1983; Davis *et al.*, 1983). Furthermore, replacement of the HA signal with the longer signal-anchor of NA permits translocation of the chimeric polypeptide (Bos *et al.*, 1984).

### 2. Anchoring of the Virus Glycoproteins in the ER Membrane

Neuraminidase becomes anchored into the lipid bilayer of the ER membrane by the hydrophobic sequences near its N-terminus (Blok *et al.*, 1982), presumably before translation and translocation of the molecule is completed. By contrast, HA and gp88 have hydrophobic transmembrane domains that begin less than 40 amino acids from the C-terminus of the polypeptide chain. Because approximately 40 newly polymerized amino acids of a nascent polypeptide are shielded within the ribosome (Blobel and Sabatini, 1970), and additional amino acids are required to span a lipid bilayer, anchoring of these proteins in the ER membrane must occur after translation has ceased. Wickner and Lodish (1985) proposed that the motive forces for the posttranslational translocation of the final 60 or 70 residues and for the co-translational insertion of the main body of a protein through the membrane may differ. Thus, it is possible that the requirement for a hydrophobic sequence to halt translocation during chain elongation may be different from the requirement for halting transfer of the terminal residues of the polypeptide after translation has ceased. Evidence consistent with this idea has recently been reported. A mutation in the hydrophobic fusion peptide at the N-terminus of the HA2 subunit of the Japan HA extends this nonpolar domain to 18 amino acids and gives it an overall hydrophobicity (Kyte and Doolittle, 1982) greater than that of a truncated (17-amino acid) transmembrane domain that still functions to anchor the mutant HA in the lipid bilayer (Doyle *et al.*, 1986). Nevertheless, this mutant fusion peptide is translocated correctly (Gething *et al.*, 1986a).

Expression of HA cDNAs that lack sequences encoding the transmembrane and cytoplasmic domain of the protein results in the efficient secretion of the truncated HA molecules from the cell (Gething and Sambrook, 1982; Sveda *et al.*, 1982), indicating that these regions are required for membrane anchoring but not for intracellular transport of the protein. Furthermore, expression of cDNAs encoding HAs that lack sequences corresponding to the hydrophilic, cytoplasmic residues results in proteins correctly anchored in cellular membranes and travel to the cell surface with kinetics very similar to those of the wild-type proteins (Doyle *et al.*, 1986). Thus, the basic residues that commonly follow the hydrophobic membrane anchor of HAs are not required for halting the transfer of the polypeptide through the lipid bilayer. Although HAs normally have hydrophobic domains longer than the 20–25 amino acids required to span the lipid bilayer, the presence of a C-terminal hydrophobic domain of as few as 17 amino acids appears to be all that is required for functional membrane association (Doyle *et al.*, 1986).

Because the amino-terminal hydrophobic region of NA serves both as a signal sequence for translocation and as a halt-transfer sequence, experiments investigating the requirements for anchoring of the protein are more complex than those carried out with HA. Markoff *et al.* (1983) demonstrated that in A/Udorn NA, an N-terminal hydrophobic region shortened from 29 to 22 amino acids retains both signal and halt-transfer functions; further truncation of this region to 9 or 7 uncharged amino acids results in proteins that accumulate in the cytoplasm. Davis *et al.* (1983) reported that the first 10 N-terminal residues of the A/WSN NA can be replaced by amino acids derived from SV40 and synthetic linker nucleic acid sequences without blocking translocation or membrane anchoring of the polypeptide. The heterologous amino acids introduced into NA included several charged residues. Thus, it is not known whether hydrophilic residues preceding the NA signal sequence-membrane anchor are required for either of the dual functions of that region.

## 3. Secondary Structure and Glycosylation of the Nascent Glycoproteins

The three-dimensional structure of HA (Wilson *et al.*, 1981) suggests that the eight extended β-strands of the globular region of the protein fold independently and conceivably could form while the remainder of the polypeptide is synthesized and translocated. The globular domain is stabilized by three intrachain disulfide bonds, one bridging residues 52 and 277; this bond and the structure to which it contributes cannot be completed until at least one half the HA polypeptide has entered the lumen of the ER. At the base of the fibrous stalk of the molecule, a five-stranded β-pleated sheet forms a major structural element. The central strand of this structure is formed by N-terminal residues, while the antiparallel strands on either side are formed by amino acids located 450 residues downstream. The last strand of this β-sheet is composed of residues from the C-terminus of the ectodomain, and thus assembly of the five-stranded β-sheet cannot occur until translocation has ceased. On the basis of their structure for HA, Wilson *et al.* (1981) suggested that a delay in the cleavage of the HA signal might serve to hold the N-terminus of the molecule close to the membrane until the C-terminal sequences that interact with it become translocated.

Evidence also exists for co-translational folding of the globular head domain of NA. The protein from the A/RI/5/57 strain contains a potential glycosylation site $Asn_{402}$-Trp-Ser that does not receive carbohydrate. In the structure of the mature NA molecule, the chain makes a bend at residue 402 that would prevent the simultaneous recognition of the side chains of $Asn_{402}$ and $Ser_{404}$, which is thought to be a requirement for the function of the oligosaccharyltransferase (reviewed by Kornfeld and Kornfeld, 1985). Colman and Ward suggested (1985) that this structure forms before glycosylation of Asn 402 can occur.

## B. Post-translational Processing in the Endoplasmic Reticulum

The RER is organized into flattened cisternae interconnected by elements of smooth transitional ER that lack ribosomes. It has been postulated that these transitional elements contain the sites from which proteins are packaged into transport vesicles for export to the Golgi apparatus (Palade, 1975). This implies that proteins move laterally within the ER as a first step in their journey to the plasma membrane. However, evidence for the movement of proteins between elements of the ER is inconclusive.

Trimming of terminal glucose residues from mannose-rich oligosaccharides, previously transferred en bloc to newly translocated polypeptides, occurs while the polypeptides are in the RER (Fig. 3). The removal of glucose residues by two glucosidases of the RER and the subsequent removal of a mannose by the ER mannosidase (Bischoff and Kornfeld, 1983) can occur either co-translationally (Atkinson and Lee, 1984) or posttranslationally (Hubbard and Robbins, 1979). Since the ribophorins that reside in the RER contain oligosaccharides of the composition $Man_6GlcNAc_2$ (Atkinson and Lee, 1984), trimming of additional mannose residues can also occur in the RER. However, it is likely that the trimming of additional mannoses from proteins such as HA that leave the ER rapidly occurs in later compartments of the pathway (Fig. 3).

Within the ER, proteins destined for transport, including viral glycoproteins that are made in large quantity, represent only a few tenths of a percent of the total protein content (Quinn et al., 1983). The resident proteins of the ER do not cycle between the ER and the Golgi apparatus (Rodriguez-Boulan et al., 1978; Amar-Costesec et al., 1984; Lewis et al., 1985; Brands et al., 1985; Yamamoto et al., 1985). Thus, between these two compartments there appears to be a gate (Palade, 1975) that functions to permit the transport of transient proteins to the Golgi apparatus while preventing the exit of the vastly more numerous ER resident proteins, including some that are not membrane bound. The precise location of this gating function is unknown, although it is thought to occur before or during the sorting of transient proteins into transport vesicles that are responsible for moving exocytic proteins to the Golgi complex (Palade, 1975). The movement of proteins from the ER to the Golgi complex requires energy (Jamieson and Palade, 1971); however, the relationship of this energy-dependent step to the gating function is not clear.

Different proteins leave the ER at different rates (Fitting and Kabat,

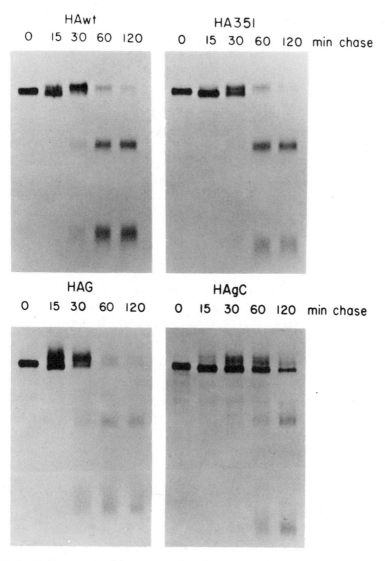

FIGURE 3. Transport of wild-type and mutant HAs through the exocytic pathway monitored by polyacrylamide electrophoresis. Simian CV-1 cells infected with SV40-HA vectors were labeled with [$^{35}$S]-methionine for 10 min and chased with nonradioactive methionine for the times shown. At the end of the chase, samples were immunoprecipitated with anti-HA antibody and analyzed by SDS–PAGE and autoradiography. After a 15-min chase, a faster migrating HA appears due to trimming of terminal mannoses from its oligosacharide side chains. A higher-molecular-weight form, due to addition of terminal sugars to proteins that have reached the *trans*-Golgi cisternae, also begins to appear by this time. For the last 15 min of each chase period, trypsin was included in the cell culture medium and HAs that reached the cell surface were cleaved into HA1 and HA2 subunits that can be seen as the two bands of lower apparent molecular weight. HAwt, wild-type HA; HA351, an HA mutant with changes in amino acids at the boundary between the ectodomain and the transmembrane domain. HAG and HAgC, chimeric HAs (see Table III). From Roth *et al.* (1986).

1982; Lodish *et al.*, 1983; Strous *et al.*, 1983; Sztul *et al.*, 1983; Rose and Bergmann, 1983; Wills *et al.*, 1984; Doyle *et al.*, 1985; Fries *et al.*, 1984; Hauri *et al.*, 1985; Williams *et al.*, 1985). These observations have been interpreted to indicate that a class of cellular receptor proteins operate the gating function for transport between the ER and the Golgi apparatus. According to this idea, receptors differ in their affinity for different exocytic proteins so that those bound at lower affinity are more slowly transported (Fitting and Kabat, 1982). Alternatively, it has been proposed that proteins destined to remain in the ER form interactions with other ER proteins that prevent entry into transport vesicles; all other proteins travel passively along the exocytic pathway (Sabatini *et al.*, 1982). According to this view, the different rates at which individual protein species exit the ER could be due to the rate at which they gain some general structural property, such as solubility, that limits their interactions with ER components and permits their export (Sabatini *et al.*, 1982; Gething *et al.*, 1985, 1986b).

## 1. Development of Quaternary Structure in Influenza Virus Glycoproteins

Hemagglutinin polypeptides must at some point assemble into trimeric, correctly folded proteins; NAs must form tetramers. Where in the biosynthetic pathway this occurs for NA is not precisely known, but several lines of evidence suggest that folding and trimerization of HA occurs within the ER before transport to the Golgi apparatus.

The timing of folding and trimerization of HAs derived from either A/Japan/305/57 (Gething *et al.*, 1986b) or the X-31 A/Aichi/68 (Copeland *et al.*, 1986) influenza viruses has been investigated. In experiments with Japan HA, the results obtained by four different methods (chemical crosslinking, velocity sedimentation, reactivity with antibodies specific for native or denatured epitopes, and protease sensitivity) were in accordance and showed that newly translocated HA is first detectable in native trimers 3 min after completion of synthesis and that the half-time of trimerization is 7–10 min (Gething *et al.*, 1986b). Similar conclusions were drawn from experiments with X-31 HA. Monomeric forms of Japan HA never become resistant to digestion by endo H, and newly formed trimers remained sensitive to endo H for 10–15 min, indicating that trimerization is complete before the protein reaches the medial Golgi cisternae. However, it is not clear that the trimerization and the folding of HA that occur in the ER necessarily complete the process of assembly of the protein. Copeland *et al.* (1986) observed that X-31 HA isolated after the protein had gained resistance to digestion with endo H was more stable than the endo H-sensitive form recognized by trimer-specific monoclonal antibodies. Thus, X-31 HA may undergo additional structural change in the Golgi apparatus at about the stage when terminal glycosylation occurs. However, no similar change in stability was seen in experiments with Japan HA (Gething *et al.*, 1986). These differences may be related to the observations that the folding and stability of the Japan HA is less dependent than that of X-31 HA on carbohydrate addition or modification (P. Gallagher, unpublished results).

Further insights into the time course of folding and oligomerization of influenza virus glycoproteins have come from analysis of the structures of the oligosaccharide side chains on the mature polypeptides. Recently, Keil *et al.* (1985) reported the composition of each of the oligosaccharide chains of FPV HA. The side chain attached to $Asn_{406}$ is predominantly of the untrimmed $Man^9GlcNAc^2$ structure, indicating that this oligosaccharide largely escapes trimming by mannosidase, an enzyme that modifies nascent polypeptides in the ER (Atkinson and Lee, 1984). $Asn_{406}$ is located in a long $\alpha$-helical region that in the mature HA trimer would be hidden at the top of the triple helical coil that forms the central structural unit stabilizing the trimer. This is consistent with trimerization of HA occurring rapidly in the RER before the mannosidase has the opportunity to trim the outer mannose residues attached to $Asn_{406}$.

Mature HAs from many different virus subtypes contain a mixture of mannose-rich and complex oligosaccharides (Schwarz *et al.*, 1977; Nakamura and Compans, 1978a, 1979; Keil *et al.*, 1979; Ward and Dopheide, 1980; Matsumoto, 1983). Of the seven oligosaccharides on X-31 HA, the only two that are clearly visible in the electron density map are of the mannose-rich type (Wilson *et al.*, 1981). One of these oligosaccharides is positioned at the interface of the globular domains; the other fills a pocket in the hinge region between the globular and stalk regions. In the trimer, both side chains would be shielded from the oligosaccharide-processing enzymes present in the Golgi apparatus. The mature forms of gp88 and the N1, N2, and type B NAs also contain at least one mannose-rich oligosaccharide (Nakamura *et al.*, 1979; Colman and Ward, 1985). These data indicate that these proteins rapidly assume a higher order structure that restricts access of processing enzymes to certain oligosaccharide side chains.

## 2. Role of Carbohydrate during Transport from the ER of Influenza Virus Glycoproteins

Such drugs as tunicamycin, which blocks the addition of carbohydrate to proteins, or swainsonine, castanospermine, or deoxynojirimycin, which interfere with enzymes responsible for ER processing of oligosaccharides, inhibit the transport of certain proteins but not others (Nakamura and Compans, 1978b; Chatis and Morrison, 1981; Peyrieras *et al.*, 1983; Lodish and Kong, 1984). The movement to the cell surface of HA and NA proteins from most strains of virus is not blocked by levels of tunicamycin that completely inhibit N-linked glycosylation (Nakamura and Compans, 1979b; Green *et al.*, 1981; Basak and Compans, 1983).

Furthermore, complete processing of oligosaccharides is not required for transport of HA or NA to the cell surface (Schwarz *et al.*, 1977; Nakamura and Compans; 1979b; Pan *et al.*, 1983; Burke *et al.*, 1984). The X-31 strain of HA is an exception, however, since in the presence of tunicamycin, newly synthesized X-31 HA does not leave the ER. Removal of each individual oligosaccharide attachment site on X-31 HA by site-directed mutagenesis does not affect the movement of the molecule to the cell surface. Thus, no single oligosaccharide plays a crucial role in transport of X-31 HA (P. Gallagher, unpublished results).

## 3. Rate of Export of Hemagglutinin from the ER

Structural alterations to the oligosaccharide side chains attached to HA cause changes in the electrophoretic mobility of the molecule (Fig. 3). Since many of these alterations occur in the Golgi apparatus, the changes in mobility can be used to monitor the rate at which newly synthesized HA arrives in and moves through the Golgi apparatus. The distal glucose residues have already been removed from the earliest form of HA that can be detected in *in vitro* pulse-labeling experiments (Burke *et al.*, 1984). At this stage, the mannose-rich oligosaccharide side chains on this form of the protein are sensitive to *in vitro* digestion by endo H; 10–20 min after synthesis of the polypeptide is complete, a form of HA appears that has a lower apparent molecular weight and that is still endo H sensitive. This form probably represents HA from which terminal mannoses have been removed by Golgi mannosidases (see Fig. 3); 15–20 min after synthesis, a heterogeneously migrating form of HA of higher molecular weight begins to appear that contains oligosaccharide side chains that are resistant to digestion with endo H and contain terminal sugars, such as fucose and galactose (Gething and Sambrook, 1982; Matlin and Simons, 1983; Doyle *et al.*, 1985). Thus, it appears that HA begins to emerge from the ER 10–15 min after completion of its synthesis. By 25–30 min after synthesis (Fig. 3), more than 90% of HA has reached the medial Golgi cisternae, where it acquires resistance to endo H.

## C. Mutants of Hemagglutinin That Are Defective in Transport from the ER

One approach to defining the requirements for transfer of newly synthesized influenza glycoproteins from the ER to the Golgi complex has been to characterize proteins defective in this process. Although viruses with temperature-sensitive (*ts*) defects in the surface expression of HA and NA have been identified, the nature of the alteration has been determined for only a few of them. Mutant HAs and NAs produced from cloned genes have been studied in much greater detail. Those HA and NA

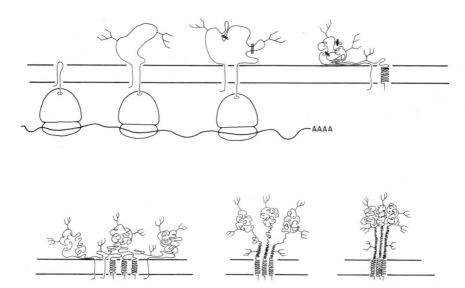

FIGURE 4. Schematic representation of the pathway of biosynthesis and folding of influenza virus HA. Biosynthesis of HA occurs on membrane-bound polysomes with membrane translocation being initiated by interaction of the amino terminal signal peptide with the membrane of the ER, here represented as occurring by the loop model (Inouye *et al.*, 1977). As the polypeptide is transferred into the lumen of the ER, oligosaccharides are added and folded structures form, in some cases stabilized by local disulphide bonds. The globular (HA1) domain may fold independently, perhaps co-translationally. However it is unlikely that the α-helices that form major structural elements of HA2 will stand away from the membrane (if they form at all) before the monomers interact to form trimers. The process of oligomerization may be initiated by association of the hydrophobic transmembrane regions. Formation of the trimer might then occur upward from the base of the spike, either before or after globular HA1 subunits form weak interactions. Finally, the correctly folded, native trimer is assembled. Data supporting this model may be found in Wilson *et al.* (1981), Gething and Sambrook (1982), and Gething *et al.* (1986). From Gething *et al.* (1986).

mutants that are defective in some particular aspect of intracellular transport are listed in Tables I, II, and III.

## 1. Temperature-Sensitive Mutants

Influenza viruses have been isolated that code for temperature-sensitive (*ts*) HA or NA glycoproteins (Scholtissek and Bowles, 1975; Ueda and Kilbourne, 1976) (see Table II). Although two mutant HAs and one mutant NA have been shown to be retained in the the ER at the nonpermissive temperature (Lohmeyer and Klenk, 1979; Rodriguez-Boulan *et al.*, 1984; Breuning and Scholtissek, 1986), the sequences of the mutant proteins have not yet been reported so that the basis for their temperature sensitivity is not understood. The ER-blocked HA protein coded by the A/WSN *ts*61 virus could not be detected in infected cells at the nonper-

TABLE II. Influenza Viruses with Temperature-Sensitive Glycoproteins[a]

| Mutant | Mutation | Surface expression | Transport defect | References |
|--------|----------|--------------------|------------------|------------|
| A/WSN/33 ts 61 | | | | |
| HA | ? | ts | Retained in ER | Ueda and Kilbourne (1976); Rodriguez-Boulan et al. (1984) |
| A/FPV/Rostock/34 | | | | |
| ts 227, ts 1 HAs | ? | ts | Retained in ER | Scholtissek and Bowles (1975); Lohmeyer and Klenk (1979) |
| ts 485, 532, 651 HAs | ? | ts | Post-Golgi | Klenk et al. (1981) |
| A/FPV/Rostock/34 | | | | |
| 113/Ho NA | ? | ts | Retained in ER | Scholtissek and Bowles (1975); Breuning and Scholtissek (1986) |

[a]ER, endoplasmic reticulum; ts, temperature sensitive.

missive temperature by use of a particular anti-HA monoclonal antibody. However, a few minutes after shifting the cells to the permissive temperature in the presence of cycloheximide, the ts61 protein could be detected with this same antibody (Rodriguez-Boulan et al., 1984). Studying the same virus in the same cell line, Rindler et al. (1984) used a mixture of two monoclonal antibodies to detect the ts61 HA in the ER at the nonpermissive temperature. These observations can now be interpreted in light of recent data that certain monoclonal antibodies can recognize only mature HA and do not recognize either the nascent polypeptide in the ER or mutant forms of the protein that are defective in transport to the Golgi apparatus (Bachi et al., 1985; Gething et al., 1986b; Copeland et al., 1986). It is thus possible that the defect in ts61 HA, like those in several of the A/Japan HA mutants made in vitro, affects both the folding and the transport of the protein.

A temperature-sensitive NA from the 113/Ho virus strain (N2) is similarly defective for transport from the ER at the nonpermissive temperature; its migration on polyacrylamide gels under nonreducing conditions is abnormal. N2 NAs normally form an interchain disulfide bond, so that each tetramer is composed of pairs of covalently linked dimers (Colman and Ward, 1985). Breuning and Scholtissek (1986) have proposed that the block in transport of the 113/Ho NA is caused by a defect that prevents dimerization and subsequent tetramerization. Supporting this proposal is the observation by Klenk and colleagues (1974) that in cells infected with FPV, NA activity could be detected in all cell fractions, including the rough microsomal fraction.

TABLE III. Influenza Virus Glycoproteins with Mutations in the Transmembrane and Cytoplasmic Domains[a]

| Mutant | Mutation | Surface expression | Transport defect | References |
|--------|----------|--------------------|------------------|------------|
| A/Japan/305/57 HA | | | | |
| HA⁻ᴬ⁻ (HAˢᵉᶜ) | TM and CD removed | + | Secreted | Gething and Sambrook (1982); Sambrook et al (1985) |
| HA11 | CD replaced by 3 aa's | ++ | Slowed transport ER to Golgi | Doyle et al (1985) |
| HA477env | RSV p15E CD added to end of HA CD | ++ | Slow ER → Golgi | Doyle et al. (1985) |
| HAxpBR | 16 aa's added to end of HA TM | − | Retained in ER | Doyle et al. (1985) |
| HA164 | CD replaced by 16 aa's | Low | Slow ER → Golgi post-Golgi block | Doyle et al. (1985) |
| HA152env | CD replaced by RSV p15E CD | ++ | No defect | Doyle et al. (1985) |
| HAG | TM and CD replaced by TM and CD of VSV G | ++ | Endocytosed | Roth et al. (1986) |
| HAgC | TM and CD replaced by TM and CD of HSV-1 gC | ++ | Slow ER → Golgi and is endocytosed | Roth et al. (1986); Gething et al. (1985) |
| A/Udorn/68 HA | | | | |
| dl-2,8, & 9 | Truncated in HA2 | + | Secreted | Sveda et al (1982) |
| dl-12 | Replaces TM with 22 nonpolar aa's | − | Retained in ER | Sveda et al. (1984) |
| A/Hong Kong/68 (X-31) HA | | | | |
| TM27 | CD deleted | ++ | Slight delay ER → Golgi | Doyle et al. (1986) |
| TM17 | TM truncated to 17 aa's, CD deleted | 20% | Slow ER → Golgi altered glycosylation | Doyle et al. (1986) |
| TM14 | TM truncated to 14 aa's, CD deleted | − | Retained in ER | Doyle et al. (1986) |
| TM9 | TM truncated to 9 aa's, CD deleted | − | Retained in ER Rapidly degraded | Doyle et al. (1986) |
| A/WSN/33 HA | | | | |
| HAG | 396 N-terminal aa's of HA fused to 152 C-terminal aa's of VSV G protein | − | Retained in ER | McQueen et al. (1984) |
| A/WSN/33 NA | | | | |
| SN10 | 10 N-terminal aa's deleted, removes CD | + | (Rate ?) polarized expression at cell surface | Davis et al. (1983); Jones et al. (1985) |

[a]aa's, amino acids; transmembrane domain.

## 2. Deletion Mutants

Studies of the biosynthesis of mutant HA or NA proteins expressed from truncated or deleted cDNA genes (Table III) have shown that the cytoplasmic domains of these polypeptides are not necessary for transport to the cell surface (Davis *et al.*, 1983; Doyle *et al.*, 1985, 1986). The transmembrane domain of HA is also not required for transport of the protein to the plasma membrane (Gething and Sambrook, 1982; Sveda and Lai, 1982). However, gross alterations to either of these domains results in structural changes that can affect the assembly or stability of the entire molecule. The transport of these deranged molecules may be much slower than usual, or it may be blocked completely. For example, deletion of the transmembrane domain of the Japan HA exactly at its junction with the large external domain results in a protein (HA-A$^-$ or HA$^{sec}$) that is glycosylated normally but that oligomerizes slowly and is delayed in movement from the ER to the Golgi apparatus (Gething and Sambrook, 1982; Gething *et al.*, 1985; Roth *et al.*, 1986). Once outside the cell, the trimers formed by this secreted HA are unstable; over time they dissociate into monomers (B. Doms, personal communication). By contrast, soluble trimers of HA (BHA) that have been cleaved from the surface of virions using bromelain are more stable (Wilson *et al.*, 1981). However, these also dissociate into monomers after treatment at low pH, whereas wild-type HA trimers extracted from virions with detergent do not (Doms *et al.*, 1986). The chief difference between BHA and HA$^{sec}$ is that BHA is derived from a molecule that is anchored during the process of trimer formation, whereas HA$^{sec}$ is not. Thus, the transmembrane domain not only serves to stabilize the protein structure after it is formed, but also must in some way facilitate formation of the correct contacts between the monomer subunits of the HA trimer.

Further examples of HA molecules whose intracellular transport is delayed or blocked include mutants with progressive truncations through the transmembrane domain. Such mutants have been expressed from HA cDNAs in which chain-termination codons have been introduced at four sites to yield proteins that lack the cytoplasmic domain and terminate at amino acids 9, 14, 17, or 27 of the wild-type hydrophobic domain (Doyle *et al.*, 1986) (Table III). Analysis of the biosynthesis and intracellular transport of these mutants shows that (1) the cytoplasmic tail is not needed for the efficient transport of HA to the cell surface, and (2) truncation of the transmembrane domain can result in drastic alterations in transport, membrane association, and stability. A hydrophobic sequence of 17 residues is sufficient to anchor HA stably in the lipid bilayer; however only a very small proportion of the mutant molecules are transported to the Golgi apparatus or the cell surface. Truncation of the hydrophobic anchor to 14 or 9 amino acids results in HA molecules that still interact in some manner with the lipid bilayer but whose transport is blocked in the ER, perhaps as a result of their inability to fold correctly.

### 3. Cytoplasmic Tail Mutants

Although the cytoplasmic domain is not required for normal intra-cellular transport of HA, mutations in this region can block or retard the transport of the protein from the ER (Doyle et al., 1985) (Table III). One mutant, HAxpBR, which contains heterologous amino acids at its C-terminus, is never transported from the ER. Several other mutants, whose cytoplasmic tails have been altered, truncated, or extended, are trans-ported slowly and asynchronously from the ER to the Golgi apparatus but then move with apparently normal kinetics to the cell surface. Because deletion of this domain has no effect on transport, the cytoplasmic tail of HA cannot contain any sequences that constitute an essential transport signal. Rather, alterations to this region produce structural changes in other parts of the protein that are important for transport (Gething et al., 1986b).

### 4. Chimeric Glycoproteins

Expression of hybrid genes that encode chimeric glycoproteins has not only yielded information about the interchangeability between differ-ent glycoproteins of signal sequences (see Table I) or transmembrane and cytoplasmic domains (Doyle et al., 1985; Gething et al., 1985; Roth et al., 1986) (see Table III), but has clarified the role of tertiary structure in transport between the ER and the Golgi apparatus as well.

Chimeric HAs [env$_s$HA-12 (Table I) and HAxpBR (Table III)] that do not move from the ER also fail to trimerize normally and give other evidence of being folded in a conformation different from wild-type HA (Gething et al., 1986b). Within 10 min of synthesis, wild-type HA forms trimers, becomes resistant to digestion by proteases added to cell lysates, and can be recognized by antibodies specific for the mature protein but not by antibodies that recognize only monomeric HA. By contrast, the mutant polypeptides remain sensitive to proteolysis and are recognized only by antibodies specific for unfolded HA. HAxpBR, which has an al-tered cytoplasmic domain, remains monomeric, whereas env$_s$HA-12, which is changed at the N-terminus of HA1, oligomerizes into aberrantly folded trimers. It is likely that the mutations in these two chimeras disrupt folding at different stages in the folding pathway (Fig. 4). Thus, trimerization, while apparently necessary for export of HA from the ER, is not sufficient for that process. Correct folding of HA into the native conformation is also required for transport of the protein to the Golgi complex.

Trimerization and folding of HA might be necessary for recognition of HA by other proteins that regulate entry of exocytic proteins into transport vesicles, or it might terminate or prevent the formation of in-teractions between HA and cellular proteins resident in the ER that hold unfolded HA molecules in that organelle. There is no evidence to support the first possibility, and there is only circumstantial evidence to support the second. During the earliest period of its biosynthesis, HA is transiently associated with a polypeptide identified as BiP (Gething et al., 1986b). For wild-type HA this interaction is brief; however, mutant HAs that are defective in export from the ER remain bound to BiP during the period that they remain in the ER (Gething et al., 1986b). Other proteins defective in transport from the ER are found associated with this protein as well (Morrison and Scharff, 1975; Haas and Wabl, 1983; Sharma et al., 1985; Bole et al., 1986). The binding of nascent or malfolded HA by BiP has the characteristics expected for an interaction that retains proteins in the ER until they have folded into a mature conformation, after which they are allowed to proceed along the exocytic pathway.

In summary, the existing data on relationship between HA structure and the rate at which HA leaves the ER support the hypothesis that trimerization and correct folding of the protein are prerequisites for export to the Golgi complex (Gething et al., 1986b). Mutations in the cytoplasmic or transmembrane domains that disturb transport of HA from the ER to the Golgi complex also change the rate of trimer formation and folding of HA. Trimerization of HA appears to be greatly facilitated by the presence of acceptable transmembrane and cytoplasmic domains, although unstable oligomers form slowly in the absence of contributions by these domains. On the basis of these observations and the structural data of Wilson et al. (1981), the following pathway for HA folding has been proposed (Gething et al., 1986b) (Fig. 4). The HA monomer is synthesized as an extended loop that first lies close to the membrane. The globular HA1 subunit folds, perhaps co-translationally, but this folding is delayed until at least one half the HA has crossed the membrane. The globular HA1 subunit is then soluble but tied to the membrane by the unfolded HA2. Completed monomers first associate through their transmembrane domains and these hydrophobic regions might form a stable structure, such as a triple-helical coil through the membrane, that ties the monomers together. This stable aggregate facilitates interactions between the subunits that permit the formation of the central α-helical coiled coil that assembles the trimer. Once assembled, the HA trimer is rapidly exported from the ER.

## III. PROTEIN TRANSPORT BETWEEN THE ENDOPLASMIC RETICULUM AND THE GOLGI APPARATUS

The movement of small vesicles carrying cellular and viral glycoproteins between the transitional elements of the ER and the Golgi apparatus has been observed by immunocytochemistry and electron microscopy (Bergmann et al., 1981; Green et al., 1981; Wehland et al., 1982; Bergmann and Singer, 1983; Saraste and Hedman, 1983; Strous et al.,

1983; Saraste and Kuismanen, 1984). The most informative experiments have employed glycoproteins that remain in the ER at a nonpermissive temperature but that become competent for export to the Golgi complex after a shift to a lower temperature (Bergmann and Singer, 1983; Saraste and Kuismanen, 1984). Immediately after the shift to the permissive temperature, such proteins become incorporated into vesicles of 50- to 100-nm diameter, some of which are coated on their cytoplasmic faces. These vesicles bud from the ER and move to a tubular network of vacuoles located at the *cis* face of the Golgi complex (Saraste and Hedman, 1983; Saraste and Kuismanen, 1984). Morphologically, the vacuoles associated with the Golgi resemble endosomes, an endocytic organelle in which the sorting of proteins is known to occur (Helenius *et al.*, 1983). Based on these observations, Saraste and Kuismanen have proposed that the pre-Golgi vacuoles are the location of the gate that controls movement between the ER and Golgi complex. At 15°C, further transport of exocytic proteins beyond these pre-Golgi vacuoles is blocked. Although HA has never been visualized in vesicles in the region between the ER and Golgi apparatus, movement of HA into the Golgi apparatus is blocked at 15°C suggesting that this glycoprotein follows a vesicular pathway to the Golgi complex that appears to be shared by all exocytic proteins.

Influenza virus HAs acquire fatty acids (Schlessinger, 1981; Schmidt, 1984) covalently bonded to Ser or Thr residues (Schmidt and Lambrecht, 1985). The NA glycoproteins are not fatty acid acylated. Several other viral glycoproteins have been reported to acquire covalently linked palmitate approximately 2 min before becoming resistant to digestion with endoglycosidase H (Schmidt, 1983; Quinn *et al.*, 1983). On the basis of these kinetic data, it has been assumed that the addition of palmitate occurs after proteins have reached the *cis*-Golgi cisternae and fatty acid acylation has been used as an early biochemical marker for the entry of proteins into the Golgi apparatus. However, several observations have been made that suggest that fatty acid acylation may not be a precise marker for entry into the Golgi complex. First, because transport of proteins across the Golgi stack can occur within 3–5 min at 37°C (Bergmann and Singer, 1983; Saraste and Kuismanen, 1984), the observed kinetics for the addition of palmitate would allow for acylation to occur while the protein is in a pre-Golgi compartment. Second, exocytic proteins can be labeled with palmitate at the nonpermissive temperature in a mutant yeast strain which has a temperature-sensitive block in transport from the ER to the Golgi apparatus (Wen and Schlesinger, 1984). Third, the HAxpBR mutant of A/Japan/305/57 HA, which fails to trimerize and shows no other biochemical or microscopic evidence of entering the Golgi complex, also becomes fatty acid acylated (Doyle *et al.*, 1985). Finally, since fatty acids appear to become attached to amino acids in the cytoplasmic domain of transmembrane proteins (Rose *et al.*, 1984), there is no reason for this modification reaction to be limited to a single compartment of the exocytic pathway.

## IV. TRANSPORT AND PROCESSING WITHIN THE GOLGI APPARATUS

### A. Organization of the Golgi Apparatus

There is extensive evidence that the Golgi apparatus is a highly compartmentalized organelle with morphologically and cytochemically distinct cisternae (reviewed by Farquhar and Palade, 1981; Dunphy and Rothman, 1985; Farquhar, 1985). Recent determination of the location and sequence of action of glycosyltransferases (Fig. 5) has led to an under-

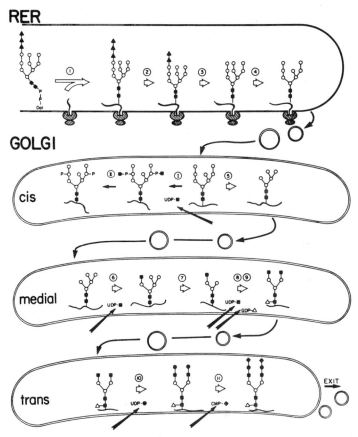

FIGURE 5. Schematic pathway of oligosaccharide processing on newly synthesized glycoproteins. Glycoproteins move vectorially through the exocytic pathway encountering the glycosidases and glycosyltransferases responsible for remodeling their oligosaccharides. The reactions are catalyzed by the following enzymes: (1) oligosacharyltransferase, (2) α-glucosidase I, (3) α-glucosidase II, (4) ER α1,2-mannosidase, (5) Golgi α-mannosidase I, (6) N-acetylglucosaminyltransferase I, (7) Golgi α-mannosidase II, (8) N-acetylglucosaminyltransferase II, (9) fucosyltransferase, (10) galactosyltransferase, (11) sialyltransferase. (■) N-acetylglucosamine; (○) mannose; (▲) glucose; (△) fucose; (●) galactose; (◆) sialic acid. From Kornfeld and Kornfeld (1985).

standing of the compartmentalization of glycosylation reactions within the Golgi complex (Kornfeld and Kornfeld, 1985), and thus of the direction and timing of movement of proteins through the Golgi stack. The trimming of terminal mannose residues that began in the ER continues as proteins move through the *cis* and into the medial Golgi cisternae. *N*-Acetylglucosamine, and in some cases fucose, is added to $Man_5GlcNac_2$ oligosaccharides in the medial cisternae. After the addition of *N*-acetylglucosamine, two terminal mannose residues may be removed, rendering the oligosaccharide insensitive to digestion by endo H. Galactose and subsequently sialic acid residues are added after the glycoprotein moves to the *trans*-Golgi cisternae.

In addition to the remodeling of N-linked oligosaccharides, other types of processing takes place in the Golgi apparatus, including specific cleavage of some proteins (Farquhar and Palade, 1981; Wills *et al.*, 1984), and the sulfation of some oligosaccharides, including those of influenza viruses (Compans and Pinter, 1975). Sulfation of tyrosines and O-linked glycosylation also occur in the Golgi complex (reviewed by Farquhar, 1985). It is likely that each of these processing events is located in a specific Golgi compartment.

The precise mechanism of movement of proteins through the Golgi apparatus has not been determined, but two major theories (reviewed by Farquhar and Palade, 1981; Farquhar, 1985) have been advanced. In the more widely held view, each Golgi cisterna is a defined organelle held in place; exocytic proteins are transferred successively through the cisternal stack in vesicles that bud from one cisterna and fuse to the next (Farquhar and Palade, 1981). In the other view, the *cis*-most Golgi cisterna arises by fusion of transport vesicles arriving from the ER; the cisterna then moves as a unit through the Golgi stack (Moore *et al.*, 1979; Saraste and Kuismanen, 1984). Since the stack has a polarity, with processing enzymes limited to subsets of the cisternae, this second mechanism requires Golgi enzymes to recycle to earlier cisternae.

## B. Movement of Influenza Virus Glycoproteins through the Golgi Apparatus

In cells derived from many different species, HAs of type A influenza viruses reach the *cis* face of the Golgi complex 10–15 min after synthesis and require 10–20 min to move through the Golgi stack and to the cell surface (Gething and Sambrook, 1982; Matlin and Simons, 1983, Alonso-Caplen and Compans, 1983; Rodriguez-Boulan *et al.*, 1984; Burke *et al.*, 1984; Doyle *et al.*, 1985). Little information is available on the rate of movement through the Golgi apparatus of NAs or the influenza C virus glycoprotein. In cells infected either with influenza viruses (Matlin and Simons, 1983; Copeland *et al.*, 1986) or SV40–HA recombinant viruses

(Gething *et al.*, 1986a), HA first becomes resistant to digestion with endo H 10–15 min after synthesis at 37°C. HA can be labeled with [³H]fucose for 15 min after protein synthesis has been blocked with puromycin (Stanley *et al.*, 1973), indicating that under those conditions most of the HA leaves the ER and reaches the medial Golgi within that period. Terminally glycosylated HAs are detected as soon as the protein begins to show resistance to digestion by endo H, indicating that movement from the medial to the *trans*-Golgi cisternae is rapid. When protein synthesis is terminated by the addition of cycloheximide, incorporation of sulfate into HA molecules continues for a further 30 min (Nakamura and Compans, 1977), indicating that a half-hour is required to move most of the HA from the RER to the *trans*-Golgi cisternae, where the sulfating enzymes are thought to be located (Farquhar, 1985). Since 15 min is required to empty the ER of HA, transport of the molecule through the Golgi stack requires an additional 15 min. This result is in fairly good agreement with two studies that used immunocytochemistry to monitor the movement of the temperature-sensitive A/WSN/33 *ts*61 HA through the Golgi complex. In these experiments, HA was first detected in the Golgi complex 10 min after a shift to 32°C (Rodriguez-Boulan *et al.*, 1984; Rindler *et al.*, 1985). Maximum labeling of the Golgi complex occurred 20 min after the temperature shift. Thus, in the MDCK cells used in this experiment, the time required to fill the Golgi stacks was approximately 10 min.

The Golgi complex maintains a pH gradient, with the *trans*-cisternae the most acidic (Anderson and Pathak, 1985; Schwartz *et al.*, 1985). The purpose of the pH gradient is not known, but it is known that agents that disrupt intracellular pH gradients cause the Golgi stack to disintegrate into swollen vacuoles (Tartakoff and Vassalli, 1977; Strous *et al.*, 1985). One such agent, the ionophore monensin, has been shown to disrupt Golgi function in the medial cisternae (Griffiths *et al.*, 1983; Quinn *et al.*, 1983). In monensin-treated cells, glycoproteins do not receive terminal sugars (Tartakoff, 1983; Alonso-Caplen and Compans, 1983; Rodriguez-Boulan *et al.*, 1984) and the transport of many secretory and membrane proteins is blocked in either the medial or the trans cisternae (Tartakoff, 1983; Griffiths *et al.*, 1983; Strous *et al.*, 1983). However, the influenza virus glycoproteins continue to appear at the cell surface in monensin-treated cells, although transport through the Golgi complex is delayed (Alonso-Caplen and Compans, 1983; Rodriguez-Boulan *et al.*, 1984). Thus, it appears unlikely that the inhibitory effect of monensin on surface expression of proteins is due to a direct block in intra-Golgi transport. In this regard, it may be important that HAs and NAs do not in general require the addition of any carbohydrate to reach the cell surface (Nakamura and Compans, 1978b; Basak and Compans, 1983), whereas the transport of certain other proteins is blocked by inhibitors of glycosylation (Leavitt *et al.*, 1977a, b; Hickman and Kornfeld, 1978; Lodish and Kong, 1984).

## C. Processing of Influenza Virus Oligosaccharides

During the transport of influenza virus glycoproteins through the Golgi apparatus, most of the mannose-rich oligosaccharide side chains are first trimmed to the $Man_5GlcNAc_2$ structure. *N*-Acetylglucosamine, fucose, galactose (Compans, 1973b; Stanley *et al.*, 1973; Nakamura and Compans, 1979a), and sialic acid (Sambrook *et al.*, 1985) residues are then added, and the oligosaccharides may become sulfated (Compans and Pinter, 1975; Nakamura and Compans, 1977). Glycoproteins of influenza viruses A and C usually contain a mixture of mannose-rich and complex oligosaccharides. However, HAs of the H2 subtype contain only the complex type (Nakamura and Compans, 1979b; Nakamura *et al.*, 1979; Ward and Colman, 1984; Keil *et al.*, 1985). The type of oligosaccharide (either mannose-rich or complex) has been mapped for a few HA and NA glycoproteins for each attachment site (Waterfield *et al.*, 1980; Brown *et al.*, 1981; Ward and Dopheide, 1980; Keil *et al.*, 1984, 1985; Ward and Colman, 1985) and these can be located on the three-dimensional structures of HA and NA. The extent to which the oligosaccharides of different HAs are processed depends on both the host cell and the HA strain (Nakamura and Compans, 1979a), and complex oligosaccharides of HA show considerable heterogeneity (Matsumoto *et al.*, 1983; Keil *et al.*, 1985; Gabel and Bergmann, personal communication). In the case of FPV HA, each attachment site has unique oligosaccharide structures, suggesting that local structural features of the HA protein influence either the accessibility of the oligosaccharide to certain processing enzymes, the rate of terminal processing, or probably both (Kiel *et al.*, 1985). Comparison of the extent of processing of two strains of FPV HA produced by the same cell type indicated that the faster-growing virus contained fewer fully processed complex oligosaccharides (Keil *et al.*, 1985). This was interpreted as indicating that the rate of movement of HA through the Golgi cisternae might control the degree to which its oligosaccharides were available for processing, providing another cause for the observed heterogeneity in glycosylation. A similar situation has been reported when HAs were expressed from cloned genes in mammalian cells. Glycoproteins synthesized in cells infected with influenza A or B viruses lack sialic acid because of the action of the viral neuraminidase (Palese *et al.*, 1974; Klenk *et al.*, 1974; Schwarz *et al.*, 1977; Nakamura and Compans, 1979b; Fuller *et al.*, 1985). However, sialic acid is added to the carbohydrate side chains on HA molecules expressed from cloned genes either in continuous murine cell lines or in simian cells infected with SV40-HA viral vectors (Sambrook *et al.*, 1985).

## D. Role of Glycosylation of Influenza Virus Glycoproteins

Although glycosylation of influenza virus glycoproteins is not required for their biosynthesis, there is substantial evidence that the addition of carbohydrate, and in certain circumstances the extent of processing of that carbohydrate in the Golgi apparatus, influences the biological functions of the proteins. The oligosaccharide side chains certainly provide protection of most HAs from extracellular proteases (Schwarz et al., 1976; Nakamura and Compans, 1978a; Basak and Compans, 1983). In some HAs, carbohydrate side chains shield potential antigenic epitopes (Wiley et al., 1981; Caton et al., 1982; Raymond et al., 1983); and the addition or loss of a glycosylation site can alter the immunogenicity of the protein. Two HAs of the H3 subtype—one produced in the laboratory and one isolated from a field strain of the virus—were found to have an additional glycosylation site at residue 63 of HA1 that blocked the binding of a monoclonal antibody (Skehel et al., 1984). Thus, it is possible that positioning of oligosaccharides plays a role in the protection of influenza viruses from the immune system of the host animal. Obviously, in addition to masking or exposing antigenic sites, oligosaccharides might shield other important surface features of influenza virus glycoproteins. This appears to be the case for the HA from the A/Chick/Penn/83 virus strain, in which the loss of an oligosaccharide attachment site that normally lies near the arginine residue that links HA1 and HA2 peptide chains has been correlated with more efficient cleavage activation of the HA molecule and with a consequent increase in the virulence of the virus (Kawaoka et al., 1984; Webster et al., 1986). A correlation has also been reported between an increase in the terminal glycosylation of oligosaccharide side chains on the HA from the A/WSN-F virus and a decrease in the affinity of the HA for host cell receptors (Deom and Schultz, 1985). Furthermore, the adaptation of influenza B viruses to growth in eggs results in HA variants that have lost the oligosaccharide attachment site at residue 196 of HA1, which is located adjacent to the sialic acid-binding site at the tip of the HA spike (Robertson et al., 1985). Presumably an oligosaccharide in this position on egg-grown HA interferes with the receptor binding required for virus infection; in mammalian cells a carbohydrate side chain in the same position does not have this effect, possibly because of a difference in the extent of modification of the oligosaccharide in the Golgi apparatus of the mammalian cell. In an experiment designed to create a similar situation through in vitro mutagenesis, a potential glycosylation site near the receptor-binding pocket on X-31 HA was added through oligonucleotide-directed mutagenesis of the cDNA gene; as expected the HA encoded by this gene is incapable of hemadsorption (P. Gallagher, personal communication).

In cell culture, HAs containing only mannose-rich oligosaccharides appear to be as biologically active as those containing complex sugar chains. Infectious influenza A virus is produced in the presence of castanospermine or of 1-deoxynojirimycin, inhibitors of the ER glucosidase I that prevent removal of terminal glucoses and subsequent processing of oligosaccharides beyond the removal of two terminal mannose residues

(Pan *et al.*, 1983; Burke *et al.*, 1984; Schwartz and Elbein, 1985). Infectious virus that contains glycoproteins lacking terminal glycosylation is also produced from cells treated with monensin (Alonso and Compans, 1981; Alonso-Caplen and Compans, 1983; Rodriguez-Boulan *et al.*, 1984). Thus, the terminal glycosylation of influenza virus glycoproteins may be relevant chiefly to the complex infectious cycle in animal hosts in which factors of the host's immune defense are present.

## V. THE *TRANS*-RETICULAR NETWORK

The *trans*-reticular network is a post-Golgi compartment consisting of a tubular network of vesicles that appears to be continuous with the final *trans*-Golgi cisterna (Novikoff, 1976, Novikoff *et al.*, 1971; Wehland *et al.*, 1982; Saraste and Kuismanen, 1984; Griffiths *et al.*, 1985; Roth *et al.*, 1985). Treatment of cells at 20°C causes an accumulation of exocytic proteins in this network and appears to bring about an expansion of this compartment (Matlin and Simons, 1983; Rodriguez-Boulan *et al.*, 1984; Pfeiffer *et al.*, 1985; Saraste and Kuismanen, 1985). The transport step that is inhibited seems to involve the formation of transport vesicles (Griffiths *et al.*, 1985). The *trans*-reticular network contains sialyl and galactosyl transferases (Pfeiffer *et al.*, 1985; Roth *et al.*, 1985) and is distinct from the endocytic pathway (Griffiths *et al.*, 1985). It can be observed in cells at 37°C; Griffiths and co-workers propose that it is identical to the structure previously designated GERL (Novikoff, 1976), which has also been shown to be structurally contiguous with secretory granules (Hand and Oliver, 1977; Novikoff *et al.*, 1977). These observations suggest that the *trans*-reticular Golgi may be the site at which the regulated secretory pathway diverges from that of the constitutive pathway, an event known to occur after proteins receive terminal glycosylation (reviewed by Farquhar, 1985).

Two events in the biosynthesis of influenza virus glycoproteins are thought to occur after completion of terminal glycosylation but before arrival of the proteins at the cell surface. In polarized epithelial cells, a sorting event directs influenza A and C glycoproteins (and presumably those of type B as well) to only one of two domains of the plasma membrane. Furthermore, in cells infected with fowl plague virus (and presumably also with other extremely virulent viruses), the cleavage of HA occurs rapidly after completion of terminal glycosylation. The intracellular compartment in which these events occur has not been defined but may involve elements of the *trans*-reticular network.

## A. Sorting of the Hemagglutinin and the Neuraminidase to the Apical Domain of Polarized Epithelial Cells

In certain cells, influenza virus glycoproteins are sorted within the cell and are displayed only at a particular domain of the plasma membrane. Epithelia are composed of polarized cells that maintain at least two distinct specializations of their plasma membranes: the apical and the basolateral domains. Once proteins arrive at one of these surface domains they are retained by tight junctions between neighboring cells that prevent mixing of membrane components through lateral diffusion (reviewed by Rodriguez-Boulan, 1983; Simons and Fuller, 1985). Polarized cells recognize and sort not only their own surface proteins but also the glycoproteins of enveloped viruses that infect them (Rodriguez-Boulan and Pendergast, 1980; Roth *et al.*, 1983a, b; Pfeiffer *et al.*, 1985; Jones *et al.*, 1985; Stephens *et al.*, 1986; Gottlieb *et al.*, 1986). This recognition does not require glycosylation (Roth *et al.*, 1979; Green *et al.*, 1981; Meiss *et al.*, 1982) and does not depend on the other viral proteins (Roth *et al.*, 1983a; Jones *et al.*, 1985; Stephens *et al.*, 1986; Gottlieb *et al.*, 1986). Thus, some aspect of the polypeptide structure serves as a recognition feature for the cellular sorting mechanisms. The glycoproteins of influenza viruses are recognized by the cell as belonging to the apical class of surface molecules (Rodriguez-Boulan and Sabatini, 1978; Herrler *et al.*, 1982). It is not clear what role, if any, the polarized expression of influenza virus glycoproteins has in the infectious cycle of the viruses. However, it is worth noting that polarized epithelial cells of the upper respiratory tract are the primary site of influenza virus infection. In polarized cells, the viruses bud from the surface that displays the viral glycoproteins, and it is possible that virions that exit from the apical cell surface and re-enter the respiratory tract early during infection are important in the spread of the virus to neighboring cells or for transmission of viruses from one animal to another.

Although the precise site within the cell where sorting of influenza glycoproteins occurs has not been determined, it must take place between the *trans*-Golgi cisternae and the cell surface. Polarity of virus maturation and glycoprotein transport is maintained in Madin–Darby canine kidney (MDCK) cells during early periods of infection with two viruses that bud from opposite surface domains, influenza virus and vesicular stomatitis (VSV) (Roth and Compans, 1982; Rindler *et al.*, 1984). Under these conditions, both the VSV G protein and the influenza virus HA have been observed in the same Golgi cisternae by immunocytochemistry (Rindler *et al.*, 1984). Biochemical data indicate that in doubly infected cells, G protein and the influenza virus NA share the transport pathway at least as far as the last compartment in which sialic acid is added to G (Fuller *et al.*, 1985). It has been shown that sialyl transferase is present in the *trans*-reticular network (Roth *et al.*, 1985), where G protein is packaged into transport vesicles and leaves the Golgi apparatus (Griffiths *et al.*, 1985). Both HA and G protein are transported directly from the *trans*-reticular network to their appropriate surface domains (Misek *et al.*, 1984; Matlin and Simons, 1984; Rindler *et al.*, 1985; Pfeiffer *et al.*, 1985; Roth *et al.*, 1986). Thus, sorting does not occur at the cell surface through endocytosis and recycling.

   The location of structural features of HA or NA that are recognized by the sorting apparatus in polarized cells was recently investigated by two groups using recombinant SV40 vectors to express cDNA copies of the HA and NA genes in a polarized rhesus kidney cell line, MA104 (Jones et al., 1985; Roth et al., 1987). A truncated HA lacking its transmembrane and cytoplasmic domains was nevertheless correctly sorted and secreted from the apical surface of the polarized cells. Furthermore, a chimeric HA, HAG (in which the transmembrane and cytoplasmic domains of HA were replaced with the homologous domains of G protein) was also expressed preferentially at the apical cell surface (Roth et al., 1987). In the case of NA, expression of a gene in which the sequences encoding the cytoplasmic domain had been replaced with sequences encoding 10 heterologous amino acids resulted in a correctly sorted protein (Jones et al., 1985). The most likely explanation for these results is that the feature responsible for sorting surface glycoproteins in polarized cells resides in the large external domain of the proteins so that direction into the correct transport pathway would occur through specific and efficient recognition of some feature or property of the ectodomain that serves as a sorting address. Exclusion of proteins from the inappropriate pathway would occur as a consequence of their efficient recruitment into the correct pathway. Proteins that lack a sorting address would be allowed into both pathways by default and would appear on both surfaces. Such a mechanism would easily explain the observation that 1–10% of viral glycoproteins expressed in polarized cells are found at the incorrect membrane domain (Rodriguez-Boulan and Pendergast, 1980; Roth et al., 1983b, 1987; Rindler et al., 1984; Misek et al., 1984; Pfeiffer et al., 1985), since a protein that is misdirected into the wrong pathway would not be prevented from reaching the wrong plasma membrane domain.

   A less likely interpretation that cannot be excluded at present is that HAs (and all proteins sorted to the apical surface) lack any feature recognized by the cell and that the feature responsible for sorting G protein (and other proteins destined for the basolateral membranes) is located in the large external domain of that protein. This is unlikely because apical proteins like HA that lack a basal membrane address would need to be recognized and efficiently excluded from the pathway to the basal surface. Furthermore, proteins found at both surface domains in polarized cells would need to contain a basal membrane address that is inefficiently recognized.

   If HA and NA contain features that are recognized by cellular sorting mechanisms, these features might be discrete, similar to the karyophilic signals identified for several proteins (Dingwall et al., 1982; Kalderon et al., 1984; Hall et al., 1984; Davey et al., 1985), or they might be some global property, such as a tendency to aggregate at a particular intracellular pH. There are few data available with which to distinguish these possibilities.

## B. Evidence for Proteolytic Processing of Certain Hemagglutinins by Cellular Proteases during Transport to the Cell Surface

Proteolytic processing is required to activate the membrane-fusion activity of HA that allows the entry of the nucleocapsid into the cytoplasm of the cell (Klenk et al., 1975; Lazarowitz et al., 1975; White et al., 1981). For some HAs, activation cleavage by proteases clearly occurs after the proteins have reached the plasma membrane (Lazarowitz et al., 1973). However, in the case of FPV HA, cleavage occurs very rapidly, perhaps within the Golgi apparatus (Klenk et al., 1974). In pulse-chase experiments in which the cleavage of radiolabelled FPV HA was measured at 37°C, HA1 and HA2 were first detected at 20 min after synthesis (Matlin and Simons, 1983). At this time, the bulk of the HA is in the Golgi complex, but some labeled protein is also beginning to arrive at the cell surface. At 20°C, a small amount of HA cleavage was observed after a 2-hr chase, an interval sufficient for the labeled HA to travel to, and accumulate in, the *trans*-reticular network. However, when the temperature was raised to 37°C at the end of the chase, quantitative cleavage occured within 15 min. This result can be interpreted either as an increase in the rate of enzyme processing at the higher temperature, or as an increased entry of HA into the compartment in which the processing enzyme is located. The latter possibility would suggest that cleavage of FPV HA occurs in transport vesicles or at the cell surface.

A proteolytic activity in MDBK cell lysates that correctly cleaves FPV HA has been reported (Klenk et al., 1985). This activity has a neutral pH optimum and requires millimolar concentrations of calcium—conditions that occur outside the cell rather than in the acidic *trans*-reticular network. Although the Golgi apparatus has been identified as the site of proteolytic processing for other viral glycoproteins (Wills et al., 1984), evidence for involvement of the Golgi apparatus in cleavage of HAs remains for the moment inconclusive.

## C. Hemagglutinins Defective in Transport from the Golgi Apparatus to the Cell Surface

Three temperature-conditional mutants of FPV have been reported that are defective in virus maturation and hemadsorption at 41°C but that synthesize HAs that receive fucose and galactose and that are cleaved into HA1 and HA2 subunits at the nonpermissive temperature (Klenk et al., 1981). Cycles of freezing and thawing of cells infected with these viruses enable them to hemadsorb, presumably by exposing intracellular HA. It is possible that these viruses contain HAs with defects in post-Golgi transport that cause them to accumulate within the cell. However, the alternative possibility that these viruses produce HAs that become defective in hemagglutination and virus assembly after reaching the cell surface has not been eliminated.

The only other influenza glycoprotein with a defect that becomes apparent late during biosynthesis is a mutant of the cloned A/Japan/305/57 HA, HA164, (see Table III) which has a novel sequence of 16 residues substituted for the wild-type cytoplasmic tail (Doyle et al., 1985). This protein collects chiefly in large vesicles that are widely distributed throughout the cytoplasm of CV-1 cells infected with a SV40-based vector. The oligosaccharides of HA164 have been isolated and characterized by serial lectin chromatography and glycosidase digestion and found to be indistinguishable from those of wild-type HA (C. Gabel and J. Bergmann, personal communication). This result indicates that HA164 migrates through the entire Golgi apparatus. In the presence of cycloheximide, HA164 does not chase from intracellular vesicles to the cell surface, nor does HA164 at the cell surface become internalized. These observations are consistent with a block in HA164 transport at some stage after exit from the Golgi complex and before arrival at the plasma membrane. The only exocytic compartment known to exist between the last Golgi compartment and the plasma membrane are the trans-reticular network and noncoated transport vesicles. The foreign sequences that comprise the cytoplasmic tail of HA164 might interfere directly with some necessary interaction between transport vesicles and a target membrane or might produce changes in the structure of the large external domain of HA164 that render the protein defective for transport, either by distorting an epitope required for recognition by cellular proteins involved in the sequestration of the glycoprotein into transport vesicles or by causing the mutant HA164 protein to aggregate or become insoluble in the particular late or post-Golgi environment. However, the acidic pH of the trans-Golgi cisterna does not appear to contribute to the HA164 defect, since raising intracellular pH with chloroquine or ammonium chloride did not change the pattern of HA164 transport (C. Doyle, unpublished results).

## VI. TRANSPORT TO THE CELL SURFACE AND INSERTION INTO THE PLASMA MEMBRANE

The entry of influenza virus glycoproteins into small vesicles that bud from the trans Golgi has not been directly observed but is presumed to be similar to that seen with other viral glycoproteins. After reaching the trans-Golgi cisterna, HAs have next been observed in smooth-surfaced vesicles of 100- to 150-nm diameter (Rodriguez-Boulan et al., 1984; Rindler et al., 1985). In experiments in which HA has been blocked in the trans-Golgi at 20°C and subsequently released by raising the temperature, transport to the cell surface requires 5–10 min (Matlin and Simons, 1983; Rodriguez-Boulan *et al.*, 1984; Rindler *et al.*, 1985).

In polarized epithelial cells, the first insertion of HA appears to occur in the center of the apical surface of the cell, directly above the Golgi complex (Rodriguez-Boulan *et al.*, 1984; Rindler *et al.*, 1985). The Golgi complex sits in a basket of microtubules (Kupfer *et al.*, 1982), and it has been suggested that during vectorial transport, vesicles move along cytoskeletal tracks to their destinations (Schnapp *et al.*, 1985). However, drugs that disrupt microtubles or actin filaments do not prevent directional transport of HA, suggesting that the specificity of delivery of HA to the apical surface is a property the transport vesicles themselves (Salas *et al.*, 1986).

Following insertion into the plasma membrane, HAs appear to be diffusely distributed over the cell surface. However, NA and the influenza C glycoprotein appear to cluster in patches (Compans and Dimmock, 1969; Compans *et al.*, 1969; Bachi *et al.*, 1969; Herrler *et al.*, 1981; Jones *et al.*, 1985). Influenza C glycoproteins form a hexagonal array on the surface of the virions, and this array is maintained on membrane sheets derived from disrupted virions and on the apparently empty virions that form at the surface of MDCK cells (Herrler *et al.*, 1981). This suggests that influenza C glycoproteins participate in lateral interactions that may be important in the formation of the virus bud. NA molecules also appear to cluster in patches on the surface of virions so that they are not randomly distributed over the viral envelope (Compans *et al.*, 1969; Murti and Webster, 1986). Thus, it is possible that NAs, unlike HAs, form higher-order arrays of oligomers at the cell surface. The location in the biosynthetic pathway in which these patches of NA or influenza C glycoproteins form has not been determined.

In cells infected with influenza viruses, the glycoproteins are rapidly recruited into budding virions. The earliest stage of virus formation that can be identified by electron microscopy is a collection of influenza glycoproteins above a thickened membrane that has viral ribonucleoprotein particles at its cytoplasmic face. Mutations in HA, matrix protein and nucleoprotein that completely block influenza virus assembly have been identified, suggesting that the process of virion assembly requires binding of all three components (reviewed by Compans and Klenk, 1979). It is tempting to speculate that the cytoplasmic domain of HA, which is not required for efficient transport of HA to the cell surface, might contain a binding site for either the matrix or the nucleoprotein, or both, since this region is highly conserved, with four of the final five residues unchanged between HAs from different virus subtypes (Gething *et al.*, 1980).

The rapidity with which influenza virus glycoproteins are recruited into virions is probably responsible for the observation that HA on the surface of cells infected with influenza virus cannot be crosslinked by multivalent antibody into an antigen cap (Basak *et al.*, 1984). Presumably, binding of HA by the viral internal components restricts its mobility, because HA expressed at the surface of permanent cell lines is readily capped by an anti-HA antibody (J. Hearing, unpublished results), as is purified HA delivered into the membrane through the fusion of liposomes (Basak *et al.*, 1984).

In cells in which HA is expressed in the absence of other influenza virus proteins, HA has a long residence time at the cell surface (Roth *et al.*, 1986; J. Hearing, personal communication). In CV-1 cells infected with SV40 expression vectors, HA appears to be specifically excluded from coated pits, even on cells expressing upwards of $10^7$ molecules of HA per cell. The rate of internalization of HA in these cells is approximately 0.08%/min (Roth *et al.*, 1986), or less than one twentieth the rate at which the plasma membrane is usually internalized through coated pits (Anderson *et al.*, 1977; Goldstein *et al.*, 1979; Steinman *et al.*, 1983; Helenius *et al.*, 1983). The exclusion of HA from coated pits is not a consequence of binding to sialic acid present on other proteins that are anchored in the plasma membrane. Two chimeric HAs, HAG and HAgC, in which the transmembrane and cytoplasmic domains of HA have been replaced with the homologous regions from either VSV G protein or HSV-1 gC protein, hemadsorb as efficiently as wild-type HA but are nevertheless rapidly internalized (Roth *et al.*, 1986). Thus it appears that HA lacks some feature required for internalization through coated pits. This is not a universal property of viral glycoproteins, since the VSV G protein is readily endocytosed when it is either implanted into cell surfaces through membrane fusion (Matlin *et al.*, 1983; Personen and Simons, 1984) or is expressed at the cell surface from a cloned gene (Gottlieb *et al.*, 1986). Nor is avoidance of endocytosis necessarily an advantage for a viral glycoprotein. Rapid recruitment of viral glycoproteins into virus buds might restrict lateral mobility in the membrane and prevent endocytosis; as in the case of HA, this process apparently prohibits antibody-induced capping. Alternatively, if a protein such as HA were endocytosed and rapidly recycled without entering a compartment with a pH below 5.5, there might be little effect on the process of virus maturation if the steady-state concentration of HA at the cell surface was sufficient for budding to occur. It is not known whether NA is effectively excluded from the endocytic pathway. However, NA has been expressed from cDNA gene using SV40 vectors, opening up the possibility of experimental investigation into this question.

## VII. SUMMARY AND PERSPECTIVES

The experiments reviewed in this chapter, together with an enormous body of work to which it has been impossible to do justice, have demonstrated great progress toward a general understanding of the cellular processes involved in protein export as well as a description of the biosynthesis, maturation, and intracellular transport of influenza virus glycoproteins in particular. A primary requirement for movement through the secretory pathway is that a nascent protein should take up a correct tertiary or quaternary structure, although how this correctness is monitored in the ER is not understood. Because all secretory or membrane proteins that are not destined to remain in the ER probably take a common route to the Golgi apparatus, specific cellular carrier proteins or receptors may not be necessary to direct their transport to the Golgi, although some mechanism must operate to separate transported proteins from those that remain in the ER. However, in or beyond the Golgi apparatus, where the pathways branch to different destinations within or without the cell, it seems highly likely that specific interactions between newly synthesized proteins and their cellular receptors must determine the direction of transport. For the influenza virus glycoproteins, it is the elucidation of the topological features that determine these interactions as well as the identification and characterization of the cellular components of the secretory apparatus that present the major challenges in the coming years. These challenges will be most efficiently met through continuing interaction and collaboration between the diverse groups of virologists, cell biologists, biochemists, and molecular biologists, whose contributions have been represented in this chapter.

## REFERENCES

Adams, G. A., and Rose, J. K., 1985, Incorporation of a charged amino acid into the membrane-spanning domain blocks cell surface expression but not membrane anchoring of a viral glycoprotein, *Mol. Cell Biol.* **5:**1442.

Adelman, M. R., Sabatini, D. D., and Blobel, G., 1973, Ribosome–membrane interaction. Nondestructive disassembly of rat liver rough microsomes into ribosomal and membranous components, *J. Cell Biol.* **56:**206.

Air, G. M., and Compans, R. W., 1983. Influenza B and Influenza C viruses, in *Genetics of Influenza Viruses* (P. Palese and D. W. Kingsbury, eds), pp. 280–304, Springer-Verlag, New York.

Alonso-Caplan, F. V., and Compans, R. W., 1983, Modulation of glycosylation and transport of viral membrane glycoproteins by a sodium ionophore, *J. Cell Biol.* **97:**659.

Amar-Costesec, A., Todd, J. A., and Kreibich, G., 1984, Segregation of the polypeptide translocation apparatus to regions of the endoplasmic reticulum containing ribophorins and ribosomes. I. Functional tests on rat liver microsomal subfractions, *J. Cell Biol.* **99:** 247.

Anderson, R. G. W., and Pathak, R. K., 1985, Vesicles and cisternae in the trans Golgi apparatus of human fibroblasts are acidic compartments, *Cell* **40:**635.

Anderson, R. G. W., Brown, M. S., and Goldstein, J. L., 1977, Role of the coated endocytic vesicle in the uptake of receptor-bound low density lipoprotein in human fibroblasts, *Cell* **10:**351.

Atkinson, P. H., and Lee J. T., 1984, Co-translational excision of α-mannose in nascent vesicular stomatitis virus G protein, *J. Cell Biol.* **98:**2245.

Bachi, T., Gerhard, W., Lindenmann, J., Muhlethaler, K., 1969, Morphogenesis of influenza A virus in Ehrlich ascites tumor cells as revealed by thin-sectioning and freeze-etching, *J. Virol.* **4:**769.

Bachi, T., Gerhard, W., and Yewdell, J. W., 1985, Monoclonal antibodies detect different forms of influenza virus during viral penetration and biosynthesis, *J. Virol.* **55:**307.

Balch, W. E., Glick, B. S., and Rothman, J. E., 1984, Sequential intermediates in the pathway of intercompartmental transport in a cell-free system, *Cell* **39**:525.

Basak, S., and Compans, 1983, Studies on the role of glycosylation in the functions and antigentic properties of influenza virus glycoproteins, *Virology* **128**:77.

Basak, S., Compans, R. W., and Oldstone, M. B. A., 1984, Restricted mobility of influenza hemagglutinin on HeLa cell plasma membranes, in: *Segmented Negative Strand Viruses* (D. H. L. Bishop and R. W. Compans, eds.), pp. 361–364, Academic, Orlando, Florida.

Bergman, L. W., and Kuehl, W. M., 1979, Formation of an intrachain disulfide bond on nascent immunoglobulin light chains, *J. Biol. Chem.* **254**:8869.

Bergmann, J. E., and Singer, S. J., 1983, Immunoelectron microscopic studies of the intracellular transport of the membrane glycoprotein (G) of vesicular stomatitis virus in infected chinese hamster ovary cells, *J. Cell Biol.* **97**:1777.

Bergmann, J. E., Tokuyasu, K. T., and Singer, S. J., 1981, Passage of an integral membrane protein, the vesicular stomatitis virus glycoprotein, through the Golgi apparatus en route to the plasma membrane, *Proc. Natl. Acad. Sci. USA* **78**:1746.

Bischoff, J., and Kornfeld, R., 1983, Evidence for an α-mannosidase in endoplasmic reticulum of rat liver, *J. Biol. Chem.* **258**:7907.

Bishop, D. H. L., Obijeski, J. F., and Simpson, R. W., 1971, Transcription of the influenza ribonucleic acid genome by a virion polymerase. II. Nature of the *in vitro* polymerase product, *J. Virol.* **8**:74.

Blobel, G., 1980, Intracellular protein topogenesis, *Proc. Natl. Acad. Sci. USA* **77**:1496.

Blobel, G., and Doberstein, B., 1975a, Transfer of proteins across membranes. I. Presence of proteolytically processed and unprocessed nascent immunoglobulin light chains on membrane-bound ribosomes of murine myeloma, *J. Cell Biol.* **67**:835.

Blobel, G., and Dobberstein, B., 1975b, Transfer of proteins across membranes. II. Reconstitution of functional rough microsomes from heterologous components, *J. Cell Biol.* **67**:852.

Blobel, G., and Sabatini, D. D., 1970, Controlled proteolysis of nascent polypeptides in rat liver cell fractions. I. Location of the polypeptides within ribosomes, *J. Cell Biol.* **45**:130.

Blobel, G., and Sabatini, D. D., 1971, Ribosome–membrane interaction in eukaryotic cells, in *Biomembranes*, Vol. 2, (L. A. Manson, ed.), pp. 193–195, Plenum, New York.

Blok, J., and Air, G. M., 1982a, Variation in the membrane-insertion and "stalk" sequences in eight subtypes of influenza A virus neuraminidase, *Biochemistry* **21**:4001.

Blok, J., and Air, G. M., 1982b, Block deletions in the neuraminidase genes from some influenza A viruses of the N1 subtype, *Virology* **118**:229.

Blok, J., Air, G. M., Laver, W. G., Ward, C. W., Lilley, G. G., Woods, E. F., Roxburgh, C. M., and Inglis, A. S., 1982c, Studies on the size, chemical composition and partial sequence of the neuraminidase (NA) from type A influenza virus show that the N-terminal region of the NA is not processed and serves to anchor the NA in the viral membrane, *Virology* **199**:109.

Bole, D. G., Hendershot, L. M., and Kearney, J. F., 1986, Posttranslational association of immunoglobulin heavy chain binding protein with nascent heavy chains in nonsecreting and secreting hybridomas, *J. Cell Biol.* **102**:1558.

Borgese, N., Mok, W., Kreibich, G., and Sabatini, D. D., 1974, Ribosome–membrane interaction: *In vitro* binding of ribosomes to microsomal membranes, *J. Mol. Biol.* **88**:559.

Bos, T. S., Davis, A. R., and Nayak, D. P., 1984, NH₂-terminal hydrophobic region of influenza virus neuraminidase provides the signal function in translocation, *Proc. Natl. Acad. Sci. USA* **81**:3976.

Brand, C. M., and Skehel, J. J., 1972, Crystalline antigen from the influenza virus envelope, *Nature New Biol.* **238**:145.

Brands, R., Snider, M. D., Hino, Y., Park, S. S., Gelboin, H. V., and Rothman, J. E., 1985, Retention of membrane proteins by the endoplasmic reticulum, *J. Cell Biol.* **101**:1724.

Breuning, A., and Scholtissek, D., 1986, A reassortant between influenza A viruses (H7N2) synthesizing an enzymatically inactive neuraminidase at 40° which is not incorporated into infectious particles, *Virology* **150**:65.

Burke, B., Matlin, K., Bause, E., Legler, G., Peyrieras, N., and Ploegh, H., 1984, Inhibition of N-linked oligosaccharide trimming does not interfere with surface expression of certain integral membrane proteins, *EMBO J.* **3**:551.

Caton, A. J., Brownlee, G. G., Yewdell, J. W., and Gerhard, W., 1982, The antigentic structure of the influenza virus A/PR/8/34 hemagglutinin (H1 subtype), *Cell* **31**:417.

Chatis, A., and Morrison, T. G,. 1981, Mutational changes in the vesicular stomatitis virus glycoprotein affect the requirement of carbohydrate in morphogenesis, *J. Virol.* **37**:307.

Colman, P. M., and Ward, C. W., 1985, Structure and diversity of influenza virus neuraminidase. *Curr. Top. Microb. Immunol.* **114**:178.

Compans, R. W., 1973a, Influenza virus proteins II. Association with components of the cytoplasm, *Virology* **51**:56.

Compans, R. W., 1973b, Distinct carbohydrate components of influenza virus glycoproteins in smooth and rough cytoplasmic membranes, *Virology* **55**:541.

Compans, R. W., and Choppin, P. W., 1971, The structure and assembly of influenza and parainfluenza viruses, in: *Comparative Virology* (K. Maramorosch and F. Kurstak, eds), pp. 407–432, Academic, New York.

Compans, R. W., and Dimmock, N. J., 1969, An electron microscopic study of single-cycle infection of chick embryo fibroblasts by influenza virus, *Virology* **39**:499.

Compans, R. W., and Klenk, H.-D., 1979, Viral membranes, in: *Comprehensive Virology* Vol. 13 (H. Fraenkel-Conrat and R. R. Wagner, eds.), pp. 293–385, Plenum, New York.

Compans, R. W., and Pinter, A., 1975, Incorporation of sulfate into influenza virus glycoproteins, *Virology* **66**:151.

Compans, R. W., Dimmock, N. J., and Meier-Ewert, H., 1969, Effect of antibody to neuraminidase on the maturation and hemagglutinating activity of an influenza virus A₂, *J. Virol.* **4**:528.

Compans, R. W., Klenk, H.-D., Caliguiri, L. A., and Choppin, P. W., 1970, Influenza virus proteins I. Analysis of polypeptides of the virion and identification of spike glycoproteins, *Virology* **42**:880.

Copeland, C., Doms, R. W., Bolzau, E. M., Webster, R. G., and Helenius, A., 1986, Assembly of influenza hemagglutinin trimers and its role in intracellular transport, *J. Cell Biol.* **103**:1179.

Daniels, R. S., Downie, J. C., Hay, A. J., Knossow, M., Skehel, J. J., Wang, M. L., and Wiley, D. C., 1985, Fusion mutants of the influenza virus hemagglutinin glycoprotein, *Cell* **40**:431.

Davey, J., Dimmock, N. J., and Colman, A., 1985, Identification of the sequence responsible for the nuclear accumulation of the influenza virus nucleoprotein in *Xenopus* oocytes, *Cell* **40**:667.

Davis, A. R., Bos, T. J., and Nayak, D. P., 1983, Active influenza virus neuraminidase is expressed in monkey cells from cDNA cloned in simian virus 40 vectors, *Proc. Natl. Acad. Sci. USA* **80**:3976.

Davis, N. G., and Model, P., 1985, An artificial anchor domain: hydrophobicity suffices to stop transfer, *Cell* **41**:607.

Deom, C. M., and Schultz, I. T., 1985, Oligosaccharide composition of an influenza virus hemagglutinin with host-determined binding protperties, *J. Biol. Chem.* **260**:14771.

Deom, C. M., Caton, A. J., and Schultz, I. T., 1986, Host cell-mediated selection of a mutant influenza A virus that has lost a complex oligosaccharide from the tip of the hemagglutinin, *Proc. Natl. Acad. Sci. USA* **83**:3771.

Dimmock, N. J., 1969, New virus-specific antigens in cells infected with influenza virus, *Virology* **39**:224.

Dingwall, C., Sharnick, S. V., and Laskey, R. A., 1982, A polypeptide domain that specifies migration of nucleoplasmin into the nucleus, *Cell* **30**:449.

Doms, R. W., Gething, M.-J., Henneberry, J. H., White, J., and Helenius, A., 1986, A variant influenza hemagglutinin that induces fusion at elevated pH, *J. Virol.* **57**:603.

Doyle, C., Roth, M. G., Sambrook, J., and Gething, M.-J., 1985, Mutations in the cytoplasmic domain of the influenza virus hemagglutinin affect different stages of intracellular transport, *J. Cell Biol.* **100**:704.

Doyle, C., Sambrook, J., and Gething, M.-J., 1986, Progressive deletions of the transmembrane and cytoplasmic domains of influenza hemagglutinin, *J. Cell Biol.* **103**:1193.

Dunphy, W. G., and Rothman, J. E., 1985, Compartmental organization of the Golgi stack, *Cell* **42**:13.

Elder, K. T., Bye, J. M., Skehel, J. J., Waterfield, M. D., and Smith, A. E., 1979, *In vitro* synthesis, glycosylation and membrane insertion of influenza virus hemagglutinin, *Virology* **95**:343.

Engelman, D. M., and Steitz, T. A, 1981, The spontaneous insertion of proteins into and across membranes the helical hairpin hypothesis, *Cell* **23**:411.

Evans, E. A., Gilmore, R., and Blobel, G., 1986, Purification of microsomal signal peptidase as a complex, *Proc. Natl. Acad. Sci. USA* **83**:581.

Farquhar, M. G., 1985, Progress in unraveling pathways of Golgi traffic, *Annu. Rev. Cell Biol.* **1**:447.

Farquhar, M. G., and Palade, G. E., 1981, The Golgi apparatus (complex)—(1954–1981)— from artifact to center stage, *J. Cell Biol.* **91**:77s.

Fields, S., Winter, G., and Brownlee, G. G., 1981, Structure of the neuraminidase in human influenza virus A/PR/8/34, *Nature (Lond.)* **290**:213.

Fitting, T., and Kabat, D., 1982, Evidence for a glycoprotein "signal" involved in transport between subcellular organelles, *J. Biol. Chem.* **257**:14011.

Fries, E. Gustafsson, L., and Peterson, P. A., 1984, Four secretory proteins synthesiszed by hepatocytes are transported from endoplasmic reticulum to Golgi complex at different rates, *EMBO J.* **3**:147.

Fuller, S. D., Bravo, R., and Simons, K., 1985, An enzymatic assay reveals that proteins destined for the apical or basolateral domains of an epithelial cell line share the same late Golgi compartments, *EMBO J.* **4**:4297.

Gething, M.-J., and Sambrook, J., 1982, Construction of influenza hemagglutinin genes that code for intracellular and secreted forms of the protein, *Nature (Lond.)* **300**:598.

Gething, M.-J., and Sambrook, J., 1983, Expression of Cloned Influenza Virus Genes, in: *Genetics of Influenza Viruses* (P. Palese and D. W. Kingsbury, eds.), pp. 169–191, Springer-Verlag, Vienna.

Gething, M.-J., Bye, J., Skehel, J. J., and Waterfield, M. D., 1980, Cloning and DNA sequence of double-stranded copies of hemagglutinin genes from H2 and H3 strains elucidates antigenic shift and drift in human influenza virus, *Nature (Lond.)* **287**:301.

Gething, M.-J., White, J. M., and Waterfield, M. D., 1978, Purification of the fusion protein of Sendai virus: Analysis of the $NH_2$-terminal sequence generated during precursor activation, *Proc. Natl. Acad. Sci. USA* **75**:2737.

Gething, M.-J., Doyle, C., Roth, M., and Sambrook, J., 1985, Mutational analysis of the structure and function of the influenza virus hemagglutinin, in: *Current Topics in Membranes and Transport*, Vol. 23, (E. A. Adelberg and C. W. Slayman, eds.), pp. 17–41, Academic, Orlando, Florida.

Gething, M.-J., Doms, R. W., York, D., and White, J., 1986a, Studies on the mechanism of membrane fusion: Site specific mutations of the hemagglutinin of influenza virus, *J. Cell Biol.* **102**:11.

Gething, M.-J., McCammon, K., and Sambrook, J., 1986b, Expression of wild-type and mutant forms of influenza hemagglutinin: the role of folding in the intracellular transport, *Cell* **46**:939.

Gilmore, R., and Blobel, G., 1983, Transient involvement of signal recognition particle and its receptor in the microsomal membrane prior to protein translocation, *Cell* **35**:677.

Gilmore, R., and Blobel, G., 1985, Translocation of secretory proteins across the microsomal membrane occurs through an environment accessible to aqueous perturbants, *Cell* **42**:497.

Gilmore, R., Blobel, G., and Walter, P., 1982a, Protein translocation across the endoplasmic reticulum. I. Detection in the microsomal membrane of a receptor for the signal recognition particle, *J. Cell Biol.* **95**:463.

Gilmore, R., Walter, P., and Blobel, G., 1982b, Protein translocation across the endoplasmic reticulum. II. Isolation and characterization of the signal recognition particle receptor, *J. Cell Biol.* **95**:470.

Glabe, C. G., Hanover, J. A., and Lennarz, W. J., 1980, Glycosylation of ovalbumin nascent chains, *J. Biol. Chem.* **255**:9236.

Goldstein, J. L., Anderson, R. G. W., and Brown, M. S., 1979, Coated pits, Coated pits, coated vesicles receptor-mediated endocytosis, *Nature (Lond.)* **279**:679.

Gottlieb, T. A., Gonzalez, A., Rizzolo, L., Rindler, M. J., Adesnik, M., and Sabatini, D. D., 1986, Sorting and endocytosis of viral glycoproteins in transfected polarized epithelial cells, *J. Cell Biol.* **102**:1242.

Gottschalk, A., 1957, Neuraminidase: The specific enzyme of influenza virus and *Vibrio cholerae, Biochem. Biophys. Acta* **23**:645.

Green, J., Griffiths, G., Louvard, D., Quinn, P., and Warren, G., 1981a, Passage of viral membrane proteins through the Golgi complex, *J. Mol. Biol.* **152**:663.

Green, R. F., Meiss, H. K., and Rodriguez-Boulan, E., 1981b, Glycosylation does not determine segregation of viral glycoproteins in the plasma membrane of epithelial cells, *J. Cell Biol.* **89**:230.

Griffith, I. P., 1975, The fine structure of influenza virus, in: *Negative Strand Viruses* (R. D. Barry and B. W. J. Mahy, eds.), pp. 121–136, Academic, Orlando, Florida.

Griffiths, G., Pfeiffer, S., Simons, K., and Matlin, K., 1985, The exit of newly synthesized membrane proteins from the trans cisternal of the Golgi complex to the plasma membrane, *J. Cell Biol.* **101**:949.

Griffiths, G., Quinn, P., and Warren, G., 1983, Dissection of the Golgi complex. I. Monensin inhibits the transport of viral membrane proteins from medial to trans Golgi cisternae in baby hamster kidney cells infected with Semliki Forest virus, *J. Cell Biol.* **96**:835.

Guan, J.-L., and Rose, J. K., 1984, Conversion of a secretory protein into a transmembrane protein results in its transport to the Golgi complex but not to the cell surface, *Cell* **37**:779.

Haas, I. G., and Wabl, M., 1983, Immunoglobulin heavy chain binding protein, *Nature (Lond.)* **306**:287.

Hall, M. N., Hereford, L., and Herskowitz, I., 1984, Targeting of *Escherichia coli* β-galactosidase to the nucleus in yeast, *Cell* **36**:1057.

Hand, A. R., and Oliver, C., 1977, Relationship between the Golgi apparatus, GERL, and secretory granules in acinar cells of the rat exorbital lacrimal gland, *J. Cell Biol.* **74**:399.

Haslam, E. A., Hampson, A. W., Egan, J. A., and White, D. O., 1970, The polypeptides of influenza virus. III. Identification of the hemagglutinin, neuraminidase and nucleocapsid proteins, *Virology* **42**:566.

Hauri, H.-P., Sterchi, E. E., Bienz, D., Fransen, J. A. M., and Marxer, A., 1985, Expression and intracellular transport of microvillus membrane hydrolases in human intestinal epithelial cells, *J. Cell Biol.* **101**:838.

Hay, A. J., 1974, Studies on the formation of the influenza virus envelope, *Virology* **60**:398.

Helenius, A., Mellman, I., Wall, D., and Hubbard, A., 1983, Endosomes, *TIBS* **8**:245.

Herrler, G., Nagele, A., Meier-Ewert, H., Bhown, A. S., and Compans, R. W., 1981, Isolation and structural analysis of influenza virus C virion glycoproteins, *Virology* **113**:439.

Hickman, S., and Kornfeld, S., 1978, Effect of tunicamycin on IgM, IgA, and IgG secretion by mouse plasmacytoma cells, *J. Immunol.* **121**:990.

Hirst, G. K., 1941, The agglutination of red cells by allantoic fluid of chick embryos infected with influenza virus, *Science* **94**:22.

Hirst, G. K., 1942, Adsorption of influenza hemagglutinins and virus by red blood cells, *J. Exp. Med.* **76**:195.

Hirst, G. K., 1943, The nature of the virus receptors of red cells. I. Evidence on the chemical nature of the virus receptors of red cells and of the existence of a closely analogous substance in normal serum, *J. Exp. Med.* **87**:301.

Horne, R. W., Waterson, A. P., Wildy, P., and Farnham, A. E., 1960, The structure and composition of the myxoviruses. I. Electron microscopic studies on the structure of myxovirus particles by negative staining techniques, *Virology* **11**:79.

Hubbard, S. C., and Ivatt, R. J., 1981, Asparagine-linked oligosaccharides, *Annu. Rev. Biochem.* **50**:555.

Hubbard, S. C., and Robbins, P. W., 1979, Synthesis and processing of protein-linked oligosaccharides in vivo, *J. Biol. Chem.* **254**:4568.

Inouye, S., Wang, S., Sekizawa, J., Halegoua, S., and Inouye, M., 1977, Amino acid sequence for the peptide extension on the prolipoprotein of *Escherichia coli* outer membrane, *Proc. Natl. Acad. Sci. USA* **74**:1004.

Jackson, R., and Blobel, G., 1977, Posttranslational cleavage of presecretory proteins with an extract of rough microsomes from dog pancreas containing signal peptidase activity, *Proc. Natl. Acad. Sci. USA* **74**:5598.

Jamieson, J. D., and Palade, G. E., 1968, Intracellular transport of secretory proteins in the pancreatic exocrine cell. IV. Metabolic requirements, *J. Cell Biol.* **39**:589.

Jones, L. V., Compans, R. W., Davis, A. R., Bos, T. J., and Nayak, D. P., 1985, Surface expression of influenza virus neuraminidase, an amino-terminally anchored viral membrane glycoprotein, in polarized epithelial cells, *Mol. Cell. Biol.* **5**:2181.

Kalderon, D., Richardson, W., Markham, A., and Smith, A, 1984, Sequence requirements for nuclear location of SV40 large-T, *Nature (Lond.)* **311**:33.

Kawaoka, Y., Naeve, C. W., and Webster, R. G., 1984, Is virulence of H5N2 influenza viruses in chickens associated with loss of carbohydrate from the hemagglutinin?, *Virology* **139**:303.

Keil, W., Klenk, H.-D., and Schwarz, R. T., 1979, Carbohydrates of influenza virus. III. Nature of oligosaccharide–protein linkage in viral glycoproteins, *J. Virol.* **31**:253.

Keil, W., Niemann, H., Schwarz, R. T., and Klenk, H.-D., 1984, Carbohydrates of influenza virus. V. Oligosaccharides attached to individual glycosylation sites of the hemagglutinin of fowl plague virus, *Virology* **133**:77.

Keil, W., Geyer, R., Dabrowski, J., Niemann, H., Stirm, S., and Klenk, H.-D., 1985, Carbohydrates of influenza virus. Structural elucidation of the individual glycans of the FPV hemagglutinin by two-dimensional $^1$H n.m.r. and methylation analysis, *EMBO J.* **4**:2711.

Klenk, H.-D., Wollert, W., Rott, R., and Scholtissek, C., 1974, Association of influenza virus proteins with cytoplasmic fractions, *Virology* **57**:28.

Klenk, H.-D., Rott, R., Orlich, M., and Blodorn, J., 1975, Activation of influenza A viruses by trypsin treatment, *Virology* **68**:426.

Klenk, H.-D., Garten, W., Keil, W., Niemann, H., Bosch, F. X., Schwarz, R. T., Scholtissek, C., and Rott, R., 1981, Processing of the hemagglutinin, in: *Genetic Variation among Influenza Viruses* (D. P. Nayak ed.), pp. 193–211, Academic, New York.

Kornfeld, R., and Kornfeld, S., 1985, Assembly of asparagine-linked oligosaccharides, *Annu. Rev. Biochem.* **54**:631.

Kreibich, G., Ulrich, B. L., and Sabatini, D. D., 1978a, Proteins of rough microsomal membranes related to ribosome binding. I. Identification of ribophorins I and II, membrane proteins characteristic of rough microsomes, *J. Cell Biol.* **77**:464.

Kreibich, G., Freienstein, C. M., Pereyra, B. N., Ulrich, B. L., and Sabatini, D. D., 1978b, Proteins of rough microsomal membranes related to ribosome binding. II. Cross-linking of bound ribosomes to specific membrane proteins exposed at the binding sites, *J. Cell Biol.* **77**:488.

Kupfer, A., Louvard, D., and Singer, S. J., 1982, Polarization of the Golgi apparatus and the microtubule-organizing center in cultured fibroblasts at the edge of an experimental wound, *Proc. Natl. Acad. Sci. USA* **79**:2603.

Kyte, J., and Doolittle, R. F., 1982, A simple method for displaying the hydropathic character of a protein, *J. Mol. Biol.* **157**:105.

Lamb, R. A., 1983, The influenza virus RNA segments and their encoded proteins, in: *Genetics of Influenza Viruses* (P. Palese and D. W. Kingsbury, eds.), pp. 21–69, Springer-Verlag, Berlin.

Laver, W. G., 1963, The structure of influenza viruses. 3. Disruption of the virus particle and separation of neuraminidase activity, *Virology* **20**:251.

Laver, W. G., 1964, Structural studies on the protein subunits from three strains of influenza virus, *J. Mol. Biol.* **9**:109.

Laver, W. G., 1971, Separation of two polypeptide chains from the hemagglutinin subunit of influenza virus, *Virology* **45**:275.

Laver, W. G., 1973, The polypeptides of influenza virus, *Adv. Virus Res.* **18**:57.

Laver, W. G., and Kilbourne, E. D., 1966, Identification in a recombinant influenza virus of structural proteins derived from both parents, *Virology* **30**:493.

Laver, W. G., and Valentine, R. C., 1969, Morphology of the isolated hemagglutinin and neuraminidase subunits of influenza virus, *Virology* **38**:105.

Laver, W. G., and Webster, R. G., 1968, Selection of antigenic mutants of influenza viruses: Isolation and peptide mapping of their hemagglutinating proteins, *Virology* **34**:193.

Lazarowitz, S. G., and Choppin, P. W., 1975, Enhancement of the infectivity of influenza A and B viruses by proteolytic cleavage of the hemagglutinin polypeptide, *Virology* **68**: 440.

Lazarowitz, S. G., Compans, R. W., and Choppin, P. W., 1971, Influenza virus structural and nonstructural proteins in infected cells and their plasma membranes, *Virology* **46**:830.

Lazarowitz, S. G., Goldberg, A. R., and Choppin, P. W., 1973, Proteolytic cleavage by plasmin of the HA polypeptide of influenza virus: Host cell activation of serum plasminogen, *Virology* **56**:172.

Leavitt, R., Schlesinger, R., and Kornfeld, S., 1977a, Tunicamycin inhibits glycosylation and multiplication of Sindbis and vesicular stomatitis viruses, *J. Virol.* **21**:375.

Leavitt, R., Schlesinger, R., and Kornfeld, S., 1977b, Impaired intracellular migration and altered solubility of nonglycosylated glycoproteins of vesicular stomatitis virus and Sinbis virus, *J. Biol. Chem.* **252**:9018.

Lewandowski, L. J., Content, J., and Leppla, S. H., 1971, Characterization of the subunit structure of the ribonucleic acid genome of influenza virus, *J. Virol.* **8**:701.

Lewis, M. J., Turco, S. J., and Green, M., 1985, Structure and assembly of the endoplasmic reticulum. Biosynthetic sorting of endoplasmic reticulum proteins, *J. Biol. Chem.* **260**: 6926.

Lingappa, V. R., Katz, F. N., Lodish H. F., and Blobel, G., 1978, A signal sequence for the insertion of a transmembrane glycoprotein. Similarities to the signals of secretory proteins in primary structure and function, *J. Biol. Chem.* **253**:8667.

Lodish, H. F., and Kong, N., 1984, Glucose removal from N-linked oligosaccharides is required for efficient maturation of certain secretory glycoproteins from the rough endoplasmic reticulum to the Golgi complex, *J. Cell Biol.* **98**:1720.

Lodish, H. F., Kong, N., Snider, M., and Strous, G. J. A. M., 1983, Hepatoma secretory proteins migrate from the rough endoplasmic reticulum to the Golgi at characteristic rates, *Nature (Lond.)* **304**:80.

Lohmeyer, J., and Klenk, H.-D., 1979, A mutant of influenza virus with a temperature-sensitive defect in the posttranslational processing of the hemagglutinin, *Virology* **93**: 134.

Marcantonio, E. E., Amar-Costesec, A. A., and Kreibich, G., 1984, Segregation of the polypeptide translocation apparatus to regions of the endoplasmic reticulum containing ribophorins and ribosomes. II. Rat liver microsomal subfractions contain equimolar amounts of ribophorins and ribosomes, *J. Cell Biol.* **99**:2254.

Markoff, L., Lin, B.-C., Sveda, M. M., and Lai, C.-J., 1983. Glycosylation and surface expression of the influenza virus neuraminidase requires the N-terminal hydrophobic region, *Mol. Cell. Biol.* **4**:8.

Matlin, K., and Simons, K., 1983, Reduced temperature prevents transfer of a membrane glycoprotein to the cell surface but does not prevent terminal glycosylation, *Cell* **34**: 233.

Matlin, K., and Simons, K., 1984, Sorting of an apical plasma membrane glycoprotein occurs before it reaches the cell surface in cultured epithelial cells, *J. Cell Biol.* **99**:2131.

Matlin, K., Bainton, D. F., Personen, M., Louvard, D., Genty, N., and Simons, K., 1983,

Transepithelial transport of a viral membrane glycoprotein implanted into the apical plasma membrane of Madin–Darby canine kidney cells. I. Morphological evidence, *J. Cell Biol.* **97**:627.

Matsumoto, A., Yoshima, H., and Kobata, A., 1983, Carbohydrates of influenza virus hemagglutinin: Structures of the whole neutral sugar chains, *Biochemistry* **22**:188 (abst.).

McCauley, J., Bye, J., Elder, K., Gething, M.-J., Skehel, J. J., Smith, A., and Waterfield, M. D., 1979, Influenza virus hemagglutinin signal sequences, *FEBS Lett.* **108**:422.

McCauley, J., Skehel, J., Elder, K., Gething, M.-J., Smith, A., and Waterfield, M., 1980, Hemagglutinin biosynthesis, in: *Structure and Variation in Influenza Virus* (G. Laver and G. Air, eds.), pp. 97–104, Elsevier/North-Holland, New York.

McClelland, L., and Hare, R., 1941, The adsorption of influenza virus by red cells and a new *in vitro* method of measuring antibodies for influenza virus, *Can. J. Public Health* **32**: 530.

Meiss, H. K., Green, R., and Rodriguez-Boulan, E. J., 1982, Lectin resistant mutants of polarized epithelial cells, *Mol. Cell Biol.* **2**:1287.

Meyer, D. I., Louvard, D., and Dobberstein, 1982, Characterization of molecules involved in protein translocation using a specific antibody, *J. Cell Biol.* **92**:579.

Milstein, C., Brownlee, G. G., Harrison, T. M., and Mathews, B. A., 1972, A possible precursor of immunoglobulin light chains, *Nature New Biol.* **239**:117.

Misek, D. E., Bard, E., and Rodriguez-Boulan, E., 1984, Biogenesis of epithelial cell polarity: Intracellular sorting and vectorial exocytosis of an apical plasma membrane glycoprotein, *Cell* **39**:537.

Moore, D. J., Kartenbeck, J., and Franke, W. W., 1979, Membrane flow and intercoversions among endomembranes, *Biochem. Biophys. Acta* **559**:71.

Morgan, C., Rose, H. M., and Moore, D. H., 1956, Structure and development of viruses observed in the electron microscope. III. Influenza virus, *J. Exp. Med.* **103**:171.

Morrison, S. L., and Scharff, M. D., 1975, Heavy chain-producing variants of a mouse myeloma cell line, *J. Immunol.* **114**:655.

Murphy, J. S., and Bang, F. B., 1952, Observations with the electron microscope on cells of the chick chorio-allantoic membrane infected with influenza virus, *J. Exp. Med.* **95**:259.

Murti, K. G., and Webster, R. G., 1986, Distribution of hemagglutinin and neuraminidase on influenza virions as revealed by immunoelectron microscopy, *Virology* **149**:36.

Nakada, S., Creager, R. S., Krystal, M., Aaronson, R. P., and Palese, P., 1984, Influenza C virus hemagglutinin: Comparison with influenza A and B virus hemagglutinins, *J. Virol.* **50**:118.

Nakamura, K., and Compans, R. W., 1977, The cellular site of sulfation of influenza virus glycoproteins, *Virology* **71**:381.

Nakamura, K., and Compans, R. W., 1978a, Glycopeptide components of influenza viral glycoproteins, *Virology* **86**:482.

Nakamura, K., and Compans, R. W., 1978b, Effects of glucosamine, 2-deoxy-D-glucose and tunicamycin on glycosylation, sulfation and assembly of influenza virus glycoproteins, *Virology* **84**:303.

Nakamura, K., and Compans, R. W., 1979a, Host cell- and virus strain-dependent differences in oligosaccharides of hemagglutinin glycoproteins of influenza A viruses, *Virology* **93**:8.

Nakamura, K., and Compans, R. W., 1979b, Biosynthesis of the oligosaccharides of influenza virus glycoproteins, *Virology* **93**:31.

Nakamura, K., Herrler, G., Petri, T., Meier-Ewert, H., and Compans, R. W., 1979, Carbohydrate components of influenza C virions, *J. Virol.* **29**:997.

Novikoff, A. B., 1976, The endoplasmic reticulum: a cytochemist's view. (Review.), *Proc. Natl. Acad. Sci. USA* **73**:2781.

Novikoff, A. B., Mori, M., Quintana, N., and Yam, A., 1977, Studies of the secretory process in the mammalian exocrine pancreas. I. The condensing vacuoles, *J. Cell Biol.* **75**:148.

Palade, G. E., 1975, Intracellular aspects of the process of protein synthesis, *Science* **189**: 347.

Palese, P., Tobita, K., Ueda, M., and Compans, R. W., 1974, Characterization of temperature sensitive influenza virus mutants defective in neuraminidase, *Virology* **61**:397.

Palmiter, R. D., Gagnon, J., Ericsson, L. H., and Walsh, K. A., 1977, Precusor of egg white lysosyme. Amino acid sequence of an $NH^2$-terminal extension, *J. Biol. Chem.* **252:** 6368.

Pan, H., Hori, H., Saul, R., Sanford, B. A., Molyneux, R. J., and Elbein, A. D., 1983, Castanospermine inhibits the processing of the oligosaccharide portion of the influenza viral hemagglutinin, *Biochemistry* **22**:3975.

Panicali, D., Davis, S. W., Weinberg, L., and Paoletti, E., 1984, Construction of live vaccines using genetically engineered poxviruses: Biological activity of recombinant vaccinia virus expressing the influenza virus hemagglutinin, *Proc. Natl. Acad. Sci. USA* **80:** 5364.

Personen, M., and Simons, K., 1984, Transcytosis of the G protein of vesicular stomatitis virus after implantation into the apical membrane of Madin-Darby canine kidney cells. I. Involvement of endosomes and lysosomes, *J. Cell Biol.* **99**:796.

Peyrieras, N., Bause, E., Lefler, G., Vasilov, R., Claesson, L., Peterson, P., and Ploegh, H., 1983, Effects of the glucosidase inhibitors nojirimycin and desoxynojirimycin on the biosynthesis of membrane and secretory glycoproteins, *EMBO J.* **2**:823.

Pfeiffer, J. B., and Compans, R. W., 1984, Structure of the influenza C glycoprotein gene as determined from cloned DNA, *Virus Res.* **1**:281.

Pfeiffer, S., Fuller, S. D., and Simons, K., 1985, Intracellular sorting and basolateral appearance of the G protein of vesicular stomatitis virus in Madin–Darby canine kidney cells. *J. Cell Biol.* **101**:470.

Pless, D. D., and Lennarz, W. J., 1977, Enzymatic conversion of proteins to glycoproteins, *Proc. Natl. Acad. Sci. USA* **74**:134.

Quinn, P., Griffiths, G., and Warren, G., 1983, Dissection of the Golgi complex. II. Density separation of specific Golgi functions in virally infected cells treated with monensin, *J. Cell Biol.* **96**:851.

Raymond, F. L., Caton, A. J., Cox, N. J., Kendall, A. P., and Brownlee, G. G., 1983, Antigenicity and evolution amongst recent influenza viruses of H1N1 subtype, *Nucleic Acids Res.* **1**:7191.

Rindler, M. J., Ivanov, I. E., Plesken, H., Rodriguez-Boulan, E. J., and Sabatini, D. D., 1984, Viral glycoproteins destined for apical or basolateral plasma membrane domains traverse the same Golgi apparatus during their intracellular transport in Madin–Darby canine kidney cells, *J. Cell Biol.* **98**:1304.

Rindler, M. J., Ivanov, I. E., Plesken, H., and Sabatini, D. D., 1985, Polarized delivery of viral glycoproteins to the apical and basolateral plasma membranes of Madin–Darby canine kidney cells infected with temperature-sensitive viruses, *J. Cell Biol.* **100**:136.

Robertson, J. S., Naeve, C. W., Webster, R. G., Bootman, J. S., Newman, R., and Schild, G. C., 1985, Alterations in hemagglutinin associated with adaptation of influenza B virus to growth in eggs, *Virology* **143**:166.

Rodriguez-Boulan, E., 1983, Membrane Biogenesis, enveloped RNA viruses and epithelial polarity, in: *Modern Cell Biology* (B. Satir, ed.), pp. 119–170, Liss, New York.

Rodriguez-Boulan, E., and Pendergast, M., 1980, Polarized distribution of viral envelope proteins in the plasma membrane of infected epithelial cells, *Cell* **20**:45.

Rodriguez-Boulan, E., and Sabatini, D., 1978, Polarized distribution of viral envelope proteins in the plasma membrane of infected epithelial cells, *Cell* **20**:45.

Rodriguez-Boulan, E., Kreibich, G., and Sabatini, D. D., 1978, Spacial orientation of glycoproteins in membranes of rat liver rough microsomes. I. Localization of lectin-binding sites in microsomal membranes, *J. Cell Biol.* **78**:874.

Rodriguez-Boulan, E., Paskiet, K. T., Salas, P. J. I., and Bard, E., 1984, Intracellular transport of influenza virus hemagglutinin to the apical surface of Madin-Darby canine kidney cells, *J. Cell Biol.* **98**:308.

Rogalski, A. A., and Singer, L. J., 1984, Associations of elements of the Golgi apparatus with microtubules, *J. Cell Biol.* **99**:1092.

Rose, J. K., Adams, G. A., and Gallione, C. J., 1984, The presence of cysteine in the cytoplasmic domain of the vesicular stomatitis virus glycoprotein is required for palmitate addition, *Proc. Natl. Acad. Sci. USA* **81**:2050.

Rose, J. K., and Bergmann, J. E., 1983, Altered cytoplasmic domains affect intracellular transport of the vesicular stomatitis virus glycoprotein, *Cell* **34**:513.

Roth, J., Taajes, D. J., Lucocq, J. M., Weinstein, J., and Paulson, J. C., 1985, Demonstration of an extensive trans-tubular network continuous with the Golgi apparatus stack that may function in glycosylation, *Cell* **43**:287.

Roth, M. G., and Compans, R. W., 1981, delayed appearance of pseudotypes between vesicular stomatitis virus and influenza virus during mixed infection of MDCK cells, *J. Virol.* **40**:848.

Roth, M. G., Fitzpatrick, J., and Compans, R. W., 1979, Polarity of influenza and vesicular stomatitis virus maturation in MDCK cells: Lack of a requirement for glycosylation of viral glycoproteins, *Proc. Natl. Acad. Sci. USA* **76**:6430.

Roth, M. G., Gething, M.-J., Sambrook, J., Giusti, L., Davis, A., Nayak, D. P., and Compans, R. W., 1983a, Influenza virus hemagglutinin expression is polarized in cells infected with recombinant SV40 viruses carrying cloned hemagglutinin DNA, *Cell* **33**:435.

Roth, M. G., Srinivas, R. V., and Compans, R. W., 1983b, Basolateral maturation of retroviruses in polarized epithelial cells, *J. Virol.* **45**:1065.

Roth, M. G., Doyle, C., Sambrook, J., and Gething, M.-J., 1986, Heterologous transmembrane and cytoplasmic domains direct functional chimeric influenza virus hemagglutinins into the endocytic pathway, *J. Cell Biol.* **102**:1271.

Roth, M. G., Gundersen, D., Patil, N., and Rodriguez-Boulan, E., 1987, The large external domain is sufficient for the correct sorting of secreted or chimeric influenza virus hemagglutinins in polarized monkey kidney cells, *J. Cell Biol.* **104**:769.

Sabatini, D. D., Kreibich, G., Morimoto, T., and Adesnik, M., 1982, Mechanisms for the incorporation of proteins in membranes and organelles, *J. Cell Biol.* **92**:1.

Salas, P. J. I., Misek, D. E., Vega-Salas, D. E., Gundersen, D., Cereijido, M., and Rodriguez-Boulan, E., 1986, Microtubules and actin filaments are not critically involved in the biogenesis of epithelial cell surface polarity, *J. Cell Biol.* **102**:1853.

Sambrook, J., Rodgers, L., White, J., and Gething, M.-J., 1985, Lines of BPV-transformed murine cells that constitutively express influenza virus hemagglutinin, *EMBO J.* **4**:91.

Saraste, J., and Hedman, K., 1983, Intracellular vesicles involved in the transport of Semliki Forest virus membrane proteins to the cell surface, *EMBO J.* **2**:2001.

Saraste, J., and Kuismanen, E., Pre- and post-Golgi vacuoles operate in the transport of Semliki Forest Virus membrane glycoproteins to the cell surface, *Cell* **38**:535.

Schatz, G., 1986, A common mechanism for different membrane systems, *Nature (Lond.)* **231**:108.

Schlesinger, M. J., 1981, Proteolipids, *Annu. Rev. Biochem.* **50**:193.

Schmidt, M. F. G., 1983, Fatty acid binding: A new kind of posttranslational modification of membrane proteins, *Curr. Top. Microbiol. Immunol.* **102**:101.

Schmidt, M. F. G., 1984, The transfer of mytistic and other fatty acids on lipid and viral protein acceptors in cultured cells infected with Semliki Forest and influenza virus, *EMBO J.* **3**:2295.

Schmidt, M. F. G., and Lambrecht, B., 1985, On the structure of the acyl linkage and the function of fatty acyl chains in the influenza virus hemagglutinin and the glycoproteins of Semliki Forest virus, *J. Gen Virol.* **66**:2635.

Schnapp, B. J., Vale, R. D., Sheetz, M. P., and Reese, T. S., 1985, Single microtubules from squid axoplasm support bidirectional movement of organelles, *Cell* **40**:455.

Scholtissek, C., and Bowles, A. L., 1975, Isolation and characterization of temperature-sensitive mutants of fowl plague virus, *Virology* **67**:576.

Schultz, I. T., 1975, The biologically active proteins of influenza virus: The hemagglutinin, in: *The Influenza Viruses and Influenza* (E. D. Kilbourne, ed.), pp. 53–82, Academic, New York.

Schwartz, A. L., Strous, G. J. A. M., Slot, J. W., and Geuze, H. J., 1985, Immunoelectron microscopic localization of acidic intracellular compartments in hepatoma cells, *EMBO J.* **4**:899.

Schwarz, R. T., Rohrschneider, J. M., and Schmidt, M. F. G., 1976, Suppression of glycoprotein formation of Semliki Forest, influenza, and avian sarcoma virus by tunicamycin, *J. Virol.* **19**:782.

Schwarz, R. T., Schmidt, M. F. G., Anwer, U., and Klenk, H.-D., 1977, Carbohydrates of influenza virus. I. Glycopeptides derived from viral glycoproteins after labeling with radioactive sugars, *J. Virol.* **23**:217.

Sekikawa, K., and Lai, C.-J., 1983, Defects in functional expression of an influenza virus hemagglutinin lacking the signal peptide sequences, *Proc. Natl. Acad. Sci. USA* **78**: 5488.

Seto, J. T., and Rott, R., 1966, Functional significance of sialidase during influenza virus multiplication, *Virology* **30**:731.

Sharma, S., Rodgers, L., Brandsma, J., Gething, M.-J., and Sambrook, J., 1985, SV40 T antigen and the exocytic pathway, *EMBO J.* **4**:1479.

Simons, K., and Fuller, S. D., 1985, Cell surface polarity in epithelia, *Annu. Rev. Cell Biol.* **1**: 243.

Skehel, J. J., 1971, Estimations of the molecular weight of the influenza virus genome, *J. Gen. Virol.* **11**:103.

Skehel, J. J., and Schild, G. C., 1971, The polypeptide composition of influenza A viruses, *Virology* **44**:396.

Skehel, J. J., Stevens, D. J., Daniels, R. S., Douglas, A. R., Knossow, M., Wilson, I. A., and Wiley, D. C., 1984, A carbohydrate side chain on hemagglutinins of Hong Kong influenza viruses inhibits recognition by a monoclonal antibody, *Proc. Natl. Acad. Sci. USA* **81**:1779.

Smith, G. L., Murphy, B. R., and Moss, B., 1983, Construction and characterization of an infectious vaccinia virus recombinant that expresses the influenza hemagglutinin gene and induces resistance to influenza virus infection in hamsters, *Proc. Natl. Acad. Sci. USA* **80**:7155–7159.

Stanley, P., and Haslam, E. A., 1971, The polypeptides of influenza virus. V. Localization of polypeptides in the virion by iodination techniques, *Virology* **46**:764.

Stanley, P., Gandhi, S. S., and White, D. O., 1973, The polypeptides of influenza virus VII. Synthesis of the hemagglutinin, *Virology* **53**:92.

Steinman, R. M., Mellman, I. S., Muller, W. A., and Cohn, Z. A., 1983, Endocytosis and the recycling of plasma membrane, *J. Cell Biol.* **96**:1.

Stephens, E. B., Compans, R. W., Earl, P., and Moss, B., 1986, Surface expression of viral glycoproteins is polarized in epithelial cells infected with recombinant vaccinia viral vectors, *EMBO J.* **5**:237.

Strous, G. J. A. M., Willemsen, R., van Kerkhof, P., Slot, J. W., Geuze, H. J., and Lodish, H. F., 1983, Vesicular stomatitis virus glycoprotein, albumin, and transferrin are transported to the cell surface via the same Golgi vesicles. *J. Cell Biol.* **97**:1815.

Strous, G. J. A. M., DuMaine, A., Zijderhand-Bleekemolen, J. E., Slot, J. W., and Schwartz, A. L., 1985, Effect of lysosomotropic amines on the secretory pathway and on the recycling of the asialoglycoprotein receptor in human hepatoma cells, *J. Cell Biol.* **101**: 531.

Struck, D. K., and Lennarz, W. J., 1980, The function of saccharide–lipids synthesis of glycoprotein, in: *The Biochemistry of Glycoproteins and Proteoglycans* (W. J. Lennarz, ed), pp. 35–84, Plenum, New York.

Sveda, M. M., Markoff, L. J., and Lai, C.-J., 1982, Cell surface expression of the influenza virus hemagglutinin requires the hydrophobic carboxy-terminal sequences, *Cell* **30**: 649.

Szczesna, E., and Boime, I., 1976, mRNA-dependent synthesis of authentic precursor to human placental lactogen: Conversion to its mature hormone form in ascites cell-free extracts, *Proc. Natl. Acad. Sci. USA* **73**:1179.

Sztul, E. S., Howell, K. E., and Palade, G. E., 1983, Intracellular and transcellular transport of secretory component and albumin in rat hepatocytes, *J. Cell Biol.* **97**:1582.

Tartakoff, A. M., 1983, Perturbation of vesicular traffic with the carboxylic ionophore monensin, *Cell* **32**:1026.

Tartakoff, A. M., and Vassalli, P., 1977, Plasma cell immunoglobulin secretion: Arrest is accompanied by alterations of the Golgi complex, *J. Exp. Med.* **146**:1332.

Tumova, B., and Pereira, H. G., 1965, Genetic interaction between influenza A viruses of human and animal origin, *Virology* **27**:253.

Ueda, M., and Kilbourne, E. D., 1976, Temperature-sensitive mutants of influenza virus: A mutation in the hemagglutinin gene, *Virology* **70**:425.

Varghese, J. N., Laver, W. G., and Colman, P. M., 1983, Structure of the influenza virus glycoprotein antigen neuraminidase at 2.9 Å resolution, *Nature (Lond.)* **303**:35.

Walter, P., and Blobel, G., 1980, Purification of a membrane-associated protein complex required for protein translocation across the endoplasmic reticulum, *Proc. Natl. Acad. Sci. USA* **77**:7112.

Walter, P., and Blobel, G., 1981, Translocation of proteins across the endoplasmic reticulum. II. Signal recognition protein (SRP) mediates the selective binding to microsomal membranes of in-vitro-assembled polysomes synthesizing secretory protein, *J. Cell Biol.* **91**:551.

Walter, P., Jackson, R. C., Marcus, M. M., Lingappa, V. R., and Blobel, G., 1979, Tryptic dissection and reconstitution of translocation activity for nascent presecretory proteins across microsomal membranes, *Proc. Natl. Acad. Sci. USA* **76**:1795.

Walter, P., Gilmore, R., and Blobel, G., 1984, Protein translocation across the endoplasmic reticulum, *Cell* **38**:5.

Ward, C. W., 1981, Structure of the influenza virus hemagglutinin, *Curr. Top. Microbiol. Immunol.* **94**:1.

Ward, C. W., and Dopheide, T. A., 1980, The Hong Kong (H3) hemagglutinin. Complete amino acid sequence and oligosaccharide distribution for the heavy chain of A/Memphis/102/72, in: *Structure and Variation in Influenza Virus* (G. Laver and G. Air, eds.), pp. 27–37, Elsevier/North-Holland, New York.

Ward, C. W., Elleman, T. C., and Azad, A. A., 1982, Amino acid sequence of the pronase-released heads of neuraminidase subtype N2 from the asian strain A/Tokyo/3/67 of influenza virus, *Biochem. J.* **207**:91.

Waterfield, M. D., Gething, M.-J., Scrace, G., and Skehel, J. J., 1980, The carbohydrate side chains and disulphide bonds of the hemagglutinin of the influenza virus A/Japan 305/57 (H2N1), in: *Structure and Variation in Influenza Virus* (G. Laver and G. Air, eds.), pp. 11–20, Elsevier/North-Holland, New York.

Webster, R. G., 1970, Estimation of the molecular weights of the polypeptide chains from the isolated hemagglutinin and neuraminidase subunits of influenza viruses, *Virology* **40**:643.

Webster, R. G., and Laver, W. G., 1967, Preparation and properties of antibody directed specifically against the neuraminidase of influenza virus, *J. Immunol.* **99**:49.

Webster, R. G., Brown, L. E., and Jackson, D. C., 1983, Changes in the antigenicity of the hemagglutinin molecule of H3 influenza virus at acidic pH, *Virology* **126**:587.

Webster, R. G., Kawaoka, Y., and Bean, W. J., 1986, Molecular changes in A/Chicken/Pennsylvanina/83 (H5N2) influenza virus associated with acquisition of virulence, *Virology* **149**:165.

Wehland, J., Willingham, M. C., Gallo, M. G., and Pastan, I., 1982, the morphologic pathway of exocytosis of the vesicular stomatitis virus G protein in cultured fibroblasts, *Cell* **28**:831.

Wen, D., and Schlesinger, M. J., 1984, Fatty acid-acylated proteins in secretory mutants of *Saccharomyces cerevisiae*, *Mol. Cell. Biol.* **4**:688.

White, J., Kielian, M., and Helenius, A., 1983, Membrane fusion proteins of enveloped animal viruses, *Q. Rev. Biophys.* **16**:151.

White, J. M., Matlin, K., and Helenius, A., 1981, Cell fusion by Semliki Forest, influenza and vesicular stomatitis virus, *J. Cell Biol.* **89**:674.

Wickner, W. T., and Lodish, H. F., 1985, Multiple mechanisms of protein insertion into and across membranes, *Science* **230**:400.

Wiley, D. C., Wilson, I. A., and Skehel, J. J., 1981, Structural identification of the antibody binding sites of the Hong Kong influenza virus hemagglutinin and their involvement in antigenic variation, *Nature (Lond.)* **289**:366.

Williams, D. B., Swiedler, S. J., and Hart, G. W., 1985, Intracellular transport of membrane glycoproteins: Two closely related histocompatibility antigens differ in their rates of transit to the cell surface, *J. Cell Biol.* **101**:725.

Wills, J. W., Srinivas, R. V., and Hunter, E., 1984, Mutations of the Rous sarcoma virus env gene that affect the transport and subcellular location of the glycoprotein products, *J. Cell Biol.* **99**:3011.

Wilson, I. A., Skehel, J. J., and Wiley, D. C., 1981, The hemagglutinin membrane glycoprotein of influenza virus: Structure at 3 Å resolution, *Nature (Lond.)* **289**:366.

Wrigley, N. G., Skehel, J. J., Charlwood, P. A., and Brand, C. M., 1973, the size and shape of influenza virus neuraminidase, *Virology* **51**:525.

Yamamoto, A., Masaki, R., and Tashiro, Y., 1985, Cytochrome P-450 transported from the endoplasmic reticulum to the Golgi apparatus in rat hepatocytes?, *J. Cell Biol.* **101**: 1733.

Yost, C. S., Hedgpeth, J., and Lingappa, V. R., 1983, A stop transfer sequence confers predictable transmembrane orientation to a previously secreted protein in cell-free systems, *Cell* **34**:759.

CHAPTER 6

# Structure of Defective-Interfering RNAs of Influenza Viruses and Their Role in Interference

DEBI P. NAYAK, THOMAS M. CHAMBERS, AND
RAMESH K. AKKINA

## I. INTRODUCTION

When viruses are passaged at high multiplicity, defective interfering (DI) particles are produced. von Magnus (1947, 1951a–c, 1952, 1954) first observed this phenomenon when he serially passaged undiluted influenza viruses in embryonated chicken eggs. He noted that although both the total amount of virus particles as assayed by hemagglutination units

Abbreviations used in this chapter: cDNA, complementary DNA; CEF, chicken embryo fibroblast cells; cRNA, complementary RNA; DI, defective interfering; DIU, defective interfering unit; DNI, defective noninterfering; $EID_{50}$, egg infectious dose (50%); HA, hemagglutinin; HAU, hemagglutinating unit; HI, hemagglutination inhibition; M, membrane protein; MDBK, Madin–Darby bovine kidney cells; MDCK, Madin–Darby canine kidney cells; MOI, multiplicity of infection; mRNA, messenger RNA; N, nucleoprotein; NA, neuraminidase; NDI, nondefective interfering; NP, nucleoprotein; NS, nonstructural protein; PAGE, polyacrylamide gel electrophoresis; PFU, plaque–forming units; Pi, postinfection; poly(A), 3' polyadenosine; RNP, ribonucleoprotein; UV, ultraviolet; vRNA, viral RNA; VSV, vesicular stomatitis virus; WSN, Wilson-Smith neurotropic.

DEBI P. NAYAK • Department of Microbiology and Immunology, Jonsson Comprehensive Cancer Center, UCLA School of Medicine, Los Angeles, California 90024. THOMAS M. CHAMBERS • Department of Virology and Molecular Biology, St. Jude Children's Research Hospital, Memphis, Tennessee 38101. RAMESH K. AKKINA • Department of Microbiology, College of Veterinary Medicine, Colorado State University, Fort Collins, Colorado 80523.

(HAU) and the amount of infectious particles as assayed by egg infectivity titer $(EID_{50})$ decreased, the ratio of infectivity to total particles $(EID_{50}/HAU)$ decreased much more precipitously during the passages at high multiplicity of infection (MOI). Clearly, during the undiluted passages, many particles were produced that were noninfectious. Subsequently, this phenomenon of multiplicity-dependent production of noninfectious virus particles has been reported with almost all animal viruses studied to date. Indeed, formation of such particles has also been observed for plant, yeast, and bacterial viruses (Kane et al., 1979; Mills et al., 1967) and probably represents a general phenomenon for all viruses. Later these noninfectious particles were called defective interfering (DI) particles in order to describe their proper phenotypic characteristics (Huang and Baltimore, 1970).

Defective interfering particles possess the following properties: first, they are defective or noninfectious, i.e., they can not replicate independently and therefore need the helper function of homologous standard virus particles for propagation; secondly, these particles are also interfering, i.e., when the same cell is coinfected with both standard and DI particles, DI particles replicate predominantly at the expense of standard viruses. Finally, these particles are noninfectious because they possess a defective genome; i.e., they do not contain the complete (standard) viral genome, but rather a shortened (deleted/altered) genome. Furthermore, in many instances, the shortened genome is altered in such a way as to give it a replicative advantage over the standard viral genome. However, although almost all DI particles appear to possess a shortened genome, it is possible to envision a DI particle containing a genome similar or even a genome larger in size than the standard genome. Alterations such as base substitution, insertion or rearrangement rather than deletion could make the viral genome both noninfectious and interfering as has been suggested for vesicular stomatitis virus (VSV) (Schubert et al., 1984).

The objective of this chapter is to review the current status of the studies on the influenza virus DI particles. This will encompass studies on the possible mode of origin and structure of the DI RNA, the function of DI RNA in the interference with standard RNA, the evolution of DI viruses and their effect on viral pathogenesis. Other comprehensive reviews on influenza virus DI particles, which deal with earlier studies on the nature of the DI genome and its role in DI virus-mediated interference, have been published (Nayak, 1980; Nayak and Sivasubramanian, 1983; Nayak et al., 1985; Barret and Dimmock, 1986).

## II. NATURE OF THE DI PARTICLE GENOME

### A. Generation and Amplification of Influenza Virus DI Particles

The study of influenza virus DI particles poses a number of unique problems. Influenza virus DI particles cannot be physically separated

from standard particles by physicochemical means, such as centrifugation and gel filtration (see Gard and von Magnus, 1947; Gard et al., 1952;
Yoshishita et al., 1959; Nayak et al., 1985). It is therefore important to
obtain virus preparations that are highly enriched in DI particles (>99%)
with only minor contamination from standard virus particles (<1%). The
next concern is that for any meaningful biochemical and biological analyses DI particles with uniform characteristics should be produced consistently in sufficient amounts and without a great deal of variation from
one preparation to another. The major problem in earlier studies was that
when DI particles were produced by repeated undiluted passages, very
few virus particles remained by the end of the fifth or sixth passages for
analysis of their genomic content, or biological or biochemical characteristics. Repeating the procedure every time using independent standard
virus preparations to generate DI particles would lead to variation due to
heterogeneity of DI particles from preparation to preparation. We have
recently reviewed the factors affecting the generation and amplification
of influenza virus DI particles (Nayak et al., 1985). We have also shown
that influenza virus DI particles can be generated, even in a single cycle of
infection, by standard virus alone and that the nature of the DI viral
genomes produced independently from separate plaques varies and cannot be predicted. Furthermore, amplification of DI particles requires co-
infection with both standard viruses as well as DI particles (Janda et al.,
1979).

The best way to consistently obtain DI particles of uniform quality is
by using an amplification procedure. In this process DI particles are generated from a number of plaques that have been passaged individually at a
high multiplicity (undiluted for three to five passages). Subsequently, one
or more of these DI preparations generated by undiluted passages can be
amplified in tissue culture by co-infecting permissive cells with both DI
particles (2–3 DIU/cell) and a standard helper virus (1–3 PFU/cell) (Janda
et al., 1979; Carter and Mahy, 1982b). The standard viruses used as helper
should be free from detectable DI particles (Nayak et al., 1978). Under a
carefully controlled condition of infection, virus preparations containing
more than 99% DI particles can be produced in relatively large quantities.
With WSN influenza virus either MDBK or MDCK cells can be used
(Nayak, 1980; Nayak and Sivasubramanian, 1983), whereas MDCK cells
appear to be the host of choice in producing DI particles of influenza B
virus (De and Nayak, 1980). In the case of fowl plague virus, either chick-
en embryos or primary chicken embryo fibroblast (CEF) cells can be used
to generate DI particles (Carter and Mahy, 1982a; Rott and Schafer, 1960).

Such a procedure usually yields a large amount of DI particles with
uniform characteristics. By screening DI particles produced from a
number of individual plaques, one or more DI preparations with the de-
sired characteristics are selected and further amplified by co-infection
with standard particles for further studies. The amount of DI particles
obtained from different plaques under these conditions may vary from 2
× 10⁶ to >10⁷ defective interfering units (DIU) per milliliter (Janda et al.,

1979; Carter and Mahy, 1982b). These DI virus preparations also contain varying amounts of standard virus, but the absolute yield of DI particles is not affected by the contaminating standard virus particles. The amplification and production of DI particles may depend on the replicating ability of a DI RNA. This may be different from its interfering ability, which instead will determine the amount of contaminating standard particles. In fact, a higher yield of DIU/ml is usually observed when the preparation also contains a relatively large amount of standard virus particles, because partial replication of standard virus is required for optimal amplification of DI virus particles.

## B. Analysis of DI RNA by PAGE

Since the DI particles are noninfectious mutant viruses, they must possess an altered genome to account for the loss or reduction of infectivity. Ada and Perry (1955, 1956) reported that the standard viruses contained approximately 0.8% nucleic acid whereas DI particles contained reduced and varying amounts of nucleic acid compared to that present in standard viruses. When the reduction in infectivity titer was plotted on a logarithmic scale, it correlated with the reduction in nucleic acid content on a linear rather than a logarithmic scale. Clearly, even these early studies indicated that although DI particles contained less RNA, the decreased infectivity was not directly proportional to the reduction of nucleic acid content. Other studies subsequently reported that DI particles contained a decreased amount of ribonucleoprotein (RNP) (Lief et al., 1956; Rott and Schafer, 1961; Rott and Scholtissek, 1963; Lenard and Compans, 1975), an overall reduction of the higher-molecular-weight RNAs and an increase in the lower-molecular weight RNAs as compared with that present in standard particles (Duesberg, 1968; Choppin and Pons, 1970; Nayak, 1969, 1972). Finally, it was demonstrated that in addition to the varying amount of the eight standard RNA segments the DI genome contained one or more novel RNA segments (DI RNA) (Fig. 1) that are not present in plaque-purified standard virus preparations and that are not required for the replication of standard virus (Nayak et al., 1978). These small RNA segments became increasingly prevalent after each serially undiluted passage. Subsequently, similar small RNA segments have been found in other influenza virus DI preparations, as well as in many standard virus preparations (Crumpton et al., 1978; Janda et al., 1979; De and Nayak, 1980; Nakajima et al., 1979), and the characteristic DI RNA pattern could be used as a marker for identity and proportional content of DI particles in a given virus preparation.

The above analyses demonstrated the following important characteristics of the influenza virus DI genome:

1. DI RNAs vary in size as well in number among different virus preparations generated and amplified independently. Each DI

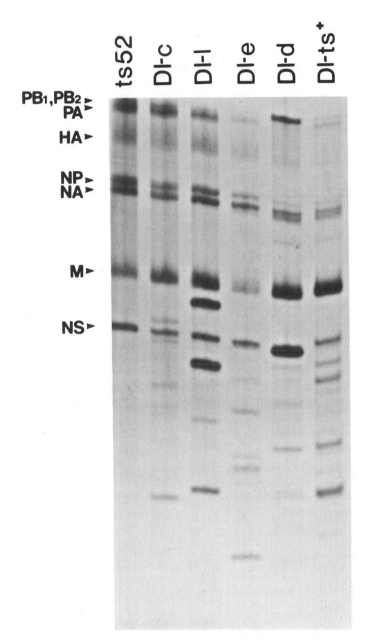

FIGURE 1. Analysis of [32]P-labeled RNAs of influenza virus DI preparations. MDBK cells were infected with WSN *ts*-52 standard virus or co-infected with standard virus and different DI virus preparations: DI-c, DI-1, DI-e, DI-d, or DI-*ts*[+], that had been obtained by serial passage of individual plaques without dilution. Standard viral RNAs are indicated at the left. Electrophoresis was in a 2.2% polyacrylamide/0.6% agarose/6M urea gel. (From Janda *et al.* 1979.)

virus preparation obtained by repeated passage of individual
plaques contains a unique set of DI RNAs that are maintained
when passaged under carefully controlled conditions. Individual
DI preparations may contain single or multiple DI RNA segments
that might not be present in equimolar ratios. Once produced,
these DI RNAs will replicate, amplify, and become the predominant RNA species in a given DI virus preparation. Upon long-term
passage, however, new DI RNA segments would be generated either independently from the progenitor standard RNA or from the
preexisting DI RNAs and would become the predominant DI RNA
species upon further passages by replacing the preexisting DI
RNAs (De and Nayak, 1980). In some cases, DI RNA species found
in later passages will come from the same progenitor gene as the
earlier DI RNA but will usually be smaller than the DI RNAs
present in earlier passages.

2. Among the standard genomic segments present in DI preparations, the amount of polymerase gene(s) is usually reduced, and
   the pattern of reduction varies among DI preparations. In some DI
   preparations, a single polymerase gene is greatly reduced in quantity, while in others two or even all three polymerase genes may
   be decreased (Janda *et al.*, 1979). In still others, no apparent reduction in quantity of any polymerase gene can be seen. Invariably,
   all DI virus preparations appear to contain one or more DI RNA
   segments. These DI RNAs that have been examined to date are of
   polymerase gene origin although subgenomic RNAs arising from
   other segments have been found (see Section II. C). We have recently proposed that the reduction of a specific polymerase gene
   occurs at the level of assembly of RNP molecules into progeny
   virus particles (see Section VI. E).

3. As with other viruses these DI RNAs are responsible for DI virus-
   mediated interference with standard influenza viruses (Janda and
   Nayak, 1979).

## C. Primary Structure of DI RNA

Once it became evident that the small RNA segments present in
influenza DI virus preparations were not the result of degradation of
standard segments but represented replicative molecules and were responsible for DI virus-mediated interference (Janda and Nayak, 1979),
investigations to determine the origin and structure of these DI RNAs
and their relationship to the standard progenitor RNA segment were
undertaken. Initial studies were done using RNA : RNA hybridization,
oligonucleotide mapping and partial direct RNA sequencing of the 5' and
3' termini. These analyses showed that all influenza virus DI RNAs are of
negative polarity and of polymerase gene origin. Furthermore, they pos-

sess both 5' and 3' genomic termini and therefore arise by internal deletions (Nayak, 1980). It was also evident that the same progenitor gene can give rise to multiple DI RNAs of varying length and that the smaller DI RNAs are not always a subset of the larger DI RNAs (Davis and Nayak, 1979; Davis *et al.*, 1980). These data suggested that DI RNAs do not arise from unique or specific sets of sequences of the progenitor RNAs, but precise data showing the exact sites of deletion(s) could not be obtained by these methods. Subsequently, nucleotide sequence analyses of cDNA clones of influenza virus standard gene segments as well as of a number of DI RNAs arising from these genes yielded detailed information about the primary structure of standard and DI viral RNAs.

## 1. DI RNA versus Subgenomic RNA

Since influenza virus DI RNAs are subgenomic in structure and since all subgenomic RNAs are not interfering in nature, influenza virus DI RNAs represent a unique subset within the subgenomic RNAs. In a given DI virus preparation, DI RNAs are often identified as a distinct RNA species that can replicate and amplify upon co-infection with standard viruses. To date approximately 31 distinct DI RNA species from a number of DI virus preparations have been examined either by oligonucleotide mapping or nucleic acid hybridization, and all appear to arise from one of the three polymerase genes. These DI RNAs were either produced in the laboratory by repeated passages of viruses (Nayak, 1980; Nakajima *et al.*, 1979; Akkina *et al.*, 1984a; Penn and Mahy, 1985; Crumpton *et al.*, 1979) or were found associated with a natural virus isolate (Chambers and Webster, 1987).

Complete nucleotide sequences of a few of the DI RNAs originating from the PB1 and PB2 genes of WSN virus are shown in Figs. 2 and 3, respectively. In addition, 35 other subgenomic RNAs of A/PR/8/34 virus have been sequenced using M13 cDNA cloning (Winter *et al.*, 1981; Jennings *et al.*, 1983). Altogether, the complete sequences of 41 subgenomic RNA segments of influenza virus, including the five DI RNAs, have been determined. These sequence analyses did not show any unique structural features that would distinguish the DI RNAs from other subgenomic RNAs. Therefore, from a structural standpoint, we have used the terms DI RNAs or subgenomic RNAs interchangeably, although all subgenomic RNAs may not have the interfering ability.

## 2. Subclasses of Subgenomic/DI RNAs

All influenza virus subgenomic/DI RNAs sequenced to date are internally deleted and belong to the 5'–3' class of DI RNAs (Lazzarini *et al.*, 1981; Nayak *et al.*, 1985), as was predicted from the earlier studies (Davis *et al.*, 1980; Nayak, 1980) in that they retain both the 3' and 5' termini of the progenitor gene. This is in contrast to the DI RNAs of vesicular

```
                   5'
WSN PB1 VRNA AGUAGAAACAAGGCAUUUUUUCAUGAAGGACAAGCUAAAUUCACUAUUUUUGCCGUCUGA       60
DI  L2B VRNA AGUAGAAACAAGGCAUUUUUUCAUGAAGGACAAGCUAAAUUCACUAUUUUUGCCGUCUGA       60
DI  L3  VRNA AGUAGAAACAAGGCAUUUUUUCAUGAAGGACAAGCUAAAUUCACUAUUUUUGCCGUCUGA       60

             GCUCUUCAAUGGUGGAACAGAUCUUCAUGAUCUCAGUGAACUCCUCUUUCUUUAUCCUUC      120
             GCUCUUCAAUGGUGGAACAGAUCUUCAUGAUCUCAGUGAACUCCUCUUUCUUUAUCCUUC      120
             GCUCUUCAAUGGUGGAACAGAUCUUCAUGAUCUCAGUGAACUCCUCUUUCUUUAUCCUUC      120

             CAGAUUCGAAAUCAAUUCGUGCAUCAAUUCGGGCUCUGGAAACCAUAGCCUCCACCAUAC      180
             CAGAUUCGAAAUCAAUUCGUGCAUCAAUUCGGGCUCUGGAAACCAUAGCCUCCACCAUAC      180
             CAGAUUCGAAAUCAAUUCGUGCAUCAAUUCGGGCUCUGGAAACCAUAGCCUCCACCAUAC      180

             UGGAUAUCCCGACUGGUCUUCUGUAUGAACUGCUGGGGAAGAAUUUUUUCAAAUAAGUUGC     240
             UGGAUAUCCCGACUGGUCUUCUGUAUGAACUGCUGGGGAAGAAUUUUUUCAAAUAAGUUGC     240
             UGGAUAUCCCGACUGGUCUUCUGUAUGAACUGCUGGGGAAGAAUUUUUUCAAAUAAGUUGC     240

             AGCACUUUUGGUACAUUUGUUCAUCUUCAAGUAUUCCUCUUUGGCUUGUAUUCAAGAUGG      300
             AGCACUUUUGGUACAUUUGUUCAUCUUCAAGUAUUCCUCUUUGGCUUGUAUUCAAGAUGG      300
             AGCA_____     244

             AUCGAUUUCUUUUGGGGAUCCAGGAGUGUGUUGUUGCAACAGCAUCAUACUCCAUGUUUU      360
             AUCGAUUUCUUUUGGGGAUCCAGGAGUGUGUUGUUGCAACAGCAUCAUACUCCAUGUUUU      360
             ----------------------------------------------------------------- ---

             UGGCUGGACCAUGUGCUGGCAUUAUCACUGCAUUGUUCACUGAUUCAAUGUCUUUAUGGU      420
             UGGCUGGACCAUGUGCUGGCAUUAUCACUGCAUUGUUCACUGAUUCAAUGUCU_____     413
             ----------------------------------------------------------------- ---

             UGACAAUGGGUUCAGUGGGUUGCAUAAACGCCCCUGGUAAUCNNNNNNNNNNNUGGGAUUC    2130
             ----------------------------------------------------------------- ---
             ----------------------------------------------------------------- ---

             CUCAAGGAAGGCCAUUGCUUCCAAUACACAAUCUGUUUGGGCAUAACCACUUGGUUCAUU    2190
                                             AUAACCACUUGGUUCAUU     431
             ----------------------------------------------------------------- ---

             GUCUUCUGGCAGUGGCCCAUCAAUCGGGUUGAGUUGCGGUGCUCCAGUUUCGGUGUUUGU    2250
             GUCUUCUGGCAGUGGCCCAUCAAUCGGGUUGAGUUGCGGUGCUCCAGUUUCGGUGUUUGU     491
                                                               UUUGU     249

             UGUCCAUCUUCCCCUUUCUGAGUACUGAUGUGUCCUGUUGACAGUAUCCAUGGUGUAUCC    2310
             UGUCCAUCUUCCCCUUUCUGAGUACUGAUGUGUCCUGUUGACAGUAUCCAUGGUGUAUCC     551
             UGUCCAUCUUCCCCUUUCUGAGUACUGAUGUGUCCUGUUGACAGUAUCCAUGGUGUAUCC     309

             UGUUCCUGUCCCAUGGCGUGUAAGGAGGGUCUCCAGUAUAAGGGAAAGUUGUGCUUAUAGC   2370
             UGUUCCUGUCCCAUGGCGUGUAAGGAGGGUCUCCAGUAUAAGGGAAAGUUGUGCUUAUAGC    611
             UGUUCCUGUCCCAUGGCGUGUAAGGAGGGUCUCCAGUAUAAGGGAAAGUUGUGCUUAUAGC    369

             AUUUUGUGCUGGCACUUUUAAGAAAAGUAAAGUCGGAUUGACAUCCAUUCAAAUGGUUUG   2430
             AUUUUGUGCUGGCACUUUUAAGAAAAGUAAAGUCGGAUUGACAUCCAUUCAAAUGGUUUG    671
             AUUUUGUGCUGGCACUUUUAAGAAAAGUAAAGUCGGAUUGACAUCCAUUCAAAUGGUUUG    429
                   3'
             CCUGCUUUCGCU                                              2442
             CCUGCUUUCGCU                                               683
             CCUGCUUUCGCU                                               441
```

FIGURE 2. Complete nucleotide sequence (negative strand polarity) of L2b and L3 DI RNAs of PB1 origin. -----, region absent in L2b and L3 DI RNAs; N-N-N-N, sequence of 1559 nucleotides in the PB1 gene. From Nayak *et al.* (1982b).

stomatitis virus (VSV) and Sendai virus, which consist of both the 5' (or copyback type) as well as 5'–3' (internally deleted or fusion) type (Kolakofsky, 1976; Huang, 1977; Perrault, 1981; Yang and Lazzarini, 1983; Amesse *et al.*, 1982). No 5' or copyback-type DI RNAs have been found with influenza viruses (see Section III).

The 5'–3' DI RNAs of influenza viruses can be further grouped into following subclasses:

1. *5′–3′ single-deletion subgenomic/DI RNA:* Subgenomic/DI RNAs that are produced by a single internal deletion in the progenitor gene. The size of the internal deletion as well as its location in the standard RNA segment may vary and would determine the size as well as the primary structure of the subgenomic/DI RNA. Most influenza virus subgenomic/DI RNAs belong to this class.
2. *5′–3′ multiple deletion subgenomic/DI RNA:* These subgenomic/DI RNAs retain both genomic termini but contain more than one internal deletion (Fig. 3). The location and the extent of the deletion may again vary.

```
PB2      cRNA  AGCGAAAGCAGGUCAAUUAUAUUCAAUAUGGAAAGAAUAAAAGAACUAAGGAAUCUAAUG    60
L2a-7    cRNA  AGCGAAAGCAGGUCAAUUAUAUUCAAUAUGGAAAGAAUAAAAGAACUAAGGAAUCUAAUG    60
L2a-17   cRNA  AGCGAAAGCAGGUCAAUUAUAUUCAAUAUGGAAAGAAUAAAAGAACUAAGGAAUCUAAUG    60

               UCGCAGUCUCGCACUCGCGAGAUACUCACAAAAACCACCGUGGACCAUAUGGCCAUAAUC   120
               UCGCAGUCUCGCACUCGCGAGAUACUCACAAAAACCACCGUGCACCAUAUGGCCAUAAUC   120
               UCGCAGUCUCGCACUCGCGAGAUACUCACAAAAACCACCGUGGACCAUAUGGCCAUAAUC   120

               AAGAAGUACACAUCAGGAAGACAGGAGAAGAACCCAGCACUUAGGAUGAAAUGGAUGAUG   180
               AAGAAGUACACAUCAGGAAGACAGGAGAAGAACCCAGCACUUAGGAUGAAAUGGAUGAUG   180
               AAGAAGUACACAUCAGGAAGACAGGAGAAGAACCCAGCACUUAGGAUGAAAUGGAUGAUG   180

               GCAAUGAAAUAUCCAAUUACAGCAGACAAGAGGAUAACGGAAAUGAUUCCUGAGAGAAAU   240
               GCAAUGAAAUAUCCAAUUACAGCAGACAAGAGGAUAACGGAAAUGAUUCCUGAGAGAAAU   240
               GCAAUGAAAUAUCCAAUUACAGCAGACAAGAGGAUAACGGAAAUGAUUCCUGAGAGAAAU   240

               GAGCAGGGACAAACUUUAUGGAGUAAAAUGAAUGACGCCGGAUCAGACCGAGUGAUGGUA   300
               GAGCAGGGACAAACUUUAUGGAGUAAAAUGAA_____   272
               GAGCAGGGACAAACUUUAUGGAGUAAAAUGAA_____   272

               NNNNNUCCUCAUUGACUAUAAAUGUGAGGGGAUCAGGAAUGAGAAUACUUGUAAGGGGCA  1981
               _____CAGGAAUGAGAAUACUUGUAAGGGGCA   299
               _____CAGGAAUGAGAAUACUUGUAAGGGGCA   299

               AUUCUCCAAUAUUCAACUACAACAAGACCACUAAAAGACUCACAGUUCUCGGAAAGGAUG  2041
               AUUCUCCAAUAUUCAACUACAACAAGACCACUAAAAGACUCACAGUUCUCGGAAAGGAUG   359
               AUUCUCCAAUAUUCAACUACAACAAGACCACUAAAAGACUCACAGUUCUCGGA_____   352

               CUGGCCCUUUAACUGAAGACCCAGAUGAAGGCACAGCUGGAGUUGAGUCCGCAGUUCUGA  2101
               CUGGCCCUUUAACUGAAGACCCAGAUGAAGGCACAGCUGGAGUUGAGUCCGCAGUUCUGA   419
               _____GUUGAGUCCGCAGUUCUGA   371

               GAGGAUUCCUCAUUCUGGGCAAAGAAGACAGGAGAUAUGGACCAGCAUUAAGCAUAAAUG  2161
               GAGAAUUCCUCAUUCUGGGCAAAGAAGACAGGAGAUAUGGACCAGCAUUAAGCAUAAAUG   419
               GAGGAUUCCUCAUUCUGGGCAAAGAAGACAGGAGAUAUGGACCAGCAUUAAGCAUAAAUG   371

               AACUGAGCAACCUUGCGAAAGGAGAGAAGGCUAAUGUGCUAAUUGGGCAAGGAGACGUGG  2221
               AACUGAGCAACCUUGCGAAAGGAGAGAAGGCUAAUGUGCUAAUUGGGCAAGGAGACGUGG   539
               AACUGAGCAACCUUGCGAAAGGAGAGAAGGCUAAUGUGCUAAUUGGGCAAGGAGACGUGG   491

               UGUUGGUAAUGAAACGGAAACGGAACUCUAGCAUACUUACUGACAGCCAGACAGCGACCA  2281
               UGUUGGUAAUGAAACGGAAACGGAACUCUAGCAUACUUACUGACAGCCAGACAGCGACCA   599
               UGUUGAUAAUGAAACGGAAACGGAACUCUAGCAUACUUACUGACAGCCAGACAGCGACCA   551

               AAAGAAUUCGGAUGGCCAUCAAUUAGUGUCGAAUAGUUUAAAAACGACCUUGUUUCUACU  2341
               AAAGAAUUCGGAUGGCCAUCAAUUAGUGUCGAAUAGUUUAAAAACGACCUUGUUUCUACU   659
               AAAGAAUUCGGAUGGCCAUCAAUUAGUGUCGAAUAGUUUAAAAACGACCUUGUUUCUACU   611
```

FIGURE 3. Complete nucleotide sequence of DI RNAs of PB2 origin. Squares indicate mismatches with the progenitor gene, _____, common nucleotide sequence (GGA) at the deletion point; -----, regions absent in the DI RNAs; N-N-N-N, sequence of 1626 nucleotides in the PB2 gene. PB2, L2a-7, and L2a-17 are of WSN virus origin. From Kaptein and Nayak, 1982; Sivasubramanian and Nayak (1983).

3. *5'–3' complex subgenomic/DI RNA:* These subgenomic/DI RNAs may contain one or more deletions and, in addition, exhibit extensive changes in the nucleic acid due to insertion, inversion, transposition, base changes, or even addition of new sequences not present in the progenitor gene (Jennings et al., 1983).

4. *5'–3' mosaic subgenomic/DI RNA:* These subgenomic/DI RNAs are produced by deletion as well as a true intersegmental recombination between two or more genomic RNA segments. Recent evidence indicates that intersegmental recombination between two different RNA molecules is possible and has been reported with picornaviruses (King et al., 1982). On the basis of the cDNA sequence data, one influenza virus subgenomic/DI RNA of mosaic type has been reported (Moss and Brownlee, 1981). However, this has not yet been confirmed by direct RNA sequencing; therefore, the possibility that this proposed mosaic structure is a cloning artifact cannot be ruled out (Buonagurio et al., 1984).

## 3. Distribution of Origin of Subgenomic/DI RNAs

Most of the subgenomic/DI RNAs sequenced to date (35 of 40, or 88%) arose from the three polymerase genes (Jennings et al., 1983; Nayak et al., 1985). Of the 35 subgenomic/DI RNAs that are of polymerase gene origin, 26 were of PB2 origin, whereas three and five originated from the PB1 and PA genes, respectively. Only six of 40 subgenomic/DI RNA segments arose from genes other than polymerase genes. Among these six RNAs, three arose from HA and one each from the NA, NP, and NS genes. However, no interfering ability of the subgenomic RNAs of non-polymerase gene origin has yet been reported. Possible reasons for this biased distribution in the origin and amplification of subgenomic/DI RNAs segments are discussed below (Sections III and IV).

## 4. Size of Subgenomic/DI RNAs

Although DI RNAs of large size have been noted in DI virus preparations (Nayak, 1980), these have not been sequenced yet. Those that have been completely sequenced fall within a range of 178–859 nucleotides, with the majority (84%) of the DI/subgenomic RNAs being 300–500 nucleotides in length. Partly, this distribution in size may have been biased because of M13 cDNA cloning, which favors insertion of cDNA within this size range. In Sendai virus, the size of the RNA appears to be an important factor in the preferential amplification of some DI RNA species (Re and Kingsbury, 1988).

## 5. Nature of the Deletion

Nearly 90% (36 of 40) of the subgenomic/DI RNAs studied arose by a single internal deletion; three possessed two internal deletions, and one each was of complex or mosaic type. The extent of the deletion in the progenitor gene varied and in some cases ranged up to 84% of the RNA segment. When the DI RNAs contain multiple deletions, the second deletion is usually smaller (approximately 50 nucleotides). Furthermore, subgenomic/DI RNAs arising from each gene also vary in size, but all subgenomic/DI RNAs appear to possess the 5'- and 3'-terminal sequences of the progenitor gene. Jennings *et al.*, (1983) further reported a relatively symmetrical contribution from the 5' and 3' ends of the progenitor gene. The range of contribution from the 5' end (positive sense) varies from 83 to 445 nucleotides and that from the 3' end is from 95 to 413 nucleotides.

## 6. Alteration in Sequence in Subgenomic/DI RNAs

Except for the internal deletion, sequence alterations among the influenza virus subgenomic DI/RNAs, unlike the DI RNAs of other viruses, are relatively uncommon. A few base mismatches have been observed, and only rarely have any drastic changes such as insertion or transposition in sequence been found. Most likely, the base mismatches as well as other alterations of nucleotide sequence in the DI virus genome occur subsequent to the generation of DI RNAs from the progenitor gene and during the amplification phase involving multiple passages of the DI virus particles. The presence of only a few base substitutions observed among the subgenomic influenza virus RNA segments sequenced by Jennings *et al.* (1983) probably reflects the recent origin of these DI RNA segments which have not undergone multiple replication cycles during repeated undiluted passages. Furthermore, absence of predominant or discrete DI RNA segments in their preparation also suggest that these subgenomic RNAs were of recent origin and did not undergo sufficient selection pressure necessary for the evolution of discrete and predominant DI RNA species. This would also explain why some subgenomic RNAs of nonpolymerase gene origin were observed in their virus preparation, whereas such subgenomic RNAs are usually absent in DI virus preparations that have undergone selection during amplification.

## 7. Maintenance of Reading Frame in Subgenomic/DI RNAs

After the point of deletion, subgenomic/DI RNAs do not show any preference for maintaining a specific reading frame over others. About one third of DI RNAs maintain the same reading frame as the progenitor gene after the deletion point and may therefore encode DI polypeptides which, if produced, are expected to retain the same $NH_2$- and COOH-termini of the parent polypeptide. The other two thirds of DI-encoded polypeptides will possess the same $NH_2$-terminus as the protein encoded by the progenitor gene but a different COOH-terminus due to a shift in the reading frame. DI-specific polypeptides have been detected with some DI preparations (see Section V.B). However, since the DI mRNAs giving rise to the DI virus-specific polypeptides have not been sequenced, it remains to be seen whether these DI-specific polypeptides were translated from DI RNAs that had the same or a different reading frame from that of the progenitor gene.

## 8. Junction Sequence

The sequences at the junction points of subgenomic/DI RNAs do not resemble the consensus splicing sequences of eukaryotic RNAs, suggesting that splicing as observed for eukaryotic mRNAs and for influenza virus M2 and NS2 mRNAs is probably not responsible for generating influenza virus subgenomic/DI RNAs (Nayak *et al.*, 1982b). Furthermore, no sequences specifying either the detachment or the reattachment of the polymerase on the template of plus or minus progenitor RNA strands have been observed. However, it is still possible that secondary structure(s) in the nucleotide sequence or some higher order structure in RNP may be responsible for causing the detachment and reattachment (see Section III, for further discussion).

## 9. Summary

Sequence analyses of a relatively large number of influenza virus subgenomic/DI RNAs indicate that all subgenomic/DI RNAs arise by internal deletion and that they predominantly arise from the polymerase genes. Furthermore, although a few subgenomic RNAs of nonpolymerase gene origin have been found, none of the DI RNAs examined to date arose from any of the five nonpolymerase genes. Sequence analyses did not provide any obvious explanation for the biased distribution observed for the origin of subgenomic/DI RNAs from the polymerase genes. One reason could be the large size of the polymerase genes, which are therefore more susceptible to error during replication. However, this cannot be the only reason because, although polymerase genes are larger, their relative molar amounts even in standard virus preparations often are less than those of smaller genes. Furthermore, intracellular replication of the polymerase RNA segments is also much lower than that of M or NS segments (Smith and Hay, 1982). Since the absolute number of nucleotides synthesized in an infected cell is the size (i.e., nucleotide number) of the RNA times the number of the molecules synthesized, it appears the total number of nucleotides of a polymerase gene synthesized may not be much different from that of the M or NS gene in infected cells. Therefore, if the generation and selection of all subgenomic (or DI) RNAs are based on random errors of the polymerase complex during the synthesis of RNA, one would expect to see a more even distribution of subgenomic/ DI RNAs of all standard RNA segments than is actually observed among the subgenomic/DI RNA species. Alternatively, subgenomic/DI RNAs may arise from other segments with equal frequency but are eventually eliminated because they are not able to replicate as efficiently as polymerase subgenomic/DI RNAs (see Section IV). However, sequence data have failed to demonstrate any unique structure(s) that may favor the generation or replication of some subgenomic/DI RNAs over others. Models that would favor the generation of DI RNAs from polymerase genes are discussed in Section III.

## III. GENERATION OF INFLUENZA VIRUS SUBGENOMIC/DI RNA

Influenza virus DI RNAs are consistently observed when the virus is passaged at high multiplicity and are present in molar excess over standard viral RNA segments in DI virus preparations. In addition, they are even present, although at much lower levels, in the clonal stocks of influenza viruses and would become amplified during subsequent passages at high multiplicity. The reason for the generation of DI RNAs, which have no apparently useful function in the biology of virus replication, with such a high frequency is not well understood (see Section VIII). The sequences of DI RNAs, as discussed above, indicate a wide variation in their structure (e.g. junction point, size) and do not show any unique feature when compared with either the progenitor RNA or other subgenomic RNAs. Furthermore, the DI RNAs that have been studied may not be the initial products of generation from the progenitor RNA but rather may represent the predominant species that have undergone evolution, survived the selection pressure, and become amplified during subsequent replicative cycles. However, these DI RNAs that are observed in a DI virus preparation may also not be the final end product of evolution either, as it has been shown that any given DI RNA may further evolve or even disappear during subsequent growth cycles of the virus (De and Nayak, 1980; D.P. Nayak, unpublished data). In fact, it is reasonable to assume that most DI RNAs studied to date may fall into the intermediate or evolving species (see Section IV).

The factors affecting the generation and evolution of influenza virus subgenomic/DI RNAs are poorly understood. It would be possible to define many variables in the generation of DI RNAs if one could generate DI RNAs in a more controllable *in vitro* replication system, but this has not yet been accomplished for influenza virus. Moreover, the detailed steps involved in the replication of the standard virus are not clearly understood. Therefore, it is difficult to propose a detailed model depicting the specific steps involved in the generation of subgenomic/DI RNAs during replication of standard RNA segments and the mechanism of subsequent amplification of a specific subset of subgenomic RNAs into discrete DI RNA species. Sequence studies rule out the normal eukaryotic splicing mechanism or post-transcriptional processing of nascent RNA in the generation DI RNAs (Nayak *et al.*, 1982b). It was suggested that an aberrant replicative event(s), in which the polymerase complex somehow skips a portion of the RNA template, is involved in the generation of DI RNAs. It appears that such a process is not restricted to generating DI RNAs alone but may produce other subgenomic RNA segments that provide altered phenotypic characteristics to the virus. Buonagurio *et al.*, (1984) reported that a 36-nucleotide internal deletion of the NS gene of influenza A/Alaska/6/77 produces a host-range variant virus. It is not known whether this NS gene with an internal deletion also possesses interfering properties.

Two general mechanisms for causing internal deletion(s) have been suggested (Nayak *et al.*, 1985). The first is the jumping polymerase model, which involves the detachment and reattachment of the polymerase complex from the template (Fig. 4a). In this scheme, during the replication of either the plus or the minus RNA strand, the polymerase complex along with the nascent RNA strand detaches from the template and reattaches downstream on the same template, giving rise to shortened RNA molecules possessing internal deletions. Such a process would require that when the polymerase complex is physically detached from the template the nascent RNA strand must remain attached to the polymerase complex. An essentially similar mechanism of detachment during the synthesis of minus strand RNA has been proposed to explain the copyback synthesis of Sendai and VSV DI RNAs (Kolakofsky, 1976; Huang, 1977). The mechanism proposed here may also be similar to the leader-primed transcription of coronavirus, in which the same primer RNA is attached to different predetermined sites on the template (Baric *et al.*, 1985). However, for the generation of influenza virus DI RNAs both the attachment and detachment sites vary. Sequence studies show that the junction and the flanking sequences of each subgenomic/DI RNA vary widely (Nayak *et al.*, 1985) and do not indicate the presence of any unique sequence or an obvious RNA secondary structure at either the junction or the flanking regions. Neither the uracil-rich regions (Fields and Winter, 1982) nor GAA and CAA sequences (Jennings *et al.*, 1983) can account for each site of detachment (see Nayak *et al.*, 1985, for discussion). Similarly, the absence of homology at the junction sites of most DI RNAs rules out the possibility that hydrogen bonding between the nascent RNA strand and the template RNA serves as a general mechanism of finding the reattachment site after the detachment of the polymerase complex from the RNA template. Therefore, if the detachment and reattachment of the polymerase on the template RNA are involved in the generation of influenza virus DI RNAs, they would occur at random or recognize some unique features of RNA in the RNP complex, which are yet to be identified.

The second scheme for generating an internal deletion can be described as a process that involves rolling over of the polymerase complex and looping out of the RNA template (rollover/loop-out model) (Fig. 4b). In this model, the polymerase complex does not completely detach from the template but instead, with the attached nascent daughter RNA

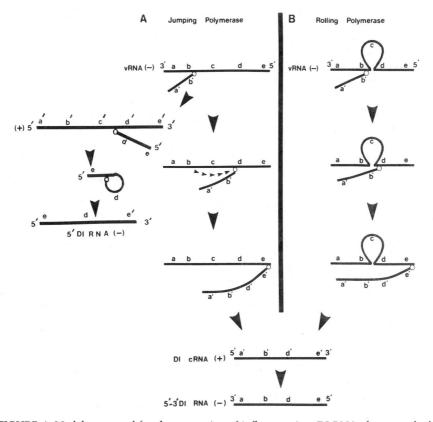

FIGURE 4. Models proposed for the generation of influenza virus DI RNAs from standard viral RNAs. Shown are negative-strand templates with sequence regions (e.g., a, b) and the corresponding positive-strand RNAs with complementary sequence regions a', b', etc. The open circles represent the functioning polymerase complex. Generation of influenza virus DI RNAs could equally occur during replication of the negative or positive-strand template. Model B is favored for the generation of influenza virus DI RNAs, as opposed to model A, which would generate both 5' (copyback type, shown at left) as well as 5'–3' DI RNAs. Modified from Nayak *et al.* (1985).

strand, rolls over to a new site on the template that is brought into juxtaposition (Nayak and Sivasubramanian, 1983; Nayak *et al.*, 1985). The fact that most, if not all, influenza virus DI RNAs are monogenic and not of polygenic origin would also argue against complete detachment of the polymerase complex containing the nascent strand from the template and reattachment to a new site but would favor a rollover/loop-out model. How the two sites on the RNA template are brought close to each other for the rollover of the polymerase complex is not clear. One possibility is that such a juxtaposition of two sites of the RNA template is caused by the formation of transient secondary structures of the RNA. The influenza virus RNP is not a rigid structure but is rather flexible; therefore, transient secondary structures in the RNA template may occur during RNA replication because of the dynamic nature of the influenza virus RNA–protein interactions. Alternatively, the supercoiled structure of RNP, as suggested by Jennings et al.(1983), may aid in bringing two sites of the RNA molecule together and would make it possible for the polymerase complex to roll over from one region of the RNA template to another. The recent evidence for the circular conformation of viral RNA due to a terminal panhandle would also support a supercoiled structure of RNP and favor the existence of multiple contact points in the template RNA (Hsu et al., 1987; Honda et al., 1987).

The above two models are likely to produce different kinds of subgenomic/DI RNAs. In the first model (Fig. 4a), which involves the detachment of the polymerase and nascent RNA from the template, most DI RNAs are likely to be 5' or copyback type, as is the case with VSV and Sendai viruses (Lazzarini et al., 1981) and only infrequently, internally deleted or 5'–3' DI RNAs will be generated when the RNA–polymerase complex reattaches to another site on the template. Internal regulatory sequences similar to polymerase initiation and termination signals may modulate the detachment and attachment of the polymerase complex on the template and consequently affect the generation of DI RNAs and DI virus particles. A number of such regulatory sequences have been proposed to be involved in the generation of DI RNAs of nonsegmented negative-strand RNA viruses, such as VSV and Sendai viruses (Herman, 1984; Re et al., 1985; Meier et al., 1984). However, since such 5' or copyback subgenomic/DI RNAs have not been found with influenza viruses, it is unlikely that influenza virus subgenomic/DI RNAs are generated by this process.

By contrast, the second model (Fig. 4b), in which the polymerase rolls over to a new site without detaching completely, would favor the generation of internally deleted influenza virus DI RNAs with conserved 5'- and 3'-termini. It would also explain why no 5' or copyback influenza virus DI RNAs have been detected because complete detachment of the polymerase complex with the nascent RNA from the template would be required for copying the 5' end of the nascent RNA. It is rather unlikely that 5' influenza virus DI RNAs are formed but are out-competed by 5'–3' DI RNAs during the amplification process because in both VSV and Sendai viruses the 5' DI RNAs have been reported to be more efficient than the 5'–3' DI RNAs in both replication as well as interference (Rao and Huang, 1982; Re and Kingsbury, 1988). Also, this model would explain why the mosaic or intersegmental DI RNAs are either absent or rare in influenza viruses which contain eight separate RNA segments. If the polymerase molecules with nascent RNA strands were completely detached from the template RNA and attached to a new template, one might expect a more frequent occurrence of mosaic DI RNAs. Taken together, current evidence suggests that rolling over of the polymerase complex and the possibility of multiple contact points within the RNA template are likely to generate 5'–3' internally deleted influenza subgenomic/DI RNAs.

The difference in the mode of generation and nature of subgenomic/DI RNAs between the segmented and the nonsegmented negative stranded RNA viruses could be attributable to a number of factors affecting the nature of the polymerase–template association. These include the following possibilities:

1. *Nature of the viral polymerase complex.* For example, since three separate polymerase molecules are involved in the process of RNA replication/transcription of influenza virus, all of them may not easily dissociate simultaneously from the template as would be required for the jumping polymerase model.

2. *Nature of the RNP template:* Unlike nonsegmented negative strand viruses, influenza virus RNP complexes are more flexible and contain exposed RNA (since the RNA in RNP is susceptible to RNase) and therefore may permit the juxtaposition of distant parts of the same RNA template. In addition, as proposed by Jennings *et al.* (1983), the coiled superstructure of influenza viral RNPs would also help in bringing distant sequences of viral RNAs together for the polymerase complex to roll over from one site on the template to another site. Flexibility in the RNP structure would also create variation in the contact points of the RNA template, accounting for the diverse junction points found in influenza virus subgenomic/DI RNAs.

3. *Nature of the P-RNP structure:* Influenza RNPs containing the P gene RNA segment may, because of their large size, form diverse loops and varying contact points and are therefore more likely to produce subgenomic/DI RNAs during replication more frequently than the smaller RNPs. Although the RNPs of nonsegmented viruses are larger, they are more rigid and lack the coiled superstructure of influenza virus RNP.

4. *Possible absence of the internal regulatory sites:* The absence of the multiple polymerase termination and initiation sites in the individual influenza viral RNA segments may prevent internal detachment and reattachment, whereas a number of such regulatory sequences present in multigenic VSV or Sendai viral RNAs would favor detachment and reattachment of the RNA–polymerase complex.

## IV. EVOLUTION OF DI RNAs

The rollover/loop-out model (fig. 4b) presented above may explain how internally deleted 5′–3′ subgenomic/DI RNAs are generated and why such a process would favor the generation of subgenomic/DI RNAs from three polymerase (P) genes. However, this would still not explain why discrete DI RNAs originating from the other five nonpolymerase gene segments have not been found in any DI virus preparation examined to date, since these nonpolymerase genes can give rise to subgenomic/DI RNAs, although infrequently (Jennings et al., 1983). Clearly, in most DI virus preparations, a relatively few subgenomic/DI RNA segments become the predominant DI RNA species. Even when multiple discrete DI RNA species are present, they are not in equimolar concentration. Therefore, only some of the subgenomic RNA species are likely to possess a replicative advantage over standard as well as other subgenomic RNA species and would become the predominant RNA species during the amplification phase. From these observations, three intriguing questions arise: (1) What are the features that provide a subgenomic RNA with a replicative advantage? (2) What are the characteristics that provide a better interfering ability? (3) How are these subgenomic/DI RNAs, with unique features of replicative advantage as well as interfering ability, selected?

Since only a few of the subgenomic RNA species become predominant in a DI preparation, DI RNAs are likely to represent a unique subclass of the large number of subgenomic RNA segments that may be randomly generated. However, sequence data have not revealed the essential characteristics of this subclass. All the predominant DI RNA species that have been examined are of polymerase origin (Nayak, 1980), suggesting that subgenomic RNAs of polymerase origin possess some advantage over other subgenomic RNA species. Possible reasons for preference for a polymerase progenitor in the generation of subgenomic RNAs have been discussed (see Section III). However, the mechanism of further selection of subgenomic/DI RNAs of polymerase gene origin over that of the nonpolymerase gene origin remains obscure.

The forces that direct the selection of influenza virus DI RNAs may be different from those operating in VSV or Sendai viruses. Re and Kingsbury (1988) recently discussed the factors that favor the survival of some Sendai virus DI RNAs over others. In their analysis, the strongest advantage is the copyback structure of the 5' DI RNAs which possess exclusively the plus-strand type of 3'-terminus and are transcriptionally inert, compared with the internally deleted structure of 5'–3' DI RNAs that possess both 5'- and 3'-termini of minus strand and are transcriptionally active. The weakest point of competitive advantage among Sendai DI RNAs is the RNA size: for DI RNAs of the same basic structure, the optimal size lies at 1200–1600 nucleotides. Larger RNAs are replicated more slowly, whereas smaller ones are probably disfavored during virus assembly and maturation. The kinetic replicational advantage of smaller RNAs over larger ones probably also holds for influenza virus DI RNAs. This may not be the sole advantage of a predominant influenza virus DI RNA species, however, since in some DI virus preparations the smaller DI RNA species are not always in molar excess over the larger ones. If any unique secondary structural features provide some DI RNAs with a replicative advantage over others, they have not been elucidated. There is currently no evidence for constraints on influenza virus RNA size during virus maturation. The majority of influenza virus DI RNAs are ~400–600 nucleotides (Section II.C.4), although DI RNAs larger than 1000 nucleotides are occasionally observed. The possible selection for specific subgenomic RNA types or sizes during virion assembly also may play a major role in the survival of influenza virus DI RNAs (see Section VI.E).

Although the specific attributes of subgenomic RNAs that are likely to become major DI RNA species are unknown, their selection must follow one of two evolutionary pathways: All subgenomic/DI RNAs, large or small, are generated directly from their progenitor genes and the ones with replicative advantage and interfering abilities will become the predominant DI RNA species. Alternatively, some of the DI RNAs do not originate directly from the progenitor genes but rather from precursor subgenomic/DI RNAs. Clearly, some of the subgenomic/DI RNAs (the first generation product) must arise directly from the progenitor gene. The largest members of DI RNAs are good candidates for this pathway. Unfortunately, complete sequences of the largest DI RNAs, particularly those generated in the first or second high-multiplicity passage, have not been determined. However, since even the largest DI RNAs studied to date do not possess all the sequences of some smaller DI RNAs (Davis and Nayak, 1979), some of the smaller DI RNAs may have originated directly from the standard progenitor genes.

Similarly, there is a good probability that some influenza virus DI RNAs are generated from a precursor DI RNA. Since two DI RNAs, L2a-7 and L2a-17 (see Fig. 3), from the same virus preparation contained one identical deletion (Sivasubramanian and Nayak, 1983), it is unlikely that they arose independently from the progenitor PB2 gene. Rather, they may have come from a single precursor DI RNA which contained the common deletion and subsequently evolved into two DI RNA species. These sequence data would indicate that influenza virus DI RNAs undergo further evolution and serve as progenitor RNAs for generating additional DI RNAs. In addition, occurrence of the progressively smaller DI RNAs after repeated viral passages would also suggest the generation of smaller DI RNAs from larger precursor DI RNAs and favor the selection of a subclass of DI RNA species with a better replicative advantage and interfering ability (De and Nayak, 1980). Clearly, certain DI RNA species possess a better replicative advantage over other DI RNAs when co-infected together (Akkina et al., 1984a; D. P. Nayak and T. M. Chambers, unpublished data). The extreme limit of the size of influenza virus DI RNA, at which they become too small to play a role in the virus replicative cycle, is unknown.

In summary, although subgenomic RNAs can be produced from all standard RNA segments, probably only a few, predominantly of polymerase origin, have the ability to become predominant DI RNA species. These DI RNA species, which possess a replicative advantage and interfering ability, represent a subclass among the subgenomic RNAs. Sequence analyses have not yet demonstrated the special attributes of this unique subclass of subgenomic RNA species. Two pathways, i.e., independent generation of DI RNA species directly from the standard progenitor gene as well as progressive evolution of subgenomic RNAs into major DI RNA species, may be operating in influenza virus replication and the selection of influenza virus DI particles. Unlike the DI RNAs of Sendai virus, the only known selective advantage of influenza virus DI RNAs is their smaller size. However, selection during the assembly of RNP into virions may affect the survival, enrichment and evolution of a specific DI RNA in a given virus preparation.

## V. REPLICATION, TRANSCRIPTION, AND TRANSLATION OF DI RNAs

Influenza virus DI RNAs, like DI RNAs of other viruses, possess two significant biological properties: replicative advantage and interfering ability. Both of these characteristics must be attributed to either the structural characteristics of the DI RNAs or their transcriptional or translational products or both. We have discussed the structural features of DI RNAs (see Sections II and III). Since DI RNAs use the same transcriptive and replicative machinery employed by the standard RNA segments, it is important to elucidate the process of transcription and replication and the nature of the viral polymerase proteins in both standard as well as DI virus-infected cells. A detailed analysis of the structure and function of the polymerase proteins and the mechanism of transcription and replication of standard viral RNAs is presented in Chapter 2, this volume. In this section, we attempt to assess the functional role of DI RNAs: (1) whether DI RNAs make functional mRNAs; (2) whether these DI mRNAs make defective proteins; (3) whether these defective proteins are likely to affect transcription and replication of standard influenza viruses.

## A. Transcription of DI RNAs

DI virus particles have been shown to contain decreased amounts of nucleic acid (Ada and Perry, 1956) and a reduced level of RNA polymerase activity (Chow and Simpson, 1971). Bean and Simpson (1976) observed that although these particles display reduced levels of polymerase activity, they transcribe all eight segments of their genome. Sequence analyses (Jennings *et al.*, 1983; Nayak *et al.*, 1982a,b; Nayak and Sivasubramanian, 1983) indicate that most DI RNAs usually retain 200–400 nucleotides from both the 3' and 5' ends of their parent gene segment (see Section II). Because the 12–13 bases at the extreme 5'- and 3'-termini are the only identified regions conserved among each of the eight standard gene segments, these regions presumably include the recognition sites for polymerase binding, transcription initiation and termination and poly (A) addition (Robertson et al., 1981). Since influenza virus DI RNAs contain the terminal regions of the progenitor genes, it was hypothesized (Nayak et al., 1982b) that, unlike the 5' DI RNAs of VSV, which are not transcribed into mRNAs, influenza virus DI RNAs should be capable of undergoing transcription into mRNAs. Chanda et al. (1983) demonstrated that all influenza virus DI RNAs examined were transcribed in vitro into poly(A)$^+$ complementary RNAs (cRNA) resembling the standard viral mRNAs. Like the standard viral mRNAs, these cRNAs could be synthesized using either globin mRNA or ApG as primer. In the in vitro system, there was no evidence for amplification of DI-specific transcripts or suppression of standard gene transcription, suggesting that interference does not occur at the level of primary transcription.

Similarly, DI-specific mRNAs (Fig. 5) were found to be present in cells co-infected with DI particles (Akkina et al., 1984a; Chambers et al., 1984). Although it has not been shown, these DI-specific poly(A)$^+$ cRNAs are presumed to possess host-derived 5' cap structures. Hybrids of specific DI RNA transcripts and their corresponding minus-strand DI RNAs are resistant to ribonuclease digestion, indicating that the DI-specific transcripts were true complements of the DI RNAs and not merely incomplete transcripts of standard RNA segments (Akkina et al., 1984a). Molar ratio analyses (Table I) indicate that in vivo, most DI RNAs are transcribed in molar excess over standard gene segments. The DI-specific RNAs in the poly(A)$^-$ cytoplasmic RNA fraction from these cells are also present in excess over the standard RNAs. Penn and Mahy (1984, 1985) using fowl plague influenza virus have also reported the synthesis in vivo of mRNAs corresponding to subgenomic RNAs of PA and PB2 origin. Since the basic structure of all influenza virus DI RNAs is the same, it is likely that all DI RNAs are capable of being transcribed.

## B. Translation of DI Virus-Specific mRNA

Because the DI RNAs are transcribed both in vitro and in vivo into poly(A)$^+$ cRNAs, it is likely that some of these mRNAs may be translated into DI-specific polypeptides. We and others (Chambers et al., 1984; Akkina et al., 1984a; Penn and Mahy, 1985) have detected several additional small polypeptides in cells co-infected with DI virus particles. Two DI virus preparations derived from WSN virus have been found to produce DI virus-specific small polypeptides (Fig. 6). Both DI virus preparations possess interfering properties as determined by infectious center reduction assays. The DI RNAs of these viruses are transcribed into poly(A)$^+$ cRNA both in vitro and in vivo (Chambers et al., 1984; Akkina et al., 1984a). One DI virus preparation contains a prominent DI RNA of ~700 nucleotides derived from the PB2 segment. MDBK cells co-infected with this DI virus preparation contain an additional polypeptide of ~22,000 molecular

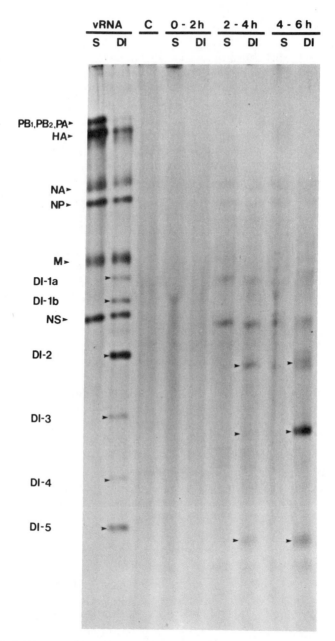

FIGURE 5. Poly(A)⁺ cytoplasmic cRNA from cells infected with standard and DI virus particles. S, RNA from infection of MDBK cells with WSN standard virus; DI, RNA from cells co-infected with standard virus and DI-*ts*⁺(Tobita); C, RNA from mock-infected cells; vRNA, marker RNA extracted from standard and DI viral particles. The indicated times are the ³²P-labeling periods postinfection. At the end of the labeling period, the cells were lysed and cytoplasmic RNA extracted. The poly(A)⁺ and poly (A)⁻ RNA fractions were separated by oligo-dT cellulose chromatography. Subsequently poly(A) was removed from the poly(A)⁺ fraction using poly(dT) and ribonuclease H (Etkind *et al.*, 1977); consequently, the mRNAs are slightly smaller and migrate further than the corresponding vRNA bands. Electrophoresis was on a 3% polyacrylamide/6 M urea gel. Arrows indicate positions of DI virus-specific poly(A)⁺ cRNAs corresponding to DI RNAs. From Akkina *et al.* (1984a).

TABLE I. Molar Ratios of *in Vivo* Poly(A)+ cRNAs of
Specific Viral and DI RNA Segments[a,b]

| RNA | Viral segments | | | | | |
| --- | --- | --- | --- | --- | --- | --- |
| | PB1 + PB2 + PA | HA | NP | NA | M | NS |
| DI-*ts*+(Tobita) viral RNA | 0.09 | 0.39 | 0.43 | 0.55 | 1.0 | .95 |
| poly(A)+ cRNA[c] | 0.07 | 0.43 | 0.34 | 0.34 | 1.0 | 1.62 |
| Standard viral RNA | 0.29 | 0.62 | 0.58 | 0.71 | 1.0 | 1.10 |
| poly(A)+ cRNA[c] | 0.09 | 1.07 | 0.63 | 0.56 | 1.0 | 1.54 |

| RNA | DI segments | | | | |
| --- | --- | --- | --- | --- | --- |
| | 1a + 1b | 2 | 3 | 4 | 5 |
| DI-*ts*+(Tobita) viral RNA | 0.73 | 1.68 | 0.65 | 0.59 | 1.04 |
| poly(A)+ cRNA[c] | 1.46 | 6.19 | 1.81 | 2.80 | 4.17 |

[a]From Chambers *et al* (1984).
[b]Molar ratios normalized to M segment:

$$\frac{(\text{segment cpm/estimated nucleotide length})}{(\text{M segment cpm/nucleotide length of M})}$$

[c]Cells labeled with $^{32}$P from 3 to 6 hr p.i.

weight (Fig. 6, lane 3). The second DI virus preparation contains minor and major DI RNAs of about 600 and 500 nucleotides derived from PB1 and PB2, respectively. Cells co-infected with this DI virus preparation contain an additional polypeptide of ~8000 molecular weight (Fig. 6, lane 4). Both polypeptides are also produced by *in vitro* translation using poly(A)+ cytoplasmic RNA from the corresponding DI virus-co-infected cells, but not from standard virus-infected cells. Furthermore, sucrose density-gradient analysis has shown that these DI virus-specific polypeptides are translated from smaller DI virus-specific mRNAs and are not the product of degradation of standard polypeptides or of premature translation-termination products of standard mRNAs (Fig. 7). Hybrid-selected translation has shown that both DI virus-specific polypeptides are of PB2 origin. These DI virus-specific polypeptides appear around 3 hr pi and reach maximal synthesis at 5 hr and continue to be synthesized even at 9 hr pi; furthermore, DI virus-specific polypeptides are also made in molar excess compared to PB2 protein to which they are related. This molar excess may reflect the more efficient replication and transcription of DI RNAs compared with that of standard viral RNAs in DI virus-co-infected cells.

## 1. DI Polypeptides and Maturation of DI Virus Particles

Mature influenza virus particles contain all three polymerase proteins responsible for primary transcription of the input template RNA. To determine whether DI virus-specific polypeptides were incorporated into DI virus particles, viral polypeptides were labeled with [$^{35}$S]meth-

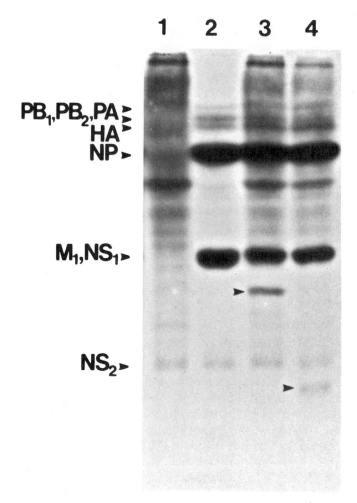

FIGURE 6. DI virus-specific polypeptides in cells coinfected with DI viruses. MDBK cells were mock infected (lane 1) or infected with WSN standard virus (lane 2), or co-infected with standard virus and DI-3 (lane 3) or DI-7 (lane 4). Cellular polypeptides were labeled with [³⁵S]methionine, 4–5 hr postinfection. Labeled cellular proteins were analyzed by electrophoresis on a 13% polyacrylamide/4 M urea gel. Arrows indicate the positions of DI virus-specific polypeptides. From Akkina *et al.* (1984a).

ionine in DI virus-co-infected cells, and the purified DI virus particles was analyzed by SDS–PAGE (Akkina *et al.*, 1984b). DI polypeptides were not detected in the virus particles, indicating that, although synthesized in large amounts in infected cells, these polypeptides are not incorporated into the mature virus particles. This finding raises the possibility that either the DI polypeptides are not transported into the nucleus or, unlike the standard PB2, they do not bind to polymerase complexes or that RNP

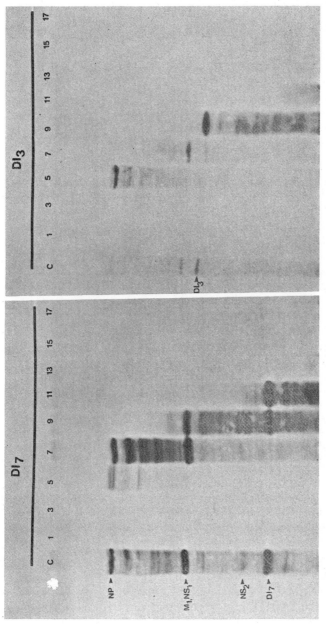

FIGURE 7. Translation *in vitro* of DI$_3$ and DI$_7$ viral mRNAs separated on sucrose velocity gradients. DI$_3$ or DI$_7$ viral mRNAs were extracted from virus-infected cells at 5 hr pi. The mRNA sample was layered onto a 5–30% (wt/vol) linear sucrose gradient and centrifuged for 18 hr at 27,000 rpm. Fractions (1 ml) were collected from the bottom. RNA was ethanol precipitated and translated *in vitro* using the wheat germ cell-free system. Panel DI$_7$, translation products of mRNA from DI$_7$-infected cells; panel DI$_3$, translation products of mRNA from DI$_3$-infected cells. Lanes C, translation products of unfractionated mRNAs. The number at the top of each lane represents the density gradient fraction number. From Akkina *et al.* (1984a).

complexes that include the DI polypeptides are not packaged into mature virus particles. Another possibility is that DI polypeptides are not stable in the cell and have a short half-life, so they may not be available for incorporation into DI virus particles.

## 2. Intracellular Location and Stability of DI Polypeptides

The PB2 protein forms a complex with the other two polymerase proteins PB1 and PA, and is transported into the nucleus (Jones et al., 1986; Akkina et al., 1987; Detjen et al., 1987; Smith et al., 1987). To determine whether PB2-derived DI virus-specific polypeptides also migrate into the nucleus, cytoplasmic and nuclear fractions of DI virus coinfected cells were analyzed by SDS–PAGE (Fig. 8b,c). Results show that DI polypeptides are also present in the nucleus. $DI_3$ polypeptide is relatively stable and only a slight decrease is noted at 2 hr of chase (Fig. 8), although some DI virus-specific polypeptides may not be stable (Penn and Mahy, 1985). $DI_3$ polypeptide appears in the nucleus within 5 min after synthesis and continues to accumulate in the nucleus. Its level in the cytoplasm decreases significantly by 2 hr. The appearance of $DI_3$ polypeptide in the nucleus correlates well with that of PB2 protein. This finding is particularly significant because polymerase proteins are active in the nucleus for both replication and transcription. Since DI polypeptides in this study are of PB2 origin, they may compete with PB2 functions, such as cap recognition or endonucleolytic activity, or both. This may also explain why in DI virus-infected cells cytopathic effect is much less pronounced and host protein synthesis is not shut off as it is in standard virus-infected cells. Furthermore, it may also explain the overall decrease in virus multiplication in cells coinfected with DI virus.

## 3. Binding of DI Polypeptides to RNP Complexes in Infected Cells

As neither of the two PB2 gene-derived DI virus polypeptides is incorporated into mature virus particles but does appear to migrate into the cell nucleus after synthesis, it is possible that both may be able to bind to RNP complexes, thereby displacing the standard PB2 protein. To see whether defective virus polypeptides associate with RNP complexes in the infected cells, labeled RNP complexes were isolated and analyzed by SDS–PAGE. Both the $DI_3$ and $DI_7$ polypeptides were found to be associated with the intracellular RNP complexes (Fig. 9). The results show that defective PB2 can form a complex with PB1 and PA and bind to RNP. It is therefore possible that some of the RNP complexes formed in the DI virus-infected cells could become functionally defective, leading to a decreased level of transcription and/or replication of viral genes. It is well established that in cells infected with DI viruses, the overall virus production as well as virus-specific macromolecular synthesis are decreased compared to that in standard virus-infected cells. Formation of defective

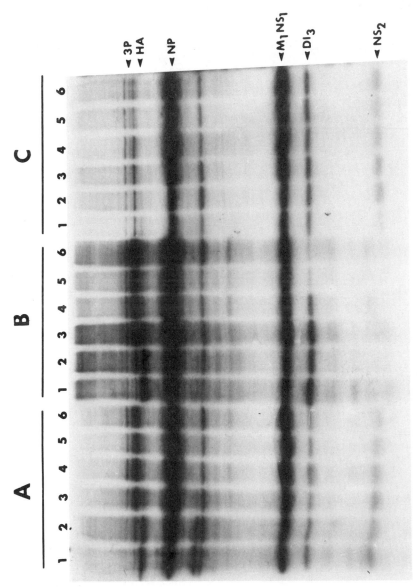

FIGURE 8. Intracellular stability and nuclear migration rate of DI$_3$ virus polypeptides in infected cells. Virus polypeptides were pulse-labeled for 5 min at 5 hr pi and chased for different times. Fractionated and unfractionated cell lysates were analyzed by 12% SDS–PAGE with 4 M urea. (a) Unfractionated cell lysates; (b) Cytoplasmic fractions; (c) Nuclear fractions. Chase periods: lane 1, 0 min; lane 2, 10 min; lane 3, 30 min; lane 4, 1 hr; lane 5, 2 hr; lane 6, 3 hr. From Akkina and Nayak (1987).

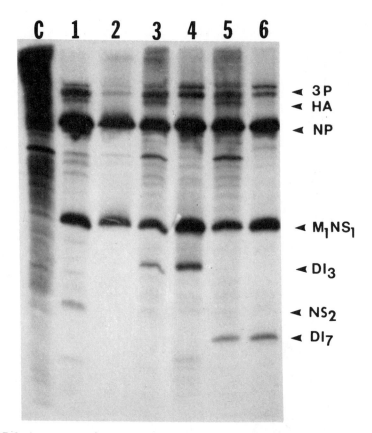

FIGURE 9. Association of DI virus polypeptides with intracellular RNP complex. Viral polypeptides in intracellular RNP complex were labeled for 1 hr, 5–6 hr pi with [$^{35}$S]methionine and were purified by sucrose-density gradient centrifugation (Akkina and Nayak, 1987). The viral polypeptides in the RNP complex were analyzed by 12% SDS–PAGE with 4 M urea. Lane 1, total standard virus-infected cell extract; lane 2, standard virus RNP; lane 3, total DI$_3$ virus co-infected cells extract; lane 4, intracellular DI$_3$ virus RNPs; lane 5, total DI$_7$ virus-co-infected cell extract; lane 6, intracellular DI$_7$ virus RNPs. C, control uninfected cell extract. From Akkina and Nayak (1987).

polymerase complexes in DI virus-infected cells may explain this phenomenon. The fact that DI polypeptides are able to bind to RNP complexes in infected cells, but are not incorporated into virus particles, would imply that defective RNPs containing DI polypeptides are somehow screened out during virus assembly. This in turn suggests some role of functional RNP–polymerase complex in virus assembly. Again, since DI proteins are made in molar excess compared to polymerase proteins, many intracellular RNPs containing DI proteins will fail to mature into complete virus particles during virus packaging and budding and would cause an overall reduction in virus production. The above findings suggest an important role for polypeptides of DI virus origin in decreasing the overall yield of virus particles and in DI virus-mediated interference.

In summary, PB2-derived DI-specific polypeptides are relatively stable in infected cells. Like the PB2 protein, these polypeptides migrate into the nucleus and form RNP–polymerase complexes, although they lack a substantial portion of PB2–polypeptide. Although they are made in large amounts in infected cells, however, they are not incorporated into mature virus particles. Furthermore, these RNP complexes containing DI virus-specific polypeptides may interfere at the level of virus assembly leading to decreased virus production. Since these DI proteins are able to form complexes with RNP and to be transported into the nucleus, they are likely to contain the structural signals required for nuclear transport as well as for interaction with NP and other polymerase proteins. Sequence analyses of these DI RNAs may elucidate the nature of the amino acid sequences involved in these functions, such as nuclear transport, complex formation, and template binding.

## VI. MECHANISM OF INTERFERENCE

In general, DI virus-mediated interference with standard viral replication possesses the following characteristics (Huang, 1975, 1977): (1) interference occurs intracellularly and not at the cell surface; (2) interference is strongest against the homologous standard virus from which the DI particles were derived, is less strong against other related viruses and nonexistent against unrelated viruses; and (3) interference by DI particles is not mediated by interferon. These features also hold for interference mediated by the DI particles of influenza virus. Although the precise mechanism of interference by DI virus particles is unknown, three effects are observed following co-infection of standard virus-infected cells with DI virus particles: (1) reduction in the yield of standard virus with amplification of DI particles; (2) overall reduction in the total yield of virus particles as well as of intracellular synthesis of viral macromolecules; and (3) reduction in cytopathic effect caused by standard virus infection.

The characteristics of influenza virus DI RNAs are substantially different from those of most other negative or positive strand DI RNAs. For example, influenza virus DI RNAs, unlike other negative strand viral DI RNAs, are formed by one or more internal deletions without any major sequence rearrangement. Since they retain both 3'- and 5'-termini, which are presumed to be polymerase-binding sites, these DI RNAs will not have any obvious replicative advantage other than their small size compared with the progenitor RNAs. Furthermore, since these DI RNAs are transcribed into mRNAs, some of which are translated into proteins, transcriptional as well as translational products of DI RNAs may also be involved in DI virus-mediated interference.

The steps in the infectious cycle in which DI RNAs and/or DI RNA-specified products act to cause interference with standard viral replication are not clear. With nonsegmented negative-strand RNA viruses like VSV, the major block appears to be decreased replication of standard RNAs compared with that of DI RNA (Blumberg and Kolakofsky, 1983; Perrault *et al.*, 1978, 1983; Huang *et al.*, 1978; Huang, 1982; Kingsbury, 1974; Leppert *et al.*, 1979; Bay and Reichmann, 1982). In a segmented virus like influenza virus, steps in which interference may occur may be multiple and more complex. For example, in addition to the steps in the replication and transcription of viral RNA, the assembly and packaging mechanism may be critically involved, because of the segmented nature of the viral genome (see Section VI.E).

## A. Effect of DI Virus Particles on the Transcription and Replication of Standard RNAs

Influenza virus DI particles after co-infection affect the mRNA transcription of the standard gene segments. Virus particles obtained after repeated high-multiplicity passages contain two- to fourfold decreased RNA transcriptase activity (Bean and Simpson, 1976; Carter and Mahy, 1982a; Chanda *et al.*, 1983; Pons, 1980). However, virus from the very first undiluted passage sometimes shows increased transcriptase activity (Carter and Mahy, 1982a; Pons, 1980). This may reflect the amplification of defective noninterfering (DNI) particles in the first passage and their subsequent replacement by DI virus particles in later passages (Carter and Mahy, 1982a). Repeated high-multiplicity passages cause reduced synthesis of total viral RNA as well as polysome-associated RNAs in infected cells (Pons and Hirst, 1969; Pons, 1980).

Synthesis of polymerase gene transcripts is markedly decreased in cells co-infected with standard and DI particles (Chambers *et al.*, 1984) and in cells infected with high-multiplicity-passaged virus (Pons and Hirst, 1969; Pons, 1980). In DI virus-co-infected cells, the DI RNAs are preferentially transcribed, accounting for about 80% of total viral transcript in these cells (see Table 1) (Chambers *et al.*, 1984). However, no evidence for preferential primary transcription of DI RNAs is seen in the *in vitro* transcription of purified DI particles; i.e., the relative levels of both standard and DI mRNAs correspond closely to those of the viral RNA template in the DI particles (Chanda *et al.*, 1983). Therefore, the reduction of standard RNA transcripts *in vivo* may reflect the effect of DI RNAs during secondary transcription. Because the early phase of secondary transcription is believed to be regulated at the level of vRNA replication (see McCauley and Mahy, 1983; Shapiro *et al.*, 1987), interference in viral mRNA transcription is probably due to DI virus-mediated interference with standard vRNA replication.

Replication of the standard viral segments is also altered after co-infection with DI particles. Again, the DI RNAs appear to be preferentially replicated over standard gene segments (Chambers *et al.*, 1984) and replication of the polymerase genes is reduced (Carter and Mahy, 1982c;

Pons, 1980). Analysis of synthesis of intracellular poly(A)$^-$ cRNAs in successive undiluted passages shows that the inhibition of polymerase gene replication occurs early even in the very first high-multiplicity passage (Carter and Mahy, 1982c), and the standard polymerase gene content of the progeny virus produced from undiluted passage is also reduced (Carter and Mahy, 1982c; Choppin and Pons, 1970; Crumpton *et al.*, 1978; Duesberg, 1968; Janda *et al.*, 1979; Pons, 1980). In some DI virus preparations, there is a preferential inhibition of a single polymerase gene such as PB1 (Ueda *et al.*, 1980), PB2 (Nakajima *et al.*, 1979; Pons, 1980), or PA (Janda *et al.*, 1979), while in others all three polymerase segments are reduced in almost equal proportions (Crumpton *et al.*, 1978, 1981; Bean and Simpson, 1976, Nayak *et al.*, 1978). Some other DI virus preparations contained reduced levels of the HA, NP, and NA genes as well as the polymerase genes (Crumpton *et al.*, 1981; Nayak *et al.*, 1978). In general, decreased synthesis and transcription of polymerase genes and polymerase mRNAs would lead to a reduction in polymerase proteins and would account for the overall reduction of viral macromolecular synthesis and virus yield.

## B. Effect of DI Particles on the Translation of Standard Polypeptides

In addition to transcription, co-infection by DI particles may also affect the levels of translation of the virus-specific proteins (Akkina *et al.*, 1984b). We have observed major changes in the synthesis of specific polymerase proteins; e.g., some DI preparations produce a decrease in the intracellular synthesis of PA and PB2 proteins, whereas others inhibit PB2 synthesis only. PB1 protein synthesis appeared almost unchanged with these DI preparations (Fig. 10). With other DI preparations, no inhibition of specific polymerase proteins could be detected. We have further shown that the decrease in the synthesis of specific polymerase proteins corresponds to the level of specific polymerase mRNAs, which, in turn, is proportional to the input ratio of polymerase genes in the DI virus preparation (Akkina *et al.*, 1984b). Lenard and Compans (1975) reported normal or slightly reduced relative levels of the polymerase proteins, reduced levels of NP and M proteins, and a twofold increase in levels of the HA and NA surface glycoproteins. However, others (Nayak *et al.*, 1978; Pons and Hirst, 1969) did not observe increased glycoprotein content in DI preparations. We have shown that some DI virus preparations exhibit 35–40% of the *in vitro* transcriptase activity compared to that present in standard viruses, and also contain a reduced amount of polymerase proteins. However, DI virus particles contained all three polymerase proteins in equimolar amounts (Akkina and Nayak, 1987), although specific polymerase proteins were reduced in DI virus-co-infected cells (Fig. 10). These results also support the hypothesis that the three polymerase proteins form a complex in $1:1:1$ ratio before or during their association

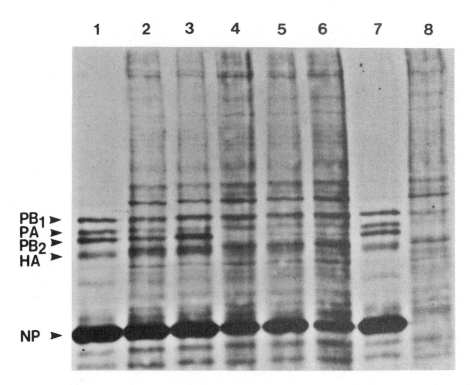

PB₁ ▶
PA ▶
PB₂ ▶
HA ▶

NP ▶

FIGURE 10. Analysis of polymerase protein synthesis in DI virus-infected cells. MDBK cells were infected with WSN standard virus (lanes 1 and 7), or co-infected with standard virus and DI-4 (lane 2) , DI-6 (lane 3), DI-7 (lane 4), DI-*ts*⁺(Tobita) (lane 5), or DI-3 (lane 6), or mock-infected (lane 8). Cellular polypeptides were labeled with [³⁵S]methionine for 1 hr at 4 hr pi. Labeled cellular proteins were analyzed by electrophoresis in an 8% polyacrylamide/4 M urea gel. Only the top portion of the gel is shown. PA is relatively decreased in lanes 3 and 5, and PB2 in lanes 4, 5, and 6. A host-cell protein (lane 8) closely co-migrates with PB1 in the DI preparations. From Akkina *et al.* (1984b).

with RNP and assembly of virus particles (Akkina *et al.*, 1984b). In the DI virus-co-infected cells, DI RNAs or their products might cause interference in one or more crucial steps involved in the biosynthesis and function of polymerase proteins (e.g., synthesis, transport, complex formation, exit from nucleus, and binding to RNP).

## C. Possible Role of DI Virus-Specific Polypeptides in Interference

Although DI virus-specific polypeptides have been detected in some DI virus preparations, their role, if any, in interference remains undetermined. The level of different standard proteins reflects the level of mRNAs rather than direct interference at the level of translation. At least some DI virus-specific proteins appear to be stable (Fig. 8) and are transported to the nucleus, where they form complexes with RNP (Fig. 9). Therefore, DI virus-specific proteins may interfere with the viral macromolecular synthesis, affecting the translation, transcription, and replication of standard viral RNAs. The polypeptides encoded by DI RNAs of polymerase genes would possess amino acid sequences such as those of polymerase proteins at the amino-terminal region up to the point of deletion, and might similarly bind to and thus block replication initiation sites.

Not every DI virus preparation produces detectable DI virus polypeptides (Chambers et al., 1984; Penn and Mahy, 1985), nor is there evidence that polypeptide-producing DI virus particles have an advantage in interference over other DI particles. All DI virus preparations we tested, whether expressing visible DI-specific polypeptides or not, possessed interfering activity. In experiments in which cells were doubly infected with equal DIU of two DI virus preparations, one of which produces DI virus-specific polypeptides, analysis of progeny particle RNA showed that the replication of the DI RNA producing a prominent DI virus polypeptide is not favored over the other DI RNA (Akkina et al., 1984a). Also, there is no evidence for the selection of DI RNAs with translation reading frames maintained in phase through the deletion point (Jennings et al., 1983). All DI RNAs are apparently transcribed into mRNAs that might produce DI virus-specific polypeptides whether detected or not, and these DI virus-specific polypeptides will possess at the amino-terminal region the same amino acid sequences as the standard polypeptide. It is quite possible that DI virus-specific polypeptides, whether detected by PAGE analysis or not, still may play an important role in interference. Hsu et al. (1985) detected putative DI virus-encoded polypeptides of Sendai virus; the importance of these polypeptides to the interference is unknown as well.

## D. Partial Reversal of DI Virus-Mediated Interference with Increasing Concentration of Standard Virus Particles

Earlier experiments have shown that when standard virus-infected MDBK cells were superinfected with DI virus particles at different times postinfection, the occurrence of interference was not restricted to the earliest stages of infection. For example, DI virus particles superinfected as late as 3 hr and 6 hr after standard virus infection inhibited the release of standard PFU by 80% and 50%, respectively (Nayak et al., 1978). Also, with influenza virus the relationship between the multiplicity of input DI particles and the yield of progeny standard virions is an inverse proportionality (Nayak et al., 1978), whereas with VSV it is a reciprocal exponential relationship (Bellett and Cooper, 1959). The relationship seen with influenza virus would support a model for interference based on

TABLE II. Dose Dependence of Standard Virus
on DI Virus-Mediated Interference[a,b]

| Infecting virus | | Progeny virus | | |
|---|---|---|---|---|
| DIU/cell | PFU/cell[c] | PFU/ml | HAU/ml | PFU/HAU ratio |
| 0 | 0.5 | $3.6 \times 10^8$ | 4096 | 90,000 |
| 4 | 0.01[d] | $1.1 \times 10^4$ | 1536 | 7 |
| 4 | 0.5 | $2.1 \times 10^4$ | 1536 | 14 |
| 4 | 1.0 | $4.3 \times 10^4$ | 2048 | 21 |
| 4 | 2.5 | $1.3 \times 10^5$ | 2048 | 63 |
| 4 | 6.25 | $3.6 \times 10^5$ | 3072 | 120 |

[a]From Akkina et al (1984b).
[b]MDBK cells were infected with standard virus and DI-Tobita. Progeny virus was harvested for the above assays at 24 hr p.i.
[c]Helper standard virus supplied.
[d]Contamination of DI preparation with residual standard virus.

competition among templates for the initiation of replication and transcription.

To determine further whether the increased multiplicity of standard virus would overcome DI virus-mediated interference (Akkina et al., 1984b), cells were co-infected with the same amount of DI particles and varying amounts of standard particles. The results (Table II) show that the DI-mediated interference with the standard virus multiplication was partially reversed with increasing amounts of standard virus and that the reversal was dose dependent. When standard virus was increased 12-fold, there was a seventeenfold increase in PFU/HAU ratio. Also, with increasing standard virus concentration, the intracellular synthesis of polymerase proteins became more equimolar as in standard virus-infected cells. These data suggest that interference by influenza DI particles is the outcome of competition for transcription and replication between the DI and standard viral genomes in the co-infected cell.

## E. Role of DI RNA in Virus Assembly

As influenza virus possesses a segmented genome, DI RNAs could affect the assembly and packaging of the different viral RNPs into virions. The mechanism of assembly and packaging of the eight different influenza virus gene segments into an infectious viral particle remains unknown. The genomic content of an individual influenza virion, standard or DI, is also unknown, except as the mean values of a heterogeneous population. Although all eight gene segments are required for infectivity, it is not known whether the eight standard RNA segments are incorporated into virions by a mechanism involving selective or random packaging. Because the DI RNAs are mutants of polymerase RNAs, it can be speculated that during the virus assembly process DI RNAs are somehow mistaken for polymerase RNAs and packaged in their place. Interference would result if the DI RNA were specifically to displace its progenitor gene from the progeny particle during assembly. Analyses of the genomic content of different DI preparations tend to favor this possibility since we and others have observed that with several DI preparations, the presence of major DI RNAs of; e.g., PB2 origin was correlated with specifically decreased levels of PB2 vRNA in that preparation (Akkina et al., 1984b; Nakajima et al., 1979). Such a mechanism would explain why a reduction of one or more specific polymerase genes is a characteristic of different influenza DI virus preparations. Furthermore, such a mechanism would suggest selective rather than random packaging in the assembly of a standard virion containing eight different RNA segments. Any competitive advantage of a given DI RNA in transcription and replication would cause a general reduction in the synthesis of standard RNA segments but would not explain the reduction of a specific polymerase gene segment in a given DI virus preparation. Smith and Hay (1982) also raised the possibility of a selective packaging mechanism, to explain the differences observed between the relative proportions of intracellular vRNAs and progeny particle vRNAs. However, whether this correlation would be a general rule in explaining the reduction of polymerase genes in influenza virus DI particles requires more careful analysis. Any advantage of the DI RNA operating during virus assembly may be critical to the perpetuation of influenza virus DI virus particles, because the competitive advantage based on size alone during transcription and replication might not be very effective in producing interference. Thus a detailed inquiry into the events leading to the genesis of DI virus particles may elucidate the process involved in assembly and budding of influenza virus particles as well as the mechanism of interference.

## F. Summary

The mechanism of interference with standard influenza virus replication by DI virus particles appears to be complex and may involve multiple steps. Both influenza virus DI RNAs as well as the DI virus-coded products (transcriptional or translational) may function in interference. Interference may occur either at the level of standard RNA replication and/or transcription since DI RNAs can fully function as both replicative and transcriptive templates. Interference at the level of transcription may also affect the production of critical enzymes such as polymerases. Furthermore, DI virus-specific proteins, because of their common amino terminal sequence, may affect the formation of functional polymerase complexes and thus, may further inhibit the total replication and transcription of viral RNAs. Finally, DI RNAs may interfere in the packaging and assembly process by selectively displacing their progenitor standard RNA segments from the virion. In all three steps, influenza virus DI RNA segments appear to function as competitors with standard RNA segments because of their smaller size but, unlike DI RNAs of other viruses, they most probably do not have an altered structure with an increased affinity for polymerase binding or replicating ability. Therefore, influenza DI virus-mediated interference as an intracellular process is rather weak and can be at least partly reversed with increasing multiplicity of standard viruses. The strength of the interference phenomenon is that DI RNAs seem to be generated easily during replication.

## VII. POSSIBLE ROLE OF INFLUENZA VIRUS DI PARTICLES IN MODULATING VIRAL PATHOGENESIS

DI virus particles are generated experimentally in the laboratory, using a controlled environment and an artificially large inoculum to provide a high multiplicity of infection. Consequently, two pertinent questions have been raised: Are DI particles produced in natural infections, and what is their role in viral pathogenesis, host response and viral evolution?

A great deal of experimental evidence, both in tissue culture and in experimental animals, as well as the ubiquity of DI virus particles, provide strong circumstantial evidence for the presence of DI particles in natural viral infections. Indeed, although rarely looked for, influenza virus strains containing subgenomic RNAs resembling DI RNAs have occasionally been isolated from natural infections (Sriram *et al.*, 1980; Bean *et al.*, 1985; Chambers and Webster, 1987). In a broad sense, DI particles represent only a small part of the total spectrum of variation observed in virus populations in nature. These variations include silent mutations without any noticeable effect on the biology of viruses, antigenic as well as host-range variations with profound effect on pathogenesis, and the generation of altered genomes, including DI RNAs, which are unable to replicate without helper viruses. Clearly, DI virus particles merely represent a group of variant viruses that lack portions of the genome encoding certain essential functions and therefore need the helper function of standard viruses for replication, but which, in turn, suppress the growth of the standard viruses. The forces that bring about such variation and that provide the selection pressure for the survival of variant viruses will greatly influence the nature of emerging variant viruses that can compete successfully. It is therefore important to determine whether the generation of DI particles is a rare or a rather common phenomenon, i.e., whether this phenomenon is restricted among some viruses or strains of viruses, or in some hosts such as cultured cells, or is rather a general phenomenon that may often occur in natural infections as well as in laboratory experiments.

First, DI particles are not restricted to any single group or strain of viruses. Since DI particles were first discovered over 40 years ago with influenza viruses, similar particles have been reported with all viruses, whether DNA or RNA viruses, or whether they contain single-stranded or double-stranded, positive sense or negative sense, unsegmented or segmented genomes, or whether they possess different biological properties: either lytic or nonlytic, oncogenic or nononcogenic (Huang, 1975; Huang and Baltimore, 1977; Perrault, 1981; Holland et al., 1980). Most highly transforming oncogenic RNA viruses are defective; i.e., they need the helper function of a nontransforming replication-competent virus, although these defective oncogenic viruses do not interfere with the growth of the helper replication-competent viruses.

Second, the generation of DI virus particles is not restricted to a group of hosts, although host cells play an important role in the generation and amplification of DI particles. Influenza virus DI virus particles have been produced in many permissive host cells such as embryonated chicken eggs as well as cultured cells (MDBK, MDCK, CEF) (von Magnus, 1951a–c, 1952; Fazekas de St. Groth and Graham, 1954a,b; Werner, 1956; Henle, 1953; Rott and Schafer, 1960, 1961; Carter and Mahy, 1982a; De and Nayak, 1980). Influenza virus DI particles have also been generated by infecting animals. Ginsberg (1954) reported the production of incomplete virus in mice by inoculating large doses of virus intranasally. Generation of DI particles both in cell cultures and in experimental animals have been reported for VSV, Sendai, reoviruses, rabies, lymphocytic choriomeningitis, and many other viruses (Huang, 1975; Huang and Baltimore, 1977; Holland et al., 1980). In short, any host system that supports the growth of a standard virus will also produce and amplify its DI particles. In fact, our studies show that the growth of DI and standard virus particles is not always mutually exclusive; rather, the growth of standard particles to a certain level helps in the optimum amplification of DI virus particles (De and Nayak, 1980).

Finally, low MOI does not prevent the generation of DI particles. It is believed that high MOI is only required for the amplification and not for the initial generation of DI virus particles (Cairns and Edney, 1952). Cells infected at very low MOI produce DI particles that are amplified in the neighboring cells during the formation of a plaque. Similarly, in naturally occurring influenza virus infections, although the initial infection occurs at a low MOI, subsequent infection of adjacent cells will occur at a high MOI. Since many cycles of infection are required before the virus can produce pathognomonic syndromes and lesions, DI virus particles generated during the early phase of infection are likely to be amplified in the neighboring cells co-infected with both standard and DI particles. These DI particles may exert their influence in modifying the outcome of the disease by interfering with the growth of the standard virus during the multicycle replication. Influenza virus DI particles are ubiquitous and are present in almost all influenza virus preparations. It is rather difficult to obtain a DI virus-free influenza virus stock and to maintain such a stock. Almost every stock of influenza A and B viruses we have examined contains visible DI RNA bands in gels and therefore, must contain a relatively large amount of DI virus particles (Janda *et al.*, 1979; De and Nayak, 1980). Clearly, our data show that DI virus particles are being continuously generated. The generation of influenza virus DI virus particles is therefore a common and natural phenomenon that almost always occurs during the replication of standard viruses. The structure of influenza virus RNA and/or the nature of the polymerase complex such as lack of editing function may be responsible for the frequent generation of DI RNAs. Therefore, although no definitive studies have been done to demonstrate and quantitate the amount of DI virus particles produced in influenza virus lesions during natural infection, DI particles are likely to be produced and amplified in individual animals or humans during the course of the disease even though the infection is transmitted at a very low multiplicity from person to person.

Although the role of DI particles in the outcome of the disease in natural infection remains undetermined, experimental evidence suggests that DI particles may have profound effects on viral pathogenesis. In cell culture, it is well known that DI particles ameliorate the effect of lytic virus by reducing cytopathic effect (Janda *et al.*, 1979; McLain *et al.*, 1988). Cells co-infected with DI virus particles can survive the cytolytic effect of standard viruses. Co-infection by DI virus particles drastically reduces the inhibitory effect of standard influenza virus on cellular protein synthesis (see Fig. 6).

Co-infection by DI particles will keep cells healthier for a longer period of time and produce fewer standard virus particles. In cell culture, the outcome of such a process may result in establishment of persistent viral infection (De and Nayak, 1980; Frielle *et al.*, 1984). In humans or animals, the effect of such a process on the outcome of the disease may be more complicated. Cave *et al.* (1985) recently reported results suggesting that in natural VSV infections DI viruses may modulate the levels of infectious virus in a host by a cycling relationship resembling the classical predator–prey relationship. Because of the healthier cells and fewer standard viruses, DI particles may limit or attenuate the disease process and favor recovery by stimulating host defenses (Gamboa *et al.*, 1975). Alternatively, such virus-infected cells may survive and evade the host immune mechanism and produce a chronic persistent viral disease (Holland *et al.*, 1980).

Undiluted A/PR/8/34 virus enriched in incomplete particles, when administered intranasally, resulted in a reduction in mortality in mice as well as decrease in virus titer in lungs as compared with the effect of standard virus produced at a low multiplicity (von Magnus, 1951b; Ginsberg, 1954; Horsfall, 1954, 1955). Similar results were reported by Doyle and Holland (1973), who also reported some reduction in viral pathogenicity with increased survival time and reduced virus titer in the lungs of mice co-infected with influenza virus DI particles. Rabinowitz and Huprikar (1979) studied the role of factors such as strain and age of mice as well as the ratio of DI particles to standard particles in the inoculum in an attempt to define further the role of DI particles in viral disease. These investigators showed that resistance to DI virus-enriched influenza virus was age dependent and varied with the mouse strain and that it was also dependent on the DI virus particle/standard particle ratio. With increased enrichment in DI particles, influenza virus grew to a lower titer and lungs contained reduced lesions as well as reduced amounts of viral antigen. In addition, mice inoculated with virus enriched in DI particles developed humoral immunity earlier in the course of the disease and to a higher antibody titer than those inoculated with standard virus, which Rabinowitz and Huprikar believed was the reason for the protective effect of DI particles.

Dimmock *et al.* (1986) reported that in C3H mice, WSN DI virus inoculated intranasally along with infectious virus protected mice against lethal infection. This DI virus-mediated protection, however, was not due to inhibition of WSN virus replication in the mouse lungs, which was unaffected. These workers found that DI virus caused the enhancement of local hemagglutination-inhibiting (HI) antibody in the lungs, but this antibody did not neutralize infectious virus. They speculated that DI virus might have ameliorated the disease state by suppressing the deleterious murine Td cell (delayed-type hypersensitivity) immune response to WSN infection, rather than by DI-mediated inhibition of standard WSN virus replication. The DI-induced production of a nonneutralizing antibody response and the relevance of this to the protective effect of DI particles in this system remain mysterious.

Chambers and Webster (1987) reported that DI virus particles are associated with a chicken influenza virus (A/chicken/Pennsylvania/83, H5N2) in nature. This virus was avirulent and contained subgenomic RNAs, whereas a closely related virus was highly virulent and lacked subgenomic RNAs. Although other evidence (Webster *et al.*, 1986) indicates that the DI particles alone were not responsible for the difference in virulence, they may have played a role in suppressing, for a time, the emergence of the virulent variant. Chickens co-infected with the avirulent DI-containing virus plus lethal doses of the virulent virus showed greatly reduced mortality, with evidence for inhibition of virulent virus multiplication before the expected onset of a specific immune response. It seems likely that the protective effect of the avirulent virus is due either to direct DI virus-mediated interference with virulent virus multiplication, or to the triggering of an early-onset nonspecific immune response. Although the mechanisms by which DI particles can be maintained in nature and can provide protection against virulent virus remain unclear, this report demonstrates that DI particles can exist in nature and may therefore have a crucial role in modifying virus infection in nature.

Clearly, DI virus particles can modulate host response in a number of ways. They can reduce growth of standard virus, help in mounting a specific immune response (Gamboa *et al.*, 1975; Rabinowitz and Huprikar, 1979; Dimmock *et al.*, 1986), and modulate expression of virus-specific antigen on cell surfaces (Welsh *et al.*, 1977). Furthermore, incomplete (or DI) influenza virus particles were found to be less toxic via either the intravenous or intracerebral route than the standard influenza virus (McKee, 1951; Manire, 1957; Schafer, 1955). Therefore, DI influenza virus particles would be better vaccines. Indeed, an ideal vaccine might consist of live viruses which, upon intranasal inoculation, would produce only incomplete (or DI) particles. Such a vaccine would not be expected to produce disease but would stimulate both local and systemic humoral antibodies as well as a cellular immune response for protection against infection.

## VIII. POSSIBLE ROLE OF DI PARTICLES IN VIRUS EVOLUTION

The importance of coexistence of DI particles in virus evolution either in experimental infection or in nature remains unknown. We assume that such a mechanism of generation and amplification of DI particles could not be built into the replication process of viruses and survive the evolutionary pressure, unless it were to provide an advantage in the evolution and survival of viruses. However, at first glance influenza virus DI particles do not appear to offer any advantage in the survival and evolution of the virus in nature. For example, any reduction of multiplication and pathogenesis of standard virus by DI virus particles would be beneficial to the host in recovery and the elimination of the virus, but clearly such an effect is not in the best interest of survival and propagation of the virus. Second, although the establishment of long-term persistent infections in humans or animals may be helpful to the survival of a virus and in enabling a virus to find another susceptible host for replication, influenza virus is not known to cause long-term persistent infections in humans or animals, except for the prolonged but asymptomatic virus replication observed in ducks and water fowl (Hinshaw *et al.*, 1980). Influenza virus infection in most animals and humans is of relatively short duration; the virus is eliminated from the host with recovery from the disease, although in an immunocompromised host the virus may persist for a prolonged period.

Third, influenza virus is maintained in the human population by evading the host immunity. It uses two mechanisms, antigenic shift and antigenic drift, to change its outer envelope proteins (HA and NA) and therefore, to escape the effect of host immunity. Antigenic shift is a much more drastic change that requires the acquisition of genes for the new envelope antigen(s). Antigenic shift is attributed to gene exchange with a nonhuman influenza strain or to the re-emergence of an older strain. DI virus particles are unlikely to have any major role in antigenic shift because of genetic reassortment except as speculated by Cane *et al.* (1987). They have reported that influenza RNAs including DI RNA can remain as nonreplicative forms for long periods of time in some cells and therefore may aid in the emergence of new virus variant upon infection with another influenza virus. However, long persistence of such non-replicative forms of influenza RNA in animals or humans has yet to be demonstrated.

By contrast, antigenic drift is a gradual change that involves mutations in the nucleic acid sequence and selection of appropriate mutants. Clearly, a number of factors, particularly host immunity, are involved in the selection of mutants responsible for the antigenic drift. Similarly, DI virus particles that interfere with the growth of homologous viruses may play a role in the selection of variants. Since the generation and perpetuation of DI particles is so common with influenza virus, it is possible that influenza virus replication is likely to be constantly subjected to the selective pressure exerted by the homologous DI particles. Therefore, DI virus particles along with immunological as well as other host and environmental factors may play an important role in the evolution of influenza viruses in nature and may aid in producing variant viruses. However, although DI virus particles have been reported to be involved in the establishment of persistent infection in cell culture (Holland et al., 1979, 1980, 1982; Rowlands et al., 1980; Weiss and Schlesinger, 1981; De and Nayak, 1980; Frielle et al., 1984) and have been implicated in the emergence of virus variants (Holland et al., 1982; Weiss and Schlesinger, 1981), there are no data on the effect of DI particles on the production of virus variants in acutely infected cells. The role of DI virus toward facilitating the emergence of virus variants in acutely infected cells or animals needs to be examined carefully.

Rott et al. (1983) reported an interesting observation which may be pertinent to the role of DI particles in virulence: Reassortant influenza viruses when passaged at high multiplicity at 41°C produced progeny viruses that were pathogenic for chickens, unlike the nonpathogenic parent viruses. These workers further showed that the gene constellations of the pathogenic viruses were the same as the parent reassortment viruses, indicating that point mutations in the parent gene(s) were responsible for the regaining of virulence. Although the mechanism of virulence reactivation remains unknown, it is possible that multicycle high-multiplicity infections, which yield DI virus particles, may have aided in the selection process. Similarly, in a natural outbreak of relatively avirulent avian influenza virus, selection pressure from DI particles may have been at least partly responsible for mutations leading to the emergence of a highly virulent variant virus (Bean et al., 1985; Chambers and Webster, 1987).

Since the influenza virus genome is segmented, it should be possible to determine the effect of a specific DI virus preparation on the changes observed in different genes and also determine which gene or gene constellation of a virus is responsible for the susceptibility or resistance of a standard virus to a given DI virus preparation. Such experiments would determine whether a specific DI virus particle produces a uniform selection pressure against all genes or whether some specific genes and gene products are more vulnerable to the interfering effect of a specific DI virus preparation.

In summary, since DI virus particles interfere with the growth of standard virus, they may exert selective pressure against the replication of homologous standard virus. Influenza virus DI particles, which are ubiquitous, may aid, along with other factors, in the generation and selection of variant viruses. In addition, the ability to generate DI particles may affect virulence.

## IX. SUMMARY AND CONCLUSION

Influenza virus DI particles are defective and interfering because they contain reduced amounts of standard viral gene segments, usually with a pronounced reduction in the level of one or more polymerase genes. This is the reason for their defective nature. In addition, the DI virus particles also contain novel small RNA segments, called DI RNAs. These DI RNAs are not required for the replication of standard viruses, and are derived from standard gene segments, predominantly from one of the polymerase genes, by one or more internal deletions. DI RNAs can undergo further evolution after they are generated producing other DI RNAs. Unlike most DI RNAs of nonsegmented negative-strand RNA viruses, influenza virus DI RNAs contain both the 5'- and 3'-termini of the parent RNAs. A possible model for the preferential generation of 5'–3' DI RNAs by rolling over of polymerase complex from one site of the template to another site of the same template is presented.

DI RNAs are specifically responsible for interference with standard viruses. DI virus-mediated interference is a complex, multistep process. DI RNAs primarily function as an effective competitor against the standard RNAs because of their smaller size. They can function as a template for both transcription and replication and are capable of producing poly(A)$^+$ mRNAs. In some preparations, DI virus-specific polypeptides, the translation products of DI viral transcripts, can be demonstrated. These properties of DI RNAs (i.e., replication, transcription, translation) are possibly involved in the mechanism of DI-mediated interference in influenza virus infection. In addition, RNPs containing these DI RNAs sometimes appear to interfere in the assembly process by exclusion of the RNPs containing progenitor RNAs. Such a mechanism would greatly enrich the proportion of DI virus particles in the final yield at the expense of standard particles and would imply a selective rather than random assembly of eight separate RNP segments to form standard infectious particles. Such an interference at the assembly level would be unique for viruses with segmented genomes. Therefore, both the generation of DI RNAs as well as the mechanism of interference by influenza virus DI particles appear to be different from that of the nonsegmented negative-strand RNA viruses.

Since DI virus particles can be generated in any host system that permits the growth of the standard virus, it is likely that they are generated in natural infections as well as in laboratory experiments. Their presence may serve to attenuate the pathogenic effects of the standard virus infection. Therefore, DI virus particles may prove useful as the basis of an effective influenza vaccine. Their occurrence may also be advantageous for the evolution of the standard virus and for the emergence of variant viruses.

ACKNOWLEDGMENTS. The work described in this review from the authors' laboratory was supported by grants R01 AI 12749 and R01 AI 16348 from the National Institute of Allergy and Infectious Diseases and by grant PCM 81-292 from the National Science Foundation. TMC was the recipient of training grants 2T32 CA09030 and 5T32 CA09030 from the National Cancer Institute and of a National Research Service Award, 1F32 AI06945-01, from the U.S. Public Health Service.

# REFERENCES

Ada, G. L., and Perry, B. T., 1955, Infectivity and nucleic acid content of influenza virus, *Nature (Lond.)* **175**:209--210.

Ada, G. L., and Perry, B. T., 1956, Influenza virus nucleic acid: Relationship between biological characteristics of the virus particle and properties of the nucleic acid, *J. Gen. Microbiol.* **14**:623--633.

Akkina, R. K., and Nayak, D. P., 1987, Interference by defective interfering influenza virus: Role of defective viral polypeptides in infected cells, in: *The Biology of Negative Strand Viruses* (B. W. J. Mahy and D. Kolakofsky, eds.), pp. 183–190, Elsevier, New York.

Akkina, R. K., Chambers, T. M., and Nayak, D. P., 1984a, Expression of defective-interfering influenza virus-specific transcripts and polypeptides in infected cells, *J. Virol.* **51**:395--403.

Akkina, R. K., Chambers, T. M., and Nayak, D. P., 1984b, Mechanism of interference by defective-interfering particles of influenza virus: Differential reduction of intracellular synthesis of specific polymerase proteins, *Virus Res.* **1**: 687--702.

Akkina, R. K., Chambers, T. M., Londo, D. R., and Nayak, D. P., 1987, Intracellular localization of cells expressing PB1 protein from cloned cDNA, *J. Virol.* **61**:2217--2224.

Amesse, L. S., Pridgen, C. L., and Kingsbury, D. W., 1982, Sendai virus DI RNA species with conserved virus genome termini and extensive internal deletions, *Virology* **118**:17--27.

Baric, R. S., Stohlman, S. A., Razavi, M. K., and Lai, M. M. C., 1985, Characterization of leader-related small RNAs in corona virus-infected cells: Further evidence of leader-primed mechanism of transcription, *Virus Res.* **3**:19--33.

Barrett, A. D. T., and Dimmock, N. J., 1986, Defective interfering viruses and infections of animals, *Curr. Top. Microbiol. Immunol.* **128**:55–84.

Bay, P. H. S., and Reichmann, M. E., 1982, *In vitro* and *in vivo* inhibition of primary transcription of vesicular stomatitis virus by a defective interfering particle, *J. Virol.* **41**: 172–182.

Bean, W. J., and Simpson, R. W., 1976, Transcriptase activity and genome composition of defective influenza virus, *J. Virol.* **18**:365–369.

Bean, W. J., Kawaoka, Y., Wood, J. M., Pearson, J. E., and Webster, R. G., 1985, Characterization of virulent and avirulent A/chicken/Pennsylvania/83 influenza A viruses: Potential role of defective interfering RNAs in nature, *J. Virol.* **54**:151–160.

Bellett, A. J. D., and Cooper, P. D., 1959, Some properties of the transmissible interfering component of vesicular stomatitis virus preparations, *J. Gen. Microbiol.* **21**:498–509.

Blumberg, B. M., and Kolakofsky, D., 1983, An analytical review of defective infections of vesicular stomatitis virus, *J. Gen. Virol.* **64:**1839-1847.

Buonagurio, D. A., Krystal, M., Palese, P., DeBorde, D. C., and Maassab, H. F., 1984, Analysis of an influenza A virus mutant with a deletion in the NS segment, *J. Virol.* **49:**418–425.

Cairns, H. J. F., and Edney, M., 1952, Quantitative aspects of influenza virus multiplication. I. Production of incomplete virus, *J. Immunol.* **69:**155–160.

Cane, C., McLain, L., and Dimmock, N. J., 1987, Intracellular stability of the interfering activity of a defective interfering influenza virus in the absence of virus multiplication. *Virol.* **159:**259–264.

Carter, M. J., and Mahy, B. W. J., 1982a, Incomplete avian influenza virus contains a defective noninterfering component, *Arch. Virol.* **71:**12–25.

Carter, M. J., and Mahy, B. W. J., 1982b, Incomplete avian influenza A virus displays anomalous interference, *Arch. Virol.* **74:**71–76.

Carter, M. J., and Mahy, B. W. J., 1982c, Synthesis of RNA segments 1–3 during generation of incomplete influenza A (fowl plague) virus, *Arch. Virol.* **73:**109–119.

Cave, D. R., Hendrickson, F. M., and Huang, A. S., 1985, Defective interfering virus particles modulate virulence, *J. Virol.* **55:**366–373.

Chambers, T. M., Akkina, R. K., and Nayak, D. P., 1984, *In vivo* transcription and translation of defective interfering particle-specific RNAs of influenza virus, in: *Segmented Negative Strand Viruses* (R. W. Compans and D. H. L. Bishop, eds.) pp. 85–91, Academic, Orlando, Florida.

Chambers, T. M., and Webster, R. G., 1987, Defective interfering virus associated with A/Chicken/Pennsylvania/83 influenza virus, *J. Virol.* **61:**1517–1523.

Chanda, P. K., Chambers, T. M., and Nayak, D. P., 1983, *In vitro* transcription of defective interfering particles of influenza virus produces poly(A) containing complementary RNAs, *J. Virol.* **45:**55–61.

Choppin, P. W., and Pons, M. W., 1970, The RNAs of infective and incomplete influenza virions grown in MDBK and HeLa cells, *Virology* **42:**603–610.

Chow, N., and Simpson, R. W., 1971, RNA-dependent RNA polymerase activity associated with virions and subviral particles of myxoviruses, *Proc Natl Acad Sci USA* **68:**752–756.

Crumpton, W. M., Dimmock, N. J., Minor, P. D., and Avery, R. J., 1978, The RNAs of defective interfering influenza virus, *Virology* **90:**370–373.

Crumpton, W. M., Clewley, J. P., Dimmock, N. J., and Avery, R. J., 1979, Origin of subgenomic RNAs in defective interfering influenza virus, *FEMS Microbiol. Lett.* **6:**431–434.

Crumpton, W. M., Avery, R. J., and Dimmock, N. J., 1981, Influence of the host cell on the genomic and subgenomic RNA content of defective interfering influenza virus, *J. Gen. Virol.* **53:**173–177.

Davis, A. R., and Nayak, D. P., 1979, Sequence relationships among defective interfering influenza viral RNAs, *Proc. Natl. Acad. Sci. USA* **76:**3092–3096.

Davis, A. R., Hiti, A. L., and Nayak, D. P., 1980, Influenza defective interfering viral RNA is formed by internal deletion of genomic RNA, *Proc. Natl. Acad. Sci. USA* **77:**215–219.

De, B. K., and Nayak, D. P., 1980, Defective interfering influenza viruses and host cells: Establishment and maintenance of persistent influenza virus infection in MDBK and HeLa cells, *J. Virol.* **36:**847–859.

Detjen, B. M., St. Angelo, C., Katze, M. G., and Krug, R. M., 1987, The three influenza viral polymerase (P) proteins not associated with viral nucleocapsids in the infected cells are in the form of a complex, *J. Virol.* **61:**16–22.

Dimmock, N. J., Beck, S., and McLain, L., 1986, Protection of mice from lethal influenza: Evidence that defective interfering virus modulates immune response and not virus multiplication, *J. Gen. Virol.* **67:**839–850.

Doyle, M., and Holland, J. J., 1973, Prophylaxis and immunization in mice by use of virus-free defective T particles to protect against intracerebral infection by vesicular stomatitis virus, *Proc. Natl. Acad. Sci. USA* **70:**2105–2108.

Duesberg, P. H., 1968, The RNAs of influenza virus, *Proc. Natl. Acad. Sci. USA* **59:**930–937.

Etkind, P. R., Buchhagen, D. L., Herz, C., Broni, B. B., and Krug, R. M., 1977, The segments of influenza viral mRNA, *J. Virol.* **22:**346–352.

Fazekas de St. Groth, S., and Graham, D. M., 1954a, The production of incomplete virus particles among influenza strains. Experiments in eggs, *Br. J. Exp. Pathol.* **35:**60–74.

Fazekas de St. Groth, S., and Graham, D. M., 1954b, Artificial production of incomplete influenza virus, *Nature (Lond.)* **173:**637–638.

Fields, S., and Winter, G., 1982, Nucleotide sequences of influenza virus segments 1 and 3 reveal mosaic structure of a small viral RNA segment, *Cell* **28:**303–313.

Frielle, D. W., Huang, D. D., and Youngner, J. S., 1984, Persistent infection with influenza A virus: Evolution of virus mutants, *Virology* **138:**103–117.

Gamboa, E. T., Harter, D. H., Daffy, P. E., and Hsu, K. C., 1975, Murine influenza virus encephalomyelitis. III. Effect of defective interfering particles, *Acta Neuropathol. (Berl.)* **34:**157–169.

Gard, S., and von Magnus, P., 1947, Studies on interference in experimental influenza. II. Purification and centrifugation experiments, *Arkiv. Kemi, Mineral Och Geol.* 24b No. 8, 1–4.

Gard, S., von Magnus, P., Svedmyr, A., and Birch-Andersen, A., 1952, Studies on the sedimentation of influenza virus, *Arch. Ges. Virusforsch.* **4:**591–611.

Ginsberg, H. S., 1954, Formation of non-infectious influenza virus in mouse lungs: Its dependence upon extensive pulmonary consolidation initiated by the viral inoculum, *J. Exp. Med.* **100:**581–603.

Henley, W., 1953, Multiplication of influenza virus in the entodermal cells of the allantois of chick embryo, *Adv. Virus Res.* **1:**141–227.

Herman, R. C., 1984, Nucleotide sequence of an aberrant glycoprotein mRNA synthesized by the internal deletion mutant of vesicular stomatitis virus, *J. Virol.* **50:**524–528.

Hinshaw, V. S., Bean, W. J., Webster, R. G., and Sriram, G., 1980, Genetic reassortment of influenza A viruses in the intestinal tract of ducks, *Virology* **102:**412–419.

Holland, J., Grabau, E. A., Jones, C. L., and Semler, B. L., 1979, Evolution of multiple genome mutations during long-term persistent infection by vesicular stomatitis virus. *Cell* **16:**495–504.

Holland, J. J., Kennedy, S. I. T., Semler, B. L., Jones, C. L., Roux, L., and Grabau, E. A., 1980, Defective interfering RNA viruses and the host cell response, in *Comprehensive Virology*, (H. Fraenkel-Conrat and R. R. Wagner, eds.), Vol. 16, pp. 137–192, Plenum, New York.

Holland, J., Spindler, K., Horodyski, F., Grabau, E., Nichol, S., and VandePol, S., 1982, Rapid evolution of RNA genomes, *Science* **215:**1577–1585.

Honda, A., Ueda, K., Nagata, K., and Ishihama, A., 1987, Identification of the RNA polymerase site on genome RNA of influenza virus. *J. Biochem.* **102:**1241–1249.

Horsfall, F. L., 1954, On the reproduction of influenza virus. Quantitative studies with procedures which enumerate infective and haemagglutinating virus particles, *J. Exp. Med.* **100:**135–161.

Horsfall, F. L., 1955, Reproduction of influenza viruses. Quantitative investigations with particle enumeration procedures on the dynamics of influenza A and B virus reproduction, *J. Exp. Med.* **102:**441–473.

Hsu, C-H, Re, G. G., Gupta, K. C., Portner, A., and Kingsbury, D. W., 1985, Expression of Sendai virus defective-interfering genomes with internal deletions, *Virology* **146:**38–49.

Hsu, M-T, Parvin, J. D., Gupta, S., Krystal, M., and Palese, P., 1987, Genomic RNAs of influenza viruses are held in a circular conformation in virions and in infected cells by a terminal panhandle, *Proc. Natl. Acad. Sci. USA* **84:**8140–8144.

Huang, A. S., 1975, Defective interfering viruses, *Annu. Rev. Microbiol.* **27:**101–117.

Huang, A. S., 1977, Viral pathogenesis and molecular biology, *Bacteriol. Rev.* **41:**811–821.

Huang, A. S., 1982, Significance of sequence rearrangements in a rhabdovirus, *ASM News* **48:**148–151.

Huang, A. S., and Baltimore, D., 1970, Defective viral particles and viral disease processes, *Nature (Lond.)* **226:**325–327.

Huang, A. S., and Baltimore, D., 1977, Defective interfering animal viruses, in: *Comprehensive Virology*, (H. Fraenkel-Conrat and R. R. Wagner, eds.), Vol. 10, pp 73–116, Plenum, New York.

Huang, A. S., Little, S. P., Oldstone, M. B. A., and Rao, D., 1978, Defective interfering particles: Their effect on gene expression and replication of vesicular stomatitis virus, in: *Persistent Viruses.* (J. G. Stevens, G. H. Todaro, and C. F. Fox, eds.), *ICN–UCLA Symposium on Molecular and Cellular Biology*, Vol VI, pp. 399–408, Academic, Orlando, Florida.

Janda, J. M., and Nayak, D. P., 1979, Defective influenza viral ribonucleoproteins cause interference, *J. Virol.* **32:**697–702.

Janda, J. M., Davis, A. R., Nayak, D. P., and De, B. K., 1979, Diversity and generation of defective interfering influenza virus particles, *Virology* **95:**48–58.

Jennings, P. A., Finch, J. T., Winter, G., and Robertson, J. S., 1983, Does the higher order structure of the influenza virus ribonucleoprotein guide sequence rearrangements in influenza viral RNA?, *Cell* **34:**619–627.

Jones, I. M., Reay, P. A., and Philpott, K. L., 1986, Nuclear location of all three polymerase proteins and a nuclear signal in polymerase PB2, *EMBO J.* **5:**2371–2376.

Kane, W. P., Pietras, D. F., and Bruenn, J. A., 1979, Evolution of defective-interfering double stranded RNAs of yeast killer virus, *J. Virol.* **32:**692–696.

Kaptein, J., and Nayak, D. P., 1982, Complete nucleotide sequence of the polymerase 3 (P3) gene of human influenza virus A/WSN/33, *J. Virol.* **42:**55–63.

King, A. M. Q., McCahon, D., Slade, W. R., and Newman, J. W. I., 1982, Recombination in RNA, *Cell* **29:**921–928.

Kingsbury, D. W., 1974, The molecular biology of paramyxoviruses, *Med. Microbiol. Immunol.* **160:**73–83.

Kolakofsky, D., 1976, Isolation and characterization of Sendai virus DI RNAs, *Cell* **8:**547–555.

Lazzarini, R. A., Keene, J. D., and Schubert, M., 1981, The origin of defective interfering particles of the negative strand RNA viruses, *Cell* **26:**145–154.

Lenard, J., and Compans, R. W., 1975, Polypeptide composition of incomplete influenza virus grown in MDBK cells, *Virology* **65:**418–426.

Leppert, M., Rittenhouse, L., Perrault, J., Summers, D., and Kolakofsky, D., 1979, Plus and minus strand leader RNAs in negative strand virus-infected cells, *Cell* **18:**735–747.

Lief, F. S., Fabiyi, A., and Henle, W., 1956, The decreased incorporation of S antigen into elementary bodies of increasing incompleteness, *Virology* **2:**782–797.

Manire, G. P., 1957, Studies on the toxicity for mice of incomplete influenza virus, *Acta Pathol. Microbiol. Scand.* **40:**501–510.

McCauley, J. W., and Mahy, B. W. J., 1983, Structure and function of the influenza virus genome, *Biochem. J.* **211:**281–294.

McLain, L., Armstrong, S. J., and Dimmock, N. J., 1988, One defective interfering particle per cell prevents influenza virus-mediated cytopathology: An efficient assay system, *J. Gen. Virol.* **69:**1415–1419.

McKee, A. P., 1951, Non-toxic influenza virus, *J. Immunol.* **66:**151–167.

Meier-Ewart, H., and Compans, R. W., 1974, Time course of synthesis and assembly of influenza virus proteins, *J. Virol.* **14:**1083–1091.

Meier, E., Harmison, G. G., Keene, J. D., and Schubert, M., 1984, Sites of copy choice replication involved in generation of vesicular stomatitis virus defective-interfering particle RNAs, *J. Virol.* **51:**515–521.

Mills, D. R., Peterson, R. I., and Spiegelman, S., 1967, An extracellular Darwinian experiment with a self-duplicating nucleic acid molecule, *Proc. Natl. Acad. Sci. USA* **58:**217–224.

Moss, B. A., and Brownlee, G. G., 1981, Sequence of DNA complementary to a small RNA segment of influenza virus A/NT/60/68, *Nucleic Acids Res* **9:**1941–1947.

Nakajima, K., Ueda, M., and Sugiura, A., 1979, Origin of small RNA in von Magnus particles of influenza virus, *J. Virol.* **29:**1142–1148.

Nayak, D. P., 1969, Influenza viruses: Structure, replication and defectiveness, *Fed. Proc.*
**28:**1858–1865.

Nayak, D. P., 1972, Defective virus RNA synthesis and production of incomplete influenza
virus in chick embryo cells, *J. Gen. Virol.* **14:**63–67.

Nayak, D. P., 1980, Defective interfering influenza viruses, *Annu. Rev. Microbiol.* **34:**619–
644.

Nayak, D. P., and Sivasubramanian, N., 1983, The structure of the influenza defective
interfering (DI) RNAs and their progenitor genes, in: *Genetics of Influenza Viruses,* (P.
Palese and D. W. Kingsbury, eds.), pp. 255–279, Springer-Verlag, Vienna.

Nayak, D. P., Tobita, K., Janda, J. M., Davis, A. R., and De, B. K., 1978, Homologous
interference mediated by defective interfering influenza virus derived from a temperature-sensitive mutant of influenza virus, *J. Virol.* **28:**375–386.

Nayak, D. P., Davis, A. R., and Cortini, R., 1982a, Defective interfering influenza viruses:
Complete sequence analysis of a DI RNA, in: *Genetic Variation Among Influenza
Viruses,* (D. P. Nayak, ed.), pp. 77–92, Academic, Orlando, Florida.

Nayak, D. P., Sivasubramanian, N., Davis, A. R., Cortini, R., and Sung, J., 1982b, Complete
sequence analyses show that two defective interfering influenza viral RNAs contain a
single internal deletion of a polymerase gene, *Proc. Natl. Acad. Sci. USA* **79:**2216–2220.

Nayak, D. P., Chambers, T. M., and Akkina, R. K., 1985, Defective interfering RNAs of
influenza viruses: Origin, structure, expression and interference, *Curr. Top. Microbiol.
Immunol.* **114:**104–151.

Penn, C. R., and Mahy, B. W. J., 1984, Expression of influenza virus subgenomic virion
RNAs in infected cells, in: *Segmented Negative Strand Viruses* (R. W. Compans and D.
H. L. Bishop, eds.), pp. 173–178, Academic, Orlando, Florida.

Penn, C. R., and Mahy, B. W. J., 1985, Novel polypeptides encoded by influenza virus
subgenomic (DI-type) virion RNAs, *Virus Res.* **3:**311–321.

Perrault, J., 1981, Origin and replication of defective interfering particles, *Curr. Top. Microbiol. Immunol.* **93:**151–207.

Perrault, J., Semler, B. W., Leavitt, R. W., and Holland, J. J., 1978, Inverted complementary
terminal sequences in defective interfering particle RNAs of vesicular stomatitis virus
and their possible role in autointerference, in: *Negative Strand Viruses and the Host
Cell* (B. W. J. Mahy and R. D. Barry, eds.), pp. 527–538, Academic, Orlando, Florida.

Perrault, J., Clinton, G. M., and McClure, M. A., 1983, RNP template of vesicular stomatitis
virus regulates transcription and replication functions, *Cell* **35:**175–185.

Pons, M. W., 1980, The genome of incomplete influenza virus, *Virology,* **100:**43–52.

Pons, M., and Hirst, G. K., 1969, The single and double-stranded RNAs and the proteins of
incomplete influenza virus, *Virology* **38:**68–72.

Rabinowitz, S. G., and Huprikar, J., 1979, Influence of defective-interfering particles of the
PR8 strain of influenza A virus on the pathogenesis of pulmonary infection in mice, *J.
Infec. Dis.* **140:**305–315.

Rao, D. D., and Huang, A. S., 1982, Interference among defective interfering particles of
vesicular stomatitis virus, *J. Virol.* **41:**210–221.

Re, G. G., and Kingsbury, D. W., 1988, Paradoxical effect of Sendai virus DI RNA size on
survival: Inefficient envelopment of small nucleocapsids, *Virology* **165:**331–337.

Re, G. G., Morgan, E. M., and Kingsbury, D. W., 1985, Nucleotide sequences responsible for
generation of internally deleted Sendai virus defective interfering genomes, *Virology*
**146:**27–37.

Robertson, J. S., Schubert, M., and Lazzarini, R. A., 1981, Polyadenylation sites for influenza
virus mRNA, *J. Virol.* **38:**157–163.

Rott, R., and Schafer, W., 1960, Untersuchungen uber die hamagglutinierenden nichtinfek-
tiosen Teilchen der Influenza-Viren. I. Die Erzeugung von "inkompletten Formen"
beim Virus der klassischen Geflugelpest (v. Magnus-Phanomen), *Z. Naturforsch.* **15b:**
691–693.

Rott, R., and Schafer, W., 1961, Untersuchungen uber die hamagglutinierenden nichtinfek-tiosen Teilchen der Influenza-Viren (German), *Z. Naturforsch.* **16b:**310–321.

Rott, R., and Scholtissek, C., 1963, Investigations about the formation of incomplete forms of fowl plague virus, *J. Gen. Microbiol.* **33:**303–312.

Rott, R., Orlich, M., and Scholtissek, C., 1983, Pathogenicity reactivation of nonpathogenic influenza virus recombinants under von Magnus conditions, *Virology* **126:**459–465.

Rowlands, D., Grabau, E., Spindler, K., Jones, C., Semler, B., and Holland, J., 1980, Virus protein changes and RNA termini alterations evolving during persistent infection, *Cell* **19:**871–880.

Schafer, W., 1955, Sero-immunologic studies on incomplete forms of the virus of classical fowl plague, *Arch. Exp. Vet. Med.* **9:**218–230 [in German].

Schubert, M., Harmison, G. G., and Meier, E., 1984, Primary structure of vesicular stom-atitis virus polymerase (L) gene: Evidence for a high frequency of mutations, *J. Virol.* **51:**505–514.

Shapiro, G. I., Gurney, T., and Krug, R. M., 1987, Influenza virus gene expression: Control mechanism at early and late times of infection and nuclear-cytoplasmic transport of virus-specific RNAs, *J. Virol.* **61:**764–773.

Sivasubramanian, N., and Nayak, D. P., 1982, Sequence analysis of the polymerase 1 gene and the secondary structure prediction of polymerase 1 protein of human influenza virus A/WSN/33, *J. Virol.* **44:**321–329.

Sivasubramanian, N., and Nayak, D. P., 1983, Defective interfering influenza RNAs of polymerase 3 gene contain single as well as multiple internal deletions, *Virology* **124:**232–237.

Smith, G. L., and Hay, A. J., 1982, Replication of the influenza virus genome, *Virology* **118:**96–108.

Smith, G. L., Levin, J. Z., Palese, P., and Moss, B., 1987, Synthesis and location of ten influenza virus polypeptides individually expressed by recombinant vaccinia virus, *Virology* **160:**336–345.

Sriram, G., Bean, W. J., Hinshaw, V. S., and Webster, R. G., 1980, Genetic diversity among avian influenza viruses, *Virology* **105:**592–599.

Ueda, M., Nakajima, K., and Sugiura, A., 1980, Extra RNAs of von Magnus particles of influenza virus cause reduction of particular polymerase genes, *J. Virol.* **34:**1–8.

von Magnus, P., 1947, Studies on interference in experimental influenza. I. Biological obser-vations, *Ark. Kemi. Mineral, Geol.* **24b:**1.

von Magnus, P., 1951a, Propagation of the PR8 strain of influenza virus in chick embryos. I. The influence of various experimental conditions on virus multiplication, *Acta Pathol. Microbiol. Scand.* **28:**250–277.

von Magnus, P., 1951b, Propagation of the PR8 strain of influenza virus in chick embryos. II. The formation of "incomplete" virus following the inoculation of large doses of seed virus, *Acta Pathol. Microbiol. Scand.* **28:**278–293.

von Magnus, P., 1951c, Propagation of the PR8 strain of influenza virus in chick embryos. III. Properties of the incomplete virus produced in serial passages of undiluted virus, *Acta Pathol. Microbiol. Scand.* **29:**157–181.

von Magnus, P., 1952, Propagation of the PR8 strain of influenza virus in chick embryos. IV. Studies on the factors involved in the formation of incomplete virus upon serial passage of undiluted virus, *Acta Pathol. Microbiol. Scand.* **30:**311–335.

von Magnus, P., 1954, Incomplete forms of influenza virus, *Adv. Virus Res.* **2:**59–78.

von Magnus, P., 1965, The *in ovo* production of incomplete virus by B/Lee and A/PR8 influenza viruses, *Arch. Ges. Virusforsch.* **17:**414–423.

Webster, R. G., Kawaoka, Y., and Bean, W. J., Jr., 1986, Molecular changes in A/chick-en/Pennsylvania/83 (H5N2) influenza virus associated with acquisition of virulence, *Virology* **149:**165–173.

Weiss, B., and Schlesinger, S., 1981, Defective interfering particles of Sindbis virus do not interfere with homologous virus obtained from persistently infected BHK cells but do interfere with Semliki Forest virus, *J. Virol.* **37:**840–844.

Welsh, R. M., Lampert, P. W., and Oldstone, M. B. A., 1977, Prevention of virus-induced cerebellar disease by defective interfering lymphocytic choriomeningitis virus, *J. Infect. Dis.* 136:391–399.

Werner, G. H., 1956, Quantitive studies on influenza virus infection of the chick embryo by the amniotic route, *J. Bacteriol.* 71:505–515.

Winter, G., Fields, S., and Ratti, G., 1981, The structure of the two subgenomic RNAs from human influenza virus A/PR/8/34, *Nucleic Acids Res.* 9:6907–6915.

Yang, F., and Lazzarini, R. A., 1983, Analysis of the recombination event generating a vesicular stomatitis virus deletion defective interfering particle, *J. Virol.* 45:766–772.

Yoshishita, T., Kawai, K., Fukai, K., and Ito, R., 1959, Analysis of infective particles in incomplete virus preparations of the von Magnus type, *Biken J.* 2:25.

CHAPTER 7

# Variation in Influenza Virus Genes
## Epidemiological, Pathogenic, and Evolutionary Consequences

FRANCES I. SMITH AND PETER PALESE

## I. INTRODUCTION

Influenza viruses can be classified into three types: A, B, and C. Strains of the same type share serologically cross-reactive matrix (M) proteins and nucleoproteins (NP), which are the major components of the virus. Recent technological advances have resulted in the cloning and sequencing of genes of many members of each type. The results indicate that the genetic information of influenza viruses is not static and that the amount and pattern of variation differs for the three types of influenza virus. The purpose of this chapter is to review the variation seen within the different types, with an emphasis on human influenza viruses. Also, we consider factors that may contribute to this variation and attempt to correlate the differences seen among the three types with differences in the biological properties and epidemiology of these viruses. Finally, an attempt is made to compare variation of influenza viruses with that seen in other RNA viruses.

## II. INFLUENZA VIRUSES IN HUMANS

The epidemiology of influenza A viruses has been well characterized owing to the association of these viruses with pandemic outbreaks of

FRANCES I. SMITH AND PETER PALESE • Department of Microbiology, Mount Sinai School of Medicine, New York, New York 10029.

disease in humans. The pandemics usually result from the appearance of a new subtype strain (antigenic shift) containing a novel hemagglutinin (HA) and/or neuraminidase (NA) that is immunologically different from that of previous circulating isolates. During this century, there appear to have been at least three occasions when new subtypes have been introduced in the human population. In 1918, a severe pandemic occurred which resulted in the death of an estimated 20 million people worldwide. The "new" viruses responsible for this widespread mortality are serologically related to viruses isolated from pigs. Consequently, these strains are referred to as the swine viruses or H1 subtype strains. Viruses of the H1 subtype continued to circulate for approximately 40 years until a second subtype shift occurred in 1957. It should be noted that the H1 strains did accumulate more subtle changes in antigenic character (antigenic drift) over the 40-year period of their circulation, so that strains isolated at the beginning of the H1 period were different from those H1 viruses isolated at the end of the subtype period. In 1957, viruses of the H2 subtype (Asian strains) appeared whose HA proteins showed little or no cross-reactivity with their counterparts in the H1 strains. For the next 11 years, H2 strains spread and gradually changed until the introduction of the new H3 subtype (Hong Kong strains) in 1968.

In 1977, a unique event occurred with respect to influenza virus epidemiology. Viruses containing surface antigens of the previous subtype, H1, reappeared and these viruses (Russian strains) cocirculated with the H3 strains prevalent at the time. The "new" H1 strains were shown to be serologically and genetically similar to H1 strains which had circulated around 1950 (Nakajima *et al.*, 1978; Kendal *et al.*, 1978; Scholtissek *et al.*, 1978b). It is unusual that the new H1 viruses did not replace the H3 strains introduced in 1968, but continue to coexist with them.

It should be noted, however, that the ratio of H1 to H3 virus isolates changes dramatically from year to year in the United States. For example, during the 1978–1979 and 1986–1987 winters, 98% and 99.3%, respectively, of all influenza virus isolates were H1 strains (Fig. 1). By contrast, 97% of all influenza virus isolates in 1984–1985 were of the H3 subtype. If one assumes that the number of virus isolates in a season reflects the prevalence in the population, the influenza virus H1 and H3 subtypes appear to be cyclical in nature over the past 10 years. Reports from other parts of the world confirm this cyclical pattern of influenza viruses. For example, in Japan most influenza cases during the 1983–1984 and 1984–1985 season were caused by H1 and type B viruses, respectively (Nakajima *et al.*, 1988). Interestingly, during the latter winter, the predominant strains in the United States were subtype 3 viruses rather than the type B viruses observed in Japan.

The epidemiology of influenza B and C viruses differs from that of influenza A viruses. Influenza B viruses usually cause less severe outbreaks of respiratory infection in man than do the A type viruses. However, the B viruses are more frequently implicated in the development of

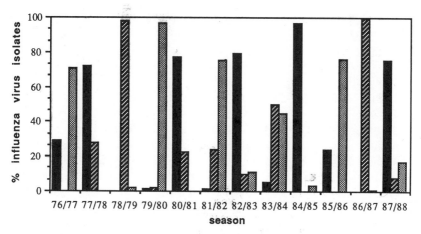

FIGURE 1. Isolation of influenza viruses in the United States, 1976–1988. Reports made to the Centers for Disease Control were collected and analyzed. The total number of influenza A and B virus isolates varied from 625 (during the 1981–1982 winter) to 2532 (during the 1987–1988 winter) per season. It is assumed that the number of isolates reported reflects the distribution in the general population. Virus isolations were reported from 49 states and the District of Columbia. The ordinate indicates H1 (▨), H3 (■), and B(▩) influenza virus isolates represented as percentage of the total number of influenza viruses identified per season. Data are from *Morbidity and Mortality Weekly Report* 37:497–503, 1988.

Reye's syndrome in children than are the A viruses (Corey *et al.*, 1976). This disease is rare, with only 0.37 and 0.88 cases per 100,000 children annually in the United States, but it is associated with a high fatality estimated to be between 22–42% (Hurwitz *et al.*, 1982). Influenza B viruses have been isolated since 1940, and there is no evidence to suggest that subtypes exist. However, there is evidence in support of antigenic change of B viruses over time (Krystal *et al.*, 1983; Bao-lan *et al.*, 1983; Oxford *et al.*, 1984; Yamashita *et al.*, 1988b). Again, as in the case of the influenza A viruses, influenza B viruses appear to go through cycles of prevalence (Fig. 1b). In some years, very few B virus isolates are obtained (e.g., 1984–1985), whereas in other years most influenza cases can be attributed to influenza B (e.g., 1985–1986).

Influenza C viruses, first isolated from man in 1947, generally cause only infrequent outbreaks of mild respiratory illness (Katagiri *et al.*, 1983). Nevertheless, a very high percentage of the young adult population possesses antibodies to these viruses (O'Callaghan *et al.*, 1980), indicating that infection occurs at a significant rate. Serological studies using hemagglutination inhibition assays of C viruses isolated over time in different geographical locations have shown that antigenic variation does occur during isolates, even though hemagglutinin subtypes of C viruses have not been observed (Chakraverty, 1978; Guo *et al.*, 1983).

In contrast to influenza A viruses, which have natural reservoirs in pigs, horses, and birds, the B and C viruses are mainly isolated from humans.

A report of influenza C virus isolations from pigs in Beijing, China, suggests however that pigs may serve as an animal reservoir for these viruses (Guo *et al.*, 1983).

## III. HOMOLOGY OF INFLUENZA A, B, AND C VIRUSES

Influenza A and B virions contain eight RNA segments, and it has been determined which RNA segment codes for which viral protein (for review, see Palese, 1977; Lamb, 1983; Air and Compans, 1983). By contrast, influenza C virions contain only seven RNA species, and only four of these—the HE, NP, M, and NS genes—have been sequenced and assigned as coding for a given viral protein(s). Although genetic exchange (reassortment) has not been observed among influenza viruses belonging to different serotypes, analysis of available data provides sufficient evidence to suggest that these three virus types are evolutionarily related.

### A. HA/HE Genes

The complete nucleotide sequence of the HA gene of influenza virus B/Lee/40 and its deduced amino acid sequence have been compared to those of influenza virus A/PR/8/34 (Krystal *et al.*, 1982). The B virus HA gene is 104 bases longer and codes for a slightly larger HA molecule. However, there are areas of identity at the nucleotide and amino acid level, and there are structural features common to these HAs. They include a hydrophobic signal peptide, hydrophobic $NH_2$- and COOH-terminals of the HA2 subunit, and an HA1/HA2 cleavage site involving an arginine residue. Alignment of the B/Lee/40 and A/PR/8/34 genes reveals that 24% of amino acids in HA1 subunits and 39% of amino acids in the HA2 subunits are conserved. The degree of relatedness between type B virus and type A virus HAs is similar to that observed among certain type A virus HAs (Table I), suggesting a close evolutionary relationship between the HAs of type A and type B influenza viruses.

In contrast with the above results, it is difficult to detect nucleotide or amino acid sequence identities between the influenza C virus surface glycoprotein referred to as HE (Vlasak *et al.*, 1988) or HEF protein (Herrler *et al.*, 1988) and the A and B virus HAs (Nakada *et al.*, 1984a). The former also possesses acetyl esterase activity, which is not found with either A or B virus HAs (Herrler, *et al.*, 1985; Vlasak *et al.*, 1988). However, there is significant structural similarity among the three surface glycoprotein molecules as a consequence of conserved hydrophobic domains and cysteine residues. Furthermore, acid-induced fusion is promoted by influenza A, B, and C viruses; this activity is most likely mediated by the HA/HE proteins, as is agglutination of erythrocytes. Thus, structural and functional similarities among the HA/HE surface glycoproteins of the

TABLE I. Sequence Identities of HA Genes
from Different Influenza Viruses

| | Nucleotide conservation[a] (%) | | Amino acid conservation[a] (%) | |
|---|---|---|---|---|
| Comparison[b] | HA1 | HA2 | HA1 | HA2 |
| H1 vs. H2 | 61 | 72 | 58 | 79 |
| H1 vs. H3 | 45 | 58 | 35 | 53 |
| H2 vs. H3 | 45 | 57 | 36 | 50 |
| B vs. H1 | 36 | 48 | 24 | 39 |

[a]Calculated as numbers of identical residues/total number of residues. The total number of residues is the average of the length of the two sequences. The signal peptide sequences are not used for this comparison. (From Krystal et al., 1982.)
[b]H1, A/PR/8/34; H2, A/Jap/305/57; H3, A/Aichi/2/68; and B, B/Lee/40. The H1, H2, and H3 sequences are from Winter et al. (1981); Gething et al. (1980); and Verhoeyen et al. (1980), respectively.

influenza A, B, and C viruses strongly suggest a common evolutionary ancestor for these genes.

## B. NP Genes

RNA segment 5 of influenza A, B, and C viruses codes for the nucleoprotein (NP), which is the major structural protein component of the nucleocapsid. Comparison of the amino acid sequence of the influenza A/PR/8/34 NP with that of influenza B/Singapore/222/79 virus (Londo et al., 1983) revealed a 37% identity over the entire length of the coding region and a similar distribution of basic amino acids. In addition, there exists a domain (amino acids 339–427) that shows an even higher degree of identity (Table II). A similar degree of sequence identity between A and B virus NPs was found using the NP sequence of B/Lee/40 virus (Briedis and Tobin, 1984). By contrast, comparison of the influenza C/Cal/78 NP (Nakada et al., 1984b) gene with those of influenza A and B viruses showed little sequence identity over the entire coding sequence (Table II). However, a significant degree of sequence conservation between A and C and B and C NPs (25% and 37%, respectively) was found within the central regions of the molecules (Table II). The results suggest that these residues are conserved because of functional requirements in protein–protein and/or protein–nucleic acid interactions.

## C. M Genes

RNA segment 7 codes for the M1 proteins of influenza A and B viruses, whereas the homologous protein of influenza C viruses is coded for by RNA 6. The M1 proteins of influenza A and B viruses possess

TABLE II. Amino Acid Sequence Identity among
Regions of Influenza A, B, and C Virus NPs

| Amino acid region | % Identity | | |
|---|---|---|---|
| | A and B | A and C | B and C |
| 1–338 | 39.4 | 9.6 | 9.7 |
| 339–427 | 43.0 | 24.7 | 37.1 |
| 428–565 | 22.0 | 16.3 | 17.0 |
| Total | 37.4 | 13.6 | 15.7 |

The numbering system and amino acid alignment are as shown in Nakada
*et al.* (1984b). The division into three domains highlights the homology
between A and C, and between B and C virus nucleoproteins. (From
Nakada *et al.* (1984b.)

approximately 25–30% sequence identity on the amino acid level, which
is compatible with similar functions (Briedis *et al.*, 1982). Between A and
C type and between B- and C-type M proteins, identities are only about
20%. The conserved amino acids are spaced throughout the molecules,
but no stretch of more than three amino acids is conserved between A-
and C- or B- and C-type M proteins. In addition, hydrophobicity patterns
of the M proteins are not superimposable (Yamashita *et al.*, 1988a).
Nevertheless, there is a clear evolutionary relationship between the M
genes of influenza C virus and those of influenza A and B viruses. For
example, eight different regions in the influenza C virus M protein are
homologous (three of four amino acids identical) with sequences in the
influenza A or B virus M proteins.

It is interesting that the splicing pattern of the influenza C virus M
gene is so different from that of influenza A and B viruses. With respect to
the latter virus types, the matrix proteins are coded for by unspliced
mRNAs; a spliced mRNA, which is not the major M gene-specific species, directs the synthesis of the M2 protein in influenza A virus-specific
cells (Briedis *et al.*, 1982; Lamb and Choppin, 1981). This contrasts with
the situation in influenza C viruses, in which the major mRNA species is
spliced and appears to code for the structural matrix protein (Yamashita
*et al.*, 1988a). There is no evidence to suggest that a second M protein is
synthesized in either influenza B or C viruses.

## D. NS Genes

In all three types of influenza virus, the shortest RNA segment codes
for two nonstructural (NS) proteins: NS1 and NS2. In all cases, the amino
acid $NH_2$-terminal coding region of NS1 and NS2 are shared, but the
COOH-terminal coding region of the NS2 is generated by a splicing event
which results in a frame shift of the mRNA molecule (for review, see
Lamb, 1983; Nakada *et al.*, 1986). There is little sequence identity among

influenza A, B and C viruses at either the nucleotide or amino acid sequence level for either the NS1 or NS2 coding regions. However, at the structural level the hydrophilicity plot of the NS2 proteins does reveal many similarities. These data suggest a possible common function for these nonstructural proteins.

## E. Conclusion

In addition to the evidence provided by the above results, the close evolutionary relatedness of influenza A and B virus genes is corroborated by new sequence data on the polymerase (P) genes of influenza B virus (Kemdirim et al., 1986; Akoto-Amanfu et al., 1987; de Borde et al., 1988). These genes code for proteins that are homologous to those of influenza A virus. No information is yet available for the P genes of influenza C viruses. Although influenza A, B, and C viruses appear to derive from a common ancestor, it is clear that sequence and structural similarities are much greater between influenza A and B viruses than between either A and C or B and C viruses. Since the evolutionary rate of influenza C virus appears to be slower than that of the A and B viruses, it seems likely that the divergence of influenza C viruses occurred before that of the influenza A and B viruses.

## IV. VARIATION IN INFLUENZA A VIRUSES

### A. Reassortants

#### 1. Reassortment of Genes Coding for Surface Proteins

Extensive changes in the HA and/or NA that lead to new subtypes of human influenza viruses are known as antigenic shifts. Sequence comparisons among the H1, H2, and H3 HA genes (Winter et al., 1981; Krystal et al., 1982) and between the N1 and N2 NA genes (Markoff and Lai, 1982; Colman and Ward, 1985) show such low levels of nucleotide homology that it is highly unlikely that these different HAs or NAs arose from one another by accumulation of point mutations during a brief period of time. Rather, it is believed that the appearance of novel surface antigens of human influenza viruses are the result of reassortment (recombination) events between previously circulating human viruses and influenza viruses of animal origin. There is ample evidence supporting reassortment between human viruses and avian strains in vivo (Webster et al., 1971), between avian strains in nature (Desselberger et al., 1978), and among human viruses in nature (Young and Palese, 1979; Bean et al., 1980; Cox et al., 1983). In addition to the three human HA subtypes (H1, H2, H3) and two NA subtypes (N1, N2), there are ten antigenically distinct HAs and seven NAs which have been identified in influenza viruses isolated from horses, pigs or birds. It is conceivable that any of these foreign HA and NA genes could be incorporated into a human virus via reassortment, resulting in a new human antigenic subtype.

The best evidence for reassortment generating new human subtypes involves the origin of the H3 HA. The Hong Kong H3N2 strain of influenza has been shown to be a reassortant (Laver and Webster, 1973). This virus contains the NA and all other genes from an Asian (H2N2) strain and an HA that is antigenically related to that of the A/duck/Ukraine/63 (H3N8) and A/equine/2/Miami/63 (H3N8) viruses (Schulman and Kilbourne, 1969; Fang et al., 1981; Ward and Dopheide, 1981a,b). The amino acid sequence identity between the HAs of the A/duck/Ukraine/63 and A/Aichi/2/68 (H3N2) viruses is 96%, suggesting that a ducklike strain donated the HA gene to the H3 virus.

### 2. Reassortment of Genes Coding for Nonsurface Proteins

Reassortment of nonsurface protein genes among cocirculating strains can lead to genetic diversity among strains of the single subtype (Young and Palese, 1979; Bean et al., 1980; Palese and Young, 1982; Cox et al., 1983). A well documented case involves strains of the reemerged H1 subtype that circulated in the winter of 1978–1979 in California (Young and Palese, 1979). Genetic analysis of these strains revealed that they were quite different from H1 viruses isolated in the previous year. A more detailed examination of the prototype A/Cal/10/78 H1 isolate using oligonucleotide mapping of isolated genes along with analysis of infected-cell polypeptides by sodium dodecyl sulfate–polyacrylamide gel electrophoresis (SDS–PAGE), suggested that the virus arose by reassortment between an earlier H1 virus and a co-circulating H3 parent. The A/Cal/10/78 strain derives its HA, NA, M and NS genes from the H1 parent and its NP and three P genes are from an H3 strain.

In addition, a second type of reassortant H1N1 virus has been described (Palese and Young, 1982). The prototype strain of this type is the A/Aberdeen/v1340/78 virus. This virus contains five H3-derived genes and an HA, NA, and NS gene from an A/USSR/90/77-like H1 strain. These data illustrate that reassortment among coexisting strains is another mechanism that can explain diversity within a subtype. Furthermore, biochemical evidence has been provided in favor of recombination among avian strains leading to new virus variants in nature (Desselberger et al., 1978; Hinshaw et al., 1980).

### B. Animal Viruses Become Virulent for Humans

New viruses may appear in humans when an animal virus acquires mutations which confer infectivity in the human population. The outbreak of influenza in 1976 among soldiers at Fort Dix, New Jersey, caused by swine viruses (H1N1) (Kendal *et al.*, 1977) and the isolation of genetically similar swine viruses from pigs and man on a farm in Wisconsin (Hinshaw *et al.*, 1978) may have resulted from such a mechanism. It should be noted, however, that in both instances these viruses were not successful in spreading through the human population, suggesting that animal influenza viruses require more than subtle changes to become epidemiologically important infectious agents of man.

## C. Re-emergence of Previously Circulating Strains

Another mechanism leading to the introduction of a new subtype of influenza A virus involves the reappearance of a subtype that had circulated at an earlier time. A well-documented example of this phenomenon is the re-emergence of H1 strains that were isolated in Anshan in northern China in May 1977 and subsequently spread to the rest of the world. Serologically, these new H1 viruses possess HA and NA antigens related to those of H1 influenza A viruses that infected man around 1950 (Kendal *et al.*, 1978). Even more remarkable, genetic analysis of the new H1 strains by RNase T1 oligonucleotide mapping, RNA : RNA hybridization studies, and sequencing (Nakajima *et al.*, 1978; Young *et al.*, 1979; Scholtissek *et al.*, 1978b; Concannon *et al.*, 1984; Buonagurio *et al.*, 1986b) revealed that the entire genomes of the reemergent H1 strains are very similar to those of H1 viruses isolated in 1950. Several mechanisms have been postulated to explain how the new H1 strains remained relatively unchanged during a 27-year period. It is possible that influenza viruses are capable of latent or persistent infection in man under conditions in which the genetic information of the virus is highly conserved, although there is no direct evidence to support this idea. The genetic information of the virus could also have been preserved by sequential passage in an animal reservoir in which influenza viruses replicate without rapid genetic change, although again all evidence so far collected suggests that genetic drift also occurs in animals. Perhaps the most likely explanation is that the 1950 virus was frozen in nature or maintained in the laboratory and then was reintroduced into man.

## D. Genetic Drift

Following the emergence of a new subtype, viruses of a single subtype show minor changes in antigenic character (antigenic drift) due to an accumulation of amino acid changing nucleotide substitutions in the genes encoding the HA and NA proteins. In addition to variation in the surface protein-coding genes, variation also occurs in the genes encoding nonsurface proteins of the influenza virus. A detailed description of genetic drift in influenza A virus RNAs is summarized in the following sections.

## 1. HA Gene

Information on variation of the HA gene within a subtype has been generated using nucleotide sequencing analysis of HA genes of viruses isolated over time. Early studies on the H3 HA (Both and Sleigh, 1981) showed that most sequence changes occur in the HA1 portion of the HA molecule, which contains the antigenic sites. The resulting amino acid changes alter the antigenic sites such that antibodies generated against an initial viral infection may not be protective against infection with a later emerging virus. A comparison of nucleotide sequences of over 14 H3 viruses, isolated between 1968 and 1980, showed that new H3 antigenic variants emerged by accumulating sequential amino acid changes within antigenic regions of the HA molecules (Both et al., 1983). By analyzing the H3 HA nucleotide sequences using the maximum parsimony procedure (Fitch, 1971), one can determine a best-fitting evolutionary tree describing their descent from a common ancestral gene. This procedure defines the best-fitting or most parsimonious tree as that which contains the minimum number of nucleotide replacements necessary to account for the evolutionary relationship among the genes (Verhoeyen et al., 1980; Sleigh et al., 1981; Hauptmann et al., 1983; Both et al., 1983; Newton et al., 1983). The most parsimonious tree for 16 HA1 genes (1068 bases analyzed per gene) contains 184 nucleotide substitutions and is illustrated in Fig. 2a. Figure 2b shows the number of nucleotide substitutions between the origin of the best tree (formed at the point where NT68 and Ai68 are joined) and the tip of each branch (Fig. 2a) plotted against the date of isolation of the viruses whose HA gene is represented by that tip. The slope of the line, derived by linear regression, indicates that 7.2 ($\pm$0.2) substitutions are occurring in the HA1 gene per year or 6.7 ($\pm$0.2) $\times 10^{-3}$ substitutions per nucleotide site per year. As can be seen from the correlation of number of substitutions with the year of isolation, this rapid uniform rate of change in the HA1 gene over time represents a good example of a molecular clock in nature.

In direct analogy with the results for the H3 HA discussed above, analysis of H1 HA variants also shows sequential amino acid substitutions. Raymond et al. (1986) compared both nucleotide and amino acid changes in the HA1 domain of the HA molecules of five H1 viruses of the 1950–1957 epidemic period and 14 isolates of the 1977–1983 H1 epidemic period. They concluded that the HAs of isolates within both periods showed sequential changes, but that the earlier and later H1 periods have followed two different evolutionary pathways. It is not surprising that divergent evolution is observed for the H1 subtype HA. In the two periods of H1 virus circulation, it is not difficult to imagine that environmental conditions were different, especially the immune status of the

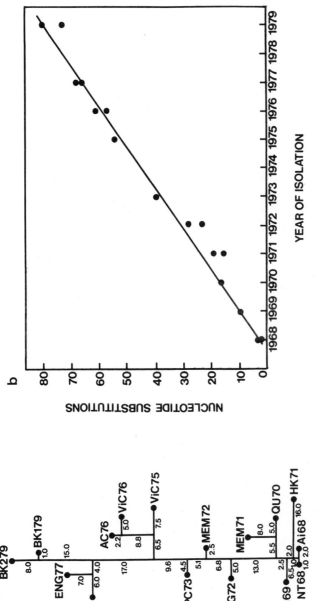

FIGURE 2. (a) Most parsimonious evolutionary tree for 16 influenza A virus H3 subtype HA1 genes. The HA nucleotide sequences have been published previously (Verhoeyen et al., 1980; Sleigh et al., 1981; Hauptmann et al., 1983; Both et al., 1983; Newton et al., 1983). The sequences were analyzed by the method of Fitch (1971). The length of the trunk and the side branches of the evolutionary tree are proportional to the minimal number of substitutions required to account for sequence differences. Nonintegral numbers arise from averaging over all possible minimal solutions. Analysis was done by W. Fitch, D. Buonagurio, and S. Nakada (unpublished data). (b) Linearity with time of number of substitutions in the H3 HA1 genes of influenza A viruses. The abscissa represents the year of isolation of the influenza A viruses used in the analysis. The ordinate indicates the number of substitutions observed in their HA1 genes between the first branching point formed by the NT68 and Ai68 sequences in (a) and the tips of all branches of the evolutionary tree. A line, generated by linear regression analysis, is drawn through the points. The slope of the line is 7.2 ± 0.2 substitutions per year.

population. During the 1950s, the H1 viruses circulated in a population possessing high levels of immunity. Then, from 1977 to the present, H1 viruses infected mostly young people born after 1957, many of whom had no pre-existing anti-H1 HA antibody. Immune pressure in the two periods probably exerted different influences on the virus, causing the H1 virus to evolve in dissimilar ways.

The observations on the evolution of the H1 subtype demonstrate the extreme flexibility of the virus in its capacity to change and what may be an unlimited potential of the influenza virus to evolve successfully in many directions and escape the host immunity. The results suggest the improbability that one can predict the molecular makeup of future epidemiologically important influenza A viruses based on information on the nature of epidemic variants of the past.

In addition to the accumulation of point mutations, insertions/deletions in the HA gene represent another mechanism that contributes to variation of influenza A viruses. Comparison of different HA gene sequences of the H3 subtype demonstrated that some strains had inserted or deleted one or more codons in their HA molecules. The A/Vic/3/75 HA, for example, contains an additional amino acid (Asn) at position 8 that is not present in any other H3 HA sequenced to date (Verhoeyen et al., 1980). Extensive deletions/insertions have also been observed in the HAs of viruses of other A subtypes ((Hiti et al., 1981).

## 2. NA Gene

Blok and Air (1980, 1982a,b) have reported on the molecular basis of antigenic drift in the NA gene through partial sequencing of the 3' end of NA genes of viruses of human subtypes N1 and N2, in addition to NA subtypes associated with nonhuman influenza viruses. The results of these studies showed that point mutations that appeared in earlier isolates were also found in later strains.

To further characterize variation in the NA gene, complete sequences have been determined and analyzed. A comparison of five N1 genes of human influenza virus isolated during a 50-year period (1933–1983) revealed almost a 20% nucleic acid and amino acid diversity, and showed a pattern of sequential mutation accumulating over time (Schreier et al., 1988). Also, comparative analysis of six complete NA sequences of natural variants of the N2 subtype isolated between 1957 and 1979 (Laver et al., 1982; Martinez et al., 1983), along with sequence information obtained from variants of the A/Tokyo/3/67 (H2N2) strain selected by monoclonal antibodies to the NA (Laver et al., 1982; Webster et al., 1982; Colman et al., 1983), revealed that antigenic drift of the N2 NA occurs via the same mechanism as that proposed for the HA polypeptide (for review, Colman and Ward, 1985). A sequential accumulation of point mutations is observed in the N2 NA genes, which alter amino acids in regions of the NA protein associated with antigenic sites.

The above-mentioned six N2 subtype NA sequences of viruses isolated between 1957 and 1979 were also examined by maximum parsimony analysis (W. Fitch, D. Buonagurio, and S. Nakada, unpublished data). From this analysis, the evolutionary rate for the N2 NA gene was calculated to be 3.2 ($\pm 0.2$) $\times$ $10^{-3}$ substitutions per nucleotide site per year. Similar to the results obtained for the H3 HA genes, a linear accumulation of changes over time is observed in the N2 NA gene, although only six strains were analyzed.

In addition, deletions/insertions have been detected in the NA genes. Partial sequencing of the 3′ end of several N1 NAs has demonstrated length variation in this region (Blok and Air, 1982b). Comparison of complete sequences of N1 and N2 NA genes also revealed deletions/insertions during the evolution of these genes (Fields *et al.*, 1981; Hiti and Nayak, 1982; Markoff and Lai, 1982). Therefore, as was observed for the HA gene, deletions/insertions also contribute to the variation seen in the NA gene.

## 3. NS Gene

The high level of genetic homology observed among NS genes of viruses belonging to different subtypes (Scholtissek *et al.*, 1978a; Hall and Air, 1981) indicates that the NS gene was conserved during the emergence of new subtype strains in this century. Early work based on partial NS sequence data from three human strains by Hall and Air (1981) suggested that sequence changes in the NS gene accumulate over time. Recently, a comparison of the complete nucleotide sequences of the NS genes of 15 influenza A viruses isolated over 53 years (1933–1985) confirms these observations, and allows an in-depth analysis of variation in the NS gene (Buonagurio *et al.*, 1986b). This sequence information was analyzed by the maximum parsimony procedure to determine the phylogenetic tree of minimum length. The best tree is illustrated in Fig. 3a. Figure 3b shows the number of nucleotide substitutions between the origin of the best tree and the tip of each branch (Fig. 3a) plotted against the date of isolation of the viruses whose NS gene is represented by that tip. The major line, derived by linear regression analysis, shows that these sequences are evolving at the steady rate of 1.7 $\pm$ 0.1 nucleotide substitutions per year, or 1.9 $\pm$ 0.1 $\times$ $10^{-3}$ substitutions per nucleotide site per year. The WSN/33 and PR/34 strains appear to have more substitutions per year than expected. Since these strains were isolated before refrigeration became available in the laboratory, we believe that continuous passaging in animal hosts and in embryonated eggs (particularly in the first 10 to 15 years after isolation of the strains) may have introduced additional mutations not present in the original isolates.

Figure 3b also shows that the group of H1 subtype strains, which reemerged in the human population in 1977 after a 27-year absence, is evolving at the same rate. These "new" H1 viruses have been cocirculat-

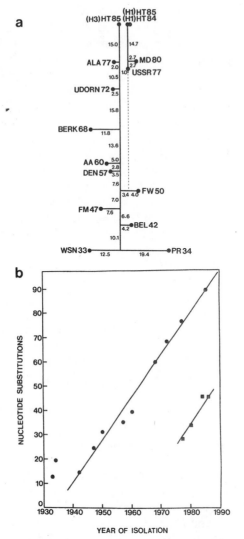

FIGURE 3. (a) Most parsimonious evolutionary tree for 15 influenza A virus NS genes. The nucleotide sequences (Buonagurio *et al.*, 1986b) were analyzed by the method of Fitch (1971). The length of the trunk and side branches of the evolutionary tree are proportional to the minimal number of substitutions required to account for the differences in sequence. The broken line represents the predicted number of additional substitutions between the NS genes of FW/50 and USSR/77 based on the calculated evolutionary rate. From Buonagurio *et al.* (1986b). (b) Linearity with time of number of substitutions in the NS genes of influenza A viruses. The abscissa represents the year of isolation of the influenza A viruses used in the analysis. The ordinate indicates the number of substitutions observed in their NS genes between the first branching point formed by the WSN/33 and PR/34 sequences in (a) and the tips of all branches of the evolutionary tree. A line, generated by linear regression analysis, is drawn through the points. The slope of the line is 1.7 ± 0.1 substitutions per year. In addition to the sequences found on the trunk of the evolutionary tree (●), the NS genes of the four new H1N1 viruses are also represented in this graph (■). From Buonagurio *et al.* (1986b).

ing with the H3 viruses since 1977 and form a separate evolutionary branch (Fig. 3a). In reality, the H1 branch should be directly connected to the FW/50 branch of the main tree, since there are only five nucleotide differences between the FW/50 and USSR/77 virus NS genes. However, the viruses were isolated 27 years apart and, on the basis of the calculated evolutionary rate of 1.7 substitutions per year, we would predict approximately 46 additional substitutions in the NS gene of USSR/77 (represented by the broken line in Fig. 3a). The observed data thus suggest a unique mechanism for the emergence of the new H1 isolates, as discussed in detail above.

The evolutionary changes in the NS genes of the 1950–1957 H1 viruses were compared with those in the NS genes of the new H1 viruses of 1977–1985 (Buonagurio et al., 1986b). The data revealed that nucleotide substitutions in the NS genes of viruses of the 1950–1957 period (A/FW/50, A/DEN/57) were quite distinct from those detected in the NS sequences of the reemerged H1 strains of 1977–1985 (A/USSR/77, A/MD/80, A/HT/84, A/HT/85). These results are consistent with those obtained for the HA genes of these isolates and suggest that the NS genes of the H1 subtype evolved along two divergent lineages in the two periods of its circulation in man.

As mentioned for the HA and NA genes, deletion/insertion events may also contribute to the variation seen in the NS gene. The NS segment of an influenza A virus host range mutant, CR43-3 (which grows in PCK but not MDCK cells), was found to differ from the NS gene of the A/ALA/6/77 parent by a deletion of 36 nucleotides in the NS1-coding region (Buonagurio et al., 1984). Also, an avian influenza A virus isolate, A/turkey/Oregon/71 (Norton et al., 1987) has been found to have a deletion in its NS1 coding region. However, although the above data suggest that a functional NS1 protein may still be formed despite certain deletion events, no natural human isolate with a deletion in the NS gene has yet been reported.

## 4. NP Gene

Immunological analysis of NP proteins has shown that antigenic differences can be detected in the molecule within and between virus subtypes (Schild et al., 1979; Van Wyke et al., 1980). In addition, migrational differences among NP proteins of virus isolates have been observed, suggesting protein variability (Oxford et al., 1981). However, genetic drift in NP genes has not been well defined. Only four complete human NP nucleotide sequences are available (Winter and Fields, 1981; Huddleston and Brownlee, 1982; Buckler-White and Murphy, 1986; Beklemishev et al., 1986). Based on the high level of nucleotide and amino acid identity (greater than 90%) among these NP genes and the deduced proteins, it appears that the NP gene has been conserved during the major antigenic shifts.

## 5. M Gene

Five complete nucleotide sequences of M genes (Winter and Fields, 1980; Lamb and Lai, 1981; Ortin *et al.*, 1983; Samokhvalov *et al.*, 1985; Markushin *et al.*, 1988) and several partial M sequences (Hall and Air, 1981; Hay *et al.*, 1987) of human influenza viruses isolated over 45 years (1934–1979) have been obtained. The viruses analyzed represent all three human antigenic subtypes. The nucleotide data illustrate that many substitutions appear to accumulate sequentially and that the M gene is highly conserved through evolution. On the amino acid level, divergence among M2 polypeptides appears to be greater than among the M1 proteins. The amino acid conservation among M1 proteins is reflected in studies using monoclonal antibodies to detect antigenic variation in M1 proteins of both human and animal viruses of different subtypes (Van Wyke et al., 1984). In this study involving 26 isolates, antigenic variation in the M1 protein could only be found in three strains.

## 6. P Genes

Information on genetic drift of the three polymerase (P) genes is extremely limited. Nucleotide sequences of the PB1 and PB2 genes of A/WSN/33 have been reported (Kapstein and Nayak, 1982; Sivasubramanian and Nayak, 1982) as well as sequences of the PB1, PB2, and PA genes of the A/PR/8/34 (Winter and Fields, 1982; Fields and Winter, 1982) and the A/NT/60/68 (Bishop *et al.*, 1982a, 1982b; Jones *et al.*, 1983). A comparison of the latter two strains alone shows that the nucleotide sequence variation per year for PB1, PB2, and PA is $4.7 \times 10^{-3}$, $2.6 \times 10^{-3}$, and $2.1 \times 10^{-3}$, respectively, and the amino acid sequences of the deduced P proteins change at a rate of $0.9 \times 10^{-3}$ (PB1), $1.2 \times 10^{-3}$ (PB2), and $1.1 \times 10^{-3}$ (PA) substitutions per site per year. It should be noted that in the case of the P genes, the number of silent mutations far exceeds the number of amino acid changing mutations, suggesting strong negative selection against the latter type of mutation (e.g., see Bishop *et al.*, 1982b).

## E. Evolution of Influenza A Viruses

The human influenza A virus genes so far examined accumulate sequential nucleotide changes over time. Therefore, phylogenetic trees can be constructed (see Figs. 2 and 3) that permit the calculation of evolutionary rates for these genes. The most sequencing information has been accumulated for the HA and NS genes, and several interesting points can be made by comparing the data obtained for these two genes. First, a striking feature of the genealogical trees of the HA and NS genes is their long, slender structure. The side branches of the trees do not greatly diverge from the main vertical axis (e.g., for the NS gene, the average age of the side branches is only 3 years). This appears to be a consequence of the short life span of any lineage other than the one that gives rise to the future generations. Most viral genes do not follow this pattern. Rather, they appear to have multiple surviving lineages undergoing slower change.

An explanation for the unusual pattern of influenza virus gene evolution may be found in positive selection of influenza A virus variants. It may be that only one influenza virus gene (most likely the HA) needs to be subject to selection. In the brief time before immunity develops to a new (antigenic) variant, that strain may sweep through the population, carrying with it whatever variants of the other genes happen by chance to be present. In this way, one gene's phylogeny may be linked (hitchhiking) to that of another gene undergoing extensive positive selection. Thus, although there is no evidence for immune surveillance of the NS gene products, the shape of its genealogical tree is similar to that of the HA gene.

However, although the shapes of the trees are similar, the nucleotide substitution rates for the H3 HA (HA1 domain) is higher than that for the NS gene ($6.7 \times 10^{-3}$ and $1.9 \times 10^{-3}$ substitutions per site per year, respectively). The observation that the nucleotide substitutions in the HA molecule are predominantly in the HA1 domain suggest that many substitutions may be selected for by immune pressure. Thus, the lower evolutionary rate of the NS gene is most likely due to factors intrinsic to influenza A virus replication. Such factors could include polymerase error rate and constraints on the number of mutations compatible with a functional protein.

Therefore, one might expect the rate of accumulation of silent mutations that is unaffected by immune pressure and functional constraints on the protein products to be similar for the HA and NS genes. In fact, as shown in Table III, these figures are very close: $3.1 \times 10^{-3}$, $1.8 \times 10^{-3}$, and $2.0 \times 10^{-3}$ silent substitutions/site/year for the HA (HA1 coding region) and the NS gene (NS1 and NS2 coding regions), respectively. We would further predict that when more sequence information becomes

TABLE III. Silent Nucleotide Substitution Rate
in Influenza A Virus Genes

| | Silent nucleotide substitutions/site/year[a] |
|---|---|
| H3 HA1 | $3.1 \times 10^{-3}$ |
| NS1 | $1.8 \times 10^{-3}$ |
| NS2 | $2.0 \times 10^{-3}$ |

[a]Estimated by summing the number of silent mutations in all sequenced strains (see Figs. 2 and 3) and dividing by the number of nucleotides in the coding regions of each gene and by the number of years over which the strains were isolated.

available the evolutionary rate of silent mutations for all influenza A virus genes would approximate that seen for the HA and NS genes.

## V. VARIATION IN INFLUENZA C VIRUSES

### A. HE Genes

Early evidence suggested that influenza C viruses are antigenically more stable over time than A type viruses (Chakraverty, 1978; Meier-Ewert et al., 1981). Furthermore, oligonucleotide mapping analysis of the genomes of different influenza C viruses (Meier-Ewert et al., 1981; Guo and Desselberger, 1984) also suggested less variation than was observed for influenza A viruses. Nucleotide sequence analysis of the HE genes of eight human influenza C virus strains isolated over approximately four decades (1947–1983) revealed that both on the nucleotide and amino acid levels changes can be observed but they do not appear to accumulate with time (Buonagurio et al., 1985). As shown in Table IV, strains isolated 31 years apart (C/AA/50 and C/YA/81) may possess almost identical HE genes whereas strains isolated only one or two years apart may differ by many changes (e.g., C/MS/80 and C/YA/81 or C/Cal/78 and C/MS/80). This suggests that the strains of influenza C viruses examined do not directly share the same evolutionary pathway with respect to the HA genes. Rather, it appears that multiple evolutionary pathways exist and C virus variants of different lineages may cocirculate in nature for extended periods of time.

Interestingly, an amino acid comparison of all HE variants reveals three clusters of amino acid substitutions in the HE1 portions of the molecules centering on regions around amino acid positions 80, 205, and 340 (Buonagurio et al., 1985). Since there are no comparable clusters of amino acid changes in the HE2 portion of the proteins, it is tempting to speculate that these three clusters in the HE1 region are associated with antigenic characteristics of the influenza C virus HEs. However, in the

TABLE IV. Nucleotide Differences between Influenza C Virus HE Genes[a]

| Virus strains | AA/50 | GL/54 | JHG/66 | CAL/78 | MS/80 | YA/81 | ENG/83 |
|---|---|---|---|---|---|---|---|
| TAY/47 | 27(1.3)[a] | 36(1.8) | 58(2.9) | 71(3.5) | 110(5.5) | 29(1.4) | 54(2.7) |
| AA/50 | | 49(2.4) | 54(2.7) | 74(3.7) | 112(5.6) | 2(0.1) | 61(3.0) |
| GL/54 | | | 73(3.6) | 49(2.4) | 112(5.6) | 49(2.4) | 36(1.8) |
| JHG/66 | | | | 91(4.5) | 128(6.3) | 56(2.8) | 84(4.2) |
| CAL/78 | | | | | 134(6.6) | 74(3.7) | 33(1.6) |
| MS/80 | | | | | | 114(5.7) | 117(5.8) |
| YA/81 | | | | | | | 61(3.0) |

[a]Nucleotides 1–58 are excluded from analysis and nucleotide deletions/insertions are counted as changes in this comparison.
[b]Number of nucleotide differences between 2 HA genes (% differences). (From Buonagurio et al., 1985.)

absence of further studies, which may include the selection of antigenic variants through the use of monoclonal antibodies, the precise antigenic structure of the C virus HE remains unknown.

## B. NS Genes

Variation in the NS genes of eight human influenza C viruses isolated during 1947–1983 has also been examined by nucleotide sequencing (Buonagurio et al., 1986a). As was found for the HE gene, nucleotide changes appearing in earlier viruses are not found in all later isolates. In addition, viruses isolated thirty years apart can either exhibit few NS differences or many changes in the NS genes. These results suggest that the NS gene of influenza C virus has multiple genetic variants belonging to different evolutionary lineages coexisting in nature. Also, since the maximum parsimony trees of the NS genes are quite different from those of the HE genes of the same viruses, it is likely that reassortment between different viral lineages occurs in nature (Buonagurio et al., 1986a).

## C. Evolution of Influenza C Viruses

The evolutionary model postulated for influenza C viruses is quite different from that suggested for influenza A viruses. For the A viruses, it has been demonstrated that antigenically dominant variants emerge with time and that successive variants accumulate mutations found in the genes of variants circulating in earlier years. In contrast, influenza C viruses do not appear to rapidly accumulate changes with time and strains of different lineages cocirculate (Kawamura et al., 1986).

What could be the nature of these differences? First, we may speculate that influenza A viruses induce a strong immune response that favors selection of antigenic variants, whereas influenza C viruses do not. This would provide an explanation for the presence of antigenically similar influenza C virus variants that continue to circulate in man. In addition, the polymerase of influenza A viruses may be more error prone than that of influenza C viruses. However, other factors may also contribute to this situation. For example, influenza A and C viruses may differ in their mode of transmission in that one or many particles are actually involved during theinfection process. If transmission of the virus is the result of infection with a single infectious particle, the virus would in effect be cloned whenever it is transmitted from one person to another. By contrast, infection with many particles during each transmission cycle would allow for a slower selection of viral variants (Brand and Palese, 1980). Alternatively, influenza C viruses may have a long replication cycle in vivo that results in a decreased number of infection cycles per unit time for C viruses. Also, the number of infected individuals worldwide may influence the genetic variation of these viruses. All or some of these factors may lead to an altered epidemiolgical pattern and to an increased number of sequence changes in the genomes of influenza A viruses as compared with those of influenza C viruses.

## VI. VARIATION IN INFLUENZA B VIRUSES

### A. HA Genes

Although antigenic shift (replacement of the HA molecule in one strain with an antigenically novel HA) has not been observed for influenza B viruses, antigenic changes do occur in B viruses (Webster and Berton, 1981; Lu et al., 1983; Oxford et al., 1984). Determination of the complete nucleotide sequences of the HA genes of five different isolates of influenza B virus has permitted detailed analysis of the variation within this gene over time (Krystal et al., 1983; Verhoeyen et al., 1983; Berton et al., 1984; Hovanec and Air, 1984).

In addition, the HA1 portions of five other strains were sequenced (Yamashita et al., 1988b). Pairwise nucleotide sequence comparisons using the B/Lee/40 HA1 gene and the HA1 genes of viruses isolated at various times between 1940 and 1987 did not show differences strictly proportional to the time between isolations. A similar pattern is observed in the deduced amino acid sequences, with many changes occurring in unique positions. A maximum parsimony analysis reflects this finding and reveals a broad evolutionary tree (Yamashita et al., 1988b).

### B. NS Genes

Recently, Yamashita et al. (1988b) determined the nucleotide sequences of the NS genes of 12 different influenza B virus isolates and compared them with those of the NS genes of B/Lee/40 (Briedis and Lamb, 1982) and B/YA/73 (Norton et al., 1987). Pairwise comparison of nucleotide and amino acid differences between the NS genes of the earliest strain, B/Lee/40 virus, and the later strains do not increase proportionally with time of isolation. A similar nonlinear relationship is seen when sequences of the B/HK/73 virus are compared with those of the strains isolated in later years. Thus, influenza B virus genes do not always show nucleotide differences proportional to the amount of time between isolations and apparently only some strains belong to the same recent evolutionary lineage.

No insertions, deletions, or changes in termination codons were observed in the NS genes after the initiation codon in positions 43–45. This is in contrast to influenza A viruses, which exhibit extreme variability in

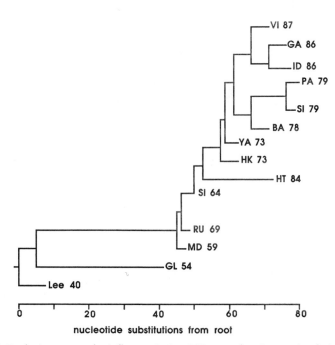

FIGURE 4. Evolutionary tree for influenza B virus NS genes showing co-circulating lineages. The nucleotide sequences (Yamashita et al., 1988b) were analyzed by the maximum parsimony method (Fitch, 1977). The length of the horizontal lines are proportional to the minimum number of substitutions required to account for the differences in sequences. Vertical lines are used to separate progeny virus lineages at the point where they branch off from the common ancestor virus lineage. The tree shown represents one of the three most parsimonious solutions. The other two solutions differ only in the branching order of the B/MD/59, B/RU/69 and B/SI/64 virus genes. From Yamashita et al. (1988b).

the size of the NS1 protein, due to changes in the termination codon in different virus strains (Parvin et al., 1983; Norton et al., 1987). In the 5′ noncoding region near the initiation codon of the B virus NS genes, several isolates have A or AA insertions and eight isolates have A to G substitutions. Consequently, many of the strains have stretches of five to seven A residues 5′ to the ATG initiation codon. A similar observation was made for RNA segments 4–8 of several influenza B isolates (Stoeckle et al., 1987). However, the significance of this A-rich sequence is not clear. Maximum parsimony analysis of the 14 above NS gene sequences revealed three trees, each having the same minimum number of substitutions (Yamashita et al., 1988b). One of these trees is shown in Fig. 4; it displays a broad multibranched pattern. The tree also shows multiple B virus lineages circulating at any one time. For example, the NS genes of the 1973–1979 strains are not direct precursors to the NS gene of the 1984 strain (HT84).

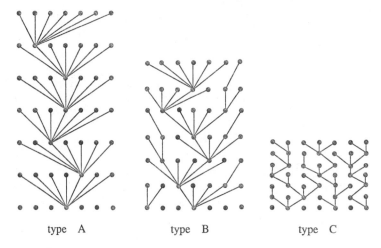

type  A                    type  B                    type  C

FIGURE 5. Evolutionary model for the propagation of influenza A, B, and C viruses in man. Dots lying on a horizontal plane represent influenza virus variants arising in the same season (year). The left part of the diagram shows influenza A virus. The variants arise from only one lineage because of the dominating effects of favorable variants. The middle and right of the diagram show, respectively, the co-circulation of multiple influenza B and C virus lineages. Variation of influenza B viruses appeared to be slower than that of influenza A viruses and faster than that of C viruses. This is illustrated by the intermediate length of the evolutionary branches of the influenza B virus pattern. (The relative length of these branches is not to scale.) For all viruses, an arbitrary number of seven seasonal cycles is shown. From Yamashita *et al.* (1988b).

## C. Comparison of Evolutionary Patterns of Influenza A, B, and C Viruses

With respect to the survival of multiple lineages, the overall evolutionary pattern of influenza B viruses follows that of influenza C viruses. However, there is a difference between the extent of sequence divergence in influenza B and C viruses. Influenza C virus genes (and perhaps the virus as a whole) may exist unchanged for decades at a time, whereas B virus gene sequences do not show such obvious sequence conservation over time. Influenza B viruses are constantly diverging, albeit at a slower rate (two- to sixfold slower) than influenza A viruses (Buonagurio *et al.*, 1986b). In summary, the evolutionary pattern of influenza B viruses is intermediate between that for human influenza A viruses and that for influenza C viruses as shown in the diagrammatic representation of Fig. 5. The reasons for these differences are not yet clear but may include differences in the antigenicity, transmission process, and mutation rate of these viruses.

## VII. GENETIC BASIS OF PATHOGENICITY

### A. Pathogenicity Is Multigenic

Most studies that have attempted to correlate pathogenicity of various strains of influenza A virus with certain gene products have, of necessity, been done in birds or mice. However, knowledge gained in these systems may also be applicable to human disease.

Two major points can be made immediately about these studies. First, it is clear that the virulence of a given strain reflects a complex set of interactions, in which both host- and virus-determined properties play a role. Although we intend to discuss in detail only results relevant to the latter, the former is also vitally important. For example, host determined factors such as immunological experience, major histocompatibility complex haplotype (Askonas et al., 1982), and Mx genotype (Haller, 1981) of the host have all been shown to have a major role in determining the outcome of influenza virus infection in the host.

Second, virulence is generally a polygenic phenomenon (reviewed by Schulman, 1983; Scholtissek, 1986), because many changes in a virus may affect its efficiency of replication. This conclusion is mainly derived from the analysis of reassortant viruses in which the parental derivation of each of the eight RNA segments has been defined. For example, Scholtissek et al. (1977) crossed ts mutants of the highly virulent fowl plague virus (FPV) with seven different human or animal influenza viruses. It was found that replacement of any FPV RNA segment, with the exception of the M gene for which no reassortants were obtained, could result in loss of pathogenicity. By contrast, other experiments show that reassortment of genes derived from two different parents may lead to a novel virus with pathogenicity traits that are more virulent than those found in either parent (Scholtissek et al., 1979). These experiments suggest that effective interaction of viral gene products is important in determining pathogenicity. The correlation between the inheritance of a constellation of gene products and virulence is further illustrated by the experiments of Rott and co-workers. These authors (Rott et al., 1979; Giesendorf et al., 1986) found that reassortant viruses obtained after mixed infection with two virulent avian viruses were, in most instances, virulent for chickens only when they derived all of the genes coding for the P and NP proteins from the same parent. All reassortant viruses in which the derivation of these genes was mixed lacked virulence.

Attenuated cold-adapted mutants have also been used to study the molecular basis of pathogenicity. Both influenza A viruses (Maassab, 1968) and influenza B viruses (Medvedeva et al., 1983) that have been adapted to grow at temperatures below those normally used for virus cultivation have been shown to become attenuated for their original hosts, and recombinants between these mutants and the wildtype strains can be prepared that are stably attenuated. Most genes of these attenuated recombinants derive from the cold-adapted parent. Nucleotide sequencing analysis shows that mutations may exist in all eight genes of the attenuated strain (Cox, 1986). Therefore, examination of revertants and single gene recombinants may help determine which of these sequence changes are responsible for attenuation in this system.

An additional procedure by which attenuation characteristics can be introduced into human influenza viruses is reassortment with temperature sensitive strains (Murphy and Chanock, 1981) or with avian strains (Murphy et al., 1982). Although in some instances transfer of the M and/or NP genes alone may confer the attenuation phenotype (Tian et al., 1985; Scholtissek et al., 1985), a general strategy to construct live virus vaccine strains may employ the exchange of all genes in the virulent strain except those coding for the HA and NA proteins.

Another multigenic trait of influenza viruses is the ability to grow to high titers in embryonated eggs (Kilbourne and Murphy, 1960). Viruses that acquire this property by reassortment can successfully be used as killed virus vaccine strains because they possess appropriate antigenic characteristics and grow to high yields in eggs (Kilbourne et al., 1971). In some instances, the latter characteristic has been shown to be associated with the NP and/or the M gene of the good growing parent strain (Schulman and Palese, 1978; Baez et al., 1980).

## B. HA Genes

The experiments cited above show that many gene products play a role in determining the pathogenicity of influenza viruses. We will now briefly review several cases in which specific changes within a single gene have been shown to dramatically affect biological properties of the virus.

Three different functions of the HA have been shown to play a role in its pathogenicity. First, the antigenic epitopes change through antigenic shift and drift, which allow the virus to successfully infect individuals who are already immune to previous strains. Genetic mechanisms responsible for these changes have already been discussed above; their effects on the three-dimensional structure of the HA were recently reviewed (Skehel and Wiley, 1988).

Second, studies on receptor-binding site variants also suggest that changes in this region of the HA molecule can lead to changes in the pathogenicity of influenza A viruses. For example, a single amino acid change in amino acid residue 226 of the H3 HA can change the receptor-binding specificity from 2-3 bound neuraminic acid to 2-6-linked neuraminic acid (Rogers et al., 1983). Subsequently, Naeve et al. (1984) reported that changes in this region are sufficient to allow replication in ducks of a reassortant virus containing a human HA gene. Additionally,

Yewdell *et al.* (1986) showed that single amino acid changes in the HA close to this site are responsible for differences in avidity for host-cell receptors. Also, differences in glycosylation of the HA, again brought about by a single amino acid change, affect receptor binding and change host preference (Deom *et al.*, 1986). Similar results have been reported for influenza B virus (Robertson *et al.*, 1985). Therefore, a single point mutation in the HA can be responsible for a change in host range as well as in specificity and avidity of receptor binding.

Third, it has been shown that the efficiency with which the HA undergoes essential host-cell-dependent posttranslational proteolytic cleavage can be correlated with changes in pathogenicity (Webster and Rott, 1987). For example, Bosch *et al.* (1981) suggest that the carboxy-termini of different HA1 molecules of avirulent H7 avian viruses contain a single arginine-glycine peptide that is susceptible to proteolytic cleavage, whereas the connecting peptide of virulent H7 strains is longer and contains additional basic amino acids. Also, studies by Webster and co-workers (Webster *et al.*, 1986; Deshpande *et al.*, 1987) on closely related avirulent and virulent strains of the influenza virus A/chicken/Pennsylvania/83 indicate that a single point mutation in the HA1 region resulting in the addition of a carbohydrate side chain may affect the cleavability and thus the virulence of the virus. However, additional experiments by the same group have shown that mutations exist in the NA gene, and may in fact exist in all genes, and that the presence of DI particles may be associated with virulence, implying that more complex factors determine pathogenicity of this virus in nature (Deshpande *et al.*, 1985; Bean *et al.*, 1985).

The pH at which the HA catalyzes fusion between viral and endosomal membranes varies among different strains of virus (White *et al.*, 1983). It is therefore also possible that such differences may play a role in determining host range and virulence. For example, Rott *et al.* (1984) showed that variants of influenza viruses that were selected for their ability to undergo activation cleavage and growth in MDCK cells have an elevated fusion pH threshold. In addition, influenza virus growth in the presence of amantadine or ammonium chloride can be correlated with changes in the HA and an altered pH optimum for fusion (Daniels *et al.*, 1985; Doms *et al.*, 1986). It should be noted that changes in the M2 protein can also be correlated with amantadine resistance.

## C. NA Genes

Influenza A/WSN/33 virus is able to undergo multicycle replication in MDBK cells. Analysis of reassortant viruses demonstrated that this property is dependent on the WSN neuraminidase (Schulman and Palese, 1977). It was further demonstrated that infectivity is dependent on a cleaved HA and that the infectivity of viruses lacking the WSN NA can be activated by *in vitro* treatment with trypsin (Schulman and Palese, 1977). In related studies, it was shown that virulence for 1-day-old chickens (Bean and Webster, 1978) and neurovirulence in mice (Sugiura and Ueda, 1980) can also be correlated with the possession of the WSN NA gene in certain reassortants.

The mechanism by which the NA facilitates cleavage of the HA has not been elucidated but may be related to differential activation or inactivation of the required host proteases and/or to an interaction between the NA and HA molecules. It is hoped that future studies will elucidate the mechanism of this interesting effect, which illustrates so well the concept that successful interaction between several different viral and host proteins may be required for pathogenicity.

## D. M Genes

Murphy and co-workers have attempted to identify sequences in the M1 and M2 proteins that are important in determining host range by combining both reassortant analyses and sequencing studies. Reassortants between the human A/Udorn/307/72 and the avian A/mallard/New York/6750/78 strains of influenza A virus were found to show the avian phenotype of restricted replication in the trachea of squirrel monkeys if they contained either the NP or M genes of the avian strain (Tian *et al.*, 1985). The M gene of the A/mallard/78 strain was then sequenced, and the deduced amino acid sequence was compared with those of two other avian and three human M proteins (Buckler-White *et al.*, 1986). Ten amino acid sites were specific for either avian or human influenza strains; i.e., a specific amino acid is conserved in the three human strains while a different amino acid is present in both avian strains at that site. It therefore seems possible that some or all of these sites may be responsible for the host-range restriction observed. Analysis of additional sequences of both human and avian influenza A virus strains and selected phenotypic mutants of these strains may help to test this prediction. It should be noted that the NP gene of two avian influenza viruses have also been shown to be able to affect host range restriction (Tian *et al.*, 1985; Snyder *et al.*, 1987).

In cell culture, low concentrations of amantadine exhibit a strain-specific inhibition of influenza A virus replication, and genetic analyses have mapped this effect to the M gene (Lubeck *et al.*, 1978) and the HA gene. Characterization of drug resistant mutants demonstrates that one target of these actions is the M2 protein (Hay *et al.*, 1985), which is a product of a spliced transcript of the M gene. Mutations that confer resistance to amantadine are restricted to four amino acids within a hydrophobic sequence, the putative membrane-associated portion of the molecule. The role of the M2 protein in viral replication is not yet clear, but these studies show that it may be important in determining the efficiency of amantadine as an antiviral drug.

In conclusion, specific amino acid changes in a single gene can be correlated with changes in biological properties, but pathogenicity of influenza viruses is, in general, a polygenic effect. A complete understanding of its diverse mechanisms can only be obtained when the details of the molecular interactions among viral and host components have been fully elucidated.

## VIII. COMPARISON OF VARIATION AMONG DIFFERENT RNA VIRUSES

### A. Tissue-Culture Studies

RNA viruses provide the opportunity to study evolutionary processes in viral genes which are manifested over a relatively short time frame (decades). By contrast, the eukaryotic chromosomes of the host cell take millions of years to achieve the same extent of molecular variation that viral genes can attain within just a few human generations. Evolutionary rates for mammalian genes are on the order of $10^{-9}$ substitutions per site per year (Li and Gojobori, 1983), a millionfold lower than that of influenza A viruses. One obvious reason for the difference in evolutionary rates is the vast difference in generation time between viruses and mammals. However, another important factor is that RNA polymerases exhibit less fidelity of genome replication than DNA polymerases, perhaps in part to a lack of proofreading enzymes. The proofreading exonuclease activity of DNA polymerase is able to remove misincorporated bases from newly synthesized DNA strands so that errors in DNA replication average as low as $10^{-8}$–$10^{-11}$ per incorporated nucleotide.

Early attempts to compare mutation rates for RNA viruses suggested that RNA viruses have high mutation rates in tissue culture (Holland *et al.*, 1982). One technique to assess mutation rates is to measure the frequency of antigenic variants resistant to neutralization by a selecting monoclonal antibody. It has been reported that the frequency of resistant (escape) mutants is similar for influenza virus, Sendai virus, vesicular stomatitis virus, poliovirus, and Coxsackie virus and is in the range of $10^{-4}$–$10^{-5}$ (Portner *et la.*, 1980; Prabhakar *et al.*, 1982; Lubeck *et al.*, 1980; Emini *et al.*, 1983; Minor *et al.*, 1983). This approach is limited, however, in that only a small region of the genome can be probed for changes using monoclonal antibody selection. The target size for mutant selection is frequently unknown and may often include more than one gene. Additionally, differences in mutation frequencies generated from resistant mutant selection experiments may be explained by differences in avidity of the discriminating monoclonal antibodies.

Analysis of variation at the nucleotide level has also been used in an attempt to determine mutation rates for different viruses. Analysis of 45 foot-and-mouth disease viral clones isolated after serial low multiplicity passage in tissue culture revealed 16 different nucleotide sequences (Sobrino et al., 1983). Also, sequencing of the VP1 gene of viral clones obtained from persistently infected cells showed extensive genetic heterogeneity (de la Torre et al., 1988). Studies on the bacteriophage QB by Domingo et al. (1978) have also shown extensive genetic heterogeneity within this viral population on serial passage, and led to a calculation of error level per genome doubling at given base positions in the RNA genome of $10^{-3}$–$10^{-4}$. Using a similar approach, Coffin et al. (1980) calculated a mutation frequency of $10^{-3}$–$10^{-4}$ for a particular point mutation in the Rous sarcoma virus RNA genome. These values are very high, but the significance of both calculations is not clear because they are mathematically based on the assumption that viral replication can be considered as a series of doubling steps. More recently, Steinhauer and Holland (1986) have quantitated polymerase error frequencies for vesicular stomatitis virus at one nucleotide site using both in vivo and in vitro assays. The extremely high frequency of base misincorporation is $\sim 10^{-4}$ substitutions per base incorporated at the site in both assays. One difficulty of this assay is that the substitution frequency at a single nucleotide site may not be representative of all the nucleotide positions of the genome RNA. In contrast to the above results, Durbin and Stollar (1986) observed much lower variability when studying reversion of a particular Sindbis virus host restricted mutant. Direct sequencing analysis of the E2 gene of revertants allowed them to estimate a mutation rate of $<10^{-6}$ errors per base incorporated.

Another approach has been taken by Parvin et al. (1986), who assayed the mutation rate in tissue culture for the NS gene of influenza A virus and for the VP1 gene of type 1 poliovirus. Each gene was directly sequenced in more than 100 randomly selected viral clones that had arisen from a single virion in one plaque generation. The VP1 gene encodes one of the structural proteins comprising the capsid of the poliovirus. Seven mutants of the influenza A virus NS gene were detected, whereas no VP1 gene poliovirus mutant was observed. The calculated mutation rates are $1.5 \times 10^{-5}$ and less than $2.1 \times 10^{-6}$ mutations per nucleotide per infectious cycle for the influenza A virus and the poliovirus-1, respectively. Sedivy et al. (1987) obtained a similar value for poliovirus ($1.5 \times 10^{-6}$ mutations per nucleotide per infectious cycle, assuming poliovirus undergoes five replication cycles per plaque generation) by using an inducible mammalian amber suppressor system to monitor the reversion frequency of a poliovirus mutant. Thus, the mutation rate of influenza A virus was found significantly higher than that of poliovirus-1.

Leider et al. (1988) employed a similar strategy to that used by Parvin et al. (1986) to measure the mutation rate of Rous sarcoma virus at $1.4 \times 10^{-4}$ mutations per nucleotide per replication cycle. Progeny descended from a single virion were collected after one replication cycle, and seven regions of the genome were analyzed for mutations by denaturing gradient gel electrophoresis. Dougherty and Temin (1988) also measured the mutation rate for single base-pair substitution of an avian retrovirus, by monitoring the expression of a suppressed gene in a spleen necrosis virus vector. They obtained a value of $2 \times 10^{-5}$ mutations per base pair per replication cycle. However, this rate represents a minimum mutation rate, because substitutions were limited to a single site, and only one change was permissible, apparently because of selection against other mutations.

In conclusion, the above results indicate that mutation rates are indeed high for RNA viruses, but that marked differences (by at least two orders of magnitude) exist between viruses. From the limited data presently available, influenza A viruses are moderately fast mutating animal viruses, slower than retroviruses, but faster than poliovirus.

## B. Variation in Nature of Human RNA Viruses

Influenza A, B, and C virus isolates in nature show both different degrees as well as different patterns of variation. To take the extreme cases, influenza C viruses isolated decades apart show little variability and there is evidence of multiple co-circulating strains, whereas influenza A viruses show a high number of nucleotide substitutions accumulating in a sequential pattern with time. Variation in nature of other human RNA viruses has generally not been investigated so thoroughly (especially at the nucleotide level). However, the examples given below suggest that markedly different patterns of variation in nature are exhibited by different human RNA viruses.

### 1. Human Immunodeficiency Virus Type 1

The retrovirus human immunodeficiency virus type 1 (HIV-1) responsible for causing acquired immune deficiency syndrome (AIDS) shows a pattern of multiple co-circulating strains of astonishingly high variability. Nucleotide sequencing of five independent isolates showed up to 8.6% total nucleotide differences (up to 24% nucleotide differences in some regions of the envelope gene), resulting largely from duplications, insertions or deletions, as well as from an accumulation of nucleotide point mutations (Starcich et al., 1986). In another study (Hahn et al., 1986) it was shown that serial virus isolates from chronically infected patients showed up to 3% nucleotide variation within the envelope gene. Although sequential isolates from any one patient did not show mutations accumulating with time, viruses from one patient were much more related to each other than to viruses obtained from other individuals. This finding suggests that serial isolates from one patient had evolved from a common progenitor within the recent past. On the basis of this assumption, the authors estimated that the rate of evolution of HIV-1 was at least $10^{-3}$ nucleotide substitutions per site per year for the env gene and $10^{-4}$ for the gag gene. The greater variation seen in the envelope gene (which codes for the surface glycoprotein of the virion) suggests either a strong positive selection for variation by immunological pressure, and/or a lower level of constraint on the amino acid sequence of the env gene product. It is not clear whether this variability is responsible for significant antigenic variation, or whether antigenic variation is necessary for survival of the HIV-1 virus within the host.

For two other retroviruses related to HIV-1, i.e., equine infectious anemia virus (EIAV) and visna virus, there is evidence that during infection changes in the envelope genes do lead to substantial changes in envelope antigenicity (Narayan et al., 1977; Salinovitch et al., 1986; for review, see Haase, 1986). For EIAV these changes appear to be sequential and are apparently directly responsible for the chronic periodic nature of the disease (Salinovitch et al., 1986). For visna virus, the biological relevance of antigenic variants is unclear, as the spread of infection seems unabated by neutralizing antibodies (Petursson et al., 1976), the parental viral strains persist longterm even after the appearance of variants (Lutley et al., 1983) and in some cases no antigenic variants are found even during advanced stages of the disease (Thomar et al., 1983). By analogy with visna virus, it is possible that HIV persistence may also be unaffected by the presence of antibody, and this would be consistent with the lack of sequentiality in the mutations of serial isolates from the same patient (vide infra).

## 2. Poliovirus

Another virus that has been extensively studied is poliovirus. In contrast to influenza, it has been possible to successfully vaccinate the population against disease caused by any known poliovirus variant by prior inoculation with three prototype strains of poliovirus. The success of this campaign suggests that poliovirus is antigenically more stable than influenza A virus. Parvin et al. (1986) presented evidence that the mutation rate of poliovirus type 1 is lower than that of influenza A virus. Also, Rico-Hesse et al. (1986) sequenced 150 nucleotides of the type 1 poliovirus genome from 45 isolates obtained from patients in five continents over a 30-year period and found that although these isolates differed from the Sabin 1 vaccine strain at up to 23% of nucleotide positions, 98% of these mutations were silent. Therefore, these results suggest that at least in this area of the genome the structure of poliovirus proteins may be subject to strong functional constraints. The combination of lower mutation rate and little structural flexibility may explain the less dramatic antigenic variation seen with this virus.

## 3. Enterovirus 70

A recently emerging RNA virus, enterovirus 70 (EV70), shows yet another interesting example of variation and its epidemiological consequences (Tanimura *et al.*, 1985; Miyamura *et al.*, 1986). EV70 is the causative agent of acute hemorrhagic conjunctivitis (AHC), which was first reported in Ghana in 1969. During the period from 1969 to 1972, EV70 spread in a pandemic fashion from Africa to Asia and Europe. After a relatively quiescent period of several years, the second pandemic of AHC occurred in 1980–1982 in parts of the world not involved in the first pandemic—India, South East Asia, South America, and southern United States, and the South Pacific Islands. Because of its dramatic and easily recognizable clinical symptoms, it is presumed that EV70 did not exist long before 1969 and that all EV70 isolates obtained over the past 17 years must trace their origin to a single ancestor. Therefore, EV70 offers the unique opportunity to study the variation that occurs in a novel virus when it is released in an immunologically naive population. Unfortunately, comparative nucleotide sequence data is not yet available. However, Takeda *et al.* (1984) have analyzed many EV70 isolates by large RNase T1-resistant oligonucleotide mapping, and find that the proportion of common spots between strains decreased as the years elapsed. This analysis suggests that the virus branched into many strains early during the first pandemic, and has evolved in a divergent fashion, yielding genetically polymorphic viruses. Miyamura *et al.* (1986) constructed a phylogenetic tree of EV70 (based on a comparison of common dots) and thereby calculated a constant and rapid rate of evolution of EV70 of $1.8 \times 10^{-3}$ nucleotide substitutions/site/year. An analysis based on RNase T1-resistant oligonucleotide mapping is subject to several assumptions and thus must await nucleotide sequence data to confirm this high evolutionary rate. Nevertheless, it seems that EV70 evolved rapidly and, because of the absence of initial immunological pressure, all antigenic variants generated had an equal chance of survival. However, it is interesting that the second pandemic did not include areas that were involved in the first pandemic, suggesting that all variants so far generated may still be antigenically cross-reactive and that prior infection by any strain confers immunity to all others. These results indicate that no more pandemics may occur in regions that have suffered in the past, unless an antigenically distinct variant emerges that again will see the previously infected population as immunologically naive.

In conclusion, considerable genetic variation has been observed in many RNA viruses. What makes influenza A virus unique is not the ability to rapidly generate new variants, but the sequential nature of mutations found in isolates over several decades. The strong positive selection of antigenic variants that is responsible for this effect is probably the consequence of the way in which the virus interacts with its hosts: high infectivity, ease of transmission, strong immunogenicity leading to high antibody responses, and neutralization by anti-HA antibodies. The ease of antigenic variant production is due to a different set of properties of the virus—its high mutation rate and structural flexibility (or lack of constraint) in the antigenic areas. It is the combination of these properties that gives influenza A virus its unique pattern of variation.

# REFERENCES

Air, G. M., and Compans, R. W., 1983, Influenza B and influenza C viruses, in: *Genetics of Influenza Viruses* (P. Palese and D. W. Kingsbury, eds.), pp. 280–304, Springer-Verlag, New York.

Akoto-Amanfu, E., Sivasubramanian, N., and Nayak, D. P., 1987, Primary structure of the polymerase acidic (PA) gene of influenza b virus (B/Sing/22/79), *Virology* **159:**147–153.

Askonas, B. A., McMichael, A. J., and Webster, R. G., 1982, The immune response to influenza viruses and the problem of protection against infection, in: *Basic and Applied Research* (A. S. Beare, ed.), pp. 157–188, CRC Press, Boca Raton, Florida.

Baez, M., Palese, P., and Kilbourne, E. D., 1980, Gene composition of high-yielding influenza vaccine strains obtained by recombination, *J. Infect. Dis.* **141:**362–369.

Bao-lan, L., Webster, R. G., Brown, L. E., and Nerome, K., 1983, Heterogeneity of influenza B viruses, *Bull. WHO* **61:**681–687.

Bean, W. J., Cox, N. J., and Kendal, A. P., 1980, Recombination of human influenza A viruses in nature, *Nature (Lond.)* **284:**638–640.

Bean, W. J., Kawaoka, Y., Wood, J. M., Pearson, J. E., and Webster, R. G., 1985, Characterization of virulent and avirulent A/chicken/Pennsylvania/83 influenza A viruses: Potential role of defective interfering RNAs in nature, *J. Virol.* **54:**151–160.

Bean, W. J., and Webster, R. G., 1978, Phenotype properties associated with influenza genome segments, in: *Negative Strand Viruses and the Host Cell* (B. W. J. Mahy and R. D. Barry, eds.), pp. 685–692, Academic, Orlando, Florida.

Beklemishev, A. B., Blynov, V. M., Vassilen, S. K., Golovin, S. Y., Karginov, V. A., Mamayev, L. V., Netesov, S. V., Petrov, N. A., and Safronov, P. F., 1986, Nucleotide sequence of a full-length DNA copy of the influenza virus A-Kiev-59-79 (H1N1)-type nucleoprotein gene, *Bioorg. Khim.* **12:**369–374.

Berton, M. T., Naeve, C. W., and Webster, R. G., 1984, Antigenic structure of the influenza B virus hemagglutinin: Nucleotide sequence analysis of antigenic variants selected with monoclonal antibodies, *J. Virol.* **52:**919–927.

Bishop, D. H. L., Huddleston, J. A., and Brownlee, G. G., 1982a, The complete sequence of RNA segment 2 of influenza A/NT/60/68 and its encoded P1 protein, *Nucl. Acids Res.* **10:**1335–1343.

Bishop, D. H. L., Jones, K. L., Huddleston, J. A., and Brownlee, G. G., 1982b, Influenza A virus evolution: Complete sequences of influenza A/NT/60/68 RNA segment 3 and its predicted acidic P polypeptide compared with those of influenza A/PR/8/34, *Virology* **120:**481–489.

Blok, J., and Air, G. M., 1980, Comparative nucleotide sequences at the 3′ end of the neuraminidase gene from eleven influenza type A viruses, *Virology* **107:**50–60.

Blok, J., and Air, G. M., 1982a, Sequence variation at the 3′ end of the neuraminidase gene from 39 influenza type A viruses. *Virology* **121:**211–229.

Blok, J., and Air, G. M., 1982b, Block deletions in the neuraminidase genes from some influenza A viruses of the N1 subtype, *Virology* **118:**229–234.

Bosch, F. X., Garten, W., Klenk, H.-D., and Rott, R., 1981, Proteolytic cleavage of influenza virus hemagglutinins: Primary structure of the connecting peptide between HA1 and

HA2 determines proteolytic cleavability and pathogenicity of avian influenza viruses, *Virology* **113**:725–735.

Both, G. W., and Sleigh, M. J., 1981, Conservation and variation in the hemagglutinins of Hong Kong subtype influenza viruses during antigenic drift, *J. Virol.* **39**:663–672.

Both, G. W., Sleigh, M. J., Cox, N. J., and Kendal, A. P., 1983, Antigenic drift in influenza virus H3 hemagglutinin from 1968–1980: Multiple evolutionary pathways and sequential amino acid changes at key antigenic sites, *J. Virol.* **48**:52–60.

Brand, C., and Palese, P., 1980, Sequential passage of influenza virus in embryonated eggs or tissue culture: Emergence of mutants, *Virology* **107**:424–433.

Briedis, D. J., and Lamb, R. A., 1982, The influenza B virus RNA segment 8 codes for two nonstructural proteins, *J. Virol.* **42**:186–193.

Briedis, D. J., and Tobin, M., 1984, Influenza B virus genome: Complete nucleotide sequence of the influenza B/Lee/40 virus genome RNA segment 5 encoding the nucleoprotein and comparison with the B/Singapore/222/79 nucleoprotein, *Virology* **133**:448–455.

Briedis, D. J., Lamb, R. A., and Choppin, P. W., 1982, Sequence of RNA segment of the influenza B virus genome: Partial amino acid homology between the membrane proteins (M1) of influenza A and B viruses and conservation of a second open reading frame, *Virology* **116**:581–588.

Buckler-White, A. J., and Murphy, B. R., 1986, Nucleotide sequence analysis of the nucleoprotein gene of an avian and a human influenza virus strain identifies two classes of nucleoproteins, *Virology* **155**:345–355.

Buckler-White, A. J., Naeve, C. W., and Murphy, B. R., 1986, Characterization of a gene coding for M proteins which is involved in host range restriction of an avian influenza A virus in monkeys, *J. Virol.* **57**:697–700.

Buonagurio, D. A., Krystal, M., Palese, P., DeBorde, D. C., and Maassab, H. F., 1984, Analysis of an influenza A virus mutant with a deletion in the NS segment, *J. Virol.* **49**:418–425.

Buonagurio, D. A., Nakada, S., Desselberger, U., Krystal, M., and Palese, P., 1985, Noncumulative sequence changes in the hemagglutinin genes of influenza C virus isolates, *Virology* **146**:221–232.

Buonagurio, D. A., Nakada, S., Fitch, W. M., and Palese, P., 1986a, Epidemiology of influenza C virus in man: Multiple evolutionary lineages and low rate of change, *Virology* **153**:12–21.

Buonagurio, D. A., Nakada, S., Parvin, J. D., Krystal, M., Palese, P., and Fitch, W. M., 1986b, Evolution of human influenza A viruses over 50 years: Rapid, uniform rate of change in NS gene. *Science* **232**:980–982.

Chakraverty, P., 1978, Antigenic relationship between influenza C viruses, *Arch. Virol.* **58**:341–348.

Coffin, J. M., Tsichlis, P. N., Barker, C. S., and Voynow, S., 1980, Variation in avian retrovirus genomes, *Ann. NY Acad. Sci.* **354**:410–425.

Colman, P. M., and Ward, C. W., 1985, Structure and diversity of influenza virus neuraminidase, *Curr. Topics Microbiol. Immunol.* **114**:177–255.

Colman, P. M., Varghese, J. N., and Laver, W. G., 1983, Structure of the catalytic and antigenic sites in influenza virus neuraminidase, *Nature (Lond.)* **303**:41–44.

Concannon, P., Cummings, I. W., and Salser, W. A., 1984, Nucleotide sequence of the influenza virus A/USSR/90/77 hemagglutinin gene, *J. Virol.* **49**:276–278.

Corey, L., Rubin, R. J., Hattwick, M. A. W., Noble, G. R., and Cassidy, E., 1976, A nationwide outbreak of Reye's syndrome: Its epidemiologic relationship to influenza B, *Am. J. Med.* **61**:615–625.

Cox, N. J., 1986, Progress and limitations in understanding the genetic basis for attenuation of live attenuated influenza vaccines, in: *Options for the Control of Influenza* (A. P. Kendal and P. A. Patriarca, eds.), pp. 207–221, Liss, New York.

Cox, N. J., Bai, Z. S., and Kendal, A. P., 1983, Laboratory-based surveillance of influenza A (H1N1) and A (H3N2) viruses in 1980–81: Antigenic and genomic analyses, *Bull. WHO* **61**:143–152.

Daniels, R. S., Downie, J. C., Hay, M., Knossow, M., Skehel, J. J., Wang, M. L., and Wiley, D. C., 1985, Fusion mutants of the influenza virus hemagglutinin glycoprotein, *Cell* **40:** 431–439.

de Borde, D. C., Donabedian, A. M., Herlocher, M. L., Naeve, C. W., and Maassab, H. F., 1988, Sequence comparison of wild type and cold-adapted B/Ann Arbor/1/66 influenza virus genes, *Virology* **163:**429–443.

de la Torre, J. C., Martinez-Salas, E., Diez, J., Villaverde, A., Gebaver, F., Rocha, E., Davila, M., and Domingo, E., 1988, Coevolution of cells and viruses in a persistent infection of foot and mouth disease virus in cell culture, *J. Virol.* **62:**2050–2058.

Deom, C. M., Caton, A. J., and Schultze, I. T., 1986, Host cell-mediated selection of a mutant influenza A virus that has lost a complex oligosaccharide from the tip of the hemagglutinin, *Proc. Natl. Acad. Sci. USA* **83:**3771–3775.

Deshpande, K. L., Fried, V. A., Ando, M., and Webster, R. G., 1987, Glycosylation affects cleavage of an H5N2 influenza virus hemagglutinin and regulates virulence, *Proc. Natl. Acad. Sci. USA* **84:**36–40.

Deshpande, K. L., Naeve, C. L., and Webster, R. G., 1985, The neuraminidase of the virulent and avirulent A/chicken/Pennsylvania/83 (H5N2) influenza A viruses: Sequence and antigenic analysis, *Virology* **147:**49–60.

Desselberger, U., Nakajima, K., Alfino, P., Pedersen, F. S., Haseltine, W. A., Hannoun, C., and Palese, P., 1978, Biochemical evidence that "new" influenza virus strains in nature may arise by recombination (reassortment), *Proc. Natl. Acad. Sci. USA* **75:**3341–3345.

Domingo, E., Sabo, D., Taniguchi, T., and Weissmann, C., 1978, Nucleotide sequence heterogeneity of an RNA phage population, *Cell* **13:**735–744.

Doms, R. W., Gething, M. -J., Henneberry, J., White, J., and Helenius, A., 1986, Variant influenza virus hemagglutinin that induces fusion at elevated pH, *J. Virol.* **57:**603–613.

Dougherty, J. P., and Temin, H. M., 1988, Determination of the rate of base-pair substitution and insertion mutations in retrovirus replication, *J. Virol.* **62:**2817–2822.

Durbin, R. K., and Stollar, V., 1986, Sequence analysis of the E2 gene of a hyperglycosylated, host restricted mutant of Sindbis virus and estimation of mutation rate from frequency of revertants, *Virology* **154:**135–143.

Emini, E. A., Kao, S.-Y., Lewis, A. J., Crainic, R., and Wimmer, E., 1983, Functional basis of poliovirus neutralization determined with monospecific neutralizing antibodies, *J. Virol.* **46:**466–474.

Fang, R., Min Jou, W., Huylebroeck, D., Devos, R., and Fiers, W., 1981, Complete structure of A/Duck/Ukraine/63 influenza hemagglutinin gene: Animal virus as progenitor of human H3 Hong Kong 1968 influenza hemagglutinin, *Cell* **25:**315–323.

Fields, S., and Winter, G., 1982, Nucleotide sequences of influenza virus segments 1 and 3 reveal mosaic structure of a small viral RNA segment, *Cell* **28:**303–313.

Fields, S., Winter, G., and Brownlee, G. G., 1981, Structure of the neuraminidase gene in human influenza virus A/PR/8/34, *Nature (Lond.)* **290:**213–217.

Fitch, W. M., 1971, Toward defining the course of evolution: Minimum change for a specific tree topology, *Syst. Zool.* **20:**406–416.

Gething, M.-J., Bye, J., Skehel, J. J., and Waterfield, M. D., 1980, Cloning and DNA sequence of double-stranded copies of hemagglutinin genes from H2 and H3 strains elucidates antigenic shift and drift in human influenza virus, *Nature (Lond.)* **287:**301–306.

Giesendorf, B., Bosch, F. X., Orlich, M., Scholtissek, C., and Rott, R., 1986, Studies on the temperature sensitivity of influenza A virus reassortants nonpathogenic for chicken, *Virus Res.* **5:**27–42.

Guo, Y. J., and Desselberger, U., 1984, Genome analysis of influenza C viruses isolated in 1981/1982 from pigs in China, *J. Gen. Virol.* **65:**1857–1872.

Guo, Y. J., Jin, F. G., Wang, P., Wang, M., and Zhu, J. M., 1983, Isolation of influenza C virus from pigs and experimental infection of pigs with influenza C virus, *J. Gen. Virol.* **64:** 177–182.

Haase, A., 1986, Pathogenesis of lentivirus infections, *Nature (Lond.)* **322:**130–136.

Hahn, B. H., Shaw, G. M., Taylor, M. E., Redfield, R. R., Markham, P. D., Salahuddin, S. Z.,

Wong-Staal, F., Gallo, R. C., Parks, E. S., and Parks, W. P., 1986, Genetic variation in HTLVIII/LAV over time in patients with AIDS or at risk for AIDS, *Science* **232:**1548–1553.

Hall, R. M., and Air, G. M., 1981, Variation in nucleotide sequences coding for the N-terminal regions of the matrix and nonstructural proteins of influenza A viruses, *Virology* **38:**1–7.

Haller, O., 1981, Inborn resistance of mice to orthomyxoviruses, *Curr. Top. Microbiol. Immunol.* **92:**25–52.

Hauptmann, R., Clarke, L. D., Mountford, R. C., Bachmayer, H., and Almond, J. W., 1983, Nucleotide sequence of the haemagglutinin gene of influenza virus A/England/321/77, *J. Gen. Virol.* **64:**215–220.

Hay, A. J., Wolstenholme, A. J., Skehel, J. J., and Smith, M. H., 1985, The molecular basis of the specific anti-influenza action of amantadine, *EMBO J.* **4:**3021–3024.

Hay, A. J., Wolstenholme, A. J., Zambon, M. C., Skehel, J. J., Smith, M. H., and Wharton, S. A., 1987, The molecular basis of the anti-influenza A action of amantadine and identification of a role for the M2 protein in influenza virus replication, in: *The Biology of Negative Strand Viruses* (B. Mahy and D. Kolakofsky, eds.), pp. 18–25, Elsevier, Amsterdam.

Herrler, G., Durkop, I., Becht, H., and Klenk, H.-D., 1988, The glycoprotein of influenza C virus is the hemagglutinin, esterase and fusion factor, *J. Gen. Virol.* **69:**839–846.

Herrler, G., Rott, R., Klenk, H.-D., Muller, H.-P., Shukla, A. K., and Schauer, R., 1985, The receptor-destroying enzyme of influenza C virus is neuraminate-O-acetylesterase, *EMBO J.* **4:**1503–1506.

Hinshaw, V. S., Bean, W. J., Jr., Webster, R. G., and Easterday, B. C., 1978, The prevalence of influenza viruses in swine and the antigenic and genetic relatedness of influenza viruses from man and swine, *Virology* **84:**51–62.

Hinshaw, V. S., Bean, W. J., Webster, R. G., and Sriram, G., 1980, Genetic reassortment of influenza A viruses in the intestinal tract of ducks, *Virology* **102:**412–419.

Hiti, A. L., and Nayak, D. P., 1982, Complete nucleotide sequence of the neuraminidase gene of human influenza virus A/WSN/33, *J. Virol.* **41:**730–734.

Hiti, A. R., David, A. R., and Nayak, D. P., 1981, Complete sequence analysis shows that the hemagglutinins of the H0 and H2 subtypes of human influenza virus are closely related, *Virology* **111:**113–124.

Holland, J., Spindler, K., Horodyski, F., Grabau, E., Nichol, S., and VandePol, S., 1982, Rapid evolution of RNA genomes, *Science* **215:**1577–1585.

Hovanec, D. L., and Air, G. M., 1984, Antigenic structure of the hemagglutinin of influenza virus B/Hong Kong/8/73 as determined from gene sequence analysis of variants selected with monoclonal antibodies, *Virology* **139:**384–392.

Huddleston, J. A., and Brownlee, G. G., 1982, The sequence of the nucleoprotein gene of human influenza A virus strain A/NT/60/68, *Nucl. Acids Res.* **10:**1029–1038.

Hurwitz, E. S., Nelson, D. B., Davis, C., Morens, D., and Schonberger, L. B., 1982, National surveillance for Reye's syndrome: A five year review, *Pediatrics* **70:**895–900.

Jones, K. L., Huddleston, J. A., and Brownlee, G. G., 1983, The sequence of RNA segment 1 of influenza virus A/NT/60/68 and its comparison with the corresponding segment of strains A/PR/8/34 and A/WSN/33, *Nucl. Acids Res.* **11:**1555–1566.

Kaptein, J. S., and Nayak, D. P., 1982, Complete nucleotide sequence of the polymerase 3 gene of human influenza virus A/WSN/33, *J. Virol.* **42:**55–63.

Katagiri, S., Ohizumi, A., and Homma, M., 1983, An outbreak of type C influenza in a children's home, *J. Infect. Dis.* **148:**51–56.

Kawamura, H., Tashiro, M., Kitame, F., Homma, M., and Nakamura, K., 1986, Genetic variation among human strains of influenza C virus isolated in Japan, *Virus Res.* **4:**275–288.

Kemdirim, S., Palefsky, J., and Briedis, D. J., 1986, Influenza B virus PB1 protein: Nucleotide sequence of the genome RNA segment predicts a high degree of structural homology with the corresponding influenza A virus polymerase protein, *Virology* **152:**126–135.

Kendal, A. P., Goldfield, M., Noble, G. R., and Dowdle, W. R., 1977, Identification and preliminary antigenic analysis of swine influenza-like virus isolated during an influenza outbreak at Fort Dix, New Jersey, *J. Infect. Dis.* **136:**381–385.

Kendal, A. P., Noble, G. R., Skehel, J. J., and Dowdle, W. R., 1978, Antigenic similarity of influenza A (H1N1) viruses from epidemics in 1977–1978 to "Scandinavian" strains isolated in epidemics of 1950–1951, *Virology* **89:**632–636.

Kilbourne, E. D., and Murphy, J. S., 1960, Genetic studies of influenza viruses. I. Viral morphology and growth capacity as exchangeable genetic traits. Rapid *in vivo* adaptation of early passage Asian strain isolates by combination with PR8, *J. Exp. Med.* **111:**387–415.

Kilbourne, E. D., Schulman, J. L., Schild, G. C., Schloer, G., Swanson, J., and Bucher, D., 1971, Correlated studies of a recombinant influenza virus vaccine. I. Derivation and characterization of virus and vaccine, *J. Infect. Dis.* **124:**449–462.

Krystal, M., Elliott, R. M., Benz, E. W., Young, J. F., and Palese, P., 1982, Evolution of influenza A and B viruses: Conservation of structural features in the hemagglutinin gene, *Proc. Natl. Acad. Sci. USA* **79:**4800–4804.

Krystal, M., Young, J. F., Palese, P., Wilson, I. A., Skehel, J. J., and Wiley, D. C., 1983, Sequential mutations in the hemagglutinins of influenza B virus isolates: Definition of antigenic domains, *Proc. Natl. Acad. Sci. USA* **80:**4527–4531.

Lamb, R. A., 1983, The influenza virus RNA segments and their encoded proteins, in: *Genetics of Influenza Viruses* (P. Palese and D. W. Kingsbury, eds.), pp. 21–69, Springer-Verlag, New York.

Lamb, R. A., and Choppin, P. W., 1981, Identification of a second protein (M2) encoded by RNA segment 7 of influenza virus, *Virology* **112:**729–737.

Lamb, R. A., and Lai, C.-J., 1981, Conservation of the influenza virus membrane protein (M1) amino acid sequence and an open reading frame of RNA segment 7 encoding a second protein (M2) in H1N1 and H3N2 strains, *Virology* **112:**746–751.

Laver, W. G., and Webster, R. G., 1973, Studies on the origin of pandemic influenza. III. Evidence implicating duck and equine influenza viruses as possible progenitors of the Hong Kong strain of human influenza, *Virology* **51:**383–391.

Laver, W. G., Air, G. M., Webster, R. G., and Markoff, L. J., 1982, Amino acid sequence changes in antigenic variants of type A influenza virus N2 neuraminidase, *Virology* **122:**450–460.

Leider, J. M., Palese, P., and Smith, F. I., 1988, Determination of the mutation rate of a retrovirus, *J. Virol.* **62:**3084–3091.

Li, W.-H., and Gojobori, T., 1983, Rapid evolution of goat and sheep globin genes following gene duplication, *Mol. Biol. Evol.* **1:**94–108.

Londo, D. R., Davis, A. R., and Nayak, D. P., 1983, Complete nucleotide sequence of the nucleoprotein gene of influenza B virus, *J. Virol.* **47:**642–648.

Lu, B. L., Webster, R. G., Brown, L. E., and Nerome, K., 1983, Heterogenicity of influenza B viruses, *Bull. WHO* **61:**681–687.

Lubeck, M. D., Schulman, J. L., and Palese, P., 1978, Susceptibility of influenza A viruses to amantadine is determined by the gene coding for M protein, *J. Virol.* **28:**710–716.

Lubeck, M. D., Schulman, J. L., and Palese, P., 1980, Antigenic variants of influenza viruses: Marked differences in the frequencies of variants selected with different monoclonal antibodies, *Virology* **102:**458–462.

Lutley, R., Petursson, C., Palsson, P. A., Georgsson, G., Klein, J., and Nathansson, N., 1983, Antigenic drift in visna: Virus variation during longterm infection of Icelandic sheep, *J. Gen. Virol.* **64:**1433–1440.

Maassab, H. F., 1968, Adaptation and growth characteristics of influenza virus at 25°C, *Nature (Lond.)* **219:**645–646.

Markoff, L., and Lai, C.-J., 1982, Sequence of the influenza A/Udorn/72 (H3N2) virus neuraminidase gene as determined from cloned full-length DNA, *Virology* **119:**288–297.

Markushin, S., Ghiasi, H., Sokolov, N., Shilov, A., Sinitsin, B., Brown, D., Klimov, A., and Nayak, D., 1988, Nucleotide sequence of RNA segment 7 and the predicted amino sequence of M1 and M2 proteins of FPV/Weybridge (H7N7) and WSN (H1N1) influenza viruses, *Virus Res.* **10**:263–272.

Martinez, C., Del Rio, L., Portela, A., Domingo, E., and Ortin, J., 1983, Evolution of the influenza virus neuraminidase gene during drift of the N2 subtype, *Virology* **130**:539–545.

Medvedeva, T. E., Gordon, M. A., Ghendon, Y. Z., Klimov, A. I., and Alexandrova, G. I., 1983, Attenuated influenza B virus recombinants obtained by crossing of B/England/2608/76 virus with a cold-adapted B/Leningrad/14/17/55 strain, *Acta Virol.* **27**:311.

Meier-Ewert, H., Petri, T., and Bishop, D. H. L., 1981, Oligonucleotide fingerprint analyses of influenza C virion RNA recovered from five different isolates, *Arch. Virol.* **67**:141–147.

Minor, P. D., Schild, G. C., Bootman, J., Evans, D. M. A., Ferguson, M., Reeve, P., Spitz, M., Stanway, G., Cann, A. J., Hauptmann, R., Clarke, L. D., Mountford, R. C., and Almond, J. W., 1983, Location and primary structure of a major antigenic site for poliovirus neutralization, *Nature (Lond.)* **310**:674–679.

Miyamura, K., Tanimura, M., Takeda, N., Kono, R., and Yamazaki, S., 1986, Evolution of enterovirus 70 in nature: All isolates were recently derived from a common ancestor, *Arch. Virol.* **89**:1–14.

Murphy, B. R., and Chanock, R. M., 1981, Genetic approaches to the prevention of influenza A virus infection, in: *Genetic Variation among Influenza Viruses* (D. Nayak, ed.), pp. 601–615, Academic, Orlando, Florida.

Murphy, B. R., Sly, D. L., Tierney, E. L., Hosier, E. L., Massicot, J. G., and Hinshaw, V. S., 1982, Influenza A reassortant virus derived from avian and human influenza A virus is attenuated and immunogenic in monkeys, *Science* **218**:1330–1332.

Naeve, C. W., Hinshaw, V. S., and Webster, R. G., 1984, Mutations in the hemagglutinin receptor-binding site can change the biological properties of an influenza virus, *J. Virol.* **51**:567–569.

Nakada, S., Creager, R. S., Krystal, M., Aaronson, R. P., and Palese, P., 1984a, Influenza C virus hemagglutinin: Comparison with influenza A and B virus hemagglutinins, *J. Virol.* **50**:118–124.

Nakada, S., Creager, R. S., Krystal, M., and Palese, P., 1984b, Complete nucleotide sequence of the influenza C/California/78 virus nucleoprotein gene, *Virus Res.* **1**:433–441.

Nakada, S., Graves, P. N., and Palese, P., 1986, The influenza C virus NS gene: Evidence for a spliced mRNA and a second NS gene product (NS2 protein), *Virus Res.* **4**:263–273.

Nakajima, S., Takeuchi, Y., and Nakajima, K., 1988, Location on the evolutionary tree of influenza H3 hemagglutinin genes of Japanese strains isolated during the 1985–86 season, *Epidemiol. Infect.* **100**:301–310.

Nakajima, K., Desselberger, U., and Palese, P., 1978, Recent human influenza A (H1N1) viruses are closely related genetically to strains isolated in 1950, *Nature (Lond.)* **274**:334–339.

Narayan, O., Griffin, D. E., and Chase, J., 1977, Antigenic shift of visna virus in persistently infected sheep, *Science* **197**:376–378.

Newton, S. E., Air, G. M., Webster, R. G., and Laver, W. G., 1983, Sequence of the hemagglutinin gene of influenza virus A/Memphis/1/71 and previously uncharacterized monoclonal antibody-derived variants, *Virology* **128**:495–501.

Norton, G. P., Tanaka, T., Tobita, K., Nakada, S., Buonagurio, D. A., Greenspan, D., Krystal, M., and Palese, P., 1987, Infectious influenza A and B virus variants with long carboxyl terminal deletions in the NS1 polypeptide, *Virology* **156**:204–213.

O'Callaghan, R. J., Gohd, R. S., and Labat, D. D., 1980, Human antibody to influenza C virus: Its age-related distribution and distinction from receptor analogs, *Infect. Immun.* **30**:500–505.

Ortin, J., Martinez, C., Del Rio, L., Davila, M., Lopez-Galindez, C., Villanueva, N., and Domingo, E., 1983, Evolution of the nucleotide sequence of influenza virus RNA segment 7 during drift of the H3N2 subtype, *Gene* **23**:233–239.

Oxford, J. S., Corcoran, T., and Schild, G. C., 1981, Intratypic electrophoretic variation of structural and non-structural polypeptides of human influenza A viruses, *J. Gen. Virol.* **56:**431–436.

Oxford, J. S., Klimov, A. I., Corcoran, T., Ghendon, Y. Z., and Schild, G. C., 1984, Biochemical and serological studies of influenza B viruses: Comparisons of historical and recent isolates, *Virus Res.* **1:**241–258.

Palese, P., 1977, The genes of influenza virus, *Cell* **10:**1–10.

Palese, P., and Young, J. F., 1982, Variation of influenza A, B, and C viruses, *Science* **215:** 1468–1474.

Parvin, J. D., Young, J. F., and Palese, P., 1983, Nonsense mutations affecting the lengths of the NS1 nonstructural proteins of influenza A virus isolates, *Virology* **128:**512–517.

Parvin, J. D., Moscona, A., Pan, W. T., Leider, J. M., and Palese, P., 1986, Measurement of the mutation rates of animal viruses: Influenza A virus and poliovirus type 1, *J. Virol.* **59:** 377–383.

Petursson, G., Nathanson, N., Georgsson, G., Panitch, H., and Palsson, P. A., 1976, Pathogenesis of visna, sequential virologic, serologic and pathologic studies, *Lab. Invest.* **35:** 402–412.

Portner, A., Webster, R. G., and Bean, W. J., 1980, Similar frequencies of antigenic variants in Sendai, vesicular stomatitis, and influenza A viruses, *Virology* **104:**235–238.

Prabhakar, B. S., Haspel, M. V., McClintock, P. R., and Notkins, A. L., 1982, High frequency of antigenic variants among naturally occurring human Coxsackie B4 virus isolates identified by monoclonal antibodies, *Nature (Lond.)* **300:**374–376.

Raymond, R. L., Caton, A. J., Cox, N. J., Kendal, A. P., and Brownlee, G. G., 1986, The antigenicity and evolution of influenza H1 haemagglutinin, from 1950–1957 and 1977–1983: Two pathways from one gene, *Virology* **148:**275–287.

Ricco-Hesse, R., Pallansch, M. A., Nottay, B. K., and Kew, O., 1986, Natural distribution of wild type 1 poliovirus genotypes, in: *Positive Strand RNA Viruses*, UCLA Symposium on Molecular and Cellular Biology, New Series, Vol. 54 (M. A. Brinton and R. Rueckert, eds.), pp. 477–486, Liss, New York.

Robertson, J. S., Naeve, C. W., Webster, R. G., Bootman, J. S., Newman, R., and Schild, G., 1985, Alterations in the hemagglutinin associated with adaptation of influenza B virus to growth in eggs, *Virology* **143:**166–174.

Rogers, G. N., Paulson, J. C., Daniels, R. S., Skehel, J. J., Wilson, I. A., and Wiley, D. C., 1983, Single amino acid substitutions in influenza hemagglutinin change receptor binding specificity, *Nature (Lond.)* **304:**76–78.

Rott, R., Orlich, M., and Scholtissek, C., 1979, Correlation of pathogenicity and gene constellation of influenza A viruses. III. Nonpathogenic recombinants derived from highly pathogenic parent strains, *J. Gen. Virol.* **44:**471–477.

Rott, R., Orlich, M., Klenk, H.-D., Wang, M. L., Skehel, J. J., and Wiley, D. C., 1984, Studies on the adaptation of influenza viruses to MDCK cells, *EMBO J.* **3:**3329–3332.

Salinovitch, O., Payne, S. L., Montelaro, R. C., Hussain, K. A., Issel, C. J., and Schnorr, K. L., 1986, Rapid emergence of novel antigenic and genetic variants of equine infectious anemia virus during persistent infection, *J. Virol.* **57:**71–80.

Samokhvalov, E. I., Karginov, V. A., Yuferov, V. P., Tschishkov, V. A., Blinov, V. M., Vasilenko, L. W., Uryvaev, L. W., and Zhdanov, V. M., 1985, Primary structure of RNA segment 7 of A/USSR/90/77 (H1N1) influenza virus, *Bioorg. Khim.* **11:**1080–1085.

Schild, G. C., Oxford, J. S., and Newman, R. W., 1979, Evidence for antigenic variation in influenza A nucleoprotein, *Virology* **93:**569–573.

Scholtissek, C., 1986, Molecular biological background of the species and organ specificity of influenza A viruses, *Angew. Chem. Int. Ed. Engl.* **25:**47–56.

Scholtissek, C., Burger, H., Kistner, O., and Shortridge, K. F., 1985, The nucleoprotein as a possible major factor in determining host specificity of influenza H3N2 viruses, *Virology* **147:**287–294.

Scholtissek, C., Rott, R., Orlich, M., Harms, E., and Rohde, W., 1977, Correlation of pathogenicity and gene contellation of an influenza A virus (fowl plague). I. Exchange of a single gene, *Virology* **81:**74–80.

Scholtissek, C., Rohde, W., Von Hoyningen, V., and Rott, R., 1978a, On the origin of the human influenza virus subtypes H2N2 and H3N2, *Virology* **87**:13–20.

Scholtissek, C., Von Hoyningen, V., and Rott, R., 1978b, Genetic relatedness between the new 1977 epidemic strains (H1N1) of influenza and human influenza strains isolated between 1947 and 1957 (H1N1), *Virology* **89**:613–617.

Scholtissek, C., Vallbracht, A., Flehmig, B., and Rott, R., 1979, Correlation of pathogenicity and gene constellation of influenza A viruses. II. Highly neurovirulent recombinants derived from non-neurovirulent or weakly neurovirulent parent virus strains, *Virology* **95**:492–500.

Schreier, E., Roeske, H., Driesel, G., Kunkel, U., Petzold, D. R., Berlinghoff, R., and Michel, S., 1988, Complete nucleotide sequence of the neuraminidase gene of the human influenza virus A/Chile/1/83 (H1N1) *Arch. Virol.* **99**:271–276.

Schulman, J. L., 1983, Virus-determined differences in the pathogenesis of influenza viruses, in: *Genetics of Influenza Viruses* (P. Palese and D. W. Kingsbury, eds.), pp. 305–320, Springer-Verlag, New York.

Schulman, J. L., and Kilbourne, E. D., 1969, Independent variation in nature of hemagglutinin and neuraminidase antigens of influenza virus: Distinctiveness of hemagglutinin antigen of Hong Kong/68 virus, *Proc. Natl. Acad. Sci. USA* **63**:326–333.

Schulman, J. L., and Palese, P., 1977, Virulence factors of influenza viruses. WSN virus neuraminidase is required for productive infection of MDBK cells, *J. Virol.* **24**:170–176.

Schulman, J. L., and Palese, P., 1978, Biological properties of influenza A/Hong Kong and PR8 viruses: Effects of genes for matrix protein and nucleoprotein on virus yield in embryonated eggs, in: *Negative Strand Viruses and the Host Cell* (B. W. J. Mahy and R. D. Barry, eds.), pp. 663–674, Academic, Orlando, Florida.

Sedivy, J. M., Capone, J. P., RajBhandary, U. L., and Sharp, P. A., 1987, An inducible mammalian amber suppressor: Propagation of a poliovirus mutant, *Cell* **50**:379–389.

Sivasubramanian, N., and Nayak, D. P., 1982, Sequence analysis of the polymerase 1 gene and the secondary structure prediction of polymerase 1 protein of human influenza virus A/WSN/33, *J. Virol.* **44**:321–329.

Skehel, J. J., and Wiley, D. C., 1988, Antigenic variation in influenza virus hemagglutinins, in: *RNA Genetics*, Vol. III: *Variability of RNA Genomes* (E. Domingo, J. J. Holland, and P. Ahlquist, eds.), pp. 139–146, CRC Press, Boca Raton, Florida.

Sleigh, M. J., Both, G. W., Underwood, P. A., and Bender, V. J., 1981, Antigenic drift in the hemagglutinin of the Hong Kong influenza subtype: Correlation of amino acid changes with alterations in viral antigenicity, *J. Virol.* **37**:845–853.

Snyder, M. H., Buckler-White, A. J., London, W. T., Tierney, E. L., and Murphy, B. R., 1987, The avian influenza virus nucleoprotein gene and a specific constellation of avian and human virus polymerase genes each specify attenuation of avian–human influenza A/pintail/79 reassortant viruses for monkeys, *J. Virol.* **61**:2857–2863.

Sobrino, F., Palma, E. L., Beck, E., Davila, M., de la Torre, J. C., Negro, P., Villanueva, N., Ortin, J., and Domingo, E., 1986, Fixation of mutations in the viral genome during an outbreak of foot-and-mouth disease: Heterogeneity and rate variations, *Gene* **50**:149–159.

Starcich, B. R., Hahn, B. H., Shaw, G. M., McNeely, P. D., Modrow, S., Wolf, H., Parks, E. S., Parks, W. P., Josephs, S. F., Gallo, R. C., and Wong-Staal, F., 1986, Identification and characterization of conserved and variable regions in the envelope gene of HTLVIII/LAV, the retrovirus of AIDS, *Cell* **45**:637–648.

Steinhauer, D. A., and Holland, J. J., 1986, Direct method for quantitation of extreme polymerase error frequencies at selected single base sites in viral RNA, *J. Virol.* **57**:219–228.

Stoeckle, M. Y., Shaw, M. W., and Choppin, P. W., 1987, Segment-specific and common nucleotide sequences in the non-coding regions of influenza B virus genome RNAs, *Proc. Natl. Acad. Sci. USA* **84**:2703–2707.

Sugiura, A., and Ueda, M., 1980, Neurovirulence of influenza virus in mice. I. Neurovirulence of recombinants between virulent and avirulent strains, *Virology* **101**:440–449.

Takeda, N., Miyamura, K., Ogino, T., Natori, K., Yamazaki, S., Sakurai, N., Nakazono, N., Ishii, K., and Kono, R., 1984, Evolution of enterovirus type 70: Oligonucleotide mapping analysis of RNA genome, *Virology* **134**:375–388.

Tanimura, M., Miyamura, K., and Takeda, N., 1985, Construction of a phylogenetic tree of enterovirus 70, *Jpn. J. Genet.* **60**:137–150.

Thormar, H., Barshatzky, M. R., Arnesen, K., and Kozlowski, P. B., 1983, The emergence of antigenic variants is a rare event in long-term visna virus infection in vivo, *J. Gen. Virol.* **64**:1427–1432.

Tian, S. F., Buckler-White, A. J., London, W. T., Reck, L. J., Chanock, R. M., and Murphy, B. R., 1985, Nucleoprotein and membrane protein genes are associated with restriction of replication of influenza A/Mallard/NY/78 virus and its reassortants in squirrel monkey respiratory tract, *J. Virol.* **53**:771–775.

Van Wyke, K. L., Yewdell, J. W., Reck, L. J., and Murphy, B. R., 1984, Antigenic characterization of influenza A virus matrix protein with monoclonal antibodies, *J. Virol.* **49**:248–252.

Van Wyke, K. L., Hinshaw, V. S., Bean, W. J., and Webster, R. G., 1980, Antigenic variation of influenza A virus nucleoprotein detected with monoclonal antibodies, *J. Virol.* **35**:24–30.

Verhoeyen, M., Fang, R., Min Jou, W., Devos, R., Huylebroeck, D., Saman, E., and Fiers, W., 1980, Antigenic drift between the haemagglutinin of the Hong Kong influenza strains A/Aichi/2/68 and A/Victoria/3/75, *Nature (Lond.)* **286**:771–776.

Verhoeyen, M., Van Rompuy, L., Min Jou, W., Huylebroeck, D., and Fiers, W., 1983, Complete nucleotide sequence of the influenza B/Singapore/222/79 virus hemagglutinin gene and comparison with the B/Lee/40 hemagglutinin, *Nucl. Acids Res.* **11**:4703–4712.

Vlasak, R., Krystal, M., Nacht, M., and Palese, P., 1987, The influenza C virus glycoprotein (HE) exhibits receptor-binding (hemagglutinin) and receptor-destroying (esterase) activities, *Virology* **160**:419–425.

Ward, C. W., and Dopheide, T. A., 1981a, Amino acid sequence and oligosaccharide distribution of the hemagglutinin from an early Hong Kong influenza virus variant A/Aichi/2/68 (X-31), *Biochem. J.* **193**:953–962.

Ward, C. W., and Dopheide, T. A., 1981b, Evolution of the Hong Kong influenza A subtype, *Biochem. J.* **195**:337–340.

Webster, R. G., and Berton, M. T., 1981, Analysis of antigenic drift in the hemagglutinin molecule of influenza B virus with monoclonal antibodies, *J. Gen. Virol.* **54**:243–251.

Webster, R. G., Campbell, C. H., and Granoff, A., 1971, The *in vivo* production of "new" influenza viruses. I. Genetic recombination between avian and mammalian influenza viruses, *Virology* **44**:317–328.

Webster, R. G., Hinshaw, V. S., and Laver, W. G., 1982, Selection and analysis of antigenic variants of the neuraminidase of N2 influenza viruses with monoclonal antibodies, *Virology* **117**:93–104.

Webster, R. G., Kawaoka, Y., and Bean, W. J., 1986, Molecular changes in A/chicken/Pennsylvania/83 (H5N2) influenza virus associated with acquisition of virulence, *Virology* **149**:165–173.

Webster, R. G., and Rott, R., 1987, Influenza virus A pathogenicity: The pivotal role of hemagglutinin, *Cell* **50**:665–666.

White, J., Kielan, M., and Helenius, A., 1983, Membrane fusion proteins of enveloped animal viruses, *Q. Rev. Biophys.* **16**:151–195.

Winter, G., and Fields, S., 1980, Cloning of influenza cDNA into M13: The sequence of the RNA segment encoding the A/PR/8/34 matrix protein, *Nucl. Acids Res.* **8**:1965–1974.

Winter, G., and Fields, S., 1981, The structure of the gene encoding the nucleoprotein of human influenza virus A/PR/8/34, *Virology* **114**:423–428.

Winter, G., and Fields, S., 1982, Nucleotide sequence of human influenza A/PR/8/34 segment 2, *Nucl. Acids Res.* **10**:2135–2143.

Winter, G., Fields, S., and Brownlee, G. G., 1981, Nucleotide sequence of the haemagglutinin of a human influenza virus H1 subtype, *Nature (Lond.)* **292**:72–75.

Yamashita, M., Krystal, M., and Palese, P., 1988a, Evidence that the matrix protein of influenza C virus is coded for by a spliced mRNA, *J. Virol.* **62**:3348–3355.

Yamashita, M., Krystal, M., Fitch, W. M., and Palese, P., 1988b, Influenza B virus evolution: Co-circulating lineages and comparison of evolutionary patterns with those of influenza A and C viruses, *Virology* **163**:112–123.

Yewdell, J. W., Caton, A. J., and Gerhard, W., 1986, Selection of influenza A virus adsorptive mutants by growth in the presence of a mixture of monoclonal anti-hemagglutinin antibodies, *J. Virol.* **57**:623–628.

Young, J. F., and Palese, P., 1979, Evolution of human influenza A viruses in nature: Recombination contributes to genetic variation of H1N1 strains, *Proc. Natl. Acad. Sci. USA* **76**:6547–6551.

Young, J. F., Desselberger, U., and Palese, P., 1979, Evolution of human influenza A viruses in nature: Sequential mutations in the genomes of new H1N1 isolates, *Cell* **18**:73–83.

CHAPTER 8

# Specificity and Function of T Lymphocytes Induced by Influenza A Viruses

JONATHAN W. YEWDELL AND CHARLES J. HACKETT

## I. INTRODUCTION

### A. Why Study T-Lymphocyte Responses to Influenza Virus?

Over the past 20 years, it has become increasingly clear that thymus-derived lymphocytes (T lymphocytes) play a pivotal role in immune responsiveness. Perhaps nowhere is this more apparent than in antiviral immunity, in which T lymphocytes provide a critical helper function in antibody responses and also function directly to reduce viral replication. Of the large number of viruses known to elicit T-lymphocyte responses, influenza virus has been the most extensively studied.

The attractiveness of influenza virus for cellular immunologists can be attributed to at least three factors: (1) influenza virus elicits vigorous responses from all defined T-lymphocyte subsets; (2) the well-known ability of influenza virus to avoid existing serological immunity has focused attention on the possibility of producing vaccines designed to elicit T-cell immunity, which is highly cross-reactive between all human influenza A viruses; and (3) a wealth of information exists concerning the structure and function of the virus and its components. Although much

JONATHAN W. YEWDELL • Laboratory of Viral Diseases, National Institute of Allergy and Infectious Diseases, National Institutes of Health, Rockville, Maryland 20852. CHARLES J. HACKETT • Wistar Institute of Anatomy and Biology, Philadelphia, Pennsylvania 19104.

remains to be learned, the broad strategies of viral penetration, replication, and assembly have been determined. All the viral genes have been cloned, sequenced, and placed in vectors that permit expression in human or mouse cells (Gething and Sambrook, 1981; Winter et al., 1981; Smith et al., 1983; Panicali et al., 1983; Young et al., 1983). A large number of virus strains with well-defined differences are available for studies of T-lymphocyte specificity, including laboratory produced variants that differ from wild-type virus by single amino acid substitutions in one of their two surface glycoproteins (Webster et al., 1982). The three-dimensional structure of each of the glycoproteins has been determined at high resolution by crystallographic means (Wilson et al., 1981; Varghese et al., 1983), and a number of their antigenic sites recognized by monoclonal antibodies have been located (Wiley et al., 1981; Caton et al., 1982). The virus further provides the researcher with a number of other proteins with different structural and biochemical properties, many of which can be obtained in large quantities. Taken together, these factors combine to allow for experimental manipulations not possible with other antigens or, for that matter, other viruses.

This chapter reviews current understanding of the specificity and function of T lymphocytes induced by influenza A virus. Particular emphasis has been placed on the nature of the viral determinants recognized by T lymphocytes, as the recent findings in this area have important implications regarding T-lymphocyte specificity, and cellular processing of exogenously and endogenously synthesized proteins for T-lymphocyte recognition. First, however, it is necessary to provide some basic information regarding T-lymphocyte recognition of foreign antigens and the structure and replication of influenza virus [for more comprehensive reviews of these topics the reader is referred respectively to Klein's comprehensive text (Klein, 1986) and to the other chapters of this volume of *The Viruses*].

## B. Background Information

### 1. Influenza Virus Structure and Function

Type A influenza viruses belong to the Orthomyxoviridae family and, along with the serologically non-cross-reactive type B influenza viruses, form the influenzavirus genus. Throughout the remainder of this chapter, influenza virus refers solely to influenza A viruses. Natural human influenza virus isolates are grouped into three subtypes; H1N1, H2N2, and H3N2. H1, H2, and H3, and N1 and N2 each designate largely non-cross-reactive serotypes of the hemagglutinin (HA) and neuraminidase (NA) molecules, respectively. Individual isolates of human influenza viruses are designated by the location, sample number, and year of isolation, and by their subtype; e.g., the eighth virus isolated in Puerto Rico in 1934 is designated PR/8/34 (H1N1).

The external virion surface consists of two integral membrane glycoproteins: the homotrimeric HA [subunit $M_r$ 63,000 (63K) nonglycosylated, 76K glycosylated], and homotetrameric NA (subunit $M_r$ 50K nonglycosylated, 60K glycosylated). These proteins form a dense layer of spikes embedded in a lipid bilayer derived from the host cell (~500 HA and 50 NA molecules per spherical virion with 100-nm diameter). Within the virion are five other viral proteins and the eight single-stranded negative sense RNA gene segments that constitute the viral genome. Matrix protein [M1 ($M_r$ 28K)] is thought to form a subenvelope shell encasing the ribonucleoprotein complex, which is composed of the RNA segments, nucleoprotein (NP) ($M_r$ 56K), and small amounts of three viral polymerases: PA ($M_r$ 83K), PB1 ($M_r$ 86K), and PB2 ($M_r$ 84K). Three other viral proteins that are largely, if not completely, excluded from virions are also synthesized by infected cells. Two of these proteins, nonstructural 1, NS1 ($M_r$ 28K) and NS2 ($M_r$ 14K), are produced from the same gene segment; the mRNA from the smaller protein arises from splicing of the NS1 mRNA (Lamb, 1983). The third, M2 ($M_r$ 11K), is an integral membrane protein produced by splicing of the mRNA that encodes M1 (Lamb et al., 1985).

The serological relationship between virus proteins derived from different virus isolates parallels their amino acid homology. The individual serotypes of the glycoproteins can exhibit as little as 25% sequence homology in their antigenically relevant portions. Within a serotype, glycoproteins can exhibit 85–99% homology; their antigenic cross-reactivity varies accordingly (Ward, 1981; Raymond et al., 1986). Although nonglycoproteins derived from different subtypes can often be distinguished using monoclonal antibodies, these proteins are highly conserved between all human isolates (>90% amino acid homology) and consequently are largely serologically cross-reactive.

The infectious cycle is initiated by binding of the virus to host-cell sialic acid residues. Release of the viral ribonucleoprotein complex into the cytoplasm is believed to occur only following internalization of virus into cellular endosomes, where the mildly acidic conditions (pH 5) induce fusion of viral and cellular membranes (Matlin et al., 1981). Synthesis of NP and NS1 can first be detected 1–2 hr following virus adsorption; 2–3 hr later synthesis of all the viral proteins can be detected. With the exception of M1 and the three integral membrane proteins, all viral proteins are present at least transiently in the nucleus of infected cells, in which transcription of viral RNA occurs. The three viral integral membrane proteins are expressed in significant quantities on infected cells surfaces (on the order of $10^6$ HA molecules per cell, 10-fold less NA and M2). Final maturation of virions occurs beginning 6–8 hr postinfection (p.i.). During this process, the M1–ribonucleoprotein complex buds through the host plasma membrane, incorporating cell-surface HA and NA while excluding M2 and host plasma membrane proteins.

As a large number of the studies discussed below are concerned with T-lymphocyte recognition of the HA, it is important to present some details of HA structure and function. The HA mediates both virus attachment to host cells and the subsequent fusion of viral and cellular membranes. The HA trimer consists of a globular head attached to the viral envelope by a fibrous tail (Wilson et al., 1981; Wiley et al., 1981). Contained within the globular head are the binding sites for host-cell sialic acid residues, as well as the epitopes recognized by antibodies that neutralize viral infectivity. HA fusion activity (and viral infectivity) is dependent on its proteolytic cleavage into disulfide-linked subunits, termed HA1 ($M_r$ 48K) and HA2 ($M_r$ 28K) (Lazarowitz and Choppin, 1975; Klenk et al., 1975). Fusion is triggered on exposure of the HA to mildly acidic conditions (Maeda and Ohnishi, 1981) that induce irreversible conformational alterations in the HA resulting in decreased interaction between the monomeric subunits (Skehel et al., 1982; Nestorowicz et al., 1985). This may permit greater access to the host cell membrane of the HA2 amino-terminal decapeptide, which is located in the fibrous tail and is believed to mediate the fusion process.

## 2. Biology of T-Lymphocyte Responses

The distinguishing characteristic of T lymphocytes is that, unlike antibodies, or antibody-producing cells (B lymphocytes), which simply recognize free antigen, T lymphocytes recognize antigen on the surface of antigen presenting cells (APCs) in conjunction with proteins encoded by the major histocompatibility complex (MHC), known as H-2 in mice and HLA in man.

Two classes of restriction molecules are encoded by the MHC. Class I molecules consist of an integral membrane glycoprotein ($M_r$ 44K) non-covalently complexed with $\beta_2$-microglobulin ($M_r$ 12K), a protein that also exists in large quantities in noncomplexed form in serum. The three-dimensional structure of a human class I molecule has been solved by crystallographic means. The most remarkable feature of the class I molecule is a groove located in the region most distal from the membrane anchor domain, that is likely to serve as an antigen binding site (Bjorkman et al. 1987a,b). Class I molecules are expressed on virtually all somatic cells. Mouse cells produce two or three, and maybe more, class I gene products that can serve as T-lymphocyte restriction elements. The loci encoding these genes are termed H-2 K, D, and L. The number of human class I restriction elements is uncertain, but there appear to be three loci encoding class I genes, termed A, B, and C. In both humans and mice, a large number of alleles exist for each class I gene. In mice, combinations of alleles that occur in certain strains, termed haplotypes, are arbitrarily assigned superscript letters (e.g., H-2$^k$), as are the alleles present at the individual loci in these mice (e.g., $K^k$) (see Festenstein and Demant, 1978).

Class II molecules consist of two noncovalently bound integral membrane glycoproteins, termed $\alpha$-($M_r$ 34K), and $\beta$-($M_r$ 28K) subunits (for review, see Mengle-Gaw and McDevitt, 1985). Unlike class I molecules, class II molecules are normally expressed on a limited number of cell types including macrophages and B lymphocytes. Class II molecules are also expressed on T lymphocytes in humans, but not in mice. In the mouse, two $\alpha$-subunits, termed A and E, and two $\beta$-subunits, also termed A and E, are known to be used as restriction elements. Subunits of the same designation associate preferentially. Genes encoding these proteins are located in the H-2 I region. In humans, there are at least six loci producing $\alpha$-subunits and seven loci producing $\beta$-subunits. These loci cluster into four regions, designated DP, DQ, DR, and DZ. The use of all but a few of the human $\alpha$- and $\beta$-subunits as restriction elements remains to be determined. As with class I molecules, a large number of alleles exist for each class II gene. The nomenclature for these alleles in mice is similar to that used for class I antigens.

T lymphocytes are divided into two major classes, termed cytotoxic T lymphocytes ($T_C$), and helper lymphocytes ($T_H$). This terminology reflects what were thought to be absolute functional differences between the classes. In recent years, however, these differences have blurred. While it was thought that $T_H$ functioned primarily to secrete lymphokines with important roles in B-cell and $T_C$ growth and differentiation, including interleukin-2 (IL-2), it is now clear that $T_H$ can also lyse target cells in an antigen-specific MHC-restricted manner (Wagner et al., 1977; Kaplan et al., 1984; Lukacher et al., 1985; Fleischer et al., 1985). Similarly, it is also known that in addition to lysing cells expressing foreign antigens, $T_C$ release lymphokines that have major biological effects, e.g., $\gamma$-interferon ($IFN_\gamma$) (Ennis, 1982; Morris et al., 1982; Klein et al., 1982). There are, however, at least two clear-cut functional differences between $T_C$ and $T_H$ subsets. First, $T_H$ require IL-1 for proliferation, while $T_C$ do not. Second, $T_C$ recognize antigen in conjunction with class I MHC molecules, while $T_H$ recognize antigen in conjunction with class II molecules. The functional distinction between $T_C$ and $T_H$ subsets is supported by their differential expression of cell surface molecules, defined by reactivity with antibodies. Both classes of cells express a pan T-cell marker, CD2. $T_H$ generally express CD4, while $T_C$ generally express CD8.

In addition to these class-specific markers, each T lymphocyte expresses a clonally unique antigen receptor, also known as T-lymphocyte idiotype structure, or Ti. This has been shown for both subclasses of T lymphocytes to be a disulfide-bonded heterodimer ($M_r$ 90K) consisting of equally sized $\alpha$- and $\beta$-subunits (for review, see Davis, 1985). Somewhat surprisingly, it appears that $T_H$ and $T_C$ use the same $\alpha$- and $\beta$-gene pool (Hedrick et al., 1985). The genomic organization of the Ti is remarkably similar to that of immunoglobulins, with coding regions corresponding in sequence homology and function to variable (V), diversity (D), joining (J), and constant (C) regions located on exons that recombine to produce the functional gene (Malissen et al., 1984).

The specific interaction of the Ti with sufficient avidity with antigen plus the MHC restricting molecule results in T lymphocyte activation. Antigen recognition alone is sufficient to shift lymphocytes from $G_0$ to $G_1$ and to induce IL-2 receptor expression and for $T_H$, IL-2 secretion (Cantrell and Smith, 1983; Imboden and Stobo, 1985). However, further progression through the cell cycle requires accessory factors (Mizel, 1982; Farrar et al., 1982). Proliferation of both $T_H$ and $T_C$ is dependent on IL-2. $T_H$ proliferation also requires IL-1 (DeFreitas et al., 1983). As a result, the clonal expansion of $T_H$ requires cells of the monocyte series (monocytes and macrophages) that secrete IL-1. By contrast, clonal expansion of $T_C$ is dependent on the presence of activated $T_H$, the only source of IL-2. (Note that the IL-2 requirement of $T_H$ is met by their own production, which is induced by antigen stimulation even in the absence of IL-1.)

The understanding of the basic growth requirements of T lymphocytes has permitted the production of permanent cloned T-lymphocyte cell lines (Schrier et al., 1980; Fathman and Frelinger, 1983). This represents a major advance in the field of cellular immunology, since it permits examination of T-lymphocyte specificity at the level of individual epitopes and T-lymphocyte function at the level of individual effector clones. Both $T_H$ and $T_C$ cloned lines can be produced from limiting dilution of cells in the presence of the proper APCs plus exogenous IL-2. In addition, it is possible to produce mouse $T_H$ hybridomas by fusion of freshly isolated $T_H$ with a mouse drug sensitive T-lymphoma cell line followed by growth in selective media (Kappler et al., 1981). These hybridomas proliferate in the absence of antigen or exogenous IL-2 but maintain the ability to secrete IL-2 upon recognition of antigen plus class II MHC molecules.

Finally, it should be mentioned that there is a third functional class of T lymphocytes that mediate supression of immune responses, termed T-suppressor lymphocytes ($T_S$) (Moller, 1976). Although the presence of T lymphocytes with suppressor activity has been detected in responses to many antigens, including influenza virus, it is uncertain whether $T_S$ represent a truly distinct T-lymphocyte lineage or whether they are derived from $T_H$ or $T_C$. These cells are generally found to express the $T_C$ surface phenotype but, in many cases, the expression of both $T_C$ and $T_H$ markers, or of $T_H$ markers alone, has been reported. Owing to difficulties in identifying $T_S$ in lymphocyte populations and producing cloned $T_S$ cell lines, it has not been possible to determine whether $T_S$ express the antigen receptor expressed by $T_C$ and $T_H$. It is also uncertain whether $T_S$ recognize foreign antigen, either alone or in an MHC-restricted manner, or are specific for unique determinants (idiotypes) on the antigen receptor of $T_C$ or $T_H$.

## 3. Nature of Antigenic Determinants Recognized by T Lymphocytes

The key feature of T-lymphocyte recognition of foreign antigens is its dependence on elements of both the foreign antigen and the appropriate self-MHC molecule. This remarkable phenomenon was first described in the early 1970s (Kindred and Shreffler, 1972; Zinkernagel and Doherty, 1974; Miller et al., 1975), and understanding its molecular basis remains one of the major quests of immunological research. Zinkernagel and Doherty (1979) originally proposed two general models for T-lymphocyte recognition that have not been superseded:

1. *Altered self:* Ti interaction solely with determinants on MHC molecules that are conformationally altered by interaction of the restriction element with foreign molecules
2. *Dual recognition:* Ti interaction with determinants on both MHC and foreign molecules; either single or multiple receptors recognizing complexed or noncomplexed MHC and foreign molecules

Current evidence indicates that recognition is mediated by a single receptor (Davis, 1985). It has been shown that foreign antigens form complexes with class II restriction molecules (Babbitt et al., 1985; Buus et al., 1986; Watts et al., 1986), probably because of interaction with a region in the class II molecule homologous to the proposed antigen binding groove in the class I molecule (Bjorkman et al., 1987b). Beyond this, there is little information regarding the molecular basis for T-lymphocyte recognition. Complete understanding of this phenomenon is ultimately dependent on determination of the three-dimensional structure of a Ti–ligand complex. This may be some years distant, as it may not be possible to crystallize Ti–ligand complexes, and a molecular definition may depend on development of new technologies able to resolve the Ti–ligand interaction when one or both components are inserted into lipid bilayers.

There has been progress, however, in analyzing the regions of MHC and foreign molecules involved in recognition. One approach is to examine the ability of monoclonal antibodies specific for MHC or foreign determinants to inhibit T-lymphocyte recognition. For both $T_H$ and $T_C$, antigen recognition can be blocked by antibodies specific for the restricting MHC molecule. Blocking with antibodies specific for foreign antigens has been observed occasionally with $T_C$, and only rarely with $T_H$. The interpretation of these studies is difficult, as it is uncertain whether blocking is due to direct steric effects on T-lymphocyte recognition, conformational alterations induced by antibody binding, or inhibition of normal cellular handling of the relevant molecules. Consequently, the information provided by antibody-blocking studies is limited essentially to defining the relevant molecules recognized by T lymphocytes.

More insight into the regions of foreign molecules that contribute to the structure recognized by the Ti has been obtained by examining T-lymphocyte recognition of progressively simpler structural forms of foreign antigens. In the case of complex antigens such as viruses, the first step in this approach is to determine which components of the virus are recognized by T lymphocytes. This is done using biochemically or genetically isolated viral components. Next, the effect of partial or total denaturation on T-lymphocyte recognition is tested. Loss of recognition indicates that the determinant is conformational, i.e., depends on elements of secondary, tertiary, or quarternary structure. At the molecular level, this could mean that the determinant is formed by the following:

1. Amino acid residues located distally in the primary structure that are brought into proximity by folding into the native structure (a discontinuous determinant)
2. Linear sequence of amino acids that adopt the proper conformation only with a very low probability following denaturation

Localization of conformational sites in the intact protein structure is generally very difficult and depends on the isolation of protein fragments maintaining native structure, or less directly, on identifying residues in the protein whose alteration affects recognition. Furthermore, it has not yet been possible to determine whether the native conformation is needed for recognition of conformational determinants or is required to prevent destruction of the determinant during processing of the protein by APCs. For these reasons, the classification of antigenic determinants as conformational is at present an operational rather than a physical description of the determinant.

If recognition is maintained upon denaturation of the antigen, it is often possible to precisely locate the region of the protein involved in recognition. This can be accomplished using small cleavage fragments of the protein, and ultimately, synthetic oligopeptides corresponding to 10–20 amino acids of the primary sequence. In this event, the determinant is termed sequential. Although this has proved a powerful approach in identifying minimal determinants required for T-lymphocyte recognition, several important caveats must be considered. First, it cannot be overstressed that the contribution of the foreign antigen to the actual structure recognized by the Ti is uncertain; evidence that even oligopeptide determinants contribute contact residues to the Ti-ligand complex is limited (Bjorkman *et al.*, 1987). Second, the ability of an oligopeptide to substitute for the intact protein does not necessarily indicate that these residues alone constitute the determinant recognized following processing of the native protein. Even in what would appear to be a much better defined case of ligand–receptor interaction (lysozyme–antibody), solution of the three-dimensional structure of the antigen–antibody complex showed the degree of interaction to be far greater than anticipated (Amit *et al.*, 1986). Third, there is evidence that suggests that T-cell responses are far more degenerate than antibody responses. This high probability of cross-reactivity makes it imperative to demonstrate not just that a given sequence from a viral protein is recognized by antiviral T lymphocytes, but that the corresponding region in the protein is recognized as well. This caveat is particularly important when screening peptide libraries for T-lymphocyte recognition. Following identification of a reactive peptide, it is critical to confirm its recognition by mutational analysis of the intact protein, *i.e.*, to demonstrate that deletions, or better yet, point mutations in the corresponding region abrogate recognition.

While the definition of antigenic sites on foreign antigens has its limitations in providing information regarding the molecular basis of T-lymphocyte recognition, it constitutes the first step in this direction. Furthermore, it provides information essential to understanding other aspects of the recognition process and the biology of T lymphocytes, including cellular processing of antigens for T-lymphocyte recognition, MHC restriction element usage, generation of Ti diversity, and *in vivo* function of T lymphocytes.

## II. GENERAL PROPERTIES OF ANTIVIRAL T-LYMPHOCYTE RESPONSES

### A. Assessing Recognition of Viral Antigens by T Lymphocytes

Anti-influenza $T_C$ activity is usually determined by the lysis of histocompatible cells expressing viral antigens, although in the case of antigen-dependent $T_C$ clones it is also possible to use proliferation as a measure of antigen recognition. Target cell lysis is generally monitored by the release of a radioactive compound retained by viable cells, usually $Na^{51}CrO_4$. The use of target cells lacking class II MHC molecules ensures that cytotoxicity is mediated by $T_C$ and not $T_H$. In circumstances in which this is not possible, e.g., human studies that must often use peripheral blood lymphocytes (PBL) as target cells, it is necessary to demonstrate class I restriction by other means, e.g., blocking of cytotoxicity using anti-class I antibodies. The level of $T_C$ activity is gauged both by the amount of radioactivity released from target cells and by the number of effector cells required to cause the release. Unlike antibodies, which demonstrate specificity over a wide range of concentrations, $T_C$ specificity occurs only over a relatively narrow ratio of effector to target cells. Using either lymphocyte populations or cloned $T_C$ lines, there is considerable "nonspecific lysis" of cells at supraoptimal effector to target cell ratios. This may merely be an artifact of *in vitro* assays. Alternatively, it is possible that this truly reflects a relative lack of specificity in $T_C$ recognition and lysis.

Anti-influenza $T_H$ activity can be measured in a variety of ways. Ideally, when using lymphocyte populations, the method used should measure unique $T_H$ functions, such as helper activity for induction of antibody or CTL responses or MHC class II-restricted target cell lysis, although it is essential to control for possible $T_C$-mediated MHC class I restricted lysis. Often, however, other methods are used that are less specific for $T_H$:

1. *Cellular proliferation assays, performed 4–6 days following in vitro stimulation with antigen:* Proliferation is assessed by incorporation of [³H]thymidine into DNA; alternatively, the number of viable cells (which is also a measure of proliferation since nonproliferating cells die under the culture conditions employed) is assessed by cleavage of a tetrazolium dye (Mossman, 1983).

2. *Delayed-type hypersensitivity (DTH) assays:* In their various forms, these assays measure inflammation at the site of injection after an 18- to 24-hr period. Although these assays do not rigorously distinguish $T_H$ from other lymphocytes, it has been found that under the proper experimental conditions, they correlate well with more specific measures of $T_H$ activity.

Stimulation of cloned $T_H$ cell lines can be measured either by proliferation or by interleukin-2 (IL-2) production (the amount of released IL-2 is determined by its ability to support proliferation of an IL-2-dependent cloned $T_C$ line). Proliferation cannot be used as an index of $T_H$ hybridoma stimulation, since these cells do not require antigen or exogenous factors for division. Instead, release of IL-2 or other lymphokines must be monitored to measure stimulation.

## B. Induction of Influenza Virus-Specific Lymphocytes

### 1. Timing of T-Lymphocyte Responses during the Course of Virus Infection

The T-lymphocyte component of the immune response is triggered early in the course of influenza virus infection. In experimental mouse infections, pulmonary virus titers reach maximal values in the first 4 days, remain at high levels for 2–3 days, and decline to below the limits of detection by 10 days postinfection (p.i.). $T_H$ activity, as measured by the *in vitro* proliferative response of nylon wool-purified lymphocytes, is detectable at least as early as 4 days p.i., with peaks of activity reported variously at days 5–13 (Butchko *et al.*, 1978; Lipscomb *et al.*, 1982; Hurwitz and Hackett, 1985). Consistent with this finding, $T_H$-dependent antibody responses are first detected 7 days p.i.

$T_C$ activity is first detectable in mouse lungs 3–4 days following intranasal infection, with peak responses occurring 2–3 days later (Cambridge *et al.*, 1976; Ennis *et al.*, 1978; Bennink *et al.*, 1978; Yap and Ada, 1978a). Following pulmonary infection, $T_C$ can also be obtained from spleen, draining lymph nodes, and to a lesser extent PBL, but the earliest and most active response occurs in the lung. The observation that splenectomized mice demonstrate no difference in lung $T_C$ activity is consistent with the idea that pulmonary $T_C$, as well as $T_H$ required for their activation, originate from antecedents present in lung lymphoid tissue (Yap and Ada, 1978c).

It is difficult to analyze the kinetics of human $T_C$ responses because of the uncertain immune status of human donors and the necessity of using PBL as a source of $T_C$. Freshly isolated PBL have been reported to demonstrate low levels of cytotoxic activity detectable 6–21 days following infection of volunteers (Daisy et al., 1981). It is usually necessary, however, to stimulate PBL in vitro with virus to observe cytotoxic activity; in this way, $T_C$ activity can be detected as early as 4 days p.i., reaching a peak 1–3 days later (Ennis et al., 1981; McMichael et al., 1983b). Note that secondary in vitro stimulation of mouse as well as human lymphocytes (Yap and Ada, 1977b; McMichael and Askonas, 1978) yields more vigorous and consistent $T_C$ responses than do lymphocytes directly isolated from immunized hosts and often must be used to detect primary responses. For this reason, in vitro-stimulated $T_C$ are often used in studies of $T_C$ specificity.

## 2. Effect of the Form of the Antigen on Induction of T-Lymphocyte Responses

$T_C$ and $T_H$ demonstrate a major difference in their response to noninfectious forms of viral antigens. While $T_H$ respond with equal efficiency to infectious and noninfectious whole and subunit virus preparations, $T_C$ responses are only poorly induced by noninfectious viral antigens (Braciale and Yap, 1978; Reiss and Shulman, 1980; Armerding et al., 1982). The high efficiency of infectious virus in $T_C$ induction is attributable to its ability to direct synthesis of viral antigens in host cells (Reiss and Shulman, 1980). The difference in $T_H$ and $T_C$ responsiveness to noninfectious viral antigens is a manifestation of a fundamental difference in the processing of viral antigens into forms recognized by class II- and class I-restricted lymphocytes (see Sections III and IV).

## 3. Effect of Preimmunization on T-Lymphocyte Responses

Preimmunization accelerates mouse $T_C$ responses to influenza infection by approximately 2 days and results in far more vigorous and consistent responses (Doherty et al., 1977; Effros et al., 1978; Bennink et al., 1978). Although comparable published data are not available for $T_H$ responses, it is common experience that a similar priming effect occurs. As with induction of primary responses, $T_H$ responses can be primed at equal efficiency using infectious or noninfectious viral antigens, while $T_C$ responses are optimally primed using infectious virus. Priming for $T_H$ and $T_C$ also differs in that secondary in vivo $T_C$ responses cannot ordinarily be induced with the same virus used for priming (Effros et al., 1977). It has been shown that this effect is caused by neutralization of viral infectivity by circulating anti-HA antibodies (Effros *et al.*, 1977; Leung *et al.*, 1980a; Greenspan and Doherty, 1982), which again emphasizes that during the course of a normal infection, the biologically relevant form of antigen for $T_C$ is produced only after infection of host cells.

The enhanced nature of secondary $T_H$ and $T_C$ responses could be accounted for by several nonmutually exclusive factors: (1) increase in precursor frequency, (2) shift of cells into a more easily triggered memory state, and (3) increased $T_H$ activity (for $T_C$ responses). The role of factors (2) and (3) in the priming process is uncertain. It remains to be determined whether memory T lymphocytes constitute a distinct functional subset. The role of $T_H$ in $T_C$ responses has been defined only at the most basic level; it is not known whether $T_H$ are limiting in primary responses. It is clear, however, that an increased precursor frequency contributes to enhanced secondary responses of both $T_H$ and $T_C$. Priming has been reported to increase precursors of antigen specific IL-2-secreting mouse lymphocytes, presumably $T_H$, by 30-fold (Miller and Reiss, 1984). In the same study, priming resulted in a 3- to 10-fold increase in the frequency of $T_H$ able to promote $T_C$ responses. Similarly, the frequency of $T_C$ precursors present in mouse spleens was shown to increase by approximately 10- to 100-fold following intraperitoneal or intranasal immunization with live virus (Askonas *et al.*, 1982; Kees and Krammer, 1984; Owen *et al.*, 1984). This increase in $T_C$ precursors has been reported to persist for the entire 2-year life span of mice (Ashman, 1982).

The longevity of the priming effect for human T-lymphocyte responses is less well established. There is evidence that priming for human secondary *in vitro* responses is relatively short lived. Ennis *et al.* reported that priming for *in vitro* $T_C$ responses lasted less than 6 months following infection of volunteers with influenza virus (Ennis *et al.*, 1981). McMichael *et al.* (1983a) found that $T_C$ *in vitro* responsiveness declined over a period of 6 years, in parallel with a decreasing prevalence of epidemic influenza. On the basis of these findings, it was proposed that human $T_C$ memory is relatively short lived and that vaccines that boost $T_C$ memory in previously infected individuals might reduce the severity of subsequent infections (McMichael *et al.*, 1983a). While this is certainly possible, it is important to recognize (1) that these studies were based on the activity of peripheral blood lymphocytes, which might not reflect the immune status of lymphocytes present either in the lung or in central lymphatic organs, and (2) that the role of $T_C$ in human immunity to influenza virus remains to be precisely defined.

## 4. Regulation of T-Lymphocyte Responses

Like virtually all immune responses, T-lymphocyte responses to influenza virus demonstrate a decline in peak activities with time and are no longer detectable without restimulation within 2–3 weeks after infection (Cambridge et al., 1976; Doherty et al., 1977). At least two factors are likely to contribute to this phenomenon. First, in the guinea pig, the decline in T-lymphocyte responses has been found to parallel decreases in viral antigen (Lipscomb et al., 1983). As most T-lymphocyte lines grown in vitro require antigenic stimulation for both activity and growth, this may indicate that the reduction of viral antigen in the host results in the death or quiescence of antigen-specific T lymphocytes.

Second, there is evidence for the active suppression of T-lymphocyte responses mediated by other T lymphocytes, summarized in Table I. Mouse DTH could be enhanced by pretreatment of mice with cyclophosphamide (Liew and Russell, 1980; Leung and Ada, 1980), a drug that has the effect of inhibiting precursors of $T_S$ (Askenase et al., 1975). Furthermore, DTH could be partially inhibited in adoptive transfer experiments by co-incubation of effector cells with cell populations containing suppressors prior to transfer. The $T_S$ in these studies were $\theta^+$, Lyt $1^+$, $2^-$ and appeared specific for the glycoproteins of the inducing strain of virus (Liew and Russell, 1983). Although this differs from the classic Lyt $1^-$, $2^+$ phenotype of most mouse $T_S$, it appears to be typical of $T_S$ that inhibit class II-restricted cells involved in DTH responses (Liew and Russell, 1980, 1983).

T-lymphocyte-mediated suppression of anti-influenza responses is not limited to DTH reactions. Leung et al. (1980) found that injection of mice with UV-irradiated noninfectious virus induced T lymphocytes that, in adoptive-transfer experiments, suppressed $T_C$ responses in a class II MHC-restricted manner. It was uncertain whether these lymphocytes suppressed $T_C$ directly, or suppressed $T_H$ needed for $T_C$ stimulation. Consistent with the latter interpretation, Miller and Reiss (1984) found that removal of Lyt $2^+$ cells greatly increased the ability of $T_H$ to help $T_C$ responses in vitro. Cyclophosphamide treatment of mice prior to initial immunization did not affect $T_H$ precursor frequency, which is consistent with the idea that $T_S$ functioned in this study by limiting the activation $T_H$, and not by decreasing their numbers. Suppression of $T_H$ has been observed in other studies as well. Mice injected intravenously 2 weeks previously with infectious virus or with isolated virus components (HA, virus cores containing NP and M1) were suppressed specifically for proliferative responses to influenza antigens (Hurwitz and Hackett, 1985). Suppression could be transferred by Lyt $2^+$ cells. Anders et al. (1981a) showed that suppressors of $T_H$ can also be induced by secondary in vitro stimulation with intact or fragmented HA.

The process of induction of influenza virus specific $T_S$ is poorly defined. Induction seems to be favored by intravenous or aerosol routes of inoculation, as well as by high or frequent doses of antigen (Liew and Russell, 1980; Anders et al., 1981a; Thompson et al., 1983; Hurwitz and Hackett, 1985). Consistent with their postulated regulatory role, $T_S$ responses typically arise later than $T_H$ responses (Liew and Russell, 1980) and are usually measured 2 weeks following immunization. The

TABLE I. Suppression of Influenza Virus-Specific Immune Responses in Mice[a]

| Induction | Assay effect on | Phenotype of Ts | Specificity | References |
|---|---|---|---|---|
| Splenocytes of primed mice cultured with high dose of virus, 4 days | Antibody production in vitro | Thy-1 | HA; HA1 1–168; HA1 266–318 | Anders et al. (1981a) |
| Aerosol infection With virus | Delayed-type Hypersensitivity | Thy-1, Lyt $1^{+2-}$ | HA subtype Specific | Liew and Russell (1980) |
| UV-treated virus i.v. | Cytotoxic T cells in vivo | Thy-1 | Type A-virus cross-reactive | Leung et al. (1980a) |
| Mice chronically immunized with UV-treated virus | Antibody production in vivo | Lyt $2^+$; radiation sensitive | Cross-reacts between H1N1 viruses A/PR/8 and A/Cam/46 | Thompson et al. (1983) |
| Virus i.p. plus or minus cylcophosphamide | Cytotoxic T cells | Lyt $2^+$ | NR | Miller and Reiss (1984) |
| Virus i.v. days 0 and 14 | T-cell proliferation | Lyt $2^+$ | Induced by HA, NP, or M1 | Hurwitz and Hackett (1985) |

[a]Abbreviations: i.v., intravenous injection; i.p., intraperitoneal injection; NR, not reported; UV, ultraviolet light.

specificity of $T_S$ is uncertain. It is not known, for example, whether $T_S$ exhibit MHC-restricted recognition of viral antigens or, for that matter, whether they recognize viral antigens at all. The latter possibility is supported by the description of a cloned human T lymphocyte that suppressed *in vitro* antibody production based on its recognition of a helper $T_H$ clone (Lamb and Feldmann, 1982). This suppressive clone was derived by stimulating autologous PBL with an irradiated M1-specific $T_H$ clone. The irradiated $T_S$ induced the stimulating $T_H$ to incorporate [$^3$H]thymidine in the absence of M1, consistent with an idiotype-specific interaction with the $T_H$ antigen receptor. These findings suggest that at least a portion of the $T_S$ response may be specific for $T_H$ idiotypes.

## III. SPECIFICITY AND FUNCTION OF ANTI-INFLUENZA VIRUS $T_H$

## A. Specificity of $T_H$

### 1. Recognition of Individual Viral Proteins

A central feature of T-lymphocyte recognition of influenza virus is its high cross-reactivity for different influenza A viruses. This phenomenon was initially discovered in studies of $T_C$ specificity (described in Section IV) and was quickly extended to $T_H$. Butchko *et al.* (1978) demonstrated that mouse T lymphocytes induced by immunization with a H3N2 virus proliferated in response to H1N1 and H2N2 viruses but not to influenza B virus. Initial evidence that recognition of the highly conserved internal virion proteins contributed to this process was provided by Russel and Liew (1979), who found that help for anti-HA antibody responses was primed by immunization with spikeless particles or with purified M1 protein. Recognition of internal viral proteins by $T_H$ populations was confirmed by Hurwitz *et al.* (1985), who found that nylon wool purified T lymphocytes derived from aerosol-infected mice proliferated in response to purified NP and M1, as well as to purified HA and NA.

$T_H$ recognition of individual viral proteins has also been examined at the clonal level. Both mouse and human clones have been obtained that are specific for each of the four major virion structural proteins (HA, NA, NP, M1) (Lamb *et al.*, 1982a,c; Hurwitz *et al.*, 1985; Fleischer *et al.*, 1985). In these studies, the proportion of clones specific for individual viral proteins varied considerably. The human clones isolated by Lamb *et al.* (1982c) were fairly evenly divided between the external and internal components, while the frequency of mouse clones isolated by Hurwitz *et al.* (1985) more reflected the weight percentage of each component. Furthermore, the mouse clones isolated by Mills *et al.* (1986b) were overwhelmingly skewed toward HA specificity. Although the reasons for these differences are uncertain, the absence of consistent trends in specificity suggests that the immunogenicity of the major viral structural proteins are not vastly different.

$T_H$ recognition of the minor viral structural proteins (polymerases) and the three nonstructural proteins has not been systematically examined. At least one $T_H$ clone isolated to date could not be assigned to one of the major virion proteins (Hurwitz et al., 1985) and might recognize one of these proteins; alternatively, it might recognize a determinant destroyed during the isolation of the individual proteins used to test specificity.

At both the clonal and population levels, the degree of cross-reactivity of $T_H$ between different influenza virus isolates has been found to reflect the degree of amino acid conservation of the virus component recognized. Generally, anti-glycoprotein $T_H$ recognize only the immunizing virus and closely related strains, while anti-internal $T_H$ recognize most, if not all, influenza A viruses (Lamb et al., 1982c; Sterkers et al., 1984a; Hurwitz et al., 1985). As might be expected, this rule is not absolute. An anti-NP clone has been described that recognizes H1N1 but not H2N2 or H3N2 viruses (Hurwitz et al., 1985). Conversely, cross-reactive anti-HA $T_H$ have been detected (Anders et al., 1981b; Katz et al., 1985b; Fleischer et al., 1985). At the population level, most of these $T_H$ appear to recognize the more highly conserved HA2 subunit. At the clonal level, however, all cross-reactive anti-HA $T_H$ isolated to date recognize the HA1 subunit (Katz et al., 1985a; Brown et al., 1987).

## 2. Fine Specificity of Anti-HA $T_H$

Studies of the fine specificity of anti-influenza $T_H$ are limited to $T_H$ specific for the HA. This does not reflect the particular importance of the HA in $T_H$ responses, since $T_H$ directed to the other viral proteins are at least of equal frequency and functional importance. Rather, the attractiveness of the HA is due to the detailed knowledge of its structure and function and the availability of mutant viruses with defined amino acid substitutions in their HAs produced in the laboratory by selection with neutralizing monoclonal antibodies (Gerhard and Webster, 1978).

### a. Studies with Immunized T-Cell Populations

Studies with $T_H$ populations showed that segments of both the HA1 and HA2 polypeptides derived from the Memphis [A/Memphis/71 (H3N2)] HA could restimulate T cells derived from virus-primed mice (Anders et al., 1981a). Cyanogen bromide fragments of the HA1 (amino acids 1–168, and all but the last eight residues of the 60-amino acid C-terminal fragment) and residues 18–115 of the HA2 were active. These findings were extended using panels of synthetic peptides. Lamb et al. (1982b) studied human $T_H$ responses using a panel of 12 peptides, 15–37 residues long, covering most of the HA1 polypeptide of the Aichi [A/

Aichi/68 (H3N2)] HA; a gap occurred at residues 249–266. Atassi and Kurisaki (1984) examined mouse responses to 12 peptides, corresponding to 11- to 18-residue-long segments present on the surface of intact HA1 and HA2 subunits of the Aichi HA. In both cases, peptides were used to restimulate T lymphocytes *in vitro* from hosts previously exposed to infectious virus. Lamb *et al.* (1982b) found that human PBL proliferated significantly in response to all peptides tested. Atassi and Kurisaki reported that 3 of the 12 peptides tested restimulated well (residues HA1 23–36, 183–199, and HA2 56–68), three were intermediate (HA1 201–218, 272–288, and 300–315), and one stimulated poorly (HA1 175–188). Five peptides failed to stimulate (HA1 124–134, 138–152, 154–167, HA2 1–11, 68–84).

These findings suggest that $T_H$ recognize sequential epitopes on the HA1 and HA2 subunits, located either on the surface, or the interior of the intact molecule. More extensive studies with $T_H$ clones have confirmed that various sites are located on the HA1, but to date, no HA2-specific clones have been isolated. More generally, lymphocyte population responses to peptides covering the HA suggest a more extensive response to the molecule than found with HA-specific clones (see Section III.A.2.b). This could mean that the process of producing $T_H$ clones may result in a biased sampling of the anti-HA repertoire. Alternatively, as neither Lamb *et al.* nor Atassi and Kurisaki examined the responsiveness of lymphocytes derived from nonimmunized hosts, the possible mitogenicity of the peptides studied was not controlled for, and responses to some of the peptides may not reflect specific recognition by HA specific $T_H$.

### b. Studies with $T_H$ Clones

Lamb and Green (1983) showed that human $T_H$ clones raised by *in vitro* restimulation with purified Aichi (H3) HA fell into several antigen-specificity groups (Table II). The major group (3 of 5 clones) recognized a 24-residue peptide (amino acids 306–329 of the HA1), another group (1 of 5) responded to peptide 105–140, while a third group (1 of 5) failed to respond to any peptide in a panel covering the entire HA1. Although truncated peptides failed to stimulate the clones, it is likely that the 24- and 36-residue peptides were in excess of the minimal size required for optimal recognition, based on findings of $T_H$ recognition of other sequential determinants (Arnon, 1981; Hackett *et al.*, 1983a; Ashwell and Schwartz, 1986).

The fine specificity of $T_H$ has been more extensively characterized using mouse $T_H$ clones. Fifteen clones that recognized the PR8 HA (H1) but not H2 or H3 HAs, were derived from 10 individual BALB/c mice immunized with PR8 (Hurwitz *et al.*, 1984; P. A. Scherle, C. J. Hackett, and W. Gerhard, unpublished data). When their specificity was examined using a panel of 43 mutant PR8 viruses having known single amino acid substitutions in the HA1 subunit (Caton *et al.*, 1982), the clones fell into

TABLE II. Recognition of Influenza HA by T$_H$ Clones[a]

| Host | In vivo[b,c] immunization | In vitro stimulation[b,c] | Number of clones of each specificity | | References |
|------|------|------|------|------|------|
| | | | HA antigen specificity | MHC restriction | |
| Human | ? | A/Texas purified HA | 3-HA1, 306-329; 1-HA1, 105-140; 1-? | NR | Lamb and Green (1983) |
| Human | A/Bangkok(H3N2); A/Brazil(H1N1); B/Singapore | A/Texas | 2-H3 HA | 1-DR7; 1-DR1 | Sterkers et al. (1984a) |
| Human | ? | A/Jap(H2N2)-infected cells | 1-H2 HA | 1-DR5 | Kaplan et al. (1984) |
| Human | ? | A/Texas | 2-H3&H4 HA | 1-DRW7; 1-DRW8 | Sterkers et al (1985) |
| Human | ? | A/USSR(H1N1) | 1-H1 thru H7 (all) | 1-Class II | Fleischer et al. (1985) |
| BALB/c mouse | A/PR8(H1N1) f.p. CFA or i.p. PBS | A/PR8 | 5-HA1, 111-119(site 1); 5-HA1, 126-138(site 2); 5-HA1, 302-313(site 3) | 5-I-E$^d$ (site 1); 5-I-A$^d$ (site 2); 5-I-E$^d$ (site 3) | Hurwitz et al. (1984); Eisenlohr et al., (1987) |
| BALB/c mouse | A/Mem-Bel(H3N1) | A/Jap-Bel(H2N1) or A/X-31 disrupted virus | 3-H3 HA1; 4-H3 and H2 HA1 | NR | Katz et al. (1985a) |
| BALB/c mouse | A/Jap | A/Jap | 2-H2 HA | 1-I-E$^d$ | Lukacher et al. (1985) |
| C57B1/6 mouse | A/PR8 f.p. CFA | A/PR8 | 1-HA1, 79-91 | 1-I-A$^b$ | Hackett and Hurwitz (unpublished data) |
| CAB mouse | A/X-31 i.n. infection | A/X-31 virus or purified HA | 12 groups, HA1-specific; one group conformational, HA1, 48-68; One group HA1, 118-138 | I-restricted I-A$^k$ | Mills et al. (1986a,b, 1988) |

[a]Abbreviations: f.p., footpad inoculation; i.p., intraperitoneal inoculation; i.n., intranasal; CFA, complete Freund's adjuvant; PBS, phosphate-buffered saline; NR, not reported; ?, unknown pre-exposure to virus.
[b]Intact virus used in immunizations, except where noted.
[c]Recombinant viruses: A/Mem-Bel = A/Memphis/102/72 × A/Bel/42; A/Jap-Bel = A/Jap/305/57 × A/Bel/42; A/X-31 = A/Hong Kong/1/68 × A/PR/8/34.

several nonoverlapping reactivity groups. Group 1 (6 of 15 clones) recognized all mutant viruses except one having a Lys for Glu substitution at residue 115 of the HA1 ($Glu_{115} \Rightarrow Lys$). Those in reactivity group 2 (5 of 15) were sensitive only to the alteration $Ser_{136} \Rightarrow Pro$, while those in group 3 (5 of 15) recognized all the mutants tested, but no other HAs derived from a number of related virus strains. More recently, analysis of T-cell clones derived by limiting dilution without prior bulk culturing demonstrated that at least three additional regions of the PR8 HA molecule are immunogenic in BALB/c mice (W. Gerhard, personal communication). These specificities are represented by clones from different individual mice, suggesting that they are not necessarily minor components of the $T_H$ response. This suggests further that preliminary culturing before fusion or cloning may bias the profile of responding clones, as cautioned by Mills *et al.* (1986b).

The antigenic determinants recognized by the three initially studied PR8 HA-specific BALB/c T cell groups have been chemically synthesized. The site 1 determinant, which is known to be of the minimal size for $T_H$ stimulation (Hackett *et al.*, 1983a; C. Moller, C. Hackett, and W. Gerhard, unpublished), is composed of residues 111–119 of the PR8 HA1. Peptides of residues 126–138 and 302–313 of the HA1 represent sites 2 and 3, respectively, although it is not known whether these may be shortened further and retain antigenic activity. Each of these peptides is also immunogenic in BALB/c mice, eliciting MHC class II-restricted T lymphocytes (Hackett *et al.*, 1985a; C. Hackett, unpublished observations). T-hybridoma clones derived from immunization with synthetic peptides respond *in vitro* to intact virus and exhibit a fine specificity of antigen recognition identical to T lymphocytes obtained by immunization with whole virus (Hackett *et al.*, 1985a). These findings indicate that at least three of the BALB/c $T_H$ sites in the PR8 HA are sequential in nature and can be recognized both antigenically and immunogenically outside the context of the intact protein.

Recognition of the Aichi (H3) HA by $T_H$ derived from CBA mice was studied by Mills *et al.* (1986b). All 27 anti-H3 HA $T_H$ clones isolated responded to purified intact HA and, with the exception of a conformation-dependent clone (described in more detail below), all clones recognized sites in the 28- to 328-amino acid tryptic fragment of HA1. The clones were placed into 12 different reactivity groups on the basis of their recognition of a panel of 13 variant HAs having 1–33 amino acid differences from the HA used to elicit the clones. The multiple changes present in many of the constituents of this panel precluded the precise definition of the regions involved in recognition by these clones. Mapping was also hindered by the failure of most clones to recognize a panel of synthetic peptides; it was possible, however, to assign groups to residues 48–68 and 118–138 of the HA1. Although it is clear that the clones exhibit considerable heterogeneity, further studies are required to determine whether each of these reactivity patterns represents separate nonoverlapping sites on the HA1 or whether they are another example of a heterogeneous $T_H$ response to a fairly limited number of determinants, as found for the H1 HA specific $T_H$ clones (see Section III.A.2.c).

It should be noted that there is a striking overlap of regions recognized by two groups of the BALB/c and human $T_H$ clones (Hurwitz et al., 1984; Lamb and Green, 1983), although the regions are on HA molecules of different subtypes and are not cross-reactively recognized by any of the clones. Specifically, the C-terminal part of the H1 subtype HA1 corresponding to the BALB/c site 3 and the region around the BALB/c site 1 were recognized by human T-lymphocyte clones to H3 HA. It is not yet clear whether this is indicative of some underlying structural feature that leads to immunodominance of the sites for human and BALB/c T cells (the H1 and H3 molecules share only 52% amino acid homology in these regions) or whether it is a fortuitous finding. The results of studies with different strains of mice and HA subtypes do suggest, however, that such overlap of $T_H$ sites between different species and HA subtypes is not a general feature of $T_H$ recognition. For example, residues 48–68 of the H3 HA recognized by the CBA clones is not one of the three major sites recognized by BALB/c anti-H1 clones. Peptide 305–328 of the H3 HA that is dominant in the human $T_H$ response does not elicit a response from mice primed with the H3 HA (L. E. Brown, University of Melbourne, personal communication). In addition, it appears that $T_H$ derived from C57B/6 mice recognize different regions of the HA than BALB/c $T_H$ (C. Hackett, J. Hurwitz, P. Scherle, and W. Gerhard, unpublished findings): a site recognized by C57B/6 $T_H$ populations and clones (residues 79–91 of HA1) is not immunogenic in BALB/c mice.

The studies discussed to this point have been limited to $T_H$ that recognize sequential determinants of the HA. A recent report suggests that $T_H$ might also recognize conformational determinants on the HA. Mills et al. (1986a) isolated CBA $T_H$ clones specific for the H3 HA that failed to recognize HAs with alterations at residue 17 or residue 208 of the HA1. Although residue 17 is located in the fibrous stem of the HA, the mutation at this site is believed to decrease the intermonomer interactions in the globular head (Rott et al., 1984). As residue 208 is located at the interface of adjacent monomers, Mills and co-workers proposed that the determinant recognized by these clones was located in the area of residue of 208 and required the native trimeric structure. Two additional findings were consistent with this interpretation. First, proliferation of these clones was greatly reduced if virus was exposed to acid conditions, thereby inducing irreversible conformational alterations in the HA and resulting in decreased interaction between monomers in the globular head. Second, the clones failed to respond to monomeric tryptic fragments of the HA (residues 1–27 or 28–328) or to a synthetic peptide of residues 1–36. These findings clearly demonstrate that recognition of the HA by the $T_H$ clones is dependent on the native structure of the HA. It is uncertain, however, whether this requirement is manifested during the process of antigen recognition itself or during the process of antigen presentation; e.g., the determinant is destroyed during processing of monomers but not of trimers.

Several conclusions can be drawn from the studies of mouse anti-HA $T_H$ specificity. First, it is clear that $T_H$ can recognize a number of distinct regions of the HA1 molecule. It would also appear that determinants on the HA2 are recognized as well, although this is less well established, since the sites recognized remain to be defined. Second, while most $T_H$ clones isolated to date recognize sequential determinants, this is possibly the result of a biased sampling of the $T_H$ repertoire. As noted by Mills *et al.* (1986b), many of these clones were induced using virus that was at least partially denatured by emulsification in adjuvant or after a preliminary *in vitro* culture with unknown effects on potential T-cell responses. Only definition of more sites recognized by $T_H$ generated by different immunization routes will reveal if sequential or conformational determinants will prove to be predominant in the $T_H$ response to influenza virus infection.

### c. Heterogeneity of Clonal $T_H$ Responses to a Single T-Cell Determinant

Detailed studies of $T_H$ clones specific for a single HA site showed that a short sequential determinant can be the target of a heterogeneous immune response. BALB/c T hybridomas specific for HA site 1 of PR8 virus (residues 111–119) were categorized into fine specificity groups (HA 1.1 or HA 1.2) by their ability to recognize a related strain of H1 virus (Hurwitz *et al.*, 1984). Initial studies with synthetic peptide analogues (Hackett *et al.*, 1985b) revealed further heterogeneity. One clone of specificity 1.1 maintained its ability to recognize a peptide having $Glu_{115} \Rightarrow$ Asp, a change sufficient to abrogate recognition by other clones of that specificity. No clone specific for site 1 could recognize the sequence change $Glu_{115} \Rightarrow$ Lys. This was not attributable to creation of a nonimmunogenic peptide caused by loss of class II binding, as is shown to occur with some substitutions of a lysozyme peptide (Allen *et al.*, 1987), since immunization with the $Lys_{115}$ analogue was able to elicit T cells restricted to the same class II element as the original peptide. $Lys_{115}$-peptide-induced T cells failed to recognize wild-type PR8 virus or peptides with $Glu_{115}$ but did recognize a variant of PR8 virus with the $Lys_{115}$ substitution. Thus, Lys versus Glu at position 115 results in recruitment of separate completely non-cross-reacting populations of T lymphocytes that recognize the identical region of the HA.

A larger panel of peptide analogues and site 1-specific T cells shows at least seven distinguishable clonotypes (A. Haberman and W. Gerhard, personal communication). Furthermore, peptides with amino acid substitutions at each position of the 111–119 sequence were able to affect recognition of at least one clone. Although nonconservative substitutions were most effective, some conservative changes (note Glu $\Rightarrow$ Asp change above) also abolished recognition.

The Ti genes used by site 1-specific $T_H$ are being characterized by A. Taylor and A. Caton (personal communication). Sequencing data show considerable diversity of both germ-line and junctional sequences in $\alpha$- and $\beta$-chain genes of the Ti. Productively rearranged receptor genes of site 1-specific cloned T-cell lines or hybridomas studied thus far have unique sequences, corroborating the functional data that a heterogeneous T-cell response is made to this peptide.

In summary, these studies demonstrate that a T-cell determinant of 9 amino acids may be recognized by more than 10 distinct $T_H$ clonotypes in the BALB/c response. Further studies are required to determine whether responses to other T-cell sites are this diverse, as well as how this may relate to immunodominance of certain regions in T-cell responses.

## 3. Class II Restriction of Anti-influenza $T_H$

Initial evidence for class II restriction of human anti-influenza $T_H$ was provided by Eckels et al. (1982), who found that homology at the HLA-D region was required for stimulation of $T_H$ clones by APCs derived from different donors. Sterkers and collaborators (1984a,b, 1985; Fischer et al., 1985) confirmed the restriction of human $T_H$ populations and clones to HLA-DR specificities. Some clones in these studies apparently were not restricted by classic HLA-D region gene products present in standard APC panels and required autologous cells as APCs. Eckels et al. (1982) observed that, in this situation, stimulation could still be inhibited by DR-specific antibodies, confirming that the response was restricted by class II gene products. Eckels and colleagues suggested that this might represent recognition of hybrid class II structures formed by trans-chain combinations in the heterozygote, possibly analogous to the unique class II antigens described in $F_1$ mice (Fischer-Lindahl and Hausmann, 1980; Kimoto and Fathman, 1980; Mengle-Gaw and McDevitt, 1985).

Observations of class II restriction of mouse anti-influenza $T_H$ were made by Leung et al. (1980b), Leung and Ada (1982), and Liew and Russell (1983) in adoptive-transfer studies of DTH. $T_H$-dependent DTH reactions required H-2I-region compatibility between donor and recipient mice. The requirement for H-2I compatibility for stimulation of mouse $T_H$ by APCs was confirmed at the clonal level by Melchers and Gerhard and their co-workers using panels of APCs derived from recombinant inbred mice to stimulate an anti-HA $T_H$ clone (Melchers et al., 1982; Gerhard et al., 1983).

More recently, the restriction of BALB/c $T_H$ clones to either $IA^d$ or $IE^d$ has been defined by stimulating clones in vitro using as APCs, L cells (an $H-2^k$ fibroblast line) transfected with either the BALB/c IA or IE genes (Lechler et al., 1986; Eisenlohr et al., 1987). $T_H$ clones specific for M1 have so far proved to be IA restricted, as have all but one of a number of NP-specific T-cell clones (P. Scherle, personal communication). As the number of sites recognized by these clones on their respective viral antigens is unknown, the significance of this finding is uncertain. When anti-HA $T_H$ clones were examined, all site 1-specific clones were found to be I-E$^d$ restricted, as were site 3-specific $T_H$, while those directed to site 2 were I-A$^d$ restricted (see Table II).

These findings are consistent with studies of $T_H$ recognition of other antigens indicating that $T_H$ recognition of individual determinants is usually limited to one of the two possible restricting elements (Schwartz, 1986). It has also been found in other systems that $T_H$ recognition of individual peptides is often limited by allelic differences in class II molecules (Heber-Katz et al., 1982). This mechanism could explain the differences between the sites recognized by BALB/c, CBA, and C57B/6 $T_H$ described in the preceding section, although it remains to be directly demonstrated that these strain-specific differences in recognition are under MHC control.

There are two explanations for the preferential recognition of foreign determinants with certain class II molecules. First, it could simply reflect the preferential association of these elements. Second, it could be due to the absence of the appropriate receptors in the $T_H$ repertoire. Current evidence from other systems favors the first possibility, since in at least three instances, $T_H$ restriction has been found to parallel the physical association of peptides with class II molecule (Babbitt et al., 1985; Buus et al., 1986; Watts et al., 1986). Given the fact that deletion of autoreactive clones is a feature of the maturation process of T lymphocytes, it might be expected that the second mechanism also contributes to this phenomenon.

## B. Antigen Processing as a Requirement for $T_H$ Recognition of Influenza Virus Components

### 1. General Considerations

Studies discussed in the preceding section indicate that (1) a substantial portion of the $T_H$ response is directed to internal viral components; and (2) even for external viral proteins such as the HA, linear determinants recognized by $T_H$ can be located in the interior of the native molecule. As $T_H$ recognition of APCs treated with intact virus is highly efficient, this suggests that APCs have the capacity for altering the structure of the virus and its individual constituents.

This problem was first addressed in studies of $T_H$ recognition of a few well-defined soluble proteins (lysozyme, cytochrome C, myoglobin) (for review, see Chesnut and Grey, 1985). These studies established that while denatured or fragmented forms of antigen can be presented to $T_H$

by metabolically inert cells (Shimonkevitz *et al.*, 1983; Allen *et al.*, 1984), or even artificial lipid bilayers containing purified class II molecules (Watts *et al.*, 1984), presentation of intact proteins requires a metabolically active cell. It was found that processing can often be inhibited by treatment with lysosomotropic agents, which prevent normal functioning of endosomal–lysosomal pathways (Chestnut *et al.*, 1983; Ziegler and Unanue 1982) or, less commonly, by certain inhibitors of lysosomal proteases (Streicher *et al.*, 1984). On the basis of these findings, it is generally believed that processing of intact soluble proteins entails internalization into a lysosomal compartment in which denaturation, and possibly fragmentation, occurs.

Although it is likely that antigen presentation *in vivo* is normally a function of cells of bone marrow lineage, since few other cells express MHC class II antigens, the cellular machinery responsible for antigen presentation is not limited to these cells. In fact by inducing expression of class II molecules on non-bone marrow-derived cells, e.g., by treating cells with interferon; introducing cloned MHC class II genes, or by isolating cells aberrantly expressing MHC class II molecules, it has been shown that a large number of cell types such as fibroblasts (Malissen *et al.*, 1984; Austin *et al.*, 1985), astrocytes (Fontana *et al.*, 1984) or even T cells (Ben-Nun *et al.*, 1985) can process complex antigens for $T_H$ recognition. There are several reports in which cells of non-bone marrow lineage have failed to present intact antigens to $T_H$, while presenting denatured or fragmented forms (Londei *et al.*, 1984; Shastri *et al.*, 1985). It is uncertain, however, whether this represents a true deficiency in processing, since in at least one instance the same cell type has been shown to process different proteins (Shastri *et al.*, 1985). Thus, it is possible that some cells exhibit a selective deficit in a portion of the processing pathway or perhaps vary in their handling of antigens such that different determinants are produced by different cell types.

## 2. Processing of Virions by APCs

Presentation of the isolated HA to $T_H$ is similar to the presentation of other soluble antigens (Eisenlohr *et al.*, 1987). Processing of intact PR8 HA (H1) by either mouse splenocytes or a cloned B-lymphoma cell line (A20) could be inhibited by treating cells with lysosomotropic agents or by mild aldehyde fixation, while these treatments failed to inhibit presentation of the site 1 peptide. These studies also demonstrated that the efficiency of presentation of the HA by metabolically active cells to $T_H$ clones specific for sites 1, 2, or 3 varied tremendously with the form of antigen provided to the APCs. Only picomolar amounts of intact virus were required to obtain a threshold stimulation, while protein fragments or synthetic peptides were needed in micromolar quantities. Similar effects were observed using $T_H$ specific for internal viral antigens. The 3–6 $\log_{10}$ difference in presentation of different forms of antigen suggests that intact virions, and not free viral proteins, are the relevant form of antigen *in vivo.*

The high efficiency of presentation of at least some determinants using intact virus was shown to be independent of biosynthesis of viral components in APCs, since presentation efficiency was not altered by ultraviolet (UV) light inactivation of viral infectivity. Rather, several types of experiments established that the high presentation efficiency of intact virus is due to specific HA-mediated attachment of virus to sialic acid residues on the APC surface. First, removal of sialic acid on the APC surface by neuraminidase treatment concomitantly prevented association of virus with the APC and presentation of viral antigens. Second, enzymatic removal of the HA from virions greatly reduced the efficiency of presentation of M1 and NP to $T_H$. Third, presentation efficiency of HA by intact virus, as well as M1 and NP, was reduced 100-fold by continuous presence of a monoclonal anti-HA antibody, which was also shown to reduce virus attachment by a similar degree.

In the same study, the association of virus with APCs was found to occur very rapidly; only 2–10 min at 20°C was sufficient to introduce virus for a maximal level of subsequent stimulation. By aldehyde fixation at various times after pulsing cells with antigen for 10 min at 25°C, it was found that at least 30 min at 37°C is required for antigen presentation following association of virus with APCs. This stage of antigen presentation was sensitive to lysosomotropic reagents, which suggests that processing of intact virus entails internalization into an endosomal compartment. Further evidence for internalization of virus was obtained by removing virus from the cell surface after pulsing by treatment with neuraminidase. It was found that shortly after association with the cell surface, virus entered a neuraminidase-insensitive compartment. This event alone was not sufficient for processing, however, as it occurred at a temperature (10°C) that completely inhibited processing. It was also noted that neuraminidase treatment of cells pulsed with infectious virus had an equivalent effect on antigen presentation and the degree of infection, as measured by synthesis of viral components. This would suggest that the same initial route is used for antigen presentation and infection.

Virus taken up by APCs should rapidly encounter low pH within endocytic vesicles, which potentially could play a role in processing of pH sensitive proteins, such as the HA. The acid-induced conformational alteration of HA (Skehel *et al.*, 1982) known to effect hemolytic activity (Yewdell *et al.*, 1983) and antibody binding to HA (Yewdell *et al.*, 1983; Doms *et al.*, 1985; Jackson *et al.*, 1986) was also shown to effect presentation of class II MHC-restricted T-cell determinants of PR8 HA (Eisenlohr *et al.*, 1988a). Virus treated briefly at pH 5 and then returned to neutrality sensitized aldehyde prefixed APCs for recognition of HA site 2 (residues 126–138) and site 3 (302–313), but not site 1 (111–119). Intact virus appeared to be the entity responsible for sensitizing the prefixed APC because (1) hemagglutination-inhibiting antibodies could block sensitization, and (2) the sensitizing activity co-sedimented with intact virus. Although it is clear that this exposure to acid pH bypasses the need for uptake by APCs to achieve functional presentation of these two sites, it cannot be entirely ruled out that some limited proteolysis or further conformational alteration on the cell membrane is also critical in formation of the structure capable of interaction with class II molecules on the APC surface. Furthermore, even though sites 2 and 3 were both presentable from without using pH 5-pretreated virus, the two sites were affected differently by the presence of the diffusible proteolytic enzyme inhibitor leupeptin, when untreated virus and viable APCs were used to present virus to T cells; i.e., site 2 presentation was abolished, while site 3 expression was enhanced fivefold by leupeptin presence, suggesting that, under ordinary conditions, both enzymatic activity and conformational changes are needed for presentation of $T_H$ sites of the HA.

While these findings strongly support the necessity for endosomal processing of the intact virus for presentation of at least a portion of determinants present on internal and external proteins, it should be recognized that a number of major issues remain unresolved. First, the involvement of fragmentation of viral antigens in presentation is uncertain. Second, it is unclear how processing of the internal virion components occurs. Somehow, these antigens must be released from the interior of the virus. In the event of fusion of virus and cellular membranes, however, these components would be on the wrong (cytoplasmic) side of the membrane. This implies that either these proteins are processed via a cytoplasmic pathway or that a mechanism for lysosomal disruption of the virus exists.

Several additional findings are relevant to these issues. Mills et al. (1986a) provided evidence that some $T_H$ recognize determinants present only on the intact HA. This finding suggests that not all determinants recognized by $T_H$ result from either denaturation or fragmentation, or both. It would also appear that endosomal processing is not required for presentation of all sequential determinants—at least those located on the surface of the intact HA—since lysosomotropic reagents did not affect presentation of intact virus to $T_H$ that recognize such a region (Mills, 1986). Curiously, in the same study, recognition of the conformational determinant was affected by these agents. These findings are consistent with the existence of multiple antigen-processing pathways, whose relative contributions vary with the type of viral protein (e.g., internal versus external) and with the nature and location of the determinant in the intact protein.

Effective presentation of T-cell antigenic determinants requires not only their ability to be processed from the native molecule and to associate with class II molecules, but also that they be expressed relatively stably on APC surfaces. Evidence has recently been obtained that T-cell determinants can exhibit distinct half-lives of expression on APCs. In these studies (Eisenlohr et al., 1988b), APCs pulsed with isolated HA or nonreplicative PR8 virus were cultured for various periods of time and then aldehyde-fixed to arrest APC activity and to preserve expression of presented determinants. T cells specific for each of the HA sites 1, 2, and 3 were used to detect presentation of the individual T-cell determinants. The individual T-cell sites showed characteristic patterns of expression over time; sites 2 and 3 achieved maximal levels of expression about 8 hr postpulse, but then declined, while site 1 continued to increase in expression over the 48-hr time course of the experiment. These observations indicate that distinct T-cell determinants, although introduced as part of the same polypeptide, can show individual patterns of expression on APC. The fact that some determinants may be much more stably expressed than others implies that a potentially important component in determining T-cell site immunodominance may be the relative stability of a given determinant on the APC surface.

## 3. Processing of Exogenous versus Endogenous Antigen

Morrison *et al.* (1986) examined the recognition of endogenously synthesized HA by two $T_H$ clones. Recognition was assessed by lysis of APCs. Although this is an unusual method of measuring $T_H$ recognition, the clones had antigen requirements typical of $T_H$, lysing target cells treated with isolated HA polypeptide or with noninfectious virus. These clones failed however, to lyse APCs expressing endogenously synthesized HA (expression of HA was achieved by infection with a recombinant vaccinia virus containing a cloned HA gene; purified vaccinia virus was used to avoid the presence of exogenous HA). The same cells were recognized by HA-specific $T_C$. $T_H$ recognition was completely abolished by incubation of target cells with chloroquine, while $T_C$ recognition was unaffected. Presumably, chlorloquine inhibition was based on the ability of this lysosomotropic amine to inhibit normal endosomal–lysosomal trafficking. These findings indicate that expression of HA in its native form as an integral membrane protein on the APC surface is not sufficient for recognition by $T_H$. More generally, they suggest that antigens recognized by $T_C$ and $T_H$ are processed by different cellular pathways that can be distinguished by their sensitivity to chloroquine. More recently, Morrison *et al.* (1988) observed that *in vitro* restimulation of virus-primed mouse splenocytes with infectious virus yielded mainly class I MHC-restricted $T_C$, while noninfectious virus under the same conditions induced primarily class II-restricted CTL. This finding suggests that viral HA newly replicated within APCs may be skewed toward association with class I MHC gene products. By contrast, using an APC line stably transfected by a recombinant retrovirus vector, Eager *et al.* (1989) have found that $T_H$ specific for sites 1, 2, and 3 in the H1 HA recognize the constitutively expressed cloned HA gene. There are at least two possibilities for this discrepancy: (1) cells constitutively expressing HA may have greater opportunity than do infected cells to recycle biosynthesized

HA, via either recycling of intact molecules expressed on the cell surface, or uptake of shed HA; and (2) the determinants recognized by clones used by Morrison *et al.* (1988) may differ qualitatively from the site 1, 2, and 3 determinants.

Not all $T_H$ specific for viral structural proteins respond to antigens present on noninfectious virions. Murine I-$E^d$-restricted T-hybridoma clones specific for the PR8 NA have been isolated that recognize infectious, but not nonreplicative virus (Eisenlohr and Hackett, 1989). Blocking of synthesis of viral components by protein synthesis inhibitors or by prior UV inactivation of viral infectivity prevented recognition by these clones. Several observations suggest that exogenous virus did not significantly contribute to formation of the NA T-cell determinant. First, $T_H$ recognition correlated with the *de novo* synthesis of viral NA within APCs. Second, recognition did not depend strictly on the amount of NA present in cultures, since high NA concentrations could be achieved by addition of nonreplicative virus without stimulating the clones. Finally, recognition of a neoantigen present on virions replicated within the host was ruled out, since, in high concentration, NA isolated from purified egg-grown virions was recognized by the clones, even if reduced and alkylated. Isolated NA was recognized when added to prefixed APCs, suggesting that this form of antigen was able to bypass the usual processing pathway of exogenous proteins. Together, these observations imply that the antigenic determinant recognized by these NA-specific $T_H$ is cryptically present on virus particles and requires cellular processing for presentation; however, the pathway followed by exogenous virus either fails to expose the determinant effectively or destroys it. The latter possibility seems more likely, since mild detergent extraction sufficiently preprocesses the determinant for recognition in context of prefixed APCs.

These data imply that endogenously synthesized proteins have access to a class II MHC-linked processing pathway that is qualitatively distinct from that predominating in the presentation of exogenous antigen. Further studies are required to determine whether such a pathway overlaps, or is identical to, that postulated for processing of class I-MHC restricted determinants. Furthermore, viral replication is seen to have an impact on the formation of certain class II as well as class I-restricted T-cell determinants, implying that vaccination strategies to obtain maximal $T_H$ activity may have to take into account this contribution as well.

## C. Role of $T_H$ in Antiviral Immunity

### 1. Requirement for $T_H$ in Antibody Responses

Evidence for the pivotal role of $T_H$ in the development of protective antibodies to influenza virus was established by studies using mice depleted of T cells either by experimental manipulation (thymectomy, lethal irradiation, and reconstitution of non-T-lymphocyte immune cells by bone marrow transfer) (Virelizier et al., 1974) or by the congenital lack of a functional thymus (nude mice, since they are also hairless) (Burns et al., 1975). Immunization of thymectomized mice with purified HA led to only poor antibody responses, and then only after repeated immunization. Despite the virtual absence of an antibody response, immunization resulted in the priming of B-lymphocyte responses, and vigorous secondary responses could be achieved following transfer of T lymphocytes (Virelizier et al., 1974). Observations with nude mice confirmed that antibody responses to influenza virus are dependent on T lymphocytes; nude mice infected with infectious influenza virus produced a transient IgM antibody response that did not demonstrate the shift to IgG isotypes characteristic of the response of normal mice (Burns et al., 1975; Sullivan et al., 1976; Lucas et al., 1978; Wells et al., 1981a,b). The requirement for $T_H$ in antiviral antibody responses was confirmed in studies of in vitro antibody responses, which also demonstrated that the antigen-specific stimulation of $T_H$ was MHC restricted (Pierce et al., 1978; Anders et al., 1979; Callard, 1979).

## 2. Requirement for $T_H$ in $T_C$ Responses

It appears that $T_C$ responses are also dependent on helper functions of T lymphocytes, but this process has been defined at only the most rudimentary levels. It is not certain whether help is delivered by the classic $T_H$ (Moll et al., 1985) or by another T-lymphocyte subset. It has been found that the addition of radioresistant T lymphocytes derived from virus-primed mice greatly augments primary in vitro mouse responses; it was necessary to irradiate transferred cells to remove $T_C$ (Ashman and Mullbacher, 1979; Askonas et al., 1982; Miller and Reiss, 1984). Lymphocytes displaying helper activity apparently display either the classical $T_H$ phenotype [CD8$^-$] or the phenotype more commonly associated with $T_C$ and $T_S$ [CD8$^+$] (Reiss and Burakoff, 1981; Ashman and Mullbacher, 1979). The helper function of these lymphocytes appears to be MHC nonrestricted. In human in vitro $T_C$ responses, a similar requirement for T-lymphocyte-mediated help has been established. Help can be provided by virus-specific CD4$^+$CD8$^-$ and CD4$^-$CD8$^-$ lymphocytes (Biddison et al., 1981). It also appears that antigen-nonspecific large granular lymphocytes play a critical role in $T_C$ induction, although it is not certain whether this is due to their interaction with $T_H$ or with $T_C$ themselves, at the level of either antigen presentation or lymphokine secretion (Burlington et al., 1984).

## 3. Delivery of Helper Function in Antibody Responses

Studies of $T_H$–B lymphocyte collaboration clearly show that both lymphocytes must be in the proper state of activation in order to obtain optimal antibody responses. $T_H$ are activated by their recognition of antigen in the presence of IL1. B lymphocyte activation appears to occur in more or less discrete stages that can be arrived at via multiple pathways. Activation generally requires crosslinking of surface immunoglobulin by antigen, the presence of factors secreted by macrophages, and the presence of antigen-specific $T_H$ (Melchers and Andersson, 1984; Melchers et al., 1985).

One of the central issues in the function of $T_H$ in antibody responses is the relationship between the specificity of $T_H$ and the B lymphocyte(s) they help. This question was initially addressed by Russell and Liew (1979), who showed that internal influenza virion antigens, particularly M1, could generate mouse $T_H$ able to help anti-HA antibody responses. The critical portion of this experiment was the demonstration that the helper effect of anti-internal $T_H$ occurred only if mice were boosted with intact virions; no effect was observed when purified preparations of M1 and HA were co-injected. This basic finding has been confirmed in both in vitro and in vivo studies. Scherle and Gerhard (1986) found that adoptive transfer of anti-NP and anti-M1 $T_H$ clones into influenza-infected nude mice resulted in a greatly enhanced anti-HA antibody response. Significantly, the antibody response to M1 and NP was not enhanced. In the same study, it was shown that $T_H$ specific for the HA can also help anti-HA responses. Fischer et al. (1982) and Lamb et al. (1982d) showed that human $T_H$ lines specific for M1 helped anti-HA antibody responses in vitro. Like Russel and Liew, Lamb et al. (1982d) found that the helper effect was observed only if M1 and HA were added to the culture in the form of intact virions. More recently, Scherle and Gerhard (1988) observed a distinct hierarchy of help for B-cell responses to influenza components by cloned T cells transferred to nude mice. B cells recognizing viral surface components HA or NA can receive help from $T_H$ clones specific for any major virion structural component. B cells specific for internal virion proteins M1 or NP are restricted to receiving help from $T_H$ clones with the same protein specificity.

These findings indicate that anti-HA or anti NA antibodies can be elicited using $T_H$ specific any of the major viral proteins. To obtain help for HA or NA by $T_H$ to other components, viral antigens must be introduced to the system as intact virions, a phenomenon termed intrastructural, intermolecular help (Russel and Liew, 1979). This phenomenon almost certainly reflects the selective presentation of internal viral antigens to the $T_H$ antigen receptor by HA- or NA-specific B lymphocytes. The basis for this could be at several levels: (1) crosslinking of surface antibody by the multivalent virus-associated HA or NA might enhance the efficiency of presentation of internal viral antigens; or (2) the presence of anti-HA or anti-NA surface antibody might significantly enhance the efficiency of virus binding or internalization over its already high levels.

The presence of primed B cells specific for HA may affect whether optimal T-cell help for NA-specific antibody responses can be achieved

(Johansson *et al.*, 1987a,b,c). Mice primed with one HA/NA subtype yielded severalfold higher titers of neuraminidase-inhibiting (NI) antibodies in response to the same NA when delivered within the context of a different HA subtype. The precise mechanism underlying this effect is unknown. Likely possibilities are that following immunization, B cells with surface immunoglobulin receptors for the predominant virion protein, HA, may (1) be present in greater abundance than NA-specific B cells, and/or (2) acquire virus particles for processing more efficiently than do NA-specific B cells.

It is important to note that delivery of the helper function does not necessarily require the direct interaction between $T_H$ and B lymphocytes. In fact, considerable evidence shows that, *in vitro*, antigen-specific help for production of human anti-HA antibodies can be delivered by soluble factors. A number of investigators have identified factors present in culture media of activated, virus specific $T_H$ that induced the specific production of antiviral antibodies (Table III). In some cases, the help provided by these factors has also been shown to be MHC restricted; in one case, the MHC restriction of the factor for individual HLA alleles was identical to the $T_H$ clone from which it was derived (Fischer *et al.*, 1981). Aside from their molecular weight, the biochemical nature of these factors is undefined. Their antigen and MHC specificity is, however, consistent with the idea that they contain at least the variable portion of the $T_H$ antigen receptor. Taking this a step further, their action would presumably result from signal transduction following binding to processed antigen preferentially expressed on anti-HA B lymphocytes.

In addition to antigen-specific factors, it is likely that $T_H$ secrete nonspecific factors necessary for B-lymphocyte stimulation. One such factor might be IL-2, which Tan *et al.* (1985) reported was required for the production of human antiviral antibodies *in vitro*. The involvement of IL-2 requires further clarification, however, as Lamb *et al.* (1983) found that $T_H$ that fail to secrete detectable amounts of IL-2 were still able to help antiviral antibody responses.

## IV. SPECIFICITY AND FUNCTION OF ANTI-INFLUENZA VIRUS $T_C$

### A. Specificity of $T_C$

#### 1. Recognition of Individual Viral Proteins

Early investigators identified two populations of influenza virus-specific mouse $T_C$—one specific for the immunizing virus or closely related strains (strain-specific), the other able to lyse cells infected with any influenza A, but not B, virus (cross-reactive) (Cambridge *et al.*, 1976; Braciale, 1977; Zweerink *et al.*, 1977b; Effros *et al.*, 1977). Soon after, it

TABLE III. Soluble Factors in Stimulated Human T Supernatants That Help B-Lymphoctye Responses to Influenza Virus in Vitro[a]

| Supernatant activity | T-cell source | T-cell stimulation | Characterization of supernatant activity | References |
|---|---|---|---|---|
| MHC nonrestricted not virus specific | Uncloned line A/X-31-specific PBL | Irradiated autologous PBL + A/X-31 | $M_r = 10,000–30,000$ | Zanders et al. (1983) |
| MHC nonrestricted antigen dependent | PBL | Autologous or allogeneic PBL + A/X-31 virus | Helped autologous or allogeneic B cells | Callard and Smith (1981) |
| MHC restricted, virus-specific | Uncloned lines A/X-31-specific (refs. 1 and 2) M1-specific clone (ref. 3) | Irradiated autologous PBL + virus (refs. 1,2); none required (ref. 3) | DR-shared helped best (ref. 1) $M_r = 50,000–70,000$ (ref. 2) | 1. Fischer et al. (1981); 2. Zanders et al. (1983); 3. Lamb et al. (1982d) |
| Virus-specific; MHC restriction not reported | Uncloned T-helper lines; type B or type A virus specific | Irradiated PBL + virus | B-virus-specific helps anti-B antibody responses; A-virus-specific helps anti-A antibody responses | Fischer et al. (1983) |
| T-cell replacing factor (TRF) | Tonsillar mononuclear cells | Phytohemagglutinin | $M_r = 35,000–43,000$; antigen required for delivery of help | Callard et al (1985) |
| Interleukin-2 (IL-2) | Tonsillar cells | Phytohemagglutinin | $M_r = 13,000$; replaced by recombinant IL-2; antibody to IL-2 receptor blocked effect | Tan et al. (1985) |

[a] Abbreviations: PBL, peripheral blood lymphocytes; A/X-31, recombinant virus A/Hong Kong × A/PR/8 (H3N2); MHC, major histocompatibility complex.

was found human $T_C$ responses are highly, if not completely, cross-reactive (McMichael and Askonas, 1978; Biddison et al., 1979). While the existence of strain-specific $T_C$ was expected based on the known serological diversity of the glycoproteins, the existence of cross-reactive $T_C$ was surprising and excited a tremendous amount of interest. It was proposed that either cross-reactive $T_C$ recognized the viral glycoproteins in a manner different from antibodies or that the highly conserved internal proteins were expressed in some form on infected cell surfaces, where they could serve as $T_C$ target structures. This problem proved to be of general importance, since cross-reactive recognition of distinct serotypes has been found to be a feature of the $T_C$ response to many viruses (Zinkernagel and Doherty, 1979).

Investigations to determine which viral components are recognized by $T_C$ can be divided into pre- and postgenetic engineering eras. Studies of the initial era provided strong evidence for the recognition of HA by strain-specific $T_C$ (Zweerink et al., 1977a; Ennis et al., 1977a,b; Braciale, 1979; Braciale et al., 1981), and also gave the first clues that $T_C$ could recognize internal viral proteins (Bennink et al., 1982; Townsend et al., 1984). These studies were limited to various degrees, however, by the inability to attain complete physical or functional separation of the individual viral components. This problem was finally overcome by cloning the individual influenza virus genes and placing them in vectors that enabled their expression in histocompatible target cells (Gething and Sambrook, 1981; Winter et al., 1981; Young et al., 1983; Smith et al., 1983, 1987; Panicali et al., 1983).

Two type of vectors have been used for the expression of cloned viral genes:

1. DNA constructs containing an influenza virus gene in conjunction with transcriptional promoters and a gene (commonly thymidine kinase or neomycin resistance) conferring a growth advantage to cells when cultured under selective conditions: These constructs are introduced into cells by direct transfection, resulting in the constitutive biosynthesis of the influenza virus gene product.
2. Recombinant vaccinia viruses (Vac) containing influenza virus genes under the control of Vac promoters: Infection of cells with Vac recombinants results in the expression of influenza virus gene products during the course of a lytic Vac infectious cycle.

Although cells expressing transfected genes have been successfully used to study $T_C$ specificity, using the currently available vectors it has proved difficult to maintain uniform levels of expression of influenza virus genes in transfected mouse cells. This problem is averted using Vac recombinants, which also offer great experimental flexibility due to their broad host range and ability to induce anti-influenza $T_C$ responses in vivo and in vitro (Bennink et al., 1984). These advantages have led to the

TABLE IV. $T_C$ Recognition of Individual Viral Proteins

| | Source of $T_C$ | | | |
| | Mice | | | |
| Protein | BALB/c (H-2$^d$) | CBA (H-2$^k$) | C57B1/6 (H-2$^b$) | Humans |
|---|---|---|---|---|
| PA | +$^a$ | −$^b$ | + | ?$^c$ |
| PB1 | + | + | ? | ? |
| PB2 | + | − | ? | + |
| HA | + | + | ? | ? |
| NP | + | + | + | + |
| NA | − | − | NT$^d$ | ? |
| M1 | − | − | − | + |
| M2 | − | − | − | ? |
| NS1 | + | + | ? | ? |
| NS2 | − | − | NT | ? |

$^a$+, $T_C$ known to recognize viral protein.
$^b$−, $T_C$ recognition not observed in extensive series of experiments.
$^c$?, $T_C$ recognition not observed in limited series of experiments.
$^d$NT, $T_C$ recognition not tested.

widespread, though not exclusive, use of Vac recombinants for studies of $T_C$ specificity.

$T_C$ recognition of all of the defined viral gene products has been examined to various degrees. These results are summarized in Table IV. Special features of $T_C$ recognition of the individual components by mouse $T_C$ are presented on a component-by-component basis. The more general conclusions that arise from these studies are then presented.

### a. HA

Braciale et al. (1984) studied recognition of transfected L cells [H-2$^k$] expressing the H2 HA, derived from A/JAP/305 (JAP) (H2N2), by $T_C$ populations and cloned $T_C$ lines derived from CBA (H-2$^k$) mice. These cells were lysed by $T_C$ populations raised against JAP and, sporadically, by $T_C$ populations raised to a H1N1 virus. While all the strain-specific cloned $T_C$ cell lines tested efficiently lysed these cells, only one of 13 cross-reactive lines demonstrated lysis, and then only at low levels compared with its recognition of virus-infected cells.

Townsend et al. (1984b) examined C3H (H-2$^k$) $T_C$ recognition of L cells expressing the H1 HA, derived from PR8. These cells were found to be specifically lysed by anti-PR8 $T_C$ populations, and to a lesser extent by anti-JAP $T_C$, but were not specifically recognized by $T_C$ induced by A/NT/60/68 (NT/60) (H3N2).

Bennink et al. (1984, 1986) examined BALB/c (H-2$^d$) $T_C$ recognition of the HA at both stimulator and target cell levels using Vac recombinants containing either the PR8 or JAP HA genes, termed H1-Vac and H2-Vac, respectively. P815 cells (H-2$^d$) infected with H2-Vac were lysed at high levels by $T_C$ elicited by priming and stimulation with JAP and at lower levels by anti-PR8 $T_C$ and were not lysed by a number of cross-reactive $T_C$ populations induced by priming and stimulation with heterologous H1N1 and H3N2 viruses. H1-Vac-infected cells were lysed by anti-PR8 $T_C$, and to a slight extent by anti-JAP $T_C$, but again were not recognized by cross-reactive $T_C$ populations induced by priming and stimulation with viruses of different subtypes. Consistent with these findings, inoculation of mice with H2-Vac or H1-Vac was found to prime their splenocytes for largely strain-specific secondary influenza $T_C$ responses upon *in vitro* stimulation with the homologous virus.

The H2-Vac recombinant was also used to establish the anti-HA specificity of 10 H-2$^d$-restricted $T_C$ clones (Braciale *et al.*, 1986). The fine specificity of these clones was then examined using cells infected with a panel of H1, H2, and H3 influenza viruses. Most clones (8 of 10) were found solely to recognize the H2 HA. Five of these clones appeared to be specific for the JAP HA, while three others also recognized cells expressing H2 HAs derived from related strains. Of the two cross-reactive clones, one recognized H1 and H2 HAs, and the other recognized H1, H2, and H3 HAs, although recognition of H1 and H3 HAs was weak relative to H2.

Considering the differences in the vectors, target cells, and mouse strains used, the results of these three studies are remarkably consistent. Taken together, they indicate that the HA is recognized largely in a strain-specific manner by $T_C$ and that the slight amount of cross-reactive recognition that occurs is limited largely to the H1 and H2 HAs. This pattern of cross-reactivity is consistent with the relatively high amino acid homologies between H1 and H2 HAs and with the low homologies between both H1 and H2 HAs with the H3 HA.

These studies also indicate that a greater $T_C$ response is induced by the JAP HA than by the PR8 HA in the mouse strains tested. Results obtained from limiting dilution assays indicate that this difference reflects the frequency of $T_C$ precursors present in secondary $T_C$ cultures; HA-specific $T_C$ were found to occur at a three- to eightfold higher frequency in JAP-primed mice than in PR8 primed mice (Andrew *et al.*, 1986; Wysocka and Bennink, 1988).

Curiously, it has not been possible to demonstrate a H3-specific $T_C$ response in a number of mouse strains using a Vac recombinant containing the H3 HA gene to either prime mice or to sensitize target cells for lysis by $T_C$ induced by H3N2 viruses (Gould *et al.* 1987; J. Bennink, J. Yewdell, and G. Smith, unpublished observations). Gould *et al.* (1987) examined $T_C$ recognition of Vac recombinants expressing a reciprocal pair of HA chimeras consisting of HA1 and HA2 chains derived from H1 and H3 HAs. The failure of H1-specific $T_C$ to recognize H3 made it possible to show that both HA1 and HA2 subunits of the H1 HA are recognized by H-2$^k$-restricted $T_C$. Complementary findings were made by Braciale *et al.* (1987) concerning the recognition of H2 HA by H-2$^d$-restricted $T_C$ clones.

## b. NP

NP was the first internal viral protein shown directly to be a $T_C$ target antigen. Townsend et al. (1984b) found that transfected L cells co-expressing the NP gene derived from NT60 and the $D^b$ gene were efficiently lysed by $T_C$ populations derived from CBA mice ($H-2^k$) induced by priming and in vitro stimulation with H1N1, H2N2, or H3N2 viruses. In addition, they were lysed by a $D^b$-restricted strain-specific $T_C$ clone, which confirmed previous evidence (Bennink et al., 1982; Townsend et al., 1984a; Kees and Krammer, 1984) that the highly conserved internal proteins can be recognized in a strain-specific manner.

Yewdell et al. (1985) found that Vac recombinant containing the PR8 NP gene (NP-Vac) sensitized P815 cells for efficient lysis by cross-reactive BALB/c $T_C$ populations induced by priming and stimulating with heterologous influenza A viruses. A major role for NP in cross-reactive $T_C$ recognition was further supported by stimulation experiments performed using NP-Vac. Inoculation of mice with NP-Vac primed their splenocytes for vigorous crossreactive secondary in vitro responses upon challenge with H1N1, H2N2, or H3N2 viruses.

## c. Polymerases

$T_C$ recognition of the polymerases has been examined using recombinant vaccinia viruses containing the PR8 PA, PB1, and PB2 genes (Bennink et al., 1987). Cross-reactive $T_C$ populations derived from BALB/c mice were found to recognize PB2-Vac infected P815 cells and, less consistently, PA-Vac infected cells. PB1-Vac-infected cells were recognized only infrequently, and then at low levels. When other target cell–mouse strain combinations were used, different patterns of recognition were obtained. Cross-reactive $T_C$ populations derived from CBA mice ($H-2^k$) consistently lysed PB1-Vac-infected L cells but failed to lyse PB2-Vac or PA-Vac-infected cells. Cross-reactive $T_C$ populations derived from C57Bl/6 mice ($H-2^b$) lysed PA-Vac-infected MC57G ($H-2^b$) target cells at higher levels than PB2-Vac-infected cells, and failed to lyse PB1-Vac-infected cells. Significantly, differential recognition of PB1 and PB2 by BALB/c and CBA $T_C$ was also found to occur using target cells derived from $H-2^k \times H-2^d$ $F_1$ mice. This indicates that the phenomenon is not due to differences in expression of PB1 and PB2 by the target cells used to assess $T_C$ activity. Using recombinant inbred mice, it was found that responsiveness to PB1 and PB2 co-segregated with $H-2^k$ and $H-2^d$ MHCs, respectively.

## d. M1 and M2

$T_C$ recognition of M1 and M2 has been examined using two Vac recombinants, one containing the full-length gene derived from PR8 (M1-Vac) and the other containing a cDNA of the M2 mRNA derived from cells infected with A/Udorn/72 (H3N2) (M2-Vac) (J. Yewdell, J. Bennink, B. Moss, and R. Lamb, unpublished observations). Note that during an influenza virus infection, M2 mRNA is produced by a splicing of the M1 mRNA. Since Vac infections occur entirely within the cytoplasm, splicing does not occur, and it was possible using the Vac recombinants to examine $T_C$ recognition of M1 and M2 separately. $T_C$ derived from BALB/c, CBA, and C57Bl/6 mice repeatedly failed to recognize M1-Vac- or M2-Vac-infected target cells. These results could not be attributed to low levels of antigen expression, since Vac recombinant-infected cells expressed similar amounts of M1 or M2 as influenza virus-infected cells. In addition, inoculation of mice with Vac recombinants failed to prime their splenocytes for secondary *in vitro* anti-influenza $T_C$ responses, although they did prime mice for secondary $T_H$ responses (C. Hackett, J. Bennink, and J. Yewdell, unpublished observations).

### e. NS1 and NS2

Using a Vac recombinant containing the PR8 NS1 gene, it was found that secondary anti-influenza $T_C$ from BALB/c, CBA, and C57Bl/6 mice recognized NS1, establishing that antiviral $T_C$ can recognize nonstructural viral proteins (Bennink *et al.*, 1987). As with the internal structural proteins, recognition was highly cross-reactive. Curiously, unlike any of the other Vac recombinants able to sensitize target cells for anti-influenza $T_C$ lysis, inoculation of BALB/c mice with NS1-Vac failed to prime their splenocytes for secondary *in vitro* anti-NS1 $T_C$ responses upon restimulation with either H1N1, H2N2, H3N2 influenza viruses or with NS1-Vac. The same virus primed perfectly well for BALB/c anti-Vac $T_C$ responses or for CBA anti-NS1 responses (Yewdell *et al.*, 1988; J. Bennink and J. Yewdell, unpublished observations). As with M1, splicing of the NS1 message results in production of an additional nonstructural protein, NS2. Using a Vac recombinant expressing PR8 NS2, it was found that mouse $T_C$ can recognize NS2, although recognition was found in only one strain (B10.M, H-2$^f$) out of seven haplotypes tested (J. Bennink, J. Yewdell, and G. Smith, unpublished observations).

### f. NA

$T_C$ recognition of NA was tested using a Vac recombinant expressing the NA from A/CAM/46 (H1N1). Splenocytes derived from NA-Vac-primed mice restimulated with PR8 failed to recognize PR8 or NA-Vac-infected cells specifically, and $T_C$ induced by PR8 priming and restimulation failed to lyse NA-Vac-infected cells in eight mouse haplotypes tested (J. Bennink, J. Yewdell, and G. Smith, unpublished observations). Although the PR8 and CAM NAs are closely related and are highly cross-reactive antigenically, it is possible that the failure to observe NA-specific $T_C$ is due to an exceptionally limited strain-specific response.

### g. General Conclusions

*i. The Vast Majority of Cross-Reactive $T_C$ Recognize Internal Viral Proteins.* This has been most clearly shown for mouse responses, but human $T_C$ have also been found to recognize target cells infected with NP-Vac, PB2-Vac, and, unlike mouse $T_C$, M1-Vac (McMichael *et al.*, 1986a; Gotch *et al.*, 1987). Nonstructural proteins as well as structural proteins can be recognized. Recognition of internal antigens is a common feature of antiviral $T_C$, as internal proteins derived from a number of other viruses, have been found to serve as $T_C$ target antigens, e.g., SV40 T antigen (Gooding, 1977; Tevethia *et al.*, 1980; Pan and Knowles, 1983), VSV nucleocapsid protein (Yewdell *et al.* 1986, Puddington *et al.*, 1986), and respiratory syncytial virus nucleocapsid protein (Bangham *et al.* 1986). It would be surprising if internal cellular antigens did not also play a prominent role in $T_C$ responses to minor histocompatibility antigens and tumor antigens.

*ii. $T_C$ Recognition of Internal Viral Proteins Is Independent of Their Relative Abundance.* Despite the fact that the polymerases are produced in very low amounts, they can be recognized with equal efficiency to NP or NS1, which are produced in 100- to 1000-fold higher amounts. It appears that even lower amounts of internal proteins can be detected in transfected cells (Townsend *et al.*, 1984b, 1985; Puddington *et al.*, 1986), suggesting that $T_C$ surveillance of cellular proteins may not often be limited by low levels of expression.

*iii. Cross-Reactive $T_C$ Recognition of Related Proteins Generally Reflects Their Degree of Amino Acid Homology.* This is largely limited to the highly conserved internal proteins. This is not to say that recognition of internal proteins is always cross-reactive; it has been shown that at least two of the internal proteins—NP (Townsend and Skehel, 1984) and PB2 (Bennink *et al.*, 1982)—can be recognized in a highly strain-specific manner. Indeed, it has been found that under some conditions, most strain-specific $T_C$ recognize internal viral proteins (Kees and Krammer, 1984).

*iv. Most, But Perhaps Not All, Viral Components Can Serve as $T_C$-Recognition Structures.* It has not been possible to demonstrate recognition of M1, M2, or NA by mouse $T_C$. If this can be shown to be related to physical properties of these proteins, it would provide valuable insight into the processing of viral antigens into $T_C$ target antigens. It is possible, however, that these antigens can only be recognized in conjunction with a very limited number of class I MHC alleles (see below), an idea favored by the recognition of M1 by human $T_C$ (Gotch *et al.*, 1987). It is also possible that the failure of mouse $T_C$ to recognize cells infected with Vac recombinants is due to interference from processes related to the vaccinia infection.

*v. The Prevalence of $T_C$ Specific for Any Given Component Depends on a Number of Factors.* One major factor is MHC haplotype, since non-responder class I MHC alleles have been identified for each viral component. Other factors also appear to influence responses to individual proteins, since even in high-responder inbred mouse strains, $T_C$ recognition of NP has been found to vary widely between individuals (Pala *et al.*, 1986). Whether this is due to specific suppression of anti-NP responses, perhaps to environmental stimuli, or to fundamental differences in individual $T_C$ repertoires remains to be determined.

*vi. Not All $T_C$ May Recognize Individual Viral Components.* Studying the specificity of PR8-specific $T_C$ clones using recombinant influenza viruses containing genes derived from PR8 and a H3N2 virus, Bennink *et al.* (1982) noted that the lytic activity of some clones could not be assigned to a single gene but was influenced by a number of genes encoding internal viral proteins. Two explanations for this finding are (1) $T_C$ recognition of complexes of viral gene products, or (2) *trans*-acting alteration of processing by some viral components. As both explanations are interesting, it will be important to reexamine this finding using currently available reagents.

*vii. Vaccinia Virus May Not Be a Perfect Expression Vector for Studies of $T_C$ Specificity.* The Vac recombinants described above all used early Vac promoters to drive transcription of foreign DNA. Coupar *et al.* (1986) examined $T_C$ recognition of cells infected with Vac recombinants containing the H1 HA gene under control of a late Vac promoter; i.e., transcription begins only after early Vac proteins are expressed and DNA replication has begun. Despite the fact that significant amounts of HA were produced, these cells were not recognized by H-$2^k$-restricted influenza-specific $T_C$. Although limiting dilution analysis (Wysocka and Bennink, 1987) suggests that most determinants are processed properly in Vac-infected cells, the findings of Coupar *et al.* serve as a warning that Vac may inhibit the processing of some determinants, even when produced under the control of early Vac promoters. By the same token, the specific ability of Vac to inhibit antigen processing is potentially useful for dissecting the cellular events that comprise antigen processing.

2. Form of Antigen Recognized by $T_C$

*a. $T_C$ Recognition of Exogenous and Endogenous Forms of Viral Antigens*

Unlike $T_H$, $T_C$ are far more efficiently induced by inoculation of animals with infectious virus than by noninfectious whole virus and subviral preparations (see Section II). This is consistent with the fact that target cells are far more efficiently sensitized for $T_C$ lysis by infectious than by noninfectious virus preparations. There is convincing evidence that $T_C$ recognize newly synthesized antigens in infected cells. First, $T_C$ lysis is inhibited by preventing biosynthesis of viral gene products, either by inactivating viral infectivity with UV irradiation (Braciale and Yap, 1978) or with β-propiolactone treatment (Hosaka et al., 1985), or by treating target cells with drugs that inhibit viral transcription or protein synthesis (Hosaka et al., 1985; Yewdell et al., 1988). Second, $T_C$ are now known to recognize at least two viral gene products (NS-1 and NS-2) that arise only during the course of the viral infection. The kinetics of $T_C$ recognition of infected cells is consistent with the recognition of newly synthesized proteins (Ada and Yap, 1977). Cells first become susceptible to lysis shortly after synthesis of viral proteins can be detected (1–2 hr p.i.). Maximal levels of recognition are reached within 4–5 hr p.i. Co-incubation of noninfected cells with infected cells does not sensitize them for lysis by $T_C$ (J. Bennink and J. Yewdell, unpublished observations). This indicates that $T_C$ do not normally recognize viral antigens leaked or secreted by infected cells, but rather that recognition is limited to cells expressing endogenously produced antigen.

Although under conditions that exist during host infection with influenza virus it is likely that $T_C$ predominantly, if not exclusively, recognize newly synthesized viral proteins, this is not an absolute requirement for recognition. It has been repeatedly demonstrated that $T_C$ responses can be induced by exogenous internal and external viral antigens. Secondary in vitro strain-specific $T_C$ responses have been elicited with purified HA and with UV irradiated virus (Zweerink et al., 1977a; Braciale and Yap, 1978; Braciale, 1979). In addition, cross-reactive in vitro responses have been elicited by UV-irradiated virus (Zweerink et al., 1977a), liposomes containing viral components (Hackett et al., 1983b), and a preparation consisting predominantly of NP (Wraith and Askonas, 1985); this preparation also induced in vivo responses. Jones et al. (1988) compared the ability of micelles and ISCOMS (highly structured micellar-like structure formed by the interaction of transmembrane domains with the glycoside Quil A) containing partially purified HA and NA to elicit $T_C$ following intranasal inoculation. ISCOMS were superior to micelles in priming for secondary in vitro responses and were able to elicit secondary responses in mice previously infected with a virus containing the same glycoproteins. Based on the subtype specificity of the response, it appeared that the response was predominantly HA specific.

It is commonly found that preparations able to induce $T_C$ fail to sensitize target cells for $T_C$ lysis. As it is likely that the $T_C$ Ti recognizes the identical determinant during induction and effector phases, this probably reflects quantitative differences in requirements for induction and measurable recognition of target cells in $^{51}Cr$ release assays. This could occur at the level of antigen concentration on the cell surface required for stimulation as compared with lysis. It might also reflect the ability of specialized antigen-handling cells, such as cells of monocyte lineage, to present antigen to $T_C$ more efficiently than cells normally used as targets in $^{51}$Cr release assays. Alternatively, it could simply reflect that stimulation of a $T_C$ response requires a relatively low percentage of cells with antigen concentration above some threshold value, while detection of $T_C$ lysis requires a higher percentage of cells expressing antigen above this value.

Further evidence that quantitative and not qualitative effects account for the differences in $T_C$ induction versus target cell sensitization comes from studies in which sensitization has been achieved using non-infectious virus. Sensitization of mouse target cells by UV-irradiated avian influenza virus was studied by Kurrle et al. (1979). $T_C$ raised to the homologous virus lysed P815 cells incubated with UV-irradiated virus nearly as efficiently as virus-infected cells. $T_C$ recognition of virus infected cells was reduced, but not totally abrogated by addition of a protein synthesis inhibitor. In both circumstances, i.e., UV virus, infectious virus plus protein synthesis inhibitors, cleavage of the HA into HA1 and HA2 subunits was required for sensitization. This would suggest that fusion of viral and cellular membranes is necessary for proper processing of exogenous antigen. But fusion is also required to initiate the infectious cycle and, as the possible biosynthesis of limited amounts of viral antigens was not examined, it cannot be excluded that in both circumstances, $T_C$ recognized endogenously biosynthesized components. As only $T_C$ elicited by the homologous virus were examined in this study, the strain-specific versus cross-reactive nature of recognition was not established, so no educated guess can be ventured as to whether internal viral components were recognized under these conditions.

Some of the variability observed in target cell sensitization using UV-irradiated virus may be related to differences in the virus strains used. A recombinant virus containing the JAP (H2) HA and A/BEL/42 (H1N1) non-HA genes was found to sensitize P815 cells following UV irradiation, while in agreement with a previous study (Braciale and Yap, 1978), UV irradiated JAP was much less effective (Ertl and Ada, 1981). Two findings suggest that sensitization was not the result of newly synthesized antigens: (1) cells were susceptible to lysis at earlier times following exposure to virus than cells infected with low doses of untreated virus, and (2) sensitized targets were recognized in a strain-specific manner. On the basis of current knowledge, this latter finding strongly suggests that $T_C$ recognition of these cells was limited to the HA, and possibly the NA. Given this conclusion, the observed difference between JAP and JAP recombinant viruses would suggest a role for non-HA genes in the presentation of HA by target cells.

Hosaka et al. (1985) reported that virus rendered noninfectious by heat treatment (30 min at 56°C) sensitized L cells for lysis by cross-reactive $T_C$. The following findings suggested that $T_C$ recognized viral antigens present in the input inoculum and not newly synthesized proteins: (1) sensitization for $T_C$ lysis occurred at earlier times with heat-treated virus than with non-heat-treated virus, and (2) $T_C$ recognition was not affected by treating target cells with agents that prevent biosynthesis of viral gene products. Additional experiments showed that target cell recognition was inhibited by choloroquine, which interferes with endosomal processing of internalized ligands, and required activation of viral fusion activity by cleavage of the HA into HA1 and HA2 subunits. This suggests that fusion of viral and cellular membranes in an endosomal compartment is required for sensitization using heat-inactivated virus. Subsequently, using cloned $T_C$ cell lines, Hosaka et al. (1988) demonstrated that L cells sensitized with heated virus were recognized by $T_C$ clones specific for HA, NP, and PB1. Each of the 11 clones of known specificity for NP and PB1 recognized cells sensitized with either infectious or noninfectious virus. By contrast, five of six HA-specific clones failed to recognize HA from heated virus. This last finding provides the best evidence that recognition of heated virus is not due to residual synthesis of viral proteins in target cells. Most importantly, it indicates that there can be differences between cellular processing of exogenous and endogenous proteins for $T_C$ recognition.

Using secondary in vitro $T_C$ populations specific for individual viral components, Yewdell et al. (1988) found independently that HA, NP, and PB1 were recognized on L cells sensitized with heated virus. To ensure that sensitization was not due to residual synthesis of viral proteins, virus was extensively UV irradiated prior to heating, and cells were continuously incubated with protein synthesis inhibitors. Strong evidence for the lack of biosynthesis of viral proteins in target cells was provided by the lack of recognition by $T_C$ specific for NS1, a protein excluded from virions, whose presence in cells is dependent on biosynthesis. In the same study, the basis for sensitization by heated virus was determined to be inactivation of viral NA activity; sensitization paralleled thermal inactivation of viral NA activity, and sensitization of targets with UV-irradiated virus was achieved by addition of a competitive inhibitor of NA activity. Yewdell et al. (1988) noted that NA inactivation greatly increased the amount of cell-associated virus and suggested that this effect accounted for its ability to sensitize cells. They acknowledged, however, that it was also possible that NA inactivation could alter cellular processing of input virus. In either event, two findings strongly suggested that sensitization was due to delivery of viral components to the cytoplasm. First, it was directly demonstrated that components of inactivated virions were released into the cytoplasm, since it was possible to visualize two karyophilic viral proteins, NP and PB2, in the nuclei of sensitized cells using monoclonal antibodies. Second, inactivation of viral fusion activity by brief exposure to acid pH (Yewdell et al., 1983), or inhibition of viral fusion by ammonium chloride, prevented target cell sensitization. If ammonium chloride was added after a short incubation time to allow for penetration of heated virus, $T_C$ recognition of cells was unaffected.

Taken together, the results with NA-inactivated virus demonstrate that reason for the great efficiency of infectious virus in eliciting $T_C$ is not its ability to direct synthesis of viral proteins in APCS *per se*, but the cytoplasmic location of the newly synthesized proteins. The failure of ammonium chloride to inhibit processing is consistent with the findings of Morrison *et al.* (1986) discussed in Section III.8.3 and provides additional evidence that processing for class I restricted recognition occurs in an extraendosomal compartment. The dependence of class I-restricted processing on the cytoplasmic location of intact proteins has been confirmed by Moore *et al.* (1988), who showed that ovalbumin was processed for $T_C$ recognition only if it was delivered to the cytoplasm of target cells, which they achieved by osmotic lysis of endosomes containing internalized ovalbumin.

Target cells have also been sensitized for lysis using subviral components. In the first of these studies, L cells treated with liposome preparations containing glycoproteins derived from A/Victoria/75 (H3N2) were found to be recognized by anti-influenza $T_C$ (Koszinowski *et al.*, 1980). Recognition was found to occur in a cross-reactive manner, which initially raised the prospect that cross-reactive $T_C$ recognized conserved regions of the glycoproteins. Based on current knowledge, it is far more likely that sensitization resulted from delivery of contaminating amounts of NP or PB1 to the cytoplasm of sensitized cells following Sendai-mediated fusion of liposome and plasma membranes.

More recently, liposomes containing glycoproteins derived from an avian influenza virus, A/FPV/Rostock (H7N1), were found to sensitize cells for $T_C$ lysis in the absence of additional fusion factors (Stitz *et al.*, 1985). In this case, recognition was limited to $T_C$ induced with the homologous virus, consistent with recognition of glycoproteins. Although it was assumed that sensitization was based on the fusion of liposomes with the plasma membrane, no direct evidence for this event was presented. This idea would, however, be consistent with $T_C$ recognition of HA from heat-treated virus. It is uncertain how, or even if, membrane-bound HA is released into the cytoplasm following the fusion of viral and cellular membranes, since it should remain oriented on the lumenal face of the endosomal membrane. It will be of great interest to determine whether membrane bound HA is processed by APCS in a manner similar to soluble internal viral components.

Although it is generally found that incorporation of viral proteins into liposomes is required for target cell sensitization, this is not an absolute requirement. As described in detail below, $T_C$ can lyse cells treated with viral determinants in their simplest form (oligopeptides). Yamada *et al.* (1985a) reported that a hybrid protein produced in bacteria consisting of the amino-terminal 81 residues of NS1 and the entire HA2 subunit of PR8 sensitized P815 cells to $T_C$ lysis. $T_C$ recognition of treated cells appeared to be limited to the HA2 portion of the protein, consistent with more recent findings that HA2 is recognized by H1-specific $T_C$ (Gould *et al.*, 1987). The same protein was found to induce *in vivo* and *in vitro* H1 HA-specific $T_C$ responses (Yamada *et al.*, 1985b). Although the mechanism of presentation was not examined in detail, Yamada and co-workers noted that the HA2 portion of the protein had two hydrophobic sequences—the fusion peptide and the carboxy-terminal membrane anchoring domain—that could directly mediate insertion of the protein into cellular membrane. It is important to determine whether sensitization is based on direct association with class I molecules on the cell surface or is due to delivery of the hybrid protein to the cytoplasm where it is subsequently processed. In the latter case, it would be of great interest to examine the role of the HA2 fusion peptide in translocation and, if active in this process, to determine its general ability to transport proteins across the plasma membrane.

### b. $T_C$ Recognition of Sequential Determinants

Until recently, it was assumed that $T_C$ recognize intact viral antigens. This assumption stifled attempts to define $T_C$ recognition structures using oligopeptides. Initial evidence for $T_C$ recognition of sequential determinants of influenza virus proteins was provided by Fan and colleagues, who found that cyanogen bromide cleavage fragments of the JAP (H2) HA could stimulate secondary *in vitro* strain-specific $T_C$ responses (Wabuke-Bunoti *et al.*, 1981; Wabuke-Bunoti and Fan, 1983). This confirmed prior observations that $T_C$ could be induced using cyanogen bromide cleavage fragments of Sendai virus fusion protein (Guertin and Fan, 1980). Subsequently it was found that *in vitro* secondary responses could also be obtained using synthetic peptides corresponding to residues 103–123 in the HA2 chain and to 181–204 in the HA1 chain (Wabuke-Bunoti *et al.*, 1984).

More direct evidence for $T_C$ recognition of fragments of viral proteins has been provided by the studies of Townsend and colleagues. They first showed that L cells (H-2$^k$) transfected with DNA containing the D$^b$ gene and various cloned fragments of the NT60 (H3N2) NP gene were efficiently recognized by cross-reactive $T_C$ populations derived from C57Bl/6 (H-2$^b$) and CBA (H-2$^k$) mice (Townsend *et al.*, 1985). $T_C$ derived from CBA mice recognized cells expressing fragments containing amino acids 1–130, 1–327, and 1–386. By contrast, C57Bl/6 $T_C$ recognized cells expressing fragments containing amino acids 1–386 and 1, 2, 328–498. Since the same pattern of reactivity was found using a strain-specific cloned $T_C$ line, known as F5, the specificity of at least a portion of the C57Bl/6 response could be assigned to amino acids 328–386.

In a subsequent study, F5 was found to recognize transfected cells expressing an NP gene containing a deletion between residues 255–339 (Townsend *et al.*, 1986). The clone was then tested for its ability to recognize synthetic peptides containing residues 345–360 or 365–380. Using as a measure of recognition either F5 proliferation, or lysis of target cells treated with peptide, F5 was found to recognize residues 345–360, but not 365–380. A series of peptides was then used to examine the

TABLE V. Peptides Recognized by Influenza-Specific $T_C$

| Protein | Location | Restriction element | Reference |
|---------|----------|--------------------|-----------|
| NP | 365–397 | H-2D$^b$ | Townsend et al. (1986) |
| NP | 335–349 | HLA-B37 | Townsend et al. (1986) |
| NP | 50–63 | H-2K$^k$ | Bastin et al. (1987) |
| NP | 147–168 | H-2K$^d$ | Bodmer et al. (1988) |
| M1 | 55–73 | HLA-A2 | Gotch et al. (1987) |
| H2 HA | 523–545 | H-2K$^d$ | Braciale et al. (1987) |
| H2 HA | 202–221 | H-2K$^d$ | Braciale et al. (1989) |

minimal sequence required for recognition. The shortest peptide able to sensitize target cells for lysis was determined to be 369–379, although the efficiency of sensitization on a molar basis decreased 10,000-fold compared with the optimally sized peptide (366–379). Three other D$^b$-restricted mouse $T_C$ clones specific for the PR8 NP were also tested for recognition of peptides. All were found to recognize a peptide consisting of residues 365–380 derived from the PR8 NP; (PR8 and NT60 differ by two amino acids in this region). Thus, all 4 D$^b$-restricted clones tested recognized the same region of NP that could be substituted by a short synthetic peptide. By contrast, a $T_C$ population derived from a human donor failed to recognize target cells treated with this peptide but did recognize target cells treated with a peptide corresponding to residues 410–425.

The work of Fan and Townsend and their colleagues provided the first indication that $T_C$ can recognize sequential determinants. Recently, a number of other investigators identified peptides recognized by influenza-specific $T_C$, these are summarized in Table V, which will, no doubt, represent only a fraction of the peptides reported by the time this chapter appears in print. Remarkably, one of the peptides recognized by $T_C$ is derived from the transmembrane domain of the HA (Braciale et al., 1987) and accounts, at least in part, for K$^d$-restricted $T_C$ recognition of the HA2 chain. $T_C$ recognition of this region indicates that the processing machinery is able to present even highly hydrophobic regions of proteins to $T_C$.

A number of peptides recognized by $T_C$ specific for other foreign antigens have also been defined (Maryanski et al., 1986; Clayberger et al., 1987; Whitton et al., 1988; Moore et al., 1988; Reddehase et al., 1989). Thus, as with $T_H$, it appears that most, if not all, epitopes recognized by $T_C$ can be substituted with synthetic oligopeptides.

## c. $T_C$ Recognition of Intact versus Fragmented Forms of Viral Antigens on Infected Cells

Following the extensive discussion of $T_C$ recognition of noninfectious viral antigens, it is perhaps important to restate that during host infection, it is likely that $T_C$ recognition is limited to cells expressing endogenously synthesized antigens. Although it is now clear that $T_C$, like

$T_H$, recognize sequential determinants on viral proteins, it is not known whether these epitopes exist on fragmented proteins, denatured intact proteins, or fully native proteins. It is also possible that more than a single form of antigen is recognized and that the relative contribution of different forms varies with the physical properties of the antigen, e.g., integral membrane protein versus soluble cytoplasmic protein.

The possibility that $T_C$ may recognize intact forms of viral antigens has some limited support. First, polyclonal and monoclonal antibodies specific for NP bind to infected cell surfaces (Virelizier et al., 1977; Yewdell et al., 1981) and are able to mediate $T_C$ lysis when chemically coupled to a monoclonal antibody specific for the T-cell antigen receptor that mediates attachment of virus-infected cells to $T_C$ (Staerz et al., 1987). Although the antigenically active moiety present on cell surfaces has not been biochemically characterized, it has been found that many of the monoclonal antibodies that bind to infected cell surfaces fail to react with NP denatured by boiling in the presence of SDS and presumably detect conformational determinants on NP (J. W. Yewdell, unpublished observations).

Second, Wraith and Vessey (1986) found that $T_C$ clones specific for NP residues 365–380 recognized cells treated with purified intact NP. Processing of antigen was not required for recognition, since antigen presentation was not affected by pretreating APCs with either ammonium chloride or paraformaldehyde. Using a $T_C$ clone of similar specificity, Bastin et al. (1987) subsequently found that target cell sensitization was 200-fold more efficient with the 365–379 peptide than with intact purified NP in either native or denatured forms. Thus, it is uncertain whether the sensitization achieved by Wraith and Vessey was due to intact protein or to small amounts of proteolytic fragments present in their NP preparation.

Third, it has been reported that anti-HA monoclonal antibodies can inhibit $T_C$ recognition of virus-infected cells (Effros et al., 1979). It is known that the antibodies used in this study recognize discontinuous antigenic determinants (Caton et al., 1982). It is not certain, however, whether blocking was truly limited to HA-specific $T_C$ in this study, as anti-viral $T_C$ populations were used. Antibody blocking of $T_C$ recognition of virus-infected target cells has also been reported in a number of other virus systems (Zinkernagel and Doherty, 1979; Pan and Knowles, 1983; Finberg et al., 1982). As these experiments can potentially provide the most conclusive evidence for recognition of conformational determinants, further studies should examine blocking of $T_C$ recognition of individual viral proteins using monoclonal antibodies known to recognize such determinants.

By contrast, arguments have been made for $T_C$ recognition of nonintact proteins. It has been found that $T_C$ can recognize transfected cells containing genes encoding NP or the vesicular stomatitis virus nucleocapsid protein, despite the absence of serologically detectable viral antigen on the cell surface (Townsend *et al.*, 1985; Puddington *et al.*, 1986). This provides only weak evidence for $T_C$ recognition of nonintact antigens however, as the minimum amount of antigen required for $T_C$ recognition is not known and could easily be beneath the level required for serological detection. Far more persuasive is the finding that $T_C$ recognize transfected cells expressing fragments of 130- and 170-amino acids from the amino and carboxy termini respectively of the NP gene (Townsend *et al.*, 1985). This unequivocally demonstrates that intact NP is not required for the eventual expression of NP epitopes on the cell surface. The approach of Townsend *et al.* has been used to map epitopes on a number of other viral proteins (Whitton *et al.*, 1988; Braciale *et al.*, 1989). It appears to be generally true then, that truncated proteins retain whatever signals are required for antigen processing and presentation to $T_C$. It remains to be determined, however, whether fragmentation of NP and other internal antigens is required for their transport to the cell surface and eventual recognition by $T_C$.

It is important to recognize that the cellular mechanism that transports cytoplasmic proteins, whether intact or fragmented, to the cellular compartment where association with class I MHC molecules occurs, is entirely undefined. Small amounts of internal proteins may be shunted into the normal exocytic pathway used for transport of integral membrane or secretory proteins:

Endoplasmic reticulum (ER) $\Rightarrow$ Golgi complex $\Rightarrow$ plasma membrane

If so, it will be of great interest to determine the organelle through which entry occurs, and the mechanism by which the protein traverses the organelle membrane. The only organelle with known transport capacity is the ER; perhaps the translocation machinery used for targeting proteins to the exocytic pathway is used for antigen processing as well.

Alternatively, the export of cytoplasmic proteins might reflect the operation of an alternative exocytic pathway. The existence of one such pathway is supported by the results of Sharma *et al.* (1985), who studied the cell-surface expression of SV40 T antigen, which, like NP, is expressed on the cell surface despite the absence of a recognizable leader sequence. The addition of an ER insertion sequence to T antigen actually decreased its cell-surface expression to below detectable levels, despite the fact the protein detectably entered the classic exocytic pathway.

Even for proteins with functional export sequences, it is possible that antigen processing occurs independently of the normal export pathway. Townsend *et al.* (1987) found that $T_C$ recognition of HA is not affected by deletion of the hydrophobic leader sequence responsible for insertion of the HA into the ER. The HA produced under these conditions was not glycosylated and could not be detected on the cell surface by a monoclonal antibody that recognizes a conformational determinant. This finding implies that under normal conditions, HA determinants are processed from a cytoplasmic HA pool which fails to enter the ER. If this were the case, it would mean that presentation of exported proteins depends on a certain inefficiency in recognition of leader sequences by the ER transport machinery. Might the poor $T_C$ recognition of NA or H3 HA result from the high efficiency of their transport into the ER?

### 3. Site of Association of Antigen with Class I Molecules

Like other integral membrane proteins, class I molecules are cotranslationally translocated into the ER and transported to the plasma membrane via the Golgi complex. Using heated virus, Yewdell *et al.* (1988) found that antigen presentation occurs in the continuous presence of protein synthesis inhibitors. This indicates that antigen association does not have to occur co-translationally with class I biosynthesis. Yewdell and Bennink (1989) examined the role of class I transport in antigen presentation using the drug brefeldin A. Brefeldin A has been recently shown to specifically inhibit the exocytosis of integral membrane and secretory proteins (Takatsuki and Tamura, 1985; Misumi *et al.*, 1986; Doms *et al.*, 1989; Lippincott-Schwartz *et al.*, 1989). Unlike other inhibitors of exocytosis, brefeldin A is believed to have minimal effects on other cellular processes.

Brefeldin A completely inhibited the presentation of biosynthesized influenza virus antigens to H-$2^k$- or H-$2^d$-restricted, anti-influenza $T_C$. Presentation of antigens from heated, noninfectious virus to H-$2^k$-restricted $T_C$ was also completely inhibited by the drug. Brefeldin A did not inhibit the presentation of preexisting MHC class I-antigen complexes since (1) the drug had no effect on $T_C$ recognition of a transfected cell line constitutively expressing the HA, and (2) brefeldin A had to be added within the first 1.5 hr following addition of virus to inhibit presentation. The effect of the drug was completely reversible. Removal of the drug for only 15 to 30 min at 37°C was sufficient to completely recover antigen presentation. Reversal of the brefeldin A blockade occurred only at temperatures greater than 20°C. A parallel effect was observed on the exocytosis of a number of integral membrane proteins, including class I MHC molecules, which is consistent with the idea that brefeldin A exerts its affect on antigen presentation by blocking class I exocytosis. Most importantly, brefeldin A failed to inhibit the presentation of NP peptides to H-$2^k$- or H-$2^d$-restricted, NP-specific $T_C$, even when target cells were pretreated for 5 hr with the drug before the addition of peptides.

Brefeldin A is the first compound identified that specifically inhibits the presentation of antigens to $T_C$. Its action is consistent with the idea that protein antigens processed from the cytoplasm associate with class I molecules during their intracellular transport to the trans-Golgi, while exogenous peptides associate with class I molecules present at the cell surface. Confirmation of this hypothesis awaits further experimentation.

## 4. Effect of MHC on Recognition of Viral Proteins

It has been known almost from the initial discovery of anti-influenza $T_C$ that recognition can be influenced by allelic differences in MHC class I molecules. In humans it was found that $T_C$ recognition of influenza-virus-infected cells did not occur in conjunction with certain HLA class I alleles (McMichael, 1978; Shaw and Biddison, 1979; Biddison and Shaw, 1979; Shaw et al., 1980). Similar findings were made in the mouse system, in which anti-influenza CTL induced by secondary in vivo stimulation in a number of mouse strains containing the $K^b$ allele failed to demonstrate $K^b$-restricted lysis of infected target cells (Doherty et al., 1978; Reiss et al., 1982). In both human and mouse systems, the phenomenon appeared to be a multigenic effect, since nonresponsiveness could be influenced by other MHC genes. In addition, even in nonresponder mouse strains, nonresponsiveness was not absolute, since responses could be obtained by varying the conditions used to elicit $T_C$, e.g., intranasal versus intraperitoneal immunization and cell type used to stimulate in vitro responses (Pala and Askonas, 1985).

These original findings indicated that anti-influenza $T_C$ can exhibit marked preference in the class I molecules used as restricting elements, but that nonresponsiveness in conjunction with certain class I alleles is not complete. When anti-influenza $T_C$ responses are examined at the level of individual viral components, however, what appears to be a simpler phenomenon occurs. Using Vac recombinants containing cloned influenza virus genes, $T_C$ responses to NP (Pala and Askonas, 1986), HA, PB2, PB1, and NS1 (Bennink and Yewdell, 1988), and PA (J. R. Bennink and J. W. Yewdell, unpublished observations), were found to occur in conjunction only with a limited number of the available class I MHC gene products expressed by mice of H-2$^k$, H-2$^d$, and H-2$^b$ haplotypes, generally a single locus in each haplotype (Table VI). Unlike $K^b$-restricted

TABLE VI. Mouse MHC Class I Responder Alleles for $T_C$
Responses to Individual Influenza Virus Proteins

| Protein | Haplotype | | |
|---------|-----------|-----------|-----------|
| | H-2$^d$ | H-2$^k$ | H-2$^b$ |
| PA | K | —[a] | ND[b] |
| PB1 | ND | K,D | ?[c] |
| PB2 | D | — | ? |
| H1 HA | K | K | ? |
| H2 HA | K | K | ? |
| NP | K | K | D |
| NS1 | L | K | ? |

[a]—, Mice of this haplotype appear to be nonresponsive to this protein.
[b]ND, Not determined.
[c]?, May not be recognized by H-2$^b$ restricted $T_C$.

antiviral responses, nonresponsiveness could not be attributed to repression mediated by responder MHC genes, since $T_C$ derived from mice containing solely nonresponder MHC alleles failed to lyse target cells infected with the appropriate Vac recombinant. As might be expected, responses to peptide determinants are more severely limited than the response to the intact molecule. For each of the six peptide determinants listed in Table V, only a single responder allele has been identified (McMichael *et al.*, 1986; Bastin *et al.*, 1987; Braciale *et al.*, 1987; Gotch *et al.*, 1987). It cannot be distinguished whether the highly restricted recognition of individual proteins and peptides is due to their intrinsic difficulties in properly associating with class I molecules or to gaps in the $T_C$ repertoire, due perhaps to deletion of self-reactive clones.

It seems likely that the highly restricted recognition of individual viral components is an important factor in the phenomenon of MHC-linked nonresponsiveness at the level of whole antiviral responses and also contributes to the forces that maintain MHC polymorphism. This phenomenon has been reported for other viruses in addition to influenza virus: in some cases, nonresponsiveness appears to be absolute and not influenced by other MHC gene products (Zinkernagel and Doherty, 1979). $K^b$-associated nonresponsiveness may be simplified when examined at the level of individual viral components, as it may reflect variability in responsiveness to one or more individual viral components.

## B. Role of $T_C$ in Antiviral Immunity

It is evident from both human and animal studies that preexisting anti-HA antibodies provide complete protection against influenza virus infections. Antibody is far less effective in eradicating established infections, however, and complete clearance of virus is probably dependent on elements of cellular immunity (Murphy and Webster, 1985). Although this section is devoted largely to $T_C$, it is important to recognize that nonspecific cellular immunity also contributes to recovery (Spencer *et al.*, 1977; Mak *et al.*, 1983). The nonspecific immune cells involved in this process have not been precisely defined, but there is no reason to exclude any of the cells normally present at the site of virus-induced inflammation, i.e., natural killer (NK) cells, activated macrophages, and granulocytes. The mode of action of these cells is similarly undefined and could include lysis of infected cells and release of IFN and possibly other antiviral factors.

The antiviral activity of $T_C$ has been studied in much greater detail and is well established with regard to influenza virus as well as number of other viruses (Bennink *et al.*, 1985; Oldstone *et al.*, 1986; Kast *et al.*, 1986). Early studies demonstrated a correlation between induction of pulmonary $T_C$ activity and reduction in virus titers. Soon after, Yap and

Ada reported that adoptive transfer of T-lymphocyte populations having class I-restricted cytotoxic activity protected naive mice against lethal influenza pneumonia, and reduced reoverable infectious virus by 100- to 10,000-fold (Yap and Ada, 1978a,b; Yap et al., 1978). The cross-reactive protection observed in these studies provided the first solid evidence for the physiological relevance of cross-reactive $T_C$ (Yap and Ada, 1978c). The antiviral activity of $T_C$ was confirmed in a number of additional studies, including those using athymic nude mice (Yap et al., 1979; Wells et al., 1981b, 1983). Definitive evidence for $T_C$-mediated protection was provided by Lin and Askonas (1981), who showed that a cross-reactive cloned $T_C$ cell line reduced viral titers and mortality when injected into naive mice. This finding has been repeated using a number of different strain-specific and cross-reactive cloned $T_C$ lines (Lukacher et al., 1984), including those with known specificity for NP (Taylor and Askonas, 1986).

There are at least three mechanisms by which $T_C$ might mediate their protective effects: (1) lysis of infected cells before progeny virus can be assembled and released; (2) direct inhibition of viral replication by releasing $IFN_\gamma$ and possibly other undefined antiviral factors; and (3) recruitment and enhancement of other immune cells with antiviral activity, e.g., NK cells and macrophages. Several studies have begun to address the relative contributions of these mechanisms to virus clearance. Lukacher et al. (1984) showed that the antiviral activity of $T_C$ is localized, since transfer of a H1 HA-specific $T_C$ clone into mice co-infected with H1N1 and H2N2 viruses resulted in enhanced clearance of only the H1N1 virus. While this result is consistent with direct lysis of infected cells, it is equally consistent with the localized effects of released factors. This latter possibility is supported by two findings. First, there appears to be a correlation between $T_C$-mediated virus clearance and secretion of $IFN_\gamma$ upon antigen recognition (Lin and Askonas, 1981; Taylor and Askonas, 1983). This correlation is based on an extremely small number of independent $T_C$ clones, however; only a single nonsecreting clone has been identified. Second, and more conclusive, Schildknecht and Ada (1985) found that treatment of mice with cyclosporine prevented adoptively transferred $T_C$ from enhancing viral clearance. Cyclosporine treatment had no measurable effect on either $T_C$ migration to the lung or lytic activity, and therefore presumably acted by its well-defined ability to inhibit $T_C$ release of lymphokines.

On the basis of these data, it would appear that release of soluble factors is critical to the effector function of $T_C$ in vivo. The mode of action of these factors is uncertain. They could act directly to inhibit viral replication or to recruit nonspecific immune cells to the site of infection. Alternatively, Schildknecht and Ada suggested that $IFN_\gamma$ is required to increase expression of MHC class I molecules on lung epithelial cells to levels that permit $T_C$ recognition. This remains to be demonstrated, however, and it is pertinent that very few cell types have been found that express low levels of class I molecules.

It is important to emphasize that the present data do not in any way exclude lysis of infected cells from being important, or even essential to virus clearance mediated by $T_C$ or $T_H$ for that matter, which can also lyse cells. Indeed, it might seem unusual, given nature's propensity for efficiency, that such a highly sophisticated mechanism would not to be used. It is possible, however, that the lytic mechanism of $T_C$, while critical in immunity to other antigens (tumor antigens, other pathogens), functions less efficiently than lymphokine release in inhibiting influenza viral replication. It is relevant that the ability of $T_C$ to limit viral replication *in vitro* has not been reported, and it is possible that release of progeny virus occurs too quickly in the infectious cycle for $T_C$-mediated lysis to be of much value.

Finally, it should not be overlooked that the evidence supporting a role for $T_C$ in protection from influenza stems solely from transfer experiments, which represent a highly artificial situation. When Andrew *et al.* (1987) examined the ability of Vac recombinants to protect CBA mice against a subsequent viral infection, they found that prior immunization with HA-Vac, but not NP-Vac, reduced mortality, morbidity, and pulmonary virus titers, despite the fact that NP-Vac primed the same mice for a more vigorous secondary $T_C$ response. Similar findings have been obtained by Bennink and Yewdell (unpublished observations), who measured the same parameters following influenza infection of BALB/c mice immunized with Vac recombinants. Of the 10 recombinants used, only HA-Vac conferred protection. These findings serve to reemphasize the importance of anti-HA antibodies in preventing influenza infection and indicate that much remains to be learned regarding the true role of $T_C$ in physiological situations. Finally, they call into question the effectiveness of vaccines designed to boost $T_C$ immunity in humans (McMichael *et al.*, 1983b).

## V. CONCLUSIONS

### A. $T_H$ versus $T_C$ Recognition; Similar Determinants, Different Processing Pathways

There is increasing evidence that a substantial, if not overwhelming, portion of both $T_H$ and $T_C$ repertoires is directed against sequential determinants. The sequences recognized by $T_H$ and $T_C$ are chemically similar and are not readily distinguished by statistical methods that weigh the chemical nature of individual residues (Berzofsky *et al.*, 1987; Berzofsky, 1988; Rothbard and Taylor, 1988). It is clear, however, that different pathways are used for the presentation of determinants to $T_C$ and $T_H$.

The endosomal–lysosomal pathway, which has been strongly implicated in presentation of antigens to $T_H$, appears to have no role in presentation of antigens to $T_C$. Whatever pathway is involved in the presentation of antigens to $T_C$, it originates in the cytoplasm of APCs and appears not to function in the presentation of antigens to $T_H$.

It seems likely that the difference in $T_H$- and $T_C$-processing pathways evolved in response to fundamental differences in the function of these cells. $T_H$ function primarily to abet the antibody response and, as such, are largely concerned with recognition of foreign proteins present in extracellular body fluids. $T_C$ function primarily to lyse aberrant cells or control intracellular pathogens by lymphokine release and, as such, are concerned with intracellular antigens.

The divergence in $T_H$ and $T_C$ antigen-processing pathways must be intimately related to differences in cellular processing and trafficking of class II, versus class I MHC, molecules, such that association with foreign antigens occurs preferentially in the cellular compartments associated with the relevant processing pathway. For example, association between foreign molecules and class II antigens may occur with highest efficiency in an endosomal/lysosomal compartment, while association with class I molecules might occur during the transit of newly synthesized class I molecules to the cell surface. This idea is supported by the finding that brefeldin A, a specific inhibitor of exocytosis, blocks the presentation of viral proteins to $T_C$ (Yewdell and Bennink, 1989). By contrast, preliminary evidence indicates that brefeldin A does not affect the presentation of exogenous proteins to $T_H$ (C. Hackett, J. Bennink, and J. Yewdell, unpublished findings). The physical state of the antigen at the time of association would be expected to play an important role in determinant selection, i.e., the process by which only specific regions of a foreign protein are selected for immune recognition.

$T_H$ and $T_C$ recognition differs in another aspect. The existence of nonresponder MHC alleles to individual viral proteins is commonplace in $T_C$ responses, while they have yet to be described in $T_H$ responses. This may indicate that fewer individual determinants on a given protein are recognized by $T_C$, due either to a less efficient system of delivering epitopes to MHC molecules in the proper cellular compartment, a less diverse repertoire (perhaps as a result of a wider universe of tolerogenic self-antigens), or to more stringent requirements for the association of determinants with class I molecules. This latter case could reflect differences either in the physical properties of antigen-binding sites in class I and class II molecules or, more likely, in the form of antigen produced by $T_H$- and $T_C$-processing pathways. These possibilities might also underlie the differences in the frequency of nonresponder alleles in $T_C$ responses for various proteins; NP is recognized in association with many human and mouse class I alleles, while recognition of other proteins is far more sporadic or even absent.

## B. T-Lymphocyte Recognition of Internal Viral Antigens Is Still Consistent with Their High Degree of Conservation

Influenza virus has the ability to mutate rapidly in response to host immunity. Despite this, internal viral proteins exhibit a degree of antigenic variation consistent with random genetic drift and/or occasional host-cell selection for functional variants. As $T_H$ and $T_C$ responses are directed against the internal components, this must mean that the T-lymphocyte response exerts, little, if any, evolutionary selection for mutants with antigenically altered internal proteins. Since, there is no apparent difference between T-lymphocyte recognition of internal and external protein, this indicates that antigenic variation in the viral glycoproteins is due solely to antibody mediated selection.

The lack of T-lymphocyte-mediated variant selection is easy to understand. First, the fact that both $T_H$ and $T_C$ responses are directed against a number of viral proteins, and in some cases a number of determinants on a single protein, means that there would be little selective advantage conferred by the failure of a single T-lymphocyte clonotype to recognize cells expressing a protein with an altered determinant. As the frequency of mutant viruses with multiple alterations is exceeding low, approximately $10^{-5n}$, where $n$ is the number of mutations (Yewdell *et al.* 1979), variant selection would have to occur in a host producing a monoclonal, or at most biclonal, T-lymphocyte response. The existence of individual with such restricted responses among populations of humans or experimental animals has yet to be described.

Second, in the case of $T_H$ there are two seemingly insurmountable additional hurdles to variant selection: (1) the vast majority of virus is produced by pulmonary epithelial cells that probably have little, if any, antigen presentation capacity, and (2) selection of antigenic variants is virtually incompatible with the likelihood that $T_H$ recognize processed input virus rather than proteins endogenously synthesized by infected cells.

Finally, even in the case of $T_C$, where selection of antigenic variants is somewhat more plausible, there are additional obstacles that make this unlikely to ever occur under normal circumstance: (1) $T_C$ would have to have the ability to lyse cells before infectious virus is produced (this remains to be established); and (2) the frequency of cells expressing a given protein solely with a variant phenotype would decrease exponentially with the multiplicity of infection (in this case, the frequency of these cells would be approximately $10^{-5n}$, where $n$ is the number of infectious virions per cell). This would mean that only cells infected with two infectious particles, at most, could act as the source of potential variants. Although the multiplicity of infection *in vivo* is unknown, it could well be higher than this.

## C. Implications of T-Lymphocyte Recognition of Influenza Antigens for Vaccine Design

One of the goals of studying T-lymphocyte responses to influenza virus is to design vaccines that would induce cross-reactive immunity to all influenza A viruses. It is now clear that the T-lymphocyte response is indeed highly cross-reactive and that, depending on the type of vaccine used, enhancement of $T_C$ responses and/or anti-HA antibody responses could be obtained. It is also apparent that if such vaccines were to be effective they would act by promoting recovery from infection and not by altering the early course of infection. The potential clinical value of such vaccines would appear to hinge on the longevity of T-lymphocyte memory. If it is true that T-lymphocyte memory lasts only 2–3 years in humans, as the available evidence suggests, then T-lymphocyte vaccines might be able to reduce either morbidity or mortality, or both. If memory is substantially longer lived than this, however, these vaccines would probably be of little use, since the absence of long lasting protection following natural infection is a hallmark of influenza.

Although the therapeutic potential of vaccines designed to elicit cellular immunity remains to be determined for influenza, they could well be useful in combating other infectious agents or in anticancer therapy. On the basis of studies of influenza virus-specific T lymphocytes it is possible to make some general suggestions regarding the composition of this type of vaccine. First, both internal and external proteins should be viewed as potential targets of T lymphocytes. Second, to avoid MHC linked nonresponsiveness, vaccines should not be based on a single protein or peptide but should include a number of structures recognized by T lymphocytes. Third, to elicit maximal $T_C$ responses, the antigen should be given in a form that ensures its expression in the cytoplasm of APCs.

ACKNOWLEDGMENTS. This work was supported by grants AI 22114 and AI 22961 from the National Institutes of Health, Bethesda. We are grateful to Lorena Brown and Jack Bennink for providing unpublished results, and to Andrew Caton, Jack Bennink, Laurence Eisenlohr, and Peggy Scherle for valuable discussions and critical reading of the manuscript.

## REFERENCES

Ada, G. L., and Yap, K. L., 1977, Matrix protein expressed at the surface of cells infected with influenza viruses, *Immunochemistry* **14**:643–651.

Allen, P. M., Strydom, D. J., and Unanue, E. R., 1984, Processing of lysozyme by macrophages: Identification of the determinant recognized by two T-cell hybridomas, *Proc. Natl. Acad. Sci. USA* **81**:2489–2493.

Amit, A. G., Mariuzza, R. A., Phillips, S. E. V., and Poljak, R. J., 1986, Three-dimensional structure of an antigen-antibody complex at 2.8 angstroms resolution, *Science* **233:** 747–753.

Anders, E. M., Peppard, P. M., Burns, W. H., and White, D. O., 1979, In vitro antibody response to influenza virus. I. T cell dependence of secondary response to hemagglutinin, *J. Immunol.* **123:**1356–1361.

Anders, E. M., Katz, J. M., Brown, L. E., Jackson, D. C., and White, D. O., 1981a, The specificity of T cells for influenza virus hemagglutinin, in: *Genetic Variation Among Influenza Viruses* (D. P. Nayak, ed.), pp. 547–565, Academic, Orlando, Florida.

Anders, E. M., Katz, J. M., Jackson, D. C., and White, D. O., 1981b, In vitro antibody response to influenza virus. II. Specificity of helper T cells recognizing hemagglutinin, *J. Immunol.* **127:**669–672.

Andrew, M. E., Coupar, B. E. H., Ada, G. L., and Boyle, D. B., 1986, Cell-mediated immune responses to influenza virus antigens expressed by vaccinia virus recombinants, *Microb. Pathogen.* **1:**443–452.

Andrew, M. E., Coupar, B. E. H., Boyle, D. B., and Ada, G. L., 1987, The roles of influenza virus haemagglutinin and nucleoprotein in protection: Analysis using vaccinia virus recombinants, *Scand. J. Immunol.* **25:**21–28.

Armerding, D., Rossiter, H., Ghazzouli, I., and Liehl, E., 1982, Evaluation of live and inactivated influenza A virus vaccines in a mouse model, *J. Infect. Dis.* **145:**320–330.

Arnon, R., 1981, Experimental allergic encephalomyelitis susceptibility and suppression, *Immunol. Rev.* **55:**5–30.

Ashman, R. B., 1982, Persistence of cell-mediated immunity to influenza A virus in mice, *Immunology* **47:**165–168.

Ashman, R. B., and Mullbacher, A., 1979, A T helper cell for anti-viral cytotoxic T cell responses, *J. Exp. Med.* **150:**1277–1282.

Ashwell, J. D., and Schwartz, R. H., 1986, T-cell recognition of antigen and the Ia molecule as a ternary complex, *Nature (Lond.)* **320:**176–179.

Askenase, P. W., Hayden, B. J., and Gershon, R. K., 1975, Augmentation of delayed-type hypersensitivity by doses of cyclophosphamide which do not affect antibody responses, *J. Exp. Med.* **141:**697–702.

Askonas, B. A., Mullbacher, A., and Ashman, R. B., 1982, Cytotoxic T-memory cells in virus infection and the specificity of helper T cells, *Immunology* **45:**79–84.

Atassi, M. Z., and Kurisaki, J-I., 1984, A novel approach for localization of the continuous protein antigenic sites by comprehensive synthetic surface scanning: Antibody and T cell activity to several influenza hemagglutinin synthetic sites, *Immunol. Commun.* **13:**539–551.

Austin, P., Trowsdale, J., Rudd, C., Bodmer, W., Feldmann, M., and Lamb, J., 1985, Functional expression of HLA-DP genes transfected into mouse fibroblasts, *Nature (Lond.)* **313:**61–64.

Babbitt, B. P., Allen, P. M., Matsueda, G., Haber, E., and Unanue, E. R., 1985, Binding of immunogenic peptides to Ia histocompatibility molecules, *Nature (Lond.)* **317:**359–361.

Bangham, C. R. M., Openshaw, P. J. M., Ball, L. A., King, A. M. Q., Wertz, G. W., and Askonas, B. A., 1986, Human and murine cytotoxic T cells specific to respiratory syncytial virus recognize the viral nucleoprotein (N), but not the major glycorptein (G), expressed by vaccinia virus recombinants, *J. Immunol.* **137:**3973–3977.

Bastin, J., Rothbard, J., Davey, J., Jones, I., and Townsend, A., 1987, Use of synthetic peptides of influenza nucleoprotein to define epitopes recognized by class I-restricted cytotoxic T lymphocytes, *J. Exp. Med.* **165:**1508–1523.

Ben-Nun, A., Strauss, W., Leeman, S. A., Cohn, L. E., Murre, C., Duby, A., Seidman, J. G., and Glimcher, L. H., 1985, An Ia-positive mouse T-cell clone is functional in presenting antigen to other T cells, *Immunogenetics* **22:**123–130.

Bennink, J. R., and Yewdell, J. W., 1988, Murine cytotoxic T lymphocyte recognition of individual influenza virus proteins: High frequency of non-responder MHC class I alleles, *J. Exp. Med.* **168:**1935–1939.

Bennink, J., Effros, R. B., and Doherty, P. C., 1978, Influenzal pneumonia: Early appearance of cross-reactive T cells in lungs of mice primed with heterologous type A viruses, *Immunology* **35**:503–509.

Bennink, J. R., Yewdell, J. W., and Gerhard, W., 1982, A viral polymerase involved in recognition of influenza virus-infected cells by a cytotoxic T-cell clone, *Nature (Lond.)* **296**:75–76.

Bennink, J. R., Yewdell, J. S., Smith, G. L., Moller, C., and Moss, B., 1984, Recombinant vaccinia primes and stimulates influenza hemagglutinin-specific cytotoxic T cells, *Nature (Lond.)* **311**:578–579.

Bennink, J. R., Yewdell, J. W., Feldman, A., Gerhard, W., and Doherty, P. C., 1985, The role of virus-specific CTL in vivo, in: *T Cell Clones* (H. Von Boehmer and W. Haas, eds.), pp. 237–242, Elsevier, New York.

Bennink, J., Yewdell, J. W., Smith, G. L., and Moss, B., 1986, Recognition of cloned influenza virus hemagglutinin gene products by cytotoxic T lymphocytes, *J. Virol.* **57**:786–791.

Bennink, J. R., Yewdell, J. W., Smith, G. L., and Moss, B., 1987, Anti-influenza cytotoxic T lymphocytes recognize the three viral polymerases and a nonstructural protein: Responsiveness to individual proteins is MHC controlled, *J. Virol.* **61**:1098–1102.

Berzofsky, J. A., 1988, The structural basis of antigen recognition by T lymphocytes: Implications for vaccines, *J. Clin. Invest.* **82**:1811–1817.

Berzofsky, J. A., Cease, K. B., Cornette, J. L., Spouge, J. L., Margalit, H., Berkower, I. J., Good, M. F., Miller, L. H., and DeLisi, C., 1987, Protein antigenic structures recognized by T cells: Potential applications to vaccine design, *Immunol. Rev.* **98**:9–52.

Biddison, W. E., and Shaw, S., 1979, Differences in HLA antigen recognition by human influenza virus-immune cytotoxic T cells, *J. Immunol.* **122**:1705–1709.

Biddison, W. E., Shaw, S., and Nelson, D. L., 1979, Virus specificity of human influenza virus-immune cytotoxic T cells, *J. Immunol.* **122**:660–664.

Biddison, W. E., Sharrow, S. O., and Shearer, G. M., 1981, T cell subpopulations required for the human cytotoxic T lymphocyte response to influenza virus: Evidence for T cell help, *J. Immunol.* **127**:487–491.

Bjorkman, P. J., Saper, M. A., Samraoui, B., Bennet, W. S., Strominger, J. L., and Wiley, D. C., 1987a, Structure of the human class I histocompatibility antigen, HLA-A2, *Nature (Lond.)* **329**:506–512.

Bjorkman, P. J., Saper, M. A., Samraoui, B., Bennet, W. S., Strominger, J. L., and Wiley, D. C., 1987b, The foreign antigen binding site and T cell recognition regions of class I histocompatibility antigen, *Nature (Lond.)* **329**:512–518.

Braciale, T. J., 1977, Immunologic recognition of influenza virus-infected cells. I. Generation of a virus-strain specific and a cross-reactive subpopulation of cytotoxic T cells in the response to type A influenza viruses of different subtypes, *Cell. Immunol.* **33**:423–436.

Braciale, T. J., 1979, Specificity of cytotoxic T cells directed to influenza virus hemagglutinin, *J. Exp. Med.* **149**:856–869.

Braciale, T. J., and Yap, K. L., 1978, Role of viral infectivity in the induction of influenza virus-specific cytotoxic T cells, *J. Exp. Med.* **147**:1236–1252.

Braciale, T. J., Andrew, M. E., and Braciale, V. L., 1981, Heterogeneity and specificity of cloned lines of influenza-virus-specific cytotoxic T lymphocytes, *J. Exp. Med.* **153**:910–919.

Braciale, T. J., Braciale, V. L., Henkel, T. J., Sambrook, J., and Gething, M.-J., 1984, Cytotoxic T lymphocyte recognition of the influenza hemagglutinin gene product expressed by DNA-mediated gene transfer, *J. Exp. Med.* **159**:341–354.

Braciale, T. J., Henkel, T. J., Lukacher, A., and Braciale, V. L., 1986, Fine specificity and antigen receptor expression among virus-specific cytolytic T lymphocyte clones, *J. Immunol.* **137**:995–1002.

Braciale, T. J., Braciale, V. L., Winkler, M., Stroynowski, I., Hood, L., Sambrook, J., and Gething, M. J., 1987, On the role of the transmembrane anchor sequence in target cell recognition by class I MHC-restricted, hemagglutinin-specific cytolytic T lymphocytes, *J. Exp. Med.* **166:**678–692.

Braciale, T. J., Sweetser, M. T., Morrison, L. A., Kittlesen, D. J., and Braciale, V. L., 1989, Class I major histocompatibility complex-restricted cytolytic T lymphocytes recognize a limited number of sites on the influenza hemagglutinin. *Proc. Natl. Acad. Sci. (USA)* **86:**277–281.

Brown, L. E., Katz, J. M., Ffrench, R. A., Anders, E. M., and White, D. O., 1987, Characterization of subtype-specific and cross-reactive helper-T-cell clones recognizing influenza virus hemagglutinin, *Cell. Immunol.* **109:**12–24.

Burlington, D. B., Djeu, J. Y., Wells, M. A., Kiley, S. C., and Quinnan, G. V., 1984, Large granular lymphocytes provide an accessory function in the *in vitro* development of influenza A virus-specific cytotoxic T cells, *J. Immunol.* **132:**3154–3158.

Burns, W. H., Billups, L. C., and Notkins, A. L., 1975, Thymus dependence of viral antigens, *Nature (Lond.)* **256:**654–655.

Butchko, G. M., Armstrong, R. B., and Ennis, F. A., 1978, Specificity studies on the proliferative response of thymus-derived lymphocytes to influenza viruses, *J. Immunol.* **121:**2381–2385.

Buus, S., Colon, S., Smith, C., Freed, J. H., Miles, C., and Grey, H. M., 1986, Interaction between a "processed" ovalbumin peptide and Ia molecules, *Proc. Natl. Acad. Sci. USA* **83:**3968–3971.

Callard, R. E., 1979, Specific *in vitro* antibody response to influenza virus by human blood lymphocytes, *Nature (Lond.)* **282:**734–736.

Callard, R. E., and Smith, C. M., 1981, Histocompatibility requirements for T cell help in specific in vitro antibody responses to influenza virus by human blood lymphocytes, *Eur. J. Immunol.* **11:**206–212.

Callard, R. E., Booth, R. J., Brown, M. H., and McCaughan, G. W., 1985, T cell-replacing factor in specific antibody responses to influenza by human blood B cells, *Eur. J. Immunol.* **15:**52–59.

Cambridge, G., Mackenzie, J. S., and Keast, D., 1976, Cell-mediated immune response to influenza virus infections in mice, *Infect. Immun.* **13:**36–43.

Cantrell, D. A., and Smith, K. A., 1983, Transient expression of interleukin 2 receptors. Consequences for T cell growth, *J. Exp. Med.* **158:**1895–1911.

Caton, A. J., Brownlee, G. G., Yewdell, J. W., and Gerhard, W., 1982, The antigenic structure of the influenza virus A/PR/8/34 hemagglutinin (H1 subtype), *Cell* **31:**417–427.

Chesnut, R. W., and Grey, H. M., 1985, Antigen presenting cells and mechanisms of antigen presentation, *CRC Crit. Rev. Immunol.* **5:**263–316.

Chesnut, R. W., Colon, S. M., and Grey, H. M., 1982, Requirements for the processing of antigens by antigen-presenting B cells. I. Functional comparison of B cell tumors and macrophages, *J. Immunol.* **129:**2382–2388.

Clayberger, C., Parham, P., Rothbard, J., Judwig, D. S., Schoolnik, G. K., and Krensky, A. M., 1987, HLA-A2 peptides can regulate cytolysis by human allogeneic T lymphocytes, *Nature (Lond.)* **330:**763–765.

Coupar, B. E. H., Andrew, M. E., Both, G. W., and Boyle, D. B., 1986, Temporal regulation of influenza hemagglutinin expression in vaccinia virus recombinants and effects on the immune response, *Eur. J. Immunol.* **16:**1479–1487.

Daisy, J. A., Tolpin, M. D., Quinnan, G. V., Rook, A. H., Murphy, B. R., Mittal, K., Clements, M. L., Mullinix, M. G., Kiley, S. C., and Ennis, F. A., 1981, Cytotoxic cellular immune responses during influenza A infection in human volunteers, in: *The Replication of Negative Strand Viruses* (D. H. L. Bishop and R. W. Compans, eds.), pp. 443–448, Elsevier, New York.

Davis, M. M., 1985, Molecular genetics of the T cell-receptor beta chain, *Annu. Rev. Immunol.* **3:**537–560.

DeFreitas, E. C., Chesnut, R. W., Grey, H. M., and Chiller, J. M., 1983, Macrophage-dependent activation of antigen-specific T cells requires antigen and a soluble monokine, *J. Immunol.* **131:**23–29.

Doherty, P. C., Effros, R. B., and Bennink, J. B., 1977, Heterogeneity of the cytotoxic response of thymus-derived lymphocytes after immunization with influenza viruses, *Proc. Natl. Acad. Sci. USA* **74:**1209–1213.

Doherty, P. C., Biddison, W. E., Bennink, J. R., and Knowles, B. B., 1978, Cytotoxic T cell responses in mice infected with influenza and vaccinia viruses vary in magnitude with H-2 genotype, *J. Exp. Med.* **148:**534–543.

Doms, R. W., Russ, G., and Yewdell, J. W., 1989, Brefeldin A redistributes resident and itinerant Golgi proteins to the endoplasmic reticulum. *J. Cell Biology* (in press).

Eager, K., Bennink, J. R., Eisenlohr, L., Ricciardi, R., and Yewdell, J., 1989, Retrovirus mediated gene transfer of the influenza virus hemagglutinin into a wide variety of murine cell lines leads to their recognition by major histocompatibility complex class I and class II restricted T lymphocytes (submitted for publication).

Eckels, D. D., Lamb, J. R., Lake, P., Woody, J. N., Johnson, A. H., and Hartzman, R. J., 1982, Antigen-specific human T-lymphocyte clones. Genetic restriction of influenza virus-specific responses to HLA-D region genes, *Hum. Immunol.* **4:**313–324.

Effros, R. B., Doherty, P. C., Gerhard, W., and Bennink, J., 1977, Generation of both cross-reactive and virus-specific T-cell populations after immunization with serologically distinct influenza A viruses, *J. Exp. Med.* **145:**557–566.

Effros, R. B., Bennink, J., and Doherty, P. C., 1978, Characteristics of secondary cytotoxic T cell responses in mice infected with influenza A viruses, *Cell. Immunol.* **36:**345–353.

Effros, R. B., Frankel, M. E., Gerhard, W., and Doherty, P. C., 1979, Inhibition of influenza-immune T cell effector function by virus-specific hybridoma antibody, *J. Immunol.* **123:**1343–1346.

Eisenlohr, L. C., and Hackett, C. J., 1989, Class II major histocompatibility complex-restricted T cells specific for a virion structural protein that do not recognize exogenous influenza virus. Evidence that presentation of labile T cell determinants is favored by endogenous antigen synthesis. *J. Exp. Med.* **169:**921–932.

Eisenlohr, L. C., Gerhard, W., and Hackett, C. J., 1987, Role of the receptor-binding activity of the viral hemagglutinin molecule in presentation of influenza virus antigens to helper T cells, *J. Virol.* **61:**1375–1383.

Eisenlohr, L. C., Gerhard, W., and Hackett, C. J., 1988a, Acid-induced conformational modification of the hemagglutinin molecule alters interaction of influenza virus with antigen-presenting cells, *J. Immunol.* **141:**1870–1876.

Eisenlohr, L. C., Gerhard, W., and Hackett, C. J., 1988b, Individual class II-restricted antigenic determinants of the same protein exhibit distinct kinetics of appearance and persistence on antigen-presenting cells, *J. Immunol.* **141:**2581–2584.

Ennis, F. A., 1982, Some newly recognized aspects of resistance against and recovery from influenza, *Arch. Virol.* **73:**207–217.

Ennis, F. A., Martin, W. J., and Verbonitz, M. W., 1977a, Cytotoxic T lymphocytes induced in mice by inactivated influenza virus vaccine, *Nature (Lond.)* **269:**418–419.

Ennis, F. A., Martin, W. J., Verbonitz, M. W., and Butchko, G. M., 1977b, Specificity studies on cytotoxic thymus derived lymphocytes reactive with influenza virus-infected cells: Evidence for dual recognition of H-2 and viral hemagglutinin antigens, *Proc. Natl. Acad. Sci. USA* **74:**3006–3010.

Ennis, F. A., Wells, M. A., Butchko, G. M., and Albrecht, P., 1978, Evidence that cytotoxic T cells are part of the host's response to influenza pneumonia, *J. Exp. Med.* **148:**1241–1250.

Ennis, F. A., Qi, Y.-H., Riley, D., Rook, A. H., Schild, G. C., Pratt, R., and Potter, C. W., 1981, HLA-restricted virus-specific cytotoxic T-lymphocyte responses to live and inactivated influenza vaccines, *Lancet* **2:**887–891.

Ertl, H., and Ada, G. L., 1981, Roles of influenza virus infectivity and glycosylation of viral antigen for recognition of target cells by cytolytic T lymphocytes, *Immunobiology* **158:**239–253.

Farrar, J. J., Benjamin, W. R., Hilfiker, M. L., Howard, M., Farrar, W. L., and Fuller-Farrar, J.,

1982, The biochemistry, biology, and role of interleukin 2 in the induction of cytotoxic T cell and antibody-forming responses, *Immunol. Rev.* **63**:129–166.

Fathman, C. G., and Frelinger, J. G., 1983, T-lymphocyte clones, *Annu. Rev. Immunol.* **1**: 633–655.

Festenstein, H., and Demant, P., 1978, HLA and H-2. Basic immunogenetics, biology, and clinical relevance, in: *Current Topics in Immunology* (J. Turk, ed.), pp. 1–212, Edward Arnold, London.

Finberg, R., Spriggs, D. R., and Fields, B. N., 1982, Host immune response to reovirus: CTL recognize the major neutralization domain of the viral hemagglutinin, *J. Immunol.* **129**: 2235–2238.

Fischer, A., Beverley, P. C. L., and Feldmann, M., 1981, Long-term human T-helper lines producing specific helper factor reactive to influenza virus, *Nature (Lond.)* **294**:166–170.

Fischer, A., Nash, S., Beverley, P. C. L., and Feldmann, M., 1982, An influenza virus matrix protein-specific human T cell line with helper activity for in vitro anti-hemagglutinin antibody production, *Eur. J. Immunol.* **12**:844–849.

Fischer, A., Sterkers, G., Charron, D., and Durandy, A., 1985, HLA class II restriction governing cell cooperation between antigen-specific helper T lymphocytes, B lymphocytes and monocytes for in vitro antibody production to influenza virus, *Eur. J. Immunol.* **15**:620–626.

Fischer Lindahl, K., and Hausmann, B., 1980, Expression of the I-E target antigen for T-cell killing requires two genes, *Immunogenetics* **11**:571–583.

Fleischer, B., Becht, H., and Rott, R., 1985, Recognition of viral antigens by human influenza A virus-specific T lymphocyte clones, *J. Immunol.* **135**:2800–2804.

Fontana, A., Fierz, W., and Wekerle, H., 1984, Astrocytes present myelin basic protein to encephalitogenic T-cell lines, *Nature (Lond.)* **307**:273–276.

Gerhard, W., and Webster, R. G., 1978, Antigenic drift in influenza A viruses. I. Selection and characterization of antigenic variants of A/PR/8/34 (H0N1) influenza virus with monoclonal antibodies, *J. Exp. Med.* **148**:383–392.

Gerhard, W., Hackett, C. J., and Melchers, F., 1983, The recognition specificity of a murine helper T cell for hemagglutinin of influenza virus A/PR/8/34, *J. Immunol.* **130**:2379–2385.

Gething, M.-J., and Sambrook, J., 1981, Cell surface expression of influenza hemagglutinin from a cloned DNA copy of the RNA gene, *Nature (Lond.)* **293**:620–625.

Gooding, L. R., 1977, Specificities of killing by cytotoxic lymphocytes generated in vivo and in vitro to syngeneic SV40 transformed cells, *J. Immunol.* **118**:920–927.

Gotch, F., McMichael, A., Smith, G., and Moss, B., 1987, Identification of the viral molecules recognized by influenza specific human cytotoxic T lymphocytes, *J. Exp. Med.* **165**:408–416.

Gotch, F., Rothbard, J., Howland, K., Townsend, A., and McMichael, A., 1987, Cytotoxic T lymphocytes recognize a fragment of influenza virus matrix protein in association with HLA-A2, *Nature (Lond.)* **326**:881–882.

Greenspan, N., and Doherty, P. C., 1982, Modification of cytotoxic T-cell response patterns by administration of hemagglutinin-specific monoclonal antibodies to mice infected with influenza A viruses, *Hybridoma* **1**:149–159.

Guertin, D. P., and Fan, D. P., 1980, Stimulation of cytolytic T cells by isolated viral peptides and HN protein coupled to agarose beads, *Nature (Lond.)* **283**:308–311.

Hackett, C. J., Dietzschold, B., Gerhard, W., Ghrist, B., Knorr, R., Gillessen, D., and Melchers, F., 1983a, Influenza virus site recognized by a murine helper T cell specific for H1 strains, *J. Exp. Med.* **158**:294–302.

Hackett, C. J., Taylor, P. M., and Askonas, B. A., 1983b, Stimulation of cytotoxic T cells by liposomes containing influenza virus or its components, *Immunology* **49**:255–263.

Hackett, C. J., Hurwitz, J. L., Dietzschold, B., and Gerhard, W., 1985a, A synthetic decapeptide of influenza virus hemagglutinin elicits helper T cells with the same fine recognition specificities as occur in response to whole virus, *J. Immunol.* **135**:1391–1394.

Hackett, C. J., Hurwitz, J. L., Moller, C., and Gerhard, W., 1985b, Fine specificity of antigen recognition by influenza hemagglutinin-specific helper T cells: Heterogeneity of clones and the role of a single amino acid position in cross-reactive responses, in *Immune Recognition of Protein Antigens* (W. G. Laver and G. M. Air, eds.), *Current Communications in Molecular Biology*, pp. 48–55, Cold Spring Harbor Laboratory Press, Cold Spring Harbor, New York.

Heber-Katz, E., Schwartz, R. H., Matis, L. A., Hannum, C. Fairwell, T., Appella, E., and Hansburg, D., 1982, Contribution of antigen-presenting cell major histocompatibility complex gene products to the specificity of antigen-induced T cell activation, *J. Exp. Med.* **155**:1086–1099.

Hedrick, S., Germain, R. N., Bevan, M. J., Dorf, M., Engel, I., Fink, P., Gascoigne, N., Heber-Katz, E., Kapp, J., Kauffmann, Y., Kaye, J., Melchers, F., Pierce, C., Schwartz, R. H., Sorensen, C., Taniguchi, M., and Davis, M. M., 1985, Rearrangement and transcription of a T-cell receptor beta-chain gene in different T-cell subsets, *Proc. Natl. Acad. Sci. USA* **82**:531–535.

Hosaka, Y., Sasao, F., and Ohara, R., 1985, Cell-mediated lysis of heat-inactivated influenza virus-coated murine targets, *Vaccine* **3**:245–252.

Hosaka, Y., Sasao, F., Yamanaka, K., Bennink, J. R., and Yewdell, J. W., 1988, Recognition of noninfectious influenza virus by class I-restricted murine cytotoxic T lymphocytes, *J. Immunol.* **140**:606–610.

Hurwitz, J. L., and Hackett, C. J., 1985, Influenza-specific suppression: Contribution of major viral proteins to the generation and function of T suppressor cells, *J. Immunol.* **135**:2134–2139.

Hurwitz, J. L., Heber-Katz, E., Hackett, C. J., and Gerhard, W., 1984, Characterization of the murine T$_H$ response to influenza virus hemagglutinin: Evidence for three major specificities, *J. Immunol.* **133**:3371–3377.

Hurwitz, J. L., Hackett, C. J., McAndrew, E. C., and Gerhard, W., 1985, Murine T$_H$ response to influenza virus: Recognition of hemagglutinin, neuraminidase, matrix, and nucleoproteins, *J. Immunol.* **134**:1994–1998.

Imboden, J. B., and Stobo, J. D., 1985, Transmembrane signalling by the T cell antigen receptor. Perturbation of the T3-antigen receptor complex generates inositol phosphates and releases calcium ions from intracellular stores, *J. Exp. Med.* **161**:446–456.

Jackson, D. C., Tang, X. I., Brown, L. E., Murray, J. M., White, D. O., and Tregear, G. W., 1986, Antigenic determinants of influenza virus hemagglutinin. XII. The epitopes of a synthetic peptide representing the C-terminus of HA1, *Virology* **155**:625–632.

Johansson, B. E., Moran, T. M., and Kilbourne, E. D., 1987a, Antigen-presenting B cells and helper T cells cooperatively mediate intravironic antigenic competition between influenza A virus surface glycoproteins, *Proc. Natl. Acad. Sci. USA* **84**:6869–6873.

Johansson, B. E., Moran, T. M., Bona, C. A., Popple, S. W., and Kilbourne, E. D., 1987b, Immunologic response to influenza virus neuraminidase is influenced by prior experience with the associated viral hemagglutinin. II. Sequential infection of mice simulates the human experience, *J. Immunol.* **139**:2010–2014.

Johansson, B. E., Moran, T. M., Bona, C. A., and Kilbourne, E. D., 1987c, Immunologic response to influenza virus neuraminidase is influenced by prior experience with the associated viral hemagglutinin. III. Reduced generation of neuraminidase-specific helper T cells in hemagglutinin-primed mice, *J. Immunol.* **139**:2015–2019.

Jones, P. D., Tha Hla, R., Morein, B., Lovgren, K., and Ada, G. L., 1988, Cellular immune responses in the murine lung to local immunization with influenza A virus glycoproteins in micelles and immunostimulatory complexes (ISCOMS), *Scand. J. Immunol.* **27**:645–652.

Kaplan, D. R., Griffith, R., Braciale, V. L., and Braciale, T. J., 1984, Influenza virus-specific human cytotoxic T cell clones: Heterogeneity in antigenic specificity and restriction by class II products, *Cell Immunol.* **88**:193–206.

Kappler, J. W., Skidmore, B., White, J., and Marrack, P., 1981, Antigen-inducible, H-2-

restricted, interleukin-2 producing T cell hybridomas. Lack of independent antigen and H-2 recognition, *J. Exp. Med.* **153:**1198–1214.

Kast, W. M., Bronkhorst, A. M., DeWaal, L. P., and Meleif, C. J. M., 1986, Cooperation between cytotoxic and helper T lymphocytes in protection against lethal Sendai virus infection. Protection by T cells is MHC-restricted and MHC-regulated; a model for MHC-disease associations, *J. Exp. Med.* **164:**723–738.

Katz, J. M., Brown, L. E., Ffrench, R. A., and White, D. O., 1985a, Murine helper T lymphocyte response to influenza virus: Recognition of haemagglutinin by subtype-specific and cross-reactive T cell clones, *Vaccine* **3:**257–262.

Katz, J. M., Laver, W. G., White, D. O., and Anders, E. M., 1985b, Recognition of influenza virus hemagglutinin by subtype-specific and cross-reactive proliferative T cells: Contribution of HA1 and HA2 polypeptide chains, *J. Immunol.* **134:**616–622.

Kees, U., and Krammer, P. H., 1984, Most influenza A virus-specific memory cytotoxic T lymphocytes react with antigenic epitopes associated with internal virus determinants, *J. Exp. Med.* **159:**365–377.

Kimoto, M., and Fathman, C. G., 1980, Antigen-reactive T cell clones. I. Transcomplementing hybrid I-A-region gene products function effectively in antigen presentation, *J. Exp. Med.* **152:**759–770.

Kindred, B., and Shreffler, D. C., 1972, H-2 dependence of co-operation between T and B cells *in vivo*, *J. Immunol.* **109:**940–943.

Klein, J., 1982, *Natural History of the Major Histocompatibility Complex*, Wiley, New York.

Klein, J. R., Raulet, D. H., Pasternack, M. S., and Bevan, M. J., 1982, Cytotoxic T lymphocytes produce immune interferon in response to antigen or mitogen, *J. Exp. Med.* **155:**1198–1203.

Klenk, H. D., Rott, R., Orlich, M., and Blodorn, J., 1975, Activation of influenza A viruses by trypsin treatment, *Virology* **68:**426–439.

Koszinowski, U. H., Allen, H., Gething, M.-J., Waterfield, M. D., and Klenk, H.-D., 1980, Recognition of viral glycoproteins by influenza A-specific cross-reactive cytolytic T lymphocytes, *J. Exp. Med.* **151:**945–958.

Kurrle, R., Wagner, H., Rollinghoff, M., and Rott, R., 1979, Influenza virus-specific T cell-mediated cytotoxicity: Integration of the virus antigen into the target cell membrane is essential for target cell formation, *Eur. J. Immunol.* **9:**107–111.

Lamb, J. R., and Feldmann, M., 1982, A human suppressor T cell clone which recognizes an autologous helper T cell clone, *Nature (Lond.)* **300:**456–458.

Lamb, J. R., and Green, N., 1983, Analysis of the antigen specificity of influenza hemagglutinin-immune human T lymphocyte clones: Identification of an immunodominant region for T cells, *Immunology* **50:**659–666.

Lamb, J. R., Eckels, D. D., Lake, P., Johnson, A. R., Hartzman, R. J., and Woody, J. N., 1982a, Antigen-specific human T lymphocyte clones: Induction, antigen specificity, and MHC restriction of influenza virus-immune clones, *J. Immunol.* **128:**233–238.

Lamb, J. R., Eckels, D. D., Lake, P., Woody, J. N., and Green, N., 1982b, Human T cell clones recognize chemically synthesized peptides of influenza hemagglutinin, *Nature (Lond.)* **300:**66–69.

Lamb, J. R., Eckels, D. D., Phelan, M., Lake, P., and Woody, J. N., 1982c, Antigen-specific human T lymphocyte clones: Viral antigen specificity of influenza virus-immune clones, *J. Immunol.* **128:**1428–1432.

Lamb, J. R., Woody, J. N., Hartzman, R. J., Eckels, D. D., 1982d, *In vitro* influenza virus-specific antibody production in man: Antigen-specific and HLA-restricted induction of helper activity mediated by cloned human T lymphocytes, *J. Immunol.* **129:**1465–1470.

Lamb, J. R., Zanders, E. D., Feldmann, M., Lake, P., Eckels, D. D., Woody, J. N., and Beverley, P. C. L., 1983, The dissociation of interleukin-2 production and antigen-specific helper activity by clonal analysis, *Immunology* **50:**397–405.

Lamb, R. A., 1983, The influenza virus RNA segments and their encoded proteins, in: *Genetics of Influenza Viruses* (P. Palese and D. W. Kingsbury, eds.), pp. 21–69, Springer-Verlag, Vienna.

Lamb, R. A., Zebedee, S. L., and Richardson, C. D., 1985, Influenza M2 protein is an integral membrane protein expressed on the infected-cell surface, *Cell* **40:**627–633.

Lazarowitz, S. G., and Choppin, P. W., 1975, Enhancement of infectivity of influenza A and B viruses by proteolytic cleavage of hemagglutinin polypeptide, *Virology* **68:**440–454.

Lechler, R. I., Ronchese, F., Braunstein, N. S., and Germain, R. N., 1986, I-A-restricted T cell antigen recognition. Analysis of the roles for A-alpha and A-beta using DNA-mediated gene transfer, *J. Exp. Med.* **163:**678–696.

Leung, K. N., and Ada, G. L., 1980, Production of DTH in the mouse to influenza virus: Comparison with conditions for stimulation of cytotoxic T cells, *Scand. J. Immunol.* **12:**129–139.

Leung, K. N., and Ada, G. L., 1982, Different functions of subsets of effector T cells in murine influenza virus infection, *Cell. Immunol.* **67:**312–324.

Leung, K.-L., Ashman, R. B., Ertl, H. C. J., and Ada, G. L., 1980a, Selective suppression of the cytotoxic T cell response to influenza virus in mice, *Eur. J. Immunol.* **10:**803–810.

Leung, K.-L., Ada, G. L., and McKenzie, I. F. C., 1980b, Specificity, Ly phenotype, and H-2 compatibility requirements of effector cells in delayed-type hypersensitivity responses to murine influenza virus infection, *J. Exp. Med.* **151:**815–826.

Liew, F. Y., and Russell, S. M., 1980, Delayed type hypersensitivity to influenza virus. Induction of antigen-specific suppressor T cells for delayed-type hypersensitivity to hemagglutinin during influenza virus infection in mice, *J. Exp. Med.* **151:**799–814.

Liew, F. Y., and Russell, S. M., 1983, Inhibition of pathogenic effect of effector T cells by specific suppressor T cells during influenza virus infection in mice, *Nature (Lond.)* **304:**541–543.

Lin, Y. L., and Askonas, B. A., 1981, Biological properties of an influenza A virus-specific killer T cell clone. Inhibition of virus replication *in vivo* and induction of delayed-type hypersensitivity reactions, *J. Exp. Med.* **154:**225–234.

Lippincott-Schwartz, J., Yuan, L. C., Bonifacino, J. S., and Klausner, R. D., 1989, Rapid redistribution of Golgi proteins into the endoplasmic reticulum in cells treated with brefeldin A: Evidence for membrane cycling from Golgi to ER, *Cell* **56:**801–813.

Lipscomb, M. F., Lyons, C. R., O'Hara, R. M., and Stein-Streilein, J., 1982, The antigen-induced selective recruitment of specific T lymphocytes to the lung, *J. Immunol.* **128:**111–115.

Lipscomb, M. F., Yeakel-Houlihan, D., Lyons, C. R., Gleason, R. R., and Stein-Streilein, J., 1983, Persistence of influenza as an immunogen in pulmonary antigen-presenting cells, *Infect. Immun.* **42:**965–972.

Londei, M., Lamb, J. R., Bottazzo, G. F., and Feldmann, M., 1984, Epithelial cells expressing aberrant MHC class II determinants can present antigen to cloned human T cells, *Nature (Lond.)* **312:**639–641.

Lucas, S. J., Barry, D. W., and Kind, P., 1978, Antibody production and protection against influenza virus in immunodeficient mice, *Infect. Immunol.* **20:**115–119.

Lukacher, A. E., Braciale, V. L., and Braciale, T. J., 1984, *In vivo* effector function of influenza virus-specific cytotoxic T lymphocyte clones is highly specific, *J. Exp. Med.* **160:**814–826.

Lukacher, A. E., Morrison, L. A., Braciale, V. L., Malissen, B., and Braciale, T. J., 1985, Expression of specific cytolytic activity by H-2I region-restricted influenza virus-specific T lymphocyte clones, *J. Exp. Med.* **162:**171–187.

Maeda, T., and Ohnishi, S., 1980, Activation of influenza virus by acidic media causes hemolysis and fusion of erythrocytes, *FEBS Lett.* **122:**283–287.

Mak, N. K., Schiltknecht, E., and Ada, G. L., 1983, Protection of mice against influenza virus infection: Enhancement of nonspecific cellular responses by *Corynebacterium parvum*, *Cell. Immunol.* **78:**314–325.

Malissen, P., Price, M. P., Goverman, J. M., McMillan, M., White, J., Kappler, J., Marrack, P., Pierres, A., Pierres, M., and Hood, L., 1984, Gene transfer of H-2 class II genes: Antigen presentation by mouse fibroblast and hamster B-cell lines, *Cell* **36:**319–327.

Malissen, M., Minard, K., Mjolsness, S., Kronenberg, M., Goverman, J., Hunkapiller, T., Prystowsky, M. B., Yoshikai, Y., Fitch, F., Mak, T. W., and Hood, L., 1984, Mouse T cell antigen receptor: Structure and organization of constant and joining gene segments encoding the beta polypeptide, *Cell* **37:**1101–1110.

Maryanski, J. L., Pala, P., Corradin, J., Jordan, B. B., and Cerrotini, J. C., 1986, H-2 restricted cytotoxic T cells specific for HLA can recognize a synthetic HLA peptide, *Nature (Lond.)* **324:**578–579.

Matlin, K. S., Reggio, H., Helenius, A., and Simons, K., 1981, Infectious entry pathway of influenza virus in a canine kidney cell line, *J. Cell Biol.* **91:**601–613.

McMichael, A., 1978, HLA restriction of human cytotoxic T lymphocytes specific for influenza virus. Poor recognition of virus associated with HLA A2, *J. Exp. Med.* **148:**1458–1467.

McMichael, A. J., and Askonas, B. A., 1978, Influenza virus-specific cytotoxic T cells in man: Induction and properties of the cytotoxic cell, *Eur. J. Immunol.* **8:**705–711.

McMichael, A. J., Dongworth, D. W., Gotch, F. M., Clark, A., and Potter, C. W., 1983a, Declining T-cell immunity to influenza, 1977–1982, *Lancet* **2:**762–765.

McMichael, A. J., Gotch, F. M., Noble, G. R., and Beare, P. A. S., 1983b, Cytotoxic T-cell immunity to influenza, *N. Engl. J. Med.* **309:**13–17.

McMichael, A. J., Gotch, F. M., and Rothbard, J., 1986a, HLA B37 determines an influenza A virus nucleoprotein epitope recognized by cytotoxic T lymphocytes, *J. Exp. Med.* **164:**1397–1406.

McMichael, A. J., Michie, C. A., Gotch, F. M., Smith, G. L., and Moss, B., 1986b, Recognition of influenza A virus nucleoprotein by human cytotoxic T lymphocytes, *J. Gen. Virol.* **67:**719–726.

Melchers, F., and Andersson, J., 1984, B cell activation: Three steps and their variations, *Cell* **37:**715–720.

Melchers, F., Zeuthen, J., and Gerhard, W., 1982, Influenza virus-specific murine T cell hybridomas which recognize virus hemagglutinin in conjunction with H-2d and display helper functions for B cells, *Curr. Top. Microbiol. Immunol.* **100:**153–163.

Melchers, F., Erdei, A., Schultz, T., and Dierich, M. P., 1985, Growth control of activated, synchronized murine B cells by the C3d fragment of human complement, *Nature (Lond.)* **317;**264–269.

Mengle-Gaw, L., and McDevitt, H. O., 1985, Genetics and expression of mouse Ia antigens, *Annu. Rev. Immunol.* **3:**367–396.

Miller, J. F. A. P., Vadas, M. A., Whitelaw, A., and Gamble, J., 1975, H-2 gene complex restricts transfer of delayed-type hypersensitivity in mice, *Proc. Natl. Acad. Sci. USA* **72:**5095–5099.

Miller, R. A., and Reiss, C. S., 1984, Limiting dilution cultures reveal latent influenza virus-specific helper T cells in virus-primed mice, *J. Mol. Cell. Immunol.* **1:**357–368.

Mills, K. H. G., 1986, Processing of viral antigens and presentation to class II-restricted T cells, *Immunol. Today* **9:**260–263.

Mills, K. H. G., Skehel, J. J., and Thomas, D. B., 1986a, Conformational dependent recognition of influenza virus hemagglutinin by murine T-helper clones, *Eur. J. Immunol.* **16:**276–280.

Mills, K. H. G., Skehel, J. J., and Thomas, D. B., 1986b, Extensive diversity in the recognition of influenza virus hemagglutinin by murine T-helper clones, *J. Exp. Med.* **163:**1477–1490.

Mills, K. H. G., Burt, D. S., Skehel, J. J., and Thomas, D. B., 1988, Fine specificity of mouse class II-restricted T cell clones for synthetic peptides of influenza virus hemagglutinin. Heterogeneity of antigen interaction with the T cell and the Ia molecule, *J. Immunol.* **140:**4083–4090.

Misumi, Y., Miki, K., Takatsuki, A., Tamura, G., and Ikehara, Y., 1986, Novel blockade by brefeldin A of intracellular transport of secretory proteins in cultured rat hepatocytes, *J. Biol. Chem.* **261:**11398–11403.

Mizel, S. B., 1982, Interleukin 1 and T cell activation, *Immunol. Rev.* **63:**51–72.

Moll, H., Eichmann, K., and Simon, M. M., 1985, Immunoregulation by mouse T-cell clones. II. The same H-Y-specific clone can provide help for the generation of cytotoxic lymphocytes and antibody-secreting cells, *Immunology* **54:**255–264.

Moller, G. (ed.), 1976, Suppressor T lymphocytes, *Transplant. Rev.* **26**:1–20.

Moore, M. W., Carbone, F. R., and Bevan, M. J., 1988, Introduction of soluble protein into the class I pathway of antigen processing and presentation, *Cell* **54**:777–780.

Morris, A. G., Lin, Y.-L., and Askonas, B. A., 1982, Immune interferon release when a cloned cytotoxic T-cell line meets its correct influenza-infected target cell, *Nature (Lond.)* **295**:150–152.

Morrison, L. A., Lukacher, A. E., Braciale, V. L., Fan, D. P., and Braciale, T. J., 1986, Differences in antigen presentation to MHC class I and class II-restricted influenza A virus-specific cytolytic T lymphocyte clones, *J. Exp. Med.* **163**:903–921.

Mossman, T., 1983, Rapid colorimetric assay for cellular growth and survival: Application to proliferation and cytotoxic assays, *J. Immunol. Methods* **65**:55–63.

Murphy, B. R., and Webster, R. G., 1985, Influenza viruses, in: *Virology* (B. N. Fields, ed.), pp. 1179–1240, Raven, New York.

Nestorowicz, A., Laver, G., and Jackson, D. C., 1985, Antigenic determinants of influenza virus haemagglutinin. X. A comparison of the physical and antigenic properties of monomeric and trimeric forms, *J. Gen. Virol.* **66**:1687–1695.

Oldstone, M. B. A., Blount, P., Southern, P. J., and Lampert, P. W., 1986, Cytoimmunotherapy for persistent virus infection reveals a unique clearance pattern from the central nervous system, *Nature (Lond.)* **321**:239–243.

Owen, J. A., Allouche, M., and Doherty, P. C., 1984, Frequency of influenza-responsive cytolytic T-lymphocyte precursors in the thymus and spleen of unprimed mice, *Cell. Immunol.* **84**:403–408.

Pala, P., and Askonas, A., 1985, Induction of K$^b$-restricted anti-influenza cytotoxic T cells in C57BL mice: Importance of stimulator cell type and immunization route, *Immunol.* **55**:601–607.

Pala, P., and Askonas, B. A., 1986, Low responder MHC alleles for T$_C$ recognition of influenza nucleoprotein, *Immunogenetics* **23**:379–384.

Pala, P., Townsend, A. R. M., and Askonas, B. A., 1986, Viral recognition by influenza A virus cross-reactive cytotoxic T (Tc) cells: The proportion of Tc cells that recognize nucleoprotein varies between individual mice, *Eur. J. Immunol.* **16**:193–198.

Pan, S., and Knowles, B. B., 1983, Monoclonal antibody to SV40 T-antigen blocks lysis of cloned cytotoxic T-cell line specific for SV40 TASA, *Virology* **125**:1–7.

Panicali, D., Davis, S. W., Weinberg, R. L., and Paoletti, E., 1983, Construction of live vaccines by using genetically engineered poxviruses: Biological activity of recombinant vaccinia virus expressing influenza hemagglutinin, *Proc. Natl. Acad. Sci. USA* **80**:5364–5368.

Pierce, S. K., Cancro, M. P., and Klinman, N. R., 1978, Individual antigen-specific T lymphocytes: Helper function in enabling the expression of multiple antibody isotypes, *J. Exp. Med.* **148**:759–765.

Puddington, L., Bevan, M. J., Rose, J. K., and Lefrançois, L., 1986, N protein is the predominant antigen recognized by vesicular stomatitis virus-specific cytotoxic T cells, *J. Virol.* **60**:708–717.

Raymond, F. L., Caton, A. J., Cox, N. J., Kendal, A. P., and Brownlee, G. G., 1986, The antigenicity and evolution of influenza H1 hemagglutinin, from 1950–1957 and 1977–1983: Two pathways from one gene, *Virology* **148**:275–287.

Reddehase, M. J., Rothbard, J. B., and Koszinowski, U. H., 1989, A pentapeptide as minimal antigen determinant for MHC class I-restricted T lymphocytes, *Nature* **337**:651–653.

Reiss, C. S., and Burakoff, S. J., 1981, Specificity of the helper T cell for the cytolytic T lymphocyte response to influenza viruses, *J. Exp. Med.* **154**:541–545.

Reiss, C. S., and Schulman, J. L., 1980, Cellular immune responses of mice to influenza virus vaccines, *J. Immunol.* **125**:2182–2188.

Reiss, C. S., Dorf, M. E., Benacerraf, B., and Burakoff, S. J., 1982, Genetic control of the cytolytic T lymphocyte response to influenza viruses: H-2D genes influence the response to H-2Kb plus virus, *J. Immunol.* **128**:2295–2299.

Rothbard, J. B., and Taylor, W. R., 1988, A sequence pattern common to T cell epitopes, *EMBO J.* **7**:93–100.

Rothbard, J. B., Lechler, R. I., Howland, K., Bal, V., Eckels, D. D., Sekaly, R., Long, E. O., Taylor, W. R., and Lamb, J. R., 1988, Structural model of HLA-DR1 restricted T cell antigen recognition, *Cell* **52**:515–523.

Rott, R., Orlich, M., Klenk, H.-D., Wang, M. L., Skehel, J. J., and Wiley, D. C., 1984, Studies on the adaptation of influenza viruses to MDCK cells, *EMBO J.* **3**:3329–3332.

Russell, S. M., and Liew, F. Y., 1979, T cells primed by influenza virion internal components can cooperate in the antibody response to hemagglutinin, *Nature (Lond.)* **280**:147–148.

Scherle, P. A., and Gerhard, W., 1986, Functional analysis of influenza-specific helper T cell clones *in vivo*. T cells specific for internal viral proteins provide cognate help for B cell responses to hemagglutinin, *J. Exp. Med.* **164**:1114–1128.

Scherle, P. A., and Gerhard, W., 1988, Differential ability of B cells specific for external versus internal influenza virus proteins to respond to help from influenza virus-specific T-cell clones *in vitro*, *Proc. Natl. Acad. Sci. USA* **85**:4446–4450.

Schiltknecht, E., and Ada, G. L., 1985, Influenza virus-specific T cells fail to reduce lung virus titers in cyclosporin-treated infected mice, *Scand. J. Immunol.* **22**:99–103.

Schrier, M. H., Iscove, N. N., Tees, R., Aarden, L., and Von Boehmer, H., 1980, Clones of killer and helper T cells: Growth requirements, specificity, and retention of function in long-term culture, *Immunol. Rev.* **51**:315–336.

Schwartz, R. H., 1986, Immune response (Ir) genes of the murine major histocompatibility complex, *Adv. Immunol.* **38**:31–201.

Sharma, S., Rogers, L., Brandsma, J., Gething, M.-J., and Sambrook, J., 1985, SV40 T antigen and the exocytic pathway, *EMBO J.* **4**:1479–1489.

Shastri, N., Malissen, B., and Hood, L., 1985, Ia-transfected L-cell fibroblasts present a lysozyme peptide but not the native protein to lysozyme-specific T cells, *Proc. Natl. Acad. Sci. USA* **82**:5885–5889.

Shaw, S., and Biddison, W. E., 1979, HLA-linked genetic control of the specificity of human cytotoxic T cell responses to influenza virus, *J. Exp. Med.* **149**:565–575.

Shaw, S., Shearer, G. M., and Biddison, W. E., 1980, Human cytotoxic T-cell responses to type A and type B influenza viruses can be restricted by different HLA antigens. Implications for HLA polymorphism and genetic regulation, *J. Exp. Med.* **151**:235–245.

Shimonkevitz, R., Kappler, J., Marrack, P., and Grey, H., 1983, Antigen recognition by H-2-restricted T cells I. Cell-free antigen processing, *J. Exp. Med.* **158**:303–316.

Skehel, J. J., Bayley, P. M., Brown, E. B., Martin, S. R., Waterfield, M. D., White, J. M., Wilson, I. A., and Wiley, D. C., 1982, Changes in the conformation of influenza virus hemagglutinin at the pH optimum of virus-mediated membrane fusion, *Proc. Natl. Acad. Sci. USA* **79**:968–972.

Smith, G. L., Murphy, B. R., and Moss, B., 1983, Construction and characterization of an infectious vaccinia virus recombinant that expresses the influenza hemagglutinin gene and induces resistance to influenza virus infection in hamsters, *Proc. Natl. Acad. Sci. USA* **80**:7155–7159.

Smith, G. L., Levin, J. Z., Palese, P., and Moss, B., 1987, Synthesis and cellular location of the ten influenza polypeptides individually expressed by recombinant vaccinia viruses, *Virology* **160**:336–345.

Spencer, J. C., Ganguly, R., and Waldman, R. H., 1977, Nonspecific protection of mice against influenza virus infection by local or systemic immunization with bacillus Calmette-Guérin, *J. Infect. Dis.* **136**:171–175.

Staerz, U. D., Yewdell, J. W., and Bevan, M. J., 1987, Hybrid antibody-mediated lysis of virus-infected cells, *Eur. J. Immunol.* **17**:571–574.

Sterkers, G., Henin, Y., Lepage, V., Fradelizzi, D., Hannoun, C., and Levy, J. P., 1984a, Influenza A hemagglutinin-specific T cell clones strictly restricted by HLA-DR1 or HLA-DR7 molecules, *Eur. J. Immunol.* **14**:125–132.

Sterkers, G., Michon, J., Lepage, V., Henin, Y., Muller, J. Y., Degos, L., and Levy, J. P., 1984b, Restriction analysis of influenza-specific cloned cell lines issued from an HLA-DRw6/DR-donor, *Immunogenetics* **20**:693–697.

Sterkers, G., Michon, J., Henin, Y., Gomard, E., Hannoun, C., and Levy, J. P., 1985, Fine specificity analysis of human influenza-specific cloned cell lines, *Cell. Immunol.* **94:** 394–405.

Stitz, L., Huang, R. T. C., Hengartner, H., Rott, R., and Zinkernagel, R., 1985, Cytotoxic T cell lysis of target cells fused with liposomes containing influenza virus hemagglutinin and neuraminidase, *J. Gen. Virol.* **66:**1333–1339.

Streicher, H. Z., Berkower, I. J., Busch, M., Gurd, F. R. N., and Berzofsky, J. Z., 1984, Antigen conformation determines processing requirements for T-cell activation, *Proc. Natl. Acad. Sci. USA* **81:**6831–6835.

Sullivan, J. L., Mayner, R. E., Barry, D. W., and Ennis, F. A., 1976, Influenza virus infection in nude mice, *J. Infect. Dis.* **133:**91–94.

Takatsuki, A., and Tamura, G., 1985, Brefeldin A, a specific inhibitor of intracellular translocation of vesicular stomatitis virus B protein: Intracellular accumulation of high mannose type B protein and inhibition of its cell surface expression, *Agric. Biol. Chem.* **49:**899–902.

Tan, P. L. J., Booth, R. J., Prestidge, R. L., Watson, J. D., Dower, S. K., and Gillis, S., 1985, Induction of antibody responses to influenza virus in human lymphocyte cultures. I. Role of interleukin 2, *J. Immunol.* **135:**2128–2133.

Taylor, P. M., and Askonas, B. A., 1983, Diversity in the biological properties of anti-influenza cytotoxic T cell clones, *Eur. J. Immunol.* **13:**707–711.

Taylor, P. J., and Askonas, B. A., 1986, Influenza nucleoprotein-specific cytotoxic T cell clones are protective *in vivo*, *Immunology* **58:**417–420.

Tevethia, S. S., Flyer, D. C., and Tjian, R., 1980, Biology of simian virus 40 (SV40) transplantation antigen (TrAg). VI. Mechanism of induction of SV40 transplantation immunity in mice by purified SV40 T antigen (D2 protein), *Virology* **107:**13–23.

Thompson, M. A., Raychaudhuri, S., and Cancro, M. P., 1983, Restricted adult clonal profiles induced by neonatal immunization. Influence of suppressor T cells, *J. Exp. Med.* **158:**112–125.

Townsend, A. R. M., and Skehel, J. J., 1984, The influenza A virus nucleoprotein gene controls the induction of both subtype specific and cross-reactive cytotoxic T cells, *J. Exp. Med.* **160:**552–563.

Townsend, A. R. M., McMichael, A. J., Carter, N. P., Huddleston, J. A., and Brownlee, G. G., 1984a, Cytotoxic T cell recognition of the influenza nucleoprotein and hemagglutinin expressed in transfected mouse L cells, *Cell* **39:**13–25.

Townsend, A. R. M., Skehel, J. J., Taylor, P. M., and Palese, P., 1984b, Recognition of influenza A virus nucleoprotein by an H-2-restricted cytotoxic T cell clone, *Virology* **133:**456–459.

Townsend, A. R., Gotch, F. M., and Davey, J., 1985, Cytotoxic T cells recognize fragments of the influenza nucleoprotein, *Cell* **42:**457–467.

Townsend, A. R. M., Rothbard, J., Gotch, F. M., Bahadur, G., Wraith, D., and McMichael, A. J., 1986, The epitopes of influenza nucleoprotein recognized by cytotoxic T lymphocytes can be defined with short synthetic peptides, *Cell* **44:**959–968.

Townsend, A. R. M., Bastin, J., Gould, K., and Brownlee, G. G., 1987, Cytotoxic T lymphocytes recognize influenza hemagglutinin lacking a signal sequence, *Nature (Lond.)* **324:** 575–577.

Varghese, J. N., Laver, W. G., and Colman, P. M., 1983, Structure of the influenza virus glycoprotein antigen neuraminidase at 2.9 Å resolution, *Nature* **303:**35–40.

Virelizier, J. L., Postlethwaite, R., Schild, G. C., and Allison, A. C., 1974, Antibody responses to antigenic determinants of influenza virus hemagglutinin. I. Thymus dependence of antibody formation and thymus independence of memory, *J. Exp. Med.* **140:** 1559–1570.

Virelizier, J. L., Allison, A., Oxford, J., and Schild, G. C., 1977, Early presence of nucleoprotein antigen on the surface of influenza virus-infected cells, *Nature (Lond.)* **266:**52–54.

Wabuke-Bunoti, M. A. N., and Fan, D. P., 1983, Isolation and characterization of a CNBr cleavage peptide of influenza viral hemagglutinin stimulatory for mouse cytolytic T lymphocytes, *J. Immunol.* **130:**2386–2391.

Wabuke-Bunoti, M., Fan, D. P., and Braciale, T. J., 1981, Stimulation of anti-influenza cytolytic T lymphocytes by CNBr cleavage fragments of the viral hemagglutinin, *J. Immunol.* **127:**1122–1125.

Wabuke-Bunoti, M. A. N., Taku, A., Fan, D. P., Kent, S., and Webster, R. G., 1984, Cytolytic T lymphocyte and antibody responses to synthetic peptides of influenza virus hemagglutinin, *J. Immunol.* **133:**2194–2201.

Wagner, H., Starzinski-Powitz, A., Jung, H., and Rollinghoff, M., 1977, Induction of I-region-restricted hapten-specific cytotoxic T lymphocytes, *J. Immunol.* **119:**1365–1368.

Ward, C. W., 1981, Structure of the influenza virus hemagglutinin, *Curr. Top. Microbiol. Immunol.* **95:**1–74.

Watts, T. H., Brian, A. A., Kappler, J. W., Marrack, P., and McConnell, H. M., 1984, Antigen presentation by supported planar membranes containing affinity-purified I-Ad, *Proc. Natl. Acad. Sci. USA* **81:**7564–7568.

Watts, T. H., Gaub, H. E., and McConnell, H. M., 1986, T-cell-mediated association of peptide antigen and major histocompatibility complex protein detected by energy transfer in an evanescent wave field, *Nature (Lond.)* **320:**179–181.

Webster, R. G., Laver, W. G., Air, G. M., and Schild, G. C., 1982, Molecular mechanisms of variation in influenza viruses, *Nature (Lond.)* **296:**115–121.

Wells, M. A., Albrecht, P., and Ennis, F. A., 1981a, Recovery from a viral respiratory infection I. Influenza pneumonia in normal and T-deficient mice, *J. Immunol.* **126:**1036–1041.

Wells, M. A., Ennis, F. A., and Albrecht, P., 1981b, Recovery from a viral respiratory infection: II. Passive transfer of immune spleen cells to mice with influenza pneumonia, *J. Immunol.* **126:**1042–1046.

Wells, M. A., Daniel, S., Djeu, J. Y., Kiley, S. C., and Ennis, F. A., 1983, Recovery from a viral respiratory tract infection. IV. Specificity of protection by cytotoxic T lymphocytes, *J. Immunol.* **130:**2908–2914.

White, J. M., and Wilson, I. A., 1987, Anti-peptide antibodies detect steps in a protein conformational change: Low-pH activation of the influenza virus hemagglutinin, *J. Cell. Biol.* **105:**2887–2896.

Whitton, J. L., Gebhard, J. R., Lewicki, H., Tishon, A., and Oldstone, M. B. A., 1988, Molecular definition of a major cytotoxic T-lymphocyte epitope in the glycoprotein of lymphocytic choriomeningitis virus, *J. Virol.* **62:**687–695.

Wiley, D. C., Wilson, I. A., and Skehel, J. J., 1981, Structural identification of the antibody-binding sites of Hong Kong influenza haemagglutinin and their involvement in antigenic variation, *Nature (Lond.)* **289:**373–378.

Wilson, I. A., Skehel, J. J., and Wiley, D. C., 1981, Structure of the haemagglutinin membrane glycoprotein of influenza virus at 3A resolution, *Nature* **289:**366–373.

Winter, G., Fields, S., and Brownlee, G. G., 1981, Nucleotide sequence of the hemagglutinin gene of a human influenza H1 subtype, *Nature (Lond.)* **292:**72–75.

Wraith, D. C., and Askonas, B. A., 1985, Induction of influenza A virus cross-reactive cytotoxic T cells by a nucleoprotein/hemagglutinin preparation, *J. Gen. Virol.* **66:**1327–1331.

Wraith, D. C., and Vessey, A. F., 1986, Influenza virus-specific cytotoxic T-cell recognition: Stimulation of nucleoprotein-specific clones with intact antigen, *Immunology* **59:**173–180.

Wysocka, M., and Bennink, J. R., 1988, Limiting dilution analysis of memory cytotoxic T lymphocytes specific for individual influenza virus gene products, *Cell. Immunol.* **112:**425–429.

Yamada, A., Young, J. F., and Ennis, F. A., 1985a, Influenza virus subtype-specific cytotoxic T lymphocytes lyse target cells coated with a protein produced in *E. coli, J. Exp. Med.* **162:**1720–1725.

Yamada, A., Ziese, M. R., Young, J. F., Yamada, Y. K., and Ennis, F. A., 1985b, Influenza virus hemagglutinin-specific cytotoxic T cell response induced by polypeptide produced in *Escherichia coli, J. Exp. Med.* **162:**663–674.

Yap, K. L., and Ada, G. L., 1977, Cytotoxic T cells specific for influenza virus-infected target cells, *Immunology* **32:**151–159.

Yap, K. L., and Ada, G. L., 1978a, Cytotoxic T cells in the lungs of mice infected with an influenza A virus, *Scand. J. Immunol.* **7:**73–80.

Yap, K. L., and Ada, G. L., 1978b, The recovery of mice from influenza A virus infection: Adoptive transfer of immunity with influenza virus-specific cytotoxic T lymphocytes recognizing a common virion antigen, *Scand. J. Immunol.* **8:**413–420.

Yap, K. L., and Ada, G. L., 1978c, The recovery of mice from influenza virus infection: Adoptive transfer of immunity with immune T lymphocytes, *Scand. J. Immunol.* **7:** 389–397.

Yap, K. L., Ada, G. L., and McKenzie, I. F. C., 1978, Transfer of specific cytotoxic T lymphocytes protects mice inoculated with influenza virus, *Nature (Lond.)* **273:**238–239.

Yap, K. L., Braciale, T. J., and Ada, G. L., 1979, Role of T cell function in recovery from murine influenza infection, *Cell. Immunol.* **43:**341–351.

Yewdell, J. W., and Bennink, J. R., 1989, Brefeldin A inhibits the presentation of protein antigens but not exogenous peptides to cytotoxic T lymphocytes, *Science* (in press).

Yewdell, J. W., Webster, R. G., and Gerhard, W., 1979, Antigenic variation in three distinct determinants of an influenza type A hemagglutinin molecule, *Nature (Lond.)* **279:**246–248.

Yewdell, J. W., Frank, E., and Gerhard, W., 1981, Expression of influenza A virus internal antigens on the surface of infected P815 cells, *J. Immunol.* **126:**1814–1819.

Yewdell, J. W., Gerhard, W., and Bachi, T., 1983, Monoclonal anti-hemagglutinin antibodies detect irreversible antigenic alterations that coincide with the acid activation of influenza virus A/PR/8/34-mediated hemolysis, *J. Virol.* **48:**239–248.

Yewdell, J. W., Bennink, J. R., Smith, G. L., and Moss, B., 1985, Influenza A virus nucleoprotein is a major target antigen for cross-reactive anti-influenza A virus cytotoxic T lymphocytes, *Proc. Natl. Acad. Sci. USA* **82:**1785–1789.

Yewdell, J. W., Bennink, J. R., Mackett, M., Lefrançois, L., Lyles, D. S., and Moss, B., 1986, Recognition of cloned vesicular stomatitis virus internal and external gene products by cytotoxic T lymphocytes, *J. Exp. Med.* **163:**1529–1538.

Yewdell, J. W., Bennink, J. R., and Hosaka, Y., 1988, Cells process exogenous proteins for recognition by cytotoxic T lymphocytes, *Science* **239:**637–640.

Young, J. F., Desselberger, U., Graves, P., Palese, P., Shatzman, A., and Rosenberg, M., 1983, Cloning and expression of influenza virus genes, in: *The Origin of Pandemic Influenza Viruses* (W. G. Laver, ed.), pp. 129–138, Elsevier, New York.

Zanders, E. D., Fischer, A., Smith, S., Beverley, P. C. L., and Feldmann, M., 1983, Antigen-specific and non-specific helper activities derived from supernatants of human influenza virus-specific T-cell lines, *Immunology* **48:**361–366.

Ziegler, H. K., and Unanue, E. R., 1982, Decrease in macrophage antigen catabolism caused by ammonia and chloroquine is associated with inhibition of antigen presentation to T cells, *Proc. Natl. Acad. Sci. USA* **79:**175–178.

Zinkernagel, R. M., and Doherty, P. C., 1974, Restriction of *in vitro* T cell-mediated cytotoxicity in lymphocytic choriomeningitis virus within a syngeneic or semi-allogeneic system, *Nature (Lond.)* **248:**701–702.

Zinkernagel, R. M., and Doherty, P. C., 1979, MHC-restricted cytotoxic T cells, *Adv. Immunol.* **27:**51–106.

Zweerink, H. J., Askonas, B. A., Millican, D., Courtneidge, S. A., and Skehel, J. J., 1977a, Cytotoxic T cells to type A influenza virus; viral hemagglutinin induces A-strain specificity while infected cells confer cross-reactive cytotoxicity, *Eur. J. Immunol.* **7:**630–635.

Zweerink, H. J., Courtneidge, S. A., Skehel, J. J., Crumpton, M. J., and Askonas, B. A., 1977b, Cytotoxic T cells kill influenza virus infected cells but do not distinguish between serologically distinct type A viruses, *Nature (Lond.)* **267:**354–356.

# Index

## DATE DUE

| | | |
|---|---|---|
| MAY 2 9 1991 | MAY 2 4 2000 | |
| MAY 2 2 1994 | | |
| OCT 1 9 1995 | MAY 2 3 2014 | |
| NOV 2 7 1995 | | |
| JUL 2 6 1997 | | |
| MAY 1 0 1998 | | |
| DEC 1 4 1999 FEB 1 1 2000 | | |
| APR 2 3 2001 | | |
| MAY 0 2 2008 | | |
| | | |
| | | |
| | | |
| | | |
| | | |
| | | |
| | | |
| | | |